LIGHTWAVE
TECHNOLOGY

LIGHTWAVE TECHNOLOGY
Components and Devices

Govind P. Agrawal

A JOHN WILEY & SONS, INC., PUBLICATION

This text is printed on acid-free paper. ∞

Copyright © 2004 by John Wiley & Sons, Inc. All rights reserved.

Published by John Wiley & Sons, Inc., Hoboken, New Jersey.
Published simultaneously in Canada.

No part of this publication may be reproduced, stored in a retrieval system, or transmitted in any form or by any means, electronic, mechanical, photocopying, recording, scanning, or otherwise, except as permitted under Section 107 or 108 of the 1976 United States Copyright Act, without either the prior written permission of the Publisher, or authorization through payment of the appropriate per-copy fee to the Copyright Clearance Center, Inc., 222 Rosewood Drive, Danvers, MA 01923, (978) 750-8400, fax (978) 646-8600, or on the web at www.copyright.com. Requests to the Publisher for permission should be addressed to the Permissions Department, John Wiley & Sons, Inc., 111 River Street, Hoboken, NJ 07030, (201) 748-6011, fax (201) 748-6008.

Limit of Liability/Disclaimer of Warranty: While the publisher and author have used their best efforts in preparing this book, they make no representations or warranties with respect to the accuracy or completeness of the contents of this book and specifically disclaim any implied warranties of merchantability or fitness for a particular purpose. No warranty may be created or extended by sales representatives or written sales materials. The advice and strategies contained herein may not be suitable for your situation. You should consult with a professional where appropriate. Neither the publisher nor author shall be liable for any loss of profit or any other commercial damages, including but not limited to special, incidental, consequential, or other damages.

For general information on our other products and services please contact our Customer Care Department within the U.S. at 877-762-2974, outside the U.S. at 317-572-3993 or fax 317-572-4002.

Wiley also publishes its books in a variety of electronic formats. Some content that appears in print, however, may not be available in electronic format.

Library of Congress Cataloging-in-Publication Data:

Agrawal, G. P. (Govind P.), 1951–
   Lightwave technology : components and devices
       p. cm
   Includes bibliographical references and index.
   ISBN 0-471-21573-2 (cloth)
    1. Electrooptics.   2. Electrooptical devices.   I. Title.

TA1750.A39 2004
621.381'045—dc22   2004044059

Printed in the United States of America.

10 9 8 7 6 5 4 3 2 1

For Anne, Sipra, Caroline, and Claire

Contents

Preface   **xiii**

1 Optical Fibers   **1**
  1.1 Basic Concepts 1
      1.1.1 Step-Index Fibers 2
      1.1.2 Graded-Index Fibers 4
  1.2 Design and Fabrication 6
      1.2.1 Silica Fibers 6
      1.2.2 Plastic Optical Fibers 9
      1.2.3 Microstructure Optical Fibers 11
      1.2.4 Cables and Connectors 12
  1.3 Fiber Modes 13
      1.3.1 Maxwell's Equations 13
      1.3.2 Modes of Step-Index Fibers 15
      1.3.3 Properties of Single-Mode Fibers 18
  1.4 Fiber Losses 22
      1.4.1 Loss Spectra of Silica Fibers 23
      1.4.2 Loss Spectra of Plastic Fibers 24
      1.4.3 Rayleigh Scattering 25
      1.4.4 Waveguide Imperfections 26
  1.5 Fiber Dispersion 27
      1.5.1 Group-Velocity Dispersion 27
      1.5.2 Material Dispersion 28
      1.5.3 Waveguide Dispersion 30
      1.5.4 Polarization-Mode Dispersion 32
      1.5.5 Pulse-Propagation Equation 34
  1.6 Fiber Nonlinearities 36
      1.6.1 Self-Phase Modulation 36
      1.6.2 Cross-Phase Modulation 38
      1.6.3 Four-Wave Mixing 39
      1.6.4 Stimulated Brillouin Scattering 40
      1.6.5 Stimulated Raman Scattering 42
  Problems 44
  References 46

2 Passive Fiber Components   49
2.1 Directional Couplers   49
- 2.1.1 Coupled-Mode Equations   50
- 2.1.2 Power-Transfer Characteristics   52
- 2.1.3 Transfer Matrix of a Coupler   53
- 2.1.4 Supermodes of a Coupler   54
- 2.1.5 Effects of Fiber Dispersion   55

2.2 Fiber Gratings   56
- 2.2.1 Bragg Diffraction   57
- 2.2.2 Photosensitivity   58
- 2.2.3 Fabrication Techniques   59
- 2.2.4 Grating Characteristics   62
- 2.2.5 Grating as an Optical Filter   67
- 2.2.6 Nonuniform Gratings   69

2.3 Fiber Interferometers   71
- 2.3.1 Fiber-Ring Resonators   72
- 2.3.2 Fabry–Perot Resonators   74
- 2.3.3 Sagnac Interferometers   76
- 2.3.4 Mach–Zehnder Interferometers   81
- 2.3.5 Michelson Interferometers   84

2.4 Isolators and Circulators   85
- 2.4.1 Faraday Effect   85
- 2.4.2 Optical Isolators   86
- 2.4.3 Optical Circulators   87

Problems   88
References   89

3 Active Fiber Components   93
3.1 Doped-Fiber Amplifiers   93
- 3.1.1 Pumping and Amplifier Gain   94
- 3.1.2 Gain Spectrum   96
- 3.1.3 Simple Theory   97
- 3.1.4 Amplifier Noise   100
- 3.1.5 Amplifier Design   102

3.2 Raman Amplifiers   105
- 3.2.1 Raman Gain and Bandwidth   105
- 3.2.2 Single-Pump Raman Amplification   107
- 3.2.3 Multiple-Pump Raman Amplification   109
- 3.2.4 Performance Limiting Factors   112

3.3 Parametric Amplifiers   118
- 3.3.1 FWM Interaction   118
- 3.3.2 Single-Pump Parametric Amplifiers   120
- 3.3.3 Dual-Pump Parametric Amplifiers   123
- 3.3.4 Performance Limiting Factors   125

3.4 Fiber Lasers   128
- 3.4.1 Cavity Design   128

		3.4.2	Continuously Operating Fiber Lasers	130
		3.4.3	Mode-Locked Fiber Lasers	132
	Problems			136
	References			137

4 Planar Waveguides   142

- 4.1 Basic Concepts . . . 142
  - 4.1.1 Modes of Planar Waveguides . . . 143
  - 4.1.2 TE and TM Modes of Symmetric Waveguides . . . 145
  - 4.1.3 TE and TM Modes of Asymmetric Waveguides . . . 147
- 4.2 Rectangular Waveguides . . . 149
  - 4.2.1 Modes of Rectangular Waveguides . . . 149
  - 4.2.2 Design of Rectangular Waveguides . . . 150
- 4.3 Materials and Waveguide Fabrication . . . 152
  - 4.3.1 Semiconductor Waveguides . . . 152
  - 4.3.2 Electro-Optic Waveguides . . . 156
  - 4.3.3 Silica Waveguides . . . 158
  - 4.3.4 Silicon-on-Insulator Technology . . . 162
  - 4.3.5 Polymer Waveguides . . . 164
- 4.4 Simple Passive Components . . . 165
  - 4.4.1 Y Junctions . . . 165
  - 4.4.2 Four-Port Couplers . . . 166
  - 4.4.3 Grating-Assisted Directional Couplers . . . 167
- 4.5 Advanced Passive Components . . . 168
  - 4.5.1 Mach–Zehnder Switches and Filters . . . 168
  - 4.5.2 Multimode Interference Couplers . . . 170
  - 4.5.3 Star Couplers . . . 171
  - 4.5.4 Arrayed-Waveguide Gratings . . . 173
- Problems . . . 175
- References . . . 177

5 Semiconductor Lasers and Amplifiers   179

- 5.1 Basic Concepts . . . 179
  - 5.1.1 Optical Gain . . . 180
  - 5.1.2 Feedback and Laser Threshold . . . 182
  - 5.1.3 Longitudinal Modes . . . 183
  - 5.1.4 Laser Structures . . . 184
- 5.2 Control of Longitudinal Modes . . . 186
  - 5.2.1 Distributed Feedback Lasers . . . 186
  - 5.2.2 Coupled-Cavity Semiconductor Lasers . . . 188
  - 5.2.3 Tunable Semiconductor Lasers . . . 190
  - 5.2.4 Vertical-Cavity Surface-Emitting Lasers . . . 191
- 5.3 Laser Characteristics . . . 192
  - 5.3.1 CW Characteristics . . . 193
  - 5.3.2 Small-Signal Modulation . . . 196
  - 5.3.3 Large-Signal Modulation . . . 198

		5.3.4	Relative Intensity Noise	200
		5.3.5	Spectral Linewidth	202
	5.4	Light-Emitting Diodes		204
		5.4.1	CW Characteristics	204
		5.4.2	Modulation Response	207
		5.4.3	LED Structures	208
	5.5	Semiconductor Optical Amplifiers		209
		5.5.1	Amplifier Design	210
		5.5.2	Amplifier Characteristics	212
		5.5.3	Practical Issues	213
		5.5.4	SOA as a Nonlinear Device	215
	Problems			218
	References			219
6	**Optical Modulators**			**223**
	6.1	Physics Behind Modulators		223
		6.1.1	Modulation Schemes	224
		6.1.2	Electroabsorption	225
		6.1.3	Electrorefraction	227
		6.1.4	Photoelastic Effect	229
	6.2	Lithium Niobate Modulators		231
		6.2.1	Phase and Amplitude Modulation	231
		6.2.2	Temporal Response	233
		6.2.3	Modulator Design	235
	6.3	Polymer-Based Modulators		237
		6.3.1	Device Fabrication	238
		6.3.2	Device Performance	241
	6.4	Electroabsorption Modulators		242
		6.4.1	Temporal Response	242
		6.4.2	Design and Performance	244
		6.4.3	Integration with Lasers	246
	Problems			250
	References			251
7	**Photodetectors**			**253**
	7.1	Basic Concepts		253
		7.1.1	Responsivity and Quantum Efficiency	253
		7.1.2	Rise Time and Bandwidth	256
	7.2	Reverse-Biased p–n Junctions		257
		7.2.1	p–n Photodiodes	257
		7.2.2	p–i–n Photodiodes	259
		7.2.3	Advanced Designs	260
	7.3	Avalanche Photodiodes		262
		7.3.1	Impact Ionization	262
		7.3.2	APD Gain	263
		7.3.3	APD Bandwidth	265

Contents   xi

			7.3.4	Advanced APD Designs	269
	7.4	MSM Photodetectors	271
	7.5	Integrated Optical Receivers	274
		7.5.1	Front End	274
		7.5.2	Linear Channel	275
		7.5.3	Decision Circuit	277
		7.5.4	Optoelectronic Integration	278
	Problems	282
	References	282

8 WDM Components   287
   8.1 Optical Filters   287
        8.1.1 Fabry–Perot Filters   289
        8.1.2 Mach–Zehnder Filters   293
        8.1.3 Bragg-Grating Filters   296
        8.1.4 Acousto-Optic Filters   300
   8.2 Multiplexers and Demultiplexers   303
        8.2.1 Grating-Based Demultiplexers   303
        8.2.2 Filter-Based Demultiplexers   305
        8.2.3 Waveguide-Grating Demultiplexers   307
   8.3 Optical Add–Drop Multiplexers   309
        8.3.1 Directional Couplers with Gratings   309
        8.3.2 Mach–Zehnder Interferometer with Gratings   311
        8.3.3 Optical Circulator with Gratings   312
        8.3.4 Microring Resonators   314
        8.3.5 Tunable Add–Drop Multiplexers   316
   8.4 WDM Transmitters and Receivers   317
        8.4.1 Optoelectronic Integrated Circuits   318
        8.4.2 Fiber-Laser Transmitters   321
        8.4.3 Spectrally Sliced WDM Transmitters   322
   Problems   325
   References   326

9 Optical Switching   334
   9.1 Switching Technologies   334
        9.1.1 Electro-Optic Switches   335
        9.1.2 Thermo-Optic Switches   338
        9.1.3 Micro-Mechanical Switches   340
        9.1.4 Liquid-Crystal Switches   342
        9.1.5 Bubble Switches   343
   9.2 Wavelength-Domain Routers   344
   9.3 Wavelength Converters   346
        9.3.1 Semiconductor-Based Devices   347
        9.3.2 Fiber-Based Devices   351
   9.4 Optical Cross-Connects   354
        9.4.1 Wavelength-Selective Cross-Connect   355

		9.4.2	Wavelength-Interchanging Cross-Connect	358
	Problems			361
	References			362

10 Time-Domain Switching — 366

- 10.1 Nonlinear Switching Schemes ... 366
    - 10.1.1 Optical Bistability ... 367
    - 10.1.2 Cross-Phase Modulation ... 370
    - 10.1.3 Four-Wave Mixing ... 373
- 10.2 Optical Flip-Flops ... 375
    - 10.2.1 Semiconductor Lasers and Amplifiers ... 375
    - 10.2.2 Passive Semiconductor Waveguides ... 378
- 10.3 Ultrafast Interferometric Switches ... 380
    - 10.3.1 Sagnac-Loop Switches ... 381
    - 10.3.2 Mach–Zehnder Switches ... 383
    - 10.3.3 Polarization-Discriminating Switches ... 386
    - 10.3.4 Lithium-Niobate Switches ... 388
- 10.4 Applications ... 389
    - 10.4.1 Time-Domain Demultiplexing ... 390
    - 10.4.2 Optical-Clock Recovery ... 396
    - 10.4.3 Packet-Switched Networks ... 397
- Problems ... 400
- References ... 401

A System of Units — 406

B Software Package — 408

C Acronyms — 410

Index — 413

Preface

The term lightwave technology was coined as a natural extension of microwave technology and refers to the developments based on the use of light in place of microwaves. The beginnings of lightwave technology can be traced to the decade of 1960s during which significant advances were made in the fields of lasers, optical fibers, and nonlinear optics. The two important milestones were realized in 1970, the year that saw the advent of low-loss optical fibers as well as room-temperature operation of semiconductor lasers. By 1980, the era of commercial lightwave transmission systems had arrived.

The first generation of fiber-optic communication systems debuting in 1980 operated at a meager bit rate of 45 Mb/s and required signal regeneration every 10 km or so. However, by 1990 further advances in lightwave technology not only increased the bit rate to 10 Gb/s (by a factor of 200), but also allowed signal regeneration after 80 km or more. The pace of innovation in all fields of lightwave technology only quickened during the 1990s, as evident from the development and commercialization of erbium-doped fiber amplifiers, fiber Bragg gratings, and wavelength-division-multiplexed lightwave systems. By 2001, the capacity of commercial terrestrial systems exceeded 1.6 Tb/s. At the same time, the capacity of transoceanic lightwave systems installed worldwide exploded. A single transpacific system could transmit information at a bit rate of more than 1 Tb/s over a distance of 10,000 km without any signal regeneration. Such a tremendous improvement was possible only because of multiple advances in all areas of lightwave technology. Although commercial development slowed down during the economic downturn that began in 2000 and was not fully over by the end of 2003, research in lightwave technology has continued to grow.

The primary objective of this two-volume set is to provide a comprehensive and up-to-date account of all major aspects of lightwave technology. This first volume is subtitled *Components and Devices*, as it is devoted to the description of a multitude of silica- and semiconductor-based optical devices. The second volume deals with the design and performance of modern transmission systems that make use of these devices. Of course, it turned out to be impossible to cover all aspects of lightwave technology in two volumes, and it was necessary to limit the scope to some extent. For example, this text does not include any material related to optical sensors mainly because several excellent books already cover the field of optical sensors. A secondary objective of this text is that it should be able to serve both as a textbook and a reference monograph. For this reason, the emphasis is on physical understanding, but engineering aspects are also discussed throughout the text. Each chapter also includes selected problems that can be assigned to students.

The primary readership for this book is likely to be of graduate students, research scientists, and professional engineers working in fields related to *lightwave technology*. An attempt is made to include as much recent material as possible so that students are exposed to the recent advances in this exciting field. The reference list at the end of each chapter is more elaborate than what is common for a typical textbook. The listing of recent research papers should be useful for researchers using this book as a reference. At the same time, students can benefit from the reference section if they are assigned problems requiring reading of original research papers. This book may be useful to students taking a graduate course devoted to fiber-optic devices or optoelectronic devices. It can also be a part of a two-semester course on optical communications or lightwave systems.

The book is divided into three major parts. Chapters 1–3 cover optical fibers and devices made with them. Chapter 4–7 are devoted to planar waveguides and devices fabricated using such waveguides. Most of the emphasis is on semiconductor waveguides employed for making lasers, photodetectors, and modulators, but silica and polymer waveguides are also discussed. Devices that are needed for modern lightwave systems are covered in Chapters 8–10. It is my hope that the ordering of 10 chapters into three major groups will help teachers in selecting the appropriate material for their classes. It should also help researchers in finding a topic they might need to consult. A detailed subject index is also provided.

The book includes a compact disk (CD) on the back cover containing software provided by RSoft, Inc. This state-of-the art software package should prove useful to readers for solving problems provided at the end of each chapter. It also contains additional problems for each chapter that may help in understanding the difficult material. Appendix B provides more details about the software. It is my hope that the CD will help in training graduate students for an industrial job.

A large number of persons have contributed to this book either directly or indirectly. It is impossible to mention all of them by name. I thank my graduate students and the students who took my course on optical communication systems and helped improve my class notes through their questions and comments. I am grateful to my colleagues and graduate students at the Institute of Optics for numerous discussions. I thank, in particular, John Marciante and Drew Maywar for reading several chapters and providing constructive feedback. Last, but not least, I thank my wife Anne and my daughters, Sipra, Caroline, and Claire, for their patience and encouragement.

Govind P. Agrawal
Rochester, NY
February 2004

Chapter 1

Optical Fibers

The phenomenon of *total internal reflection*, responsible for guiding light in optical fibers, has been known for more than 150 years [1]. Although glass fibers were made in the 1920s, their use became practical only in the 1950s, when the use of a cladding layer led to considerable improvement in their guiding characteristics [2]–[4]. Before 1970, optical fibers were used mainly for medical imaging over short distances [5]. Their use for communication purposes was considered impractical because of high losses ($\sim$1,000 dB/km). However, the situation changed drastically in 1970 when, following an earlier suggestion [6], the loss of optical fibers was reduced to below 20 dB/km [7]. Further progress resulted by 1979 in a loss of only 0.2 dB/km near the 1.55-μm spectral region [8]. The availability of low-loss fibers led to a revolution in the field of lightwave technology and started the era of fiber-optic communications. By the end of 1990s, fibers were being used worldwide not only for signal transmission but also for making many useful optical devices in the form of couplers, gratings, amplifiers, and sensors.

This chapter is organized as follows. Section 1.1 employs geometrical optics to explain the guiding mechanism for both the step-index and graded-index fibers. Section 1.2 focuses on the design and fabrication of silica, plastic, and microstructure fibers. Maxwell's equations are solved in Section 1.3 to introduce the concept of fiber modes and discuss the properties of single-mode fibers. The loss mechanisms for optical fibers are considered in Section 1.4, whereas Section 1.5 focuses on the dispersion characteristics of single-mode fibers. In the last section we discuss various nonlinear effects that often degrade lightwave systems but can also be employed for making a variety of useful fiber-based devices.

1.1 Basic Concepts

In its simplest form an optical fiber consists of a cylindrical core (diameter <1 mm) made of silica glass or an organic polymer (plastic), which is surrounded by a cladding whose refractive index is lower than that of the core. Because of an abrupt index change at the core–cladding interface, such fibers are called *step-index fibers*. In a different

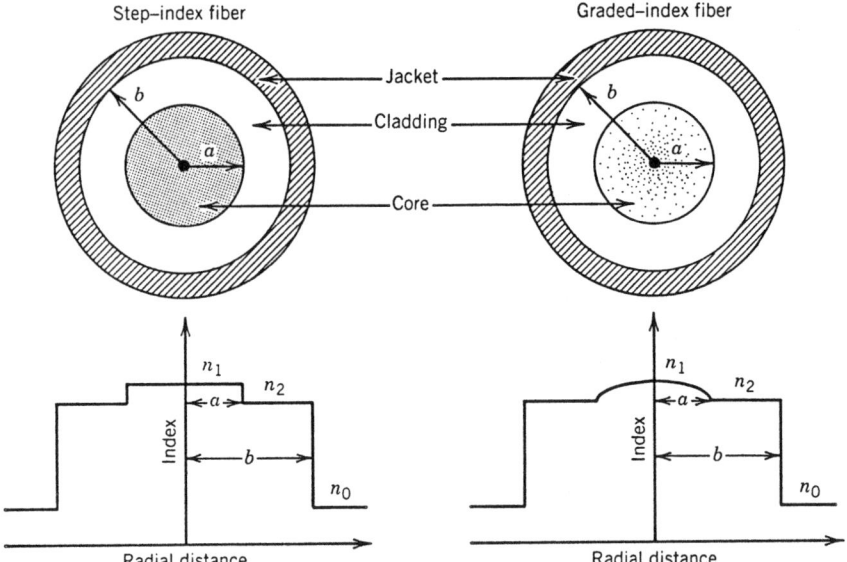

Figure 1.1: Cross section and refractive-index profile for step-index and graded-index fibers.

type of fiber, known as *graded-index fiber*, the refractive index decreases gradually inside the core. Figure 1.1 shows schematically the index profile and the cross section for the two kinds of fibers.

The first question that must be answered is what makes an optical fiber to confine light to its core in spite of a curved path. Considerable insight in the guiding properties of optical fibers can be gained by using a ray picture based on geometrical optics. The geometrical-optics description, although approximate, is valid when the core radius a is much larger than the light wavelength λ. When the two become comparable, it is necessary to use the modal description of Section 1.3 based on Maxwell's equations [9]–[16]. Since the guiding mechanism is somewhat different for the step-index and graded-index fibers, we consider the two types of fibers in separate subsections.

1.1.1 Step-Index Fibers

Consider the geometry of Figure 1.2, where a ray making an angle θ_i with the fiber axis is incident at the core center. Because of refraction at the fiber–air interface, the ray bends toward the normal. The angle θ_r of the refracted ray is given by [17]

$$n_0 \sin \theta_i = n_1 \sin \theta_r, \tag{1.1.1}$$

where n_1 and n_0 are the refractive indices of the fiber core and air, respectively. The refracted ray hits the core–cladding interface and is refracted again. However, refraction is possible only for an angle of incidence ϕ such that $\sin \phi < n_2/n_1$, where n_2 is the cladding index. For angles larger than a *critical angle* ϕ_c, defined as [17]

$$\sin \phi_c = n_2/n_1, \tag{1.1.2}$$

1.1 Basic Concepts

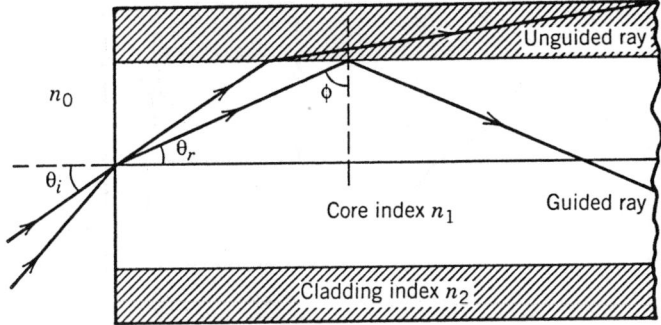

Figure 1.2: Light confinement through total internal reflection in step-index fibers. Rays for which $\phi < \phi_c$ are refracted out of the core.

the ray experiences total internal reflection at the core–cladding interface. Since such reflections occur throughout the fiber length, all rays with $\phi > \phi_c$ remain confined to the fiber core. This is the basic mechanism behind light confinement in optical fibers.

We can use Eqs. (1.1.1) and (1.1.2) to find the maximum angle that the incident ray should make with the fiber axis to remain confined inside the core. Noting that $\theta_r = \pi/2 - \phi_c$ for such a ray and substituting it in Eq. (1.1.1), we obtain

$$n_0 \sin \theta_i = n_1 \cos \phi_c = (n_1^2 - n_2^2)^{1/2}. \tag{1.1.3}$$

In analogy with lenses, $n_0 \sin \theta_i$ is known as the *numerical aperture* (NA) of the fiber. It represents the light-gathering capacity of an optical fiber. For $n_1 \simeq n_2$ the numerical aperture can be approximated by

$$\mathrm{NA} = n_1 (2\Delta)^{1/2}, \qquad \Delta = (n_1 - n_2)/n_1, \tag{1.1.4}$$

where Δ is the fractional index change at the core–cladding interface. Clearly, Δ should be made as large as possible in order to couple maximum light into the fiber. However, such fibers are not useful for the purpose of optical communications because of a phenomenon known as multipath dispersion, also called *modal dispersion*.

Multipath dispersion can be understood by referring to Figure 1.2, where different rays travel along paths of different lengths. As a result, these rays disperse in time at the output end of the fiber even if they were coincident at the input end and traveled at the same speed inside the fiber. A short pulse (called an *impulse*) would broaden considerably as a result of different path lengths. One can estimate the extent of pulse broadening simply by considering the shortest and longest ray paths. The shortest path occurs for $\theta_i = 0$ and is just equal to the fiber length L. The longest path occurs for θ_i given by Eq. (1.1.3) and has a length $L/\sin\phi_c$. If we use the speed of ray propagation $v = c/n_1$, the time delay is given by

$$\Delta T = \frac{n_1}{c} \left(\frac{L}{\sin \phi_c} - L \right) = \frac{L n_1^2}{c\, n_2} \Delta. \tag{1.1.5}$$

The time delay between the two rays taking the shortest and longest paths is a measure of broadening experienced by an impulse launched at the fiber input.

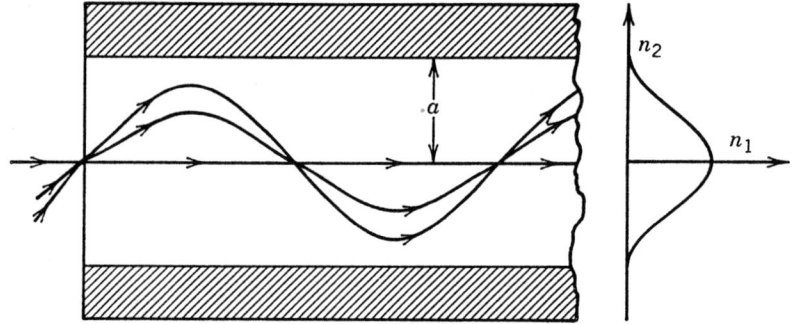

Figure 1.3: Ray trajectories in a graded-index fiber.

We can relate ΔT to the information-carrying capacity of the fiber measured through the bit rate B. Although a precise relation between B and ΔT depends on many details, such as the pulse shape, it is clear intuitively that ΔT should be less than the allocated bit slot ($T_B = 1/B$). Thus, an order-of-magnitude estimate of the bit rate is obtained from the condition $B \Delta T < 1$. By using Eq. (1.1.5), we obtain

$$BL < \frac{n_2}{n_1^2} \frac{c}{\Delta}. \tag{1.1.6}$$

This condition provides a rough estimate of a fundamental limitation of step-index fibers. As an illustration, consider an unclad glass fiber with $n_1 = 1.5$ and $n_2 = 1$. The bit rate–distance product of such a fiber is limited to quite small values since $BL < 0.4$ (Mb/s)-km. Considerable improvement occurs for cladded fibers with a small index step. Most fibers for communication applications are designed with $\Delta < 0.01$. As an example, $BL < 100$ (Mb/s)-km for $\Delta = 2 \times 10^{-3}$. Such fibers can transmit data at a bit rate of 10 Mb/s over distances up to 10 km. As discussed next, the problem of multipath dispersion can be solved to a large extent by using graded-index fibers.

1.1.2 Graded-Index Fibers

The refractive index of the core in graded-index fibers is not constant but decreases gradually from its maximum value n_1 at the core center to its minimum value n_2 at the core–cladding interface. Most graded-index fibers are designed to have a nearly quadratic decrease and are analyzed by using the so-called α-profile of the form

$$n(\rho) = \begin{cases} n_1[1 - \Delta(\rho/a)^\alpha]; & \rho < a, \\ n_1(1 - \Delta) \equiv n_2 \ ; & \rho \geq a, \end{cases} \tag{1.1.7}$$

where a is the core radius. The parameter α determines the index profile. A step-index profile is approached in the limit of large α, while a *parabolic-index fiber* is realized for $\alpha = 2$.

It is easy to understand qualitatively why multipath dispersion is reduced for graded-index fibers. Figure 1.3 shows schematically paths for three different rays. Similar to

1.1 Basic Concepts

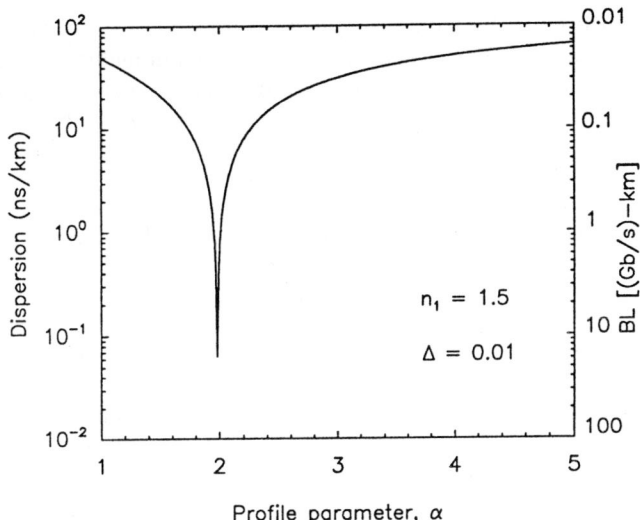

Figure 1.4: Variation of intermodal dispersion $\Delta T/L$ with the profile parameter α for a graded-index fiber. The scale on the right shows the corresponding bit rate–distance product.

the case of step-index fibers, the path is longer for more oblique rays. However, the ray velocity changes along the path because of variations in the refractive index. More specifically, the ray propagating along the fiber axis takes the shortest path but travels most slowly as the index is largest along this path. Oblique rays have a large part of their path in a medium of lower refractive index, where they travel faster. It is therefore possible for all rays to arrive together at the fiber output by a suitable choice of the refractive-index profile.

Geometrical optics can be used to show that a parabolic-index profile leads to nondispersive pulse propagation within the *paraxial approximation*. The trajectory of a paraxial ray is obtained by solving [17]

$$\frac{d^2\rho}{dz^2} = \frac{1}{n}\frac{dn}{d\rho}, \tag{1.1.8}$$

where ρ is the radial distance of the ray from the axis. By using Eq. (1.1.7) for $\rho < a$ with $\alpha = 2$, Eq. (1.1.8) reduces to an equation of harmonic oscillator and has the general solution

$$\rho = \rho_0 \cos(pz) + (\rho_0'/p)\sin(pz), \tag{1.1.9}$$

where $p = (2\Delta/a^2)^{1/2}$ and ρ_0 and ρ_0' are the position and the direction of the input ray, respectively. Equation (1.1.9) shows that all rays recover their initial positions and directions periodically at distances $z = 2m\pi/p$, where m is an integer (see Figure 1.3). Such a complete restoration of the input implies that a parabolic-index fiber does not suffer from multipath dispersion even though different rays take different paths.

The conclusion above holds only within the paraxial and the geometrical-optics approximations, both of which must be relaxed for practical fibers. Modal dispersion in graded-index fibers has been studied extensively by using wave-propagation techniques [10]–[13]. The quantity $\Delta T/L$, where ΔT is the maximum multipath delay in a fiber of length L, is found to vary considerably with α. Figure 1.4 shows this variation for $n_1 = 1.5$ and $\Delta = 0.01$. The minimum dispersion occurs for $\alpha = 2(1-\Delta)$ and depends on Δ as [15]

$$\Delta T/L = n_1 \Delta^2/(8c). \tag{1.1.10}$$

The limiting bit rate–distance product is obtained by using the criterion $\Delta T < 1/B$ and is given by

$$BL < 8c/(n_1 \Delta^2). \tag{1.1.11}$$

The right scale in Figure 1.4 shows the BL product as a function of α. Graded-index fibers with a suitably optimized index profile can transmit data at a bit rate of 100 Mb/s over distances up to 100 km. The BL product of such fibers is improved by nearly three orders of magnitude over that of step-index fibers. Indeed, the first generation of lightwave systems used graded-index fibers. Further improvement is possible only by using single-mode fibers in which core radius is made comparable to the light wavelength. Geometrical optics cannot be used for such fibers.

Although graded-index silica fibers are rarely used for long-haul links, the use of graded-index *plastic* fibers attracted considerable attention for data-link applications during the 1990s and has become quite common with the advent of the Gigabit Ethernet. The next section is devoted to the design and fabrication issues related to silica and plastic optical fibers.

1.2 Design and Fabrication

In this section we discuss the engineering aspects of optical fibers made using either silica glass or a suitable plastic material. Manufacturing of fiber cables, suitable for use in an actual lightwave system, involves sophisticated technology with attention to many practical details available in several books [18]–[20]. We begin with silica fibers and then consider plastic fibers. Both types of materials have been used in recent years to make microstructure fibers discussed in a separate subsection.

1.2.1 Silica Fibers

In the case of silica fibers, both the core and the cladding are made using silicon dioxide (SiO_2) or silica as the base material. The difference in their refractive indices is realized by doping the core, or the cladding, or both with a suitable material. Dopants such as GeO_2 and P_2O_5 increase the refractive index of silica and are suitable for the core. On the other hand, dopants such as B_2O_3 and fluorine decrease the refractive index of silica and are suitable for the cladding. The major design issues are related to the refractive-index profile, the amount of dopants, and the core and cladding dimensions [21]–[25]. The diameter of the outermost cladding layer has the standard value of 125 μm for all communication-grade silica fibers.

Figure 1.5 shows typical index profiles that have been used for different kinds of fibers. The top row corresponds to standard fibers that are designed to have minimum

1.2 Design and Fabrication

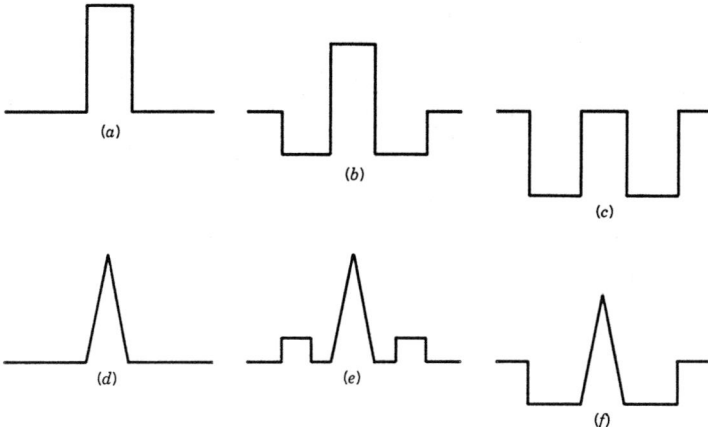

Figure 1.5: Several index profiles used in the design of single-mode fibers. Upper and lower rows correspond to standard and dispersion-shifted fibers, respectively.

dispersion near 1.3 μm. The simplest design [Figure 1.5(a)] consists of a pure-silica cladding and a core doped with GeO_2 to obtain $\Delta \approx 3 \times 10^{-3}$. A commonly used variation [Figure 1.5(b)] lowers the cladding index over a region adjacent to the core by doping it with fluorine. It is also possible to have an undoped core by using a design shown in Figure 1.5(c). The fibers of this kind are referred to as doubly clad or *depressed-cladding fibers* [21]. They are also called W fibers, reflecting the shape of the index profile. The bottom row in Figure 1.5 shows three index profiles used for dispersion-shifted fibers for which the zero-dispersion wavelength is chosen in the range 1.45–1.60 μm. A triangular index profile with a depressed or raised cladding is often used for this purpose [22]–[24]. The refractive indices and the thickness of different layers are optimized to design a fiber with desirable dispersion characteristics [25]. Sometimes as many as four cladding layers are used for controlling the dispersion of a silica fiber.

Fabrication of telecommunication-grade silica fibers involves two stages. In the first stage a vapor-deposition method is used to make a *cylindrical preform* with the desired refractive-index profile. The preform is typically 1 m long and 2 cm in diameter and contains core and cladding layers with correct relative dimensions. In the second stage, the preform is drawn into a fiber by using a precision-feed mechanism that feeds the preform into a furnace at the proper speed.

Several methods can be used to make the preform. The three commonly used methods [18] are known as modified chemical vapor deposition (MCVD), outside-vapor deposition (OVD), and vapor-axial deposition (VAD). Figure 1.6 shows a schematic diagram of the MCVD process. In this process, successive layers of SiO_2 are deposited on the inside of a fused silica tube by mixing the vapors of $SiCl_4$ and O_2 at a temperature of about 1800°C. To ensure uniformity, a multiburner torch is moved back and forth across the tube length using an automatic translation stage. The refractive index of the cladding layers is controlled by adding fluorine to the tube. When a sufficient cladding thickness has been deposited, the core is formed by adding the vapors

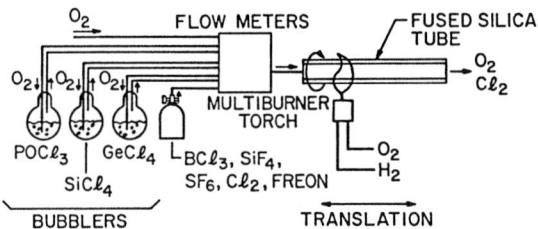

Figure 1.6: MCVD process commonly used for fiber fabrication.

of $GeCl_4$ or $POCl_3$. The underlying chemical reactions forming the core and cladding can be written as

$$SiCl_4 + O_2 \rightarrow SiO_2 + 2Cl_2,$$
$$GeCl_4 + O_2 \rightarrow GeO_2 + 2Cl_2,$$
$$4POCl_3 + 3O_2 \rightarrow 2P_2O_5 + 6Cl_2.$$

The flow rate of $GeCl_4$ or $POCl_3$ determines the amount of dopant and the corresponding increase in the refractive index of the core. A triangular-index core can be fabricated simply by varying the flow rate from layer to layer. When all layers forming the core have been deposited, the torch temperature is raised to collapse the tube into a solid rod of preform.

The MCVD process is also known as the *inner-vapor-deposition method*, as the core and cladding layers are deposited inside a silica tube. In a related process, known as the *plasma-activated chemical vapor deposition* process [26], the chemical reaction is initiated by a microwave plasma. By contrast, in the OVD and VAD processes the core and cladding layers are deposited on the outside of a rotating mandrel by using the technique of *flame hydrolysis*. The mandrel is removed prior to sintering. The porous soot boule is then placed in a sintering furnace to form a glass boule. The central hole allows an efficient way of reducing water vapors through dehydration in a controlled atmosphere of Cl_2–He mixture, although it results in a central dip in the index profile. The dip can be minimized by closing the hole during sintering.

The fiber drawing step is essentially the same irrespective of the process used to make the preform [18]. Figure 1.7 shows the drawing apparatus schematically. The preform is fed into a furnace in a controlled manner where it is heated to a temperature of about 2000°C. The melted preform is drawn into a fiber by using a precision-feed mechanism. The fiber diameter is monitored optically by diffracting light emitted by a laser from the fiber. A change in the diameter changes the diffraction pattern, which in turn changes the photodiode current. This current change acts as a signal for a servocontrol mechanism that adjusts the winding rate of the fiber. The fiber diameter can be kept constant to within 0.1% by this technique. A polymer coating is applied to the fiber during the drawing step. It serves a dual purpose, as it provides mechanical protection and preserves the transmission properties of the fiber. The diameter of the coated fiber is typically 250 μm, although it can be as large as 900 μm when multiple coatings are used. The tensile strength of the fiber is monitored during its winding

1.2 Design and Fabrication

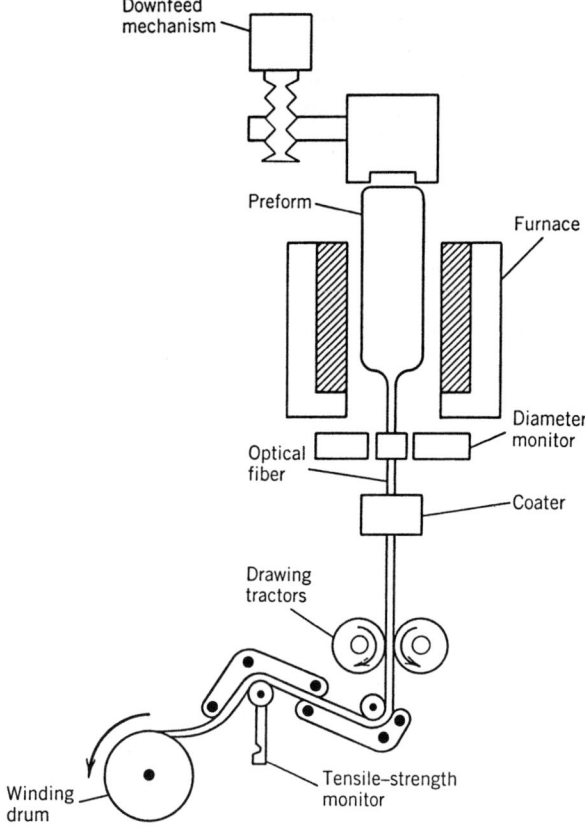

Figure 1.7: Apparatus used for fiber drawing.

on the drum. The winding rate is typically 0.2–0.5 m/s. Several hours are required to convert a single preform into a fiber of about 5 km length. This brief discussion is intended to give a general idea. Fabrication of optical fibers generally requires attention to a large number of engineering details [19].

1.2.2 Plastic Optical Fibers

The interest in plastic optical fibers grew during the 1990s as the need for cheaper fibers capable of transmitting data over short distances (typically <1 km) became evident [27]–[33]. Such fibers have a relatively large core (diameter as large as 1 mm), resulting in a high numerical aperture and high coupling efficiency, but they exhibit high losses (typically exceeding 50 dB/km). For this reason, they are used to transmit data at bit rates of up to 10 Gb/s over short distances (1 km or less). In a 1996 demonstration, a 10-Gb/s signal was transmitted over 0.5 km with a bit-error rate of less than 10^{-11} [28]. Graded-index plastic optical fibers provide an ideal solution for transferring data among computers and are becoming increasingly important for the Gigabit Ethernet and other Internet-related applications requiring bit rates in excess of 1 Gb/s.

As the name implies, plastic optical fibers use plastics in the form of organic polymers for making both the core and the cladding. The commonly used polymers for this purpose are polymethyl methacrylate (PMMA), polystyrene, polycarbonate, and an amorphous fluorinated polymer poly(perfluoro-butenylvinyl ether), or PFBVE, known commercially as CYTOP [32]. The PMMA plastic was used to make step-index fibers as early as 1968. By 1995, the technology had advanced enough that it was possible to make graded-index plastic fibers with a relatively large bandwidth [27]. Since then, considerable progress has been made in making new types of plastic fibers with relatively low losses even in the wavelength region near 1.3 μm [31]–[33]. The core diameter of plastic fibers can vary from 10 μm to 1 mm depending on the application. In the case of low-cost applications, the core size is typically 120 μm, while the cladding diameter approaches 200 μm.

Manufacturing of modern plastic fibers follows the same two-step process used for silica fibers in the sense that a preform is prepared first with the correct refractive-index profile and is then converted into the fiber form. An important technique used for making the preform for graded-index plastic fibers is known as the interfacial gel polymerization method [27]. In this technique, one begins with a hollow cylinder made of the polymer (such as PMMA) destined to be used for the cladding. This hollow cylinder is filled with a mixture of the monomer from which the cladding polymer was made, a dopant with higher refractive index than that of the cladding polymer, a chemical compound that helps in initiating the polymerization process, and another chemical known as the chain-transfer agent. The filled cylinder is heated to a temperature close to 95°C and rotated on its axis for a period of up to 24 hours.

The polymerization of the core begins near the inner wall of the cylinder because of the so-called gel effect and then gradually moves toward the center of the tube. More specifically, the monomer penetrates the inner tube wall and produces a swollen "gel" phase in which the process of polymerization begins. Since larger dopant molecules are partially excluded from the gel phase, the dopant concentration increases gradually toward the core center. At the end of the polymerization process, one ends up with a graded-index preform in the form of a solid cylinder. Many dopants can be used to increase the refractive index of the core depending on the design. For example, when PMMA with a refractive index of 1.492 is used for the cladding, one can employ diphenyl sulfure ($n = 1.633$), bromobenzen ($n = 1.60$), benzyl benzoate ($n = 1.568$), benzyl n-buytl phthalate ($n = 1.54$), or triphenyl phosphate ($n = 1.536$). A diffusion technique is also used for making preforms when a flourinated polymer is used. The index gradient in this case is produced by diffusing an index-raising dopant into a solid cylinder made of the cladding polymer.

The fiber-drawing step is identical to that used for silica fibers. The drawing apparatus similar to that shown in Figure 1.7 is used for this purpose. The main difference is that the melting temperature of plastics is much lower than that of silica (about 200°C in place of 1800°C). At a temperature near 200°C, a plastic material acquires enough viscosity that it begins to extrude downward. The extrusion velocity and the speed at which the winding drum rotates are adjusted using a computer-controlled mechanism so that the drawn fiber has the desired constant diameter. During the drawing process,

1.2 Design and Fabrication

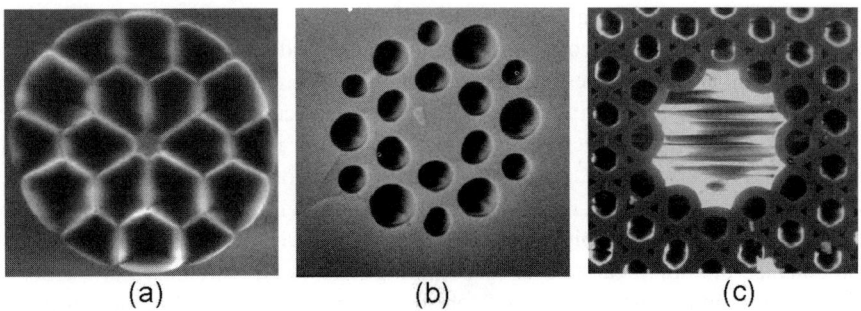

Figure 1.8: Scanning electron micrographs of three microstructure fibers.

the relative ratio between the core and the cladding is maintained. The fiber diameter is continuously monitored using a suitable optical technique, and another plastic coating is applied to the fiber. The top plastic coating protects the fiber against microbending and facilitates its handling. When the coating has hardened, the tensile strength of the fiber is tested by passing the coated fiber through several pulleys.

1.2.3 Microstructure Optical Fibers

During the 1990s, new types of fibers were developed in which the cladding region contained multiple tiny air holes [34]–[44]. The hole diameter was typically ~ 1 μm but can be made smaller. Such fibers are known by several names such as holey fibers, microstructure fibers, or photonic-crystal fibers. They can be divided into two main categories. In the first category, light is still guided by total internal reflection, and the air holes are mostly used to reduce the index of the cladding region adjacent of the core, resulting in a tighter mode confinement. In the second category, the air holes form a periodic pattern, resulting in a two-dimensional photonic crystal, which can confine the light to the core region even when the core region has a lower refractive index then the cladding material. In the extreme case, the core may only contain air in which light remains confined because of the existence of a photonic bandgap resulting from the periodic variation of the refractive index in the cladding.

Figure 1.8 shows three examples of microstructure fibers. In part (a), the air holes surrounding a narrow silica core are so large in size that the core is mostly surrounded by air. This design is sometimes referred to as the "grapefruit" structure because of its appearance. In part (b), the air holes surrounding the silica core are nonuniform in size and do not extend much further from the core. Both of these designs correspond to the first category because the waveguiding is provided by the larger refractive index of the central silica core compared with the cladding. In contrast, the part (c) shows a photonic-crystal fiber in which air holes surrounding the central low-index core follow a regular periodic pattern, and it is this periodicity that guides any light propagating inside the air core.

A simple technique for fabricating microstructure fibers consists of stacking multiple capillary tubes of pure silica (diameter about 1 mm) in a hexagonal pattern around a solid silica rod and drawing such a preform into a fiber form using a fiber-drawing apparatus [34]. A polymer coating is added on the outside to protect the resulting microstructure fiber. When viewed under a scanning electron microscope, such a fiber shows a two-dimensional pattern of air holes around the central region acting as a core.

The interest in microstructure fibers stems from their interesting nonlinear and dispersive properties [36]–[38]. As discussed later in Section 1.5, traditional silica fibers exhibit "normal" dispersion in the wavelength region below 1.3 μm. In contrast, microstructure fibers can exhibit anomalous dispersion at visible wavelengths. They can also enhance the nonlinear effects by confining light to an extremely narrow central core region (about 1 μm in size). Such fibers are sometimes called highly nonlinear fibers, although this name is misleading because the intrinsic nonlinearity of silica does not change. Rather, it is the tighter confinement of light that enhances the optical intensity.

The photonic-crystal fibers with an air or vacuum core [see Figure 1.8(c)] are attracting attention because they provide an environment in which light can propagate without experiencing much losses. At the same time, the dispersive and nonlinear effects are likely to be drastically reduced or even completely eliminated. Such fibers offer the possibility of transmitting high-power signals over long lengths. They can be made using the same technique as other microstructure fibers with a silica core, that is, by stacking capillary tubes of pure silica in a hexagonal pattern. In the design shown in Figure 1.8(c) the central hole was created by removing seven capillary tubes (one at the center and six surrounding it) before the preform was drawn into a fiber. Such a fiber was found to transmit light along the central core in several frequency bands in the visible and infrared regions [40]. This transmission is attributed to an optical mode that is created by the two-dimensional periodicity and confined to the central air hole.

1.2.4 Cables and Connectors

Cabling of optical fibers is necessary to protect them from deterioration during transportation and installation [45]. Cable design depends on the type of application. For some applications it may be enough to buffer the fiber by placing it inside a plastic jacket. For others the cable must be made mechanically strong by using strengthening elements such as steel rods.

A light-duty cable is made by surrounding the fiber by a buffer jacket of hard plastic. A tight jacket can be provided by applying a buffer plastic coating of 0.5–1 mm thickness on top of the primary coating applied during the drawing process. In an alternative approach the fiber lies loosely inside a plastic tube. Microbending losses are nearly eliminated in this loose-tube construction, since the fiber can adjust itself within the tube. This construction can also be used to make multifiber cables by using a slotted tube with a different slot for each fiber.

Heavy-duty cables use steel or a strong polymer such as Kevlar to provide the mechanical strength. Figure 1.9 shows schematically three kinds of cables. In the loose-tube construction, fiberglass rods embedded in polyurethane and a Kevlar jacket provide the necessary mechanical strength (left drawing). The same design can be extended to multifiber cables by placing several loose-tube fibers around a central steel core (middle drawing). When a large number of fibers need to be placed inside a single

1.3 Fiber Modes

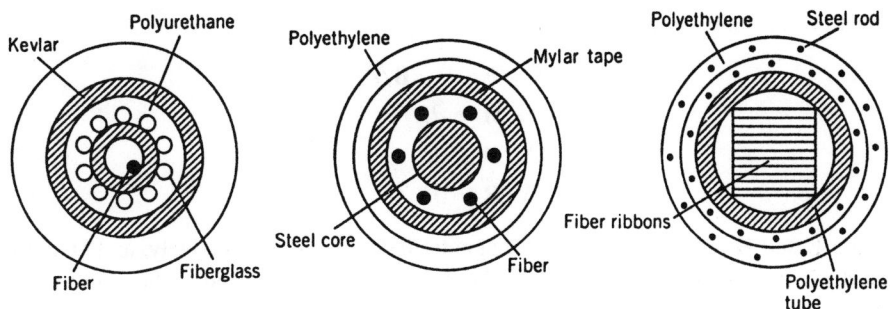

Figure 1.9: Typical designs for heavy-duty fiber cables.

cable, a ribbon cable is used (right drawing). The ribbon is manufactured by packaging typically 12 fibers between two polyester tapes. Several ribbons are then stacked into a rectangular array that is placed inside a polyethylene tube. The mechanical strength is provided by using steel rods in the two outermost polyethylene jackets. The outer diameter of such fiber cables is about 1–1.5 cm.

Connectors are needed to use optical fibers in any actual lightwave system. They can be divided into two categories. A permanent joint between two fibers is known as a fiber splice, and a detachable connection between them is realized by using a fiber connector. Connectors are used to link fiber cable with the transmitter (or the receiver), while splices are used to join two fiber segments permanently. The main issue in the use of splices and connectors is related to the loss. Some power is always lost, as the two fiber ends are never perfectly aligned in practice. Splice losses below 0.1 dB are routinely realized by using the technique of fusion splicing [46]. Connector losses are generally larger. State-of-the-art connectors provide an average loss of about 0.3 dB [47]. The technology behind the design of splices and connectors is quite sophisticated. For details, the reader is referred to Ref. [48], a book devoted entirely to this issue.

1.3 Fiber Modes

To understand the optical properties of fibers, it is first important to understand the concept of fiber modes. In this section we introduce this concept by solving the Maxwell equations for step-index fibers. It turns out that a single parameter V governs the number of modes supported by a fiber. By controlling the numerical value of this parameter, one can design a step-index fiber such that it supports only a single mode. Since single-mode fibers are used often in practice, we discuss the properties of such fibers in detail.

1.3.1 Maxwell's Equations

Like all electromagnetic phenomena, propagation of optical fields in fibers is governed by Maxwell's equations. For a nonconducting medium without free charges, these equations take the form (in MKS units)

$$\nabla \times \mathbf{E} = -\partial \mathbf{B}/\partial t, \tag{1.3.1}$$
$$\nabla \times \mathbf{H} = \partial \mathbf{D}/\partial t, \tag{1.3.2}$$
$$\nabla \cdot \mathbf{D} = 0, \tag{1.3.3}$$
$$\nabla \cdot \mathbf{B} = 0, \tag{1.3.4}$$

where $\mathbf{E}$ and $\mathbf{H}$ are the electric and magnetic field vectors, respectively, and $\mathbf{D}$ and $\mathbf{B}$ are the corresponding flux densities. The flux densities are related to the field vectors by the constitutive relations [49]

$$\mathbf{D} = \varepsilon_0 \mathbf{E} + \mathbf{P}, \tag{1.3.5}$$
$$\mathbf{B} = \mu_0 \mathbf{H} + \mathbf{M}, \tag{1.3.6}$$

where ε_0 is the vacuum permittivity, μ_0 is the vacuum permeability, and $\mathbf{P}$ and $\mathbf{M}$ are the induced electric and magnetic polarizations, respectively. For optical fibers $\mathbf{M} = 0$ because of the nonmagnetic nature of silica glass.

Evaluation of the electric polarization $\mathbf{P}$ requires a microscopic quantum-mechanical approach. Although such an approach is essential when the optical frequency is near a medium resonance, a phenomenological relation between $\mathbf{P}$ and $\mathbf{E}$ can be used far from medium resonances. This is the case for silica fibers in the wavelength region 0.5–2 μm, a range that covers the low-loss region that is of interest for lightwave systems. In general, the relation between $\mathbf{P}$ and $\mathbf{E}$ can be nonlinear. Although the nonlinear effects in optical fibers are of considerable interest [50] and are discussed in Section 1.6, they can be ignored in a discussion of fiber modes. In the linear case, $\mathbf{P}$ is then related to $\mathbf{E}$ by the relation

$$\mathbf{P}(\mathbf{r},t) = \varepsilon_0 \int_{-\infty}^{\infty} \chi(\mathbf{r}, t-t') \mathbf{E}(\mathbf{r},t') \, dt'. \tag{1.3.7}$$

Linear susceptibility χ is, in general, a second-rank tensor but reduces to a scalar for an isotropic medium such as silica glass. Optical fibers become slightly birefringent because of unintentional variations in the core shape or in local strain; such birefringent effects are considered later in this section. Equation (1.3.7) includes the delayed nature of the temporal response, a feature that has important implications for optical fibers because it leads to chromatic dispersion.

Equations (1.3.1) through (1.3.7) provide a general formalism for studying wave propagation in optical fibers. In practice, it is convenient to use a single field variable $\mathbf{E}$. By taking the curl of Eq. (1.3.1) and using Eqs. (1.3.2), (1.3.5), and (1.3.6), we obtain the wave equation

$$\nabla \times \nabla \times \mathbf{E} = -\frac{1}{c^2} \frac{\partial^2 \mathbf{E}}{\partial t^2} - \mu_0 \frac{\partial^2 \mathbf{P}}{\partial t^2}, \tag{1.3.8}$$

where the speed of light in vacuum is defined as usual by $c = (\mu_0 \varepsilon_0)^{-1/2}$. By introducing the Fourier transform of $\mathbf{E}(\mathbf{r},t)$ through the relation

$$\tilde{\mathbf{E}}(\mathbf{r},\omega) = \int_{-\infty}^{\infty} \mathbf{E}(\mathbf{r},t) \exp(i\omega t) \, dt, \tag{1.3.9}$$

1.3 Fiber Modes

as well as a similar relation for $\mathbf{P}(\mathbf{r},t)$ in Eq. (1.3.7), Eq. (1.3.8) can be written in the frequency domain as

$$\nabla \times \nabla \times \tilde{\mathbf{E}} = -\varepsilon(\mathbf{r},\omega)(\omega^2/c^2)\tilde{\mathbf{E}}, \tag{1.3.10}$$

where the frequency-dependent *dielectric constant* is defined as

$$\varepsilon(\mathbf{r},\omega) = 1 + \tilde{\chi}(\mathbf{r},\omega), \tag{1.3.11}$$

and $\tilde{\chi}(\mathbf{r},\omega)$ is the Fourier transform of $\chi(\mathbf{r},t)$. In general, $\varepsilon(\mathbf{r},\omega)$ is complex. Its real and imaginary parts are related to the *refractive index* n and the *absorption coefficient* α by the definition

$$\varepsilon = (n + i\alpha c/2\omega)^2. \tag{1.3.12}$$

From Eqs. (1.3.11) and (1.3.12), n and α are related to $\tilde{\chi}$ as

$$n = (1 + \operatorname{Re}\tilde{\chi})^{1/2}, \tag{1.3.13}$$

$$\alpha = (\omega/nc)\operatorname{Im}\tilde{\chi}, \tag{1.3.14}$$

where Re and Im stand for the real and imaginary parts, respectively. Both n and α are frequency dependent. The frequency dependence of n is referred to as *chromatic dispersion* or simply as material dispersion.

Two further simplifications can be made before solving Eq. (1.3.10). First, ε can be taken to be real and replaced by n^2 because of low optical losses in silica fibers. Second, since $n(\mathbf{r},\omega)$ is independent of the spatial coordinate $\mathbf{r}$ in both the core and the cladding of a step-index fiber, one can use the identity

$$\nabla \times \nabla \times \tilde{\mathbf{E}} \equiv \nabla(\nabla \cdot \tilde{\mathbf{E}}) - \nabla^2 \tilde{\mathbf{E}} = -\nabla^2 \tilde{\mathbf{E}}, \tag{1.3.15}$$

where we used Eq. (1.3.3) and the relation $\tilde{\mathbf{D}} = \varepsilon \tilde{\mathbf{E}}$ to set $\nabla \cdot \tilde{\mathbf{E}} = 0$. This simplification is made even for graded-index fibers. Equation (1.3.15) then holds approximately as long as the index changes occur over a length scale much longer than the wavelength. By using Eq. (1.3.15) in Eq. (1.3.10), we obtain the Helmholtz equation,

$$\nabla^2 \tilde{\mathbf{E}} + n^2(\omega)k_0^2 \tilde{\mathbf{E}} = 0, \tag{1.3.16}$$

where the free-space wave number k_0 is defined as

$$k_0 = \omega/c = 2\pi/\lambda, \tag{1.3.17}$$

and λ is the vacuum wavelength of the optical field oscillating at the frequency ω. Equation (1.3.16) is solved next to obtain the optical modes of step-index fibers.

1.3.2 Modes of Step-Index Fibers

An *optical mode* refers to a specific solution of the wave equation (1.3.16) that satisfies the appropriate boundary conditions and has the property that its spatial distribution does not change with propagation. The fiber modes can be classified as guided modes, leaky modes, and radiation modes [11]. As one might expect, signal transmission takes place through the guided modes only. The following discussion focuses exclusively on the guided modes of a step-index fiber.

To take advantage of the cylindrical symmetry, Eq. (1.3.16) should be written in the cylindrical coordinates ρ, ϕ, and z as

$$\frac{\partial^2 E_z}{\partial \rho^2} + \frac{1}{\rho}\frac{\partial E_z}{\partial \rho} + \frac{1}{\rho^2}\frac{\partial^2 E_z}{\partial \phi^2} + \frac{\partial^2 E_z}{\partial z^2} + n^2 k_0^2 E_z = 0, \quad (1.3.18)$$

where for a step-index fiber of core radius a, the refractive index n is of the form

$$n = \begin{cases} n_1; & \rho \leq a, \\ n_2; & \rho > a. \end{cases} \quad (1.3.19)$$

For simplicity of notation, the tilde over $\tilde{E}$ has been dropped and the frequency dependence of all variables is implicitly understood. Equation (1.3.18) is written for the axial component E_z of the electric field vector. Similar equations can be written for the other five components of **E** and **H**. However, it is not necessary to solve all six equations since only two components out of six are independent. It is customary to choose E_z and H_z as the independent components and obtain E_ρ, E_ϕ, H_ρ, and H_ϕ in terms of them. Equation (1.3.18) is easily solved by using the method of separation of variables and writing E_z as

$$E_z(\rho, \phi, z) = F(\rho)\Phi(\phi)Z(z). \quad (1.3.20)$$

By using Eq. (1.3.20) in Eq. (1.3.18), we obtain the following three ordinary differential equations:

$$d^2Z/dz^2 + \beta^2 Z = 0, \quad (1.3.21)$$

$$d^2\Phi/d\phi^2 + m^2 \Phi = 0, \quad (1.3.22)$$

$$\frac{d^2 F}{d\rho^2} + \frac{1}{\rho}\frac{dF}{d\rho} + \left(n^2 k_0^2 - \beta^2 - \frac{m^2}{\rho^2}\right) F = 0. \quad (1.3.23)$$

Equation (1.3.21) has a solution of the form $Z = \exp(i\beta z)$, where β has the physical significance of the propagation constant. Similarly, Eq. (1.3.22) has a solution $\Phi = \exp(im\phi)$, but the constant m is restricted to take only integer values since the field must be periodic in ϕ with a period of 2π.

Equation (1.3.23) is the well-known differential equation satisfied by the Bessel functions [51]. Its general solution in the core and cladding regions can be written as

$$F(\rho) = \begin{cases} AJ_m(p\rho) + A'Y_m(p\rho); & \rho \leq a, \\ CK_m(q\rho) + C'I_m(q\rho); & \rho > a, \end{cases} \quad (1.3.24)$$

where A, A', C, and C' are constants and J_m, Y_m, K_m, and I_m are different kinds of Bessel functions [51]. The parameters p and q are defined by

$$p^2 = n_1^2 k_0^2 - \beta^2, \quad (1.3.25)$$

$$q^2 = \beta^2 - n_2^2 k_0^2. \quad (1.3.26)$$

Considerable simplification occurs when we use the boundary condition that the optical field for a guided mode should be finite at $\rho = 0$ and decay to zero at $\rho = \infty$. Since

1.3 Fiber Modes

$Y_m(p\rho)$ has a singularity at $\rho = 0$, $F(0)$ can remain finite only if $A' = 0$. Similarly $F(\rho)$ vanishes at infinity only if $C' = 0$. The general solution of Eq. (1.3.18) is thus of the form

$$E_z = \begin{cases} AJ_m(p\rho)\exp(im\phi)\exp(i\beta z); & \rho \leq a, \\ CK_m(q\rho)\exp(im\phi)\exp(i\beta z); & \rho > a. \end{cases} \quad (1.3.27)$$

The same method can be used to obtain H_z which also satisfies Eq. (1.3.18). Indeed, the solution is the same but with different constants B and D, that is,

$$H_z = \begin{cases} BJ_m(p\rho)\exp(im\phi)\exp(i\beta z); & \rho \leq a, \\ DK_m(q\rho)\exp(im\phi)\exp(i\beta z); & \rho > a. \end{cases} \quad (1.3.28)$$

The other four components E_ρ, E_ϕ, H_ρ, and H_ϕ can be expressed in terms of E_z and H_z by using the Maxwell equations. In the core region, we obtain

$$E_\rho = \frac{i}{p^2}\left(\beta\frac{\partial E_z}{\partial \rho} + \mu_0\frac{\omega}{\rho}\frac{\partial H_z}{\partial \phi}\right), \quad (1.3.29)$$

$$E_\phi = \frac{i}{p^2}\left(\frac{\beta}{\rho}\frac{\partial E_z}{\partial \phi} - \mu_0\omega\frac{\partial H_z}{\partial \rho}\right), \quad (1.3.30)$$

$$H_\rho = \frac{i}{p^2}\left(\beta\frac{\partial H_z}{\partial \rho} - \varepsilon_0 n^2\frac{\omega}{\rho}\frac{\partial E_z}{\partial \phi}\right), \quad (1.3.31)$$

$$H_\phi = \frac{i}{p^2}\left(\frac{\beta}{\rho}\frac{\partial H_z}{\partial \phi} + \varepsilon_0 n^2\omega\frac{\partial E_z}{\partial \rho}\right). \quad (1.3.32)$$

These equations can be used in the cladding region after replacing p^2 by $-q^2$.

Equations (1.3.27) through (1.3.32) express the electromagnetic field in the core and cladding regions of an optical fiber in terms of four constants A, B, C, and D. These constants are determined by applying the boundary condition that the tangential components of **E** and **H** be continuous across the core–cladding interface. By requiring the continuity of E_z, H_z, E_ϕ, and H_ϕ at $\rho = a$, we obtain a set of four homogeneous equations satisfied by A, B, C, and D [14]. These equations have a nontrivial solution only if the determinant of the coefficient matrix vanishes. After considerable algebraic details, this condition leads us to the following eigenvalue equation [14]–[16]:

$$\left[\frac{J'_m(pa)}{pJ_m(pa)} + \frac{K'_m(qa)}{qK_m(qa)}\right]\left[\frac{J'_m(pa)}{pJ_m(pa)} + \frac{n_2^2}{n_1^2}\frac{K'_m(qa)}{qK_m(qa)}\right] = \frac{m^2}{a^2}\left(\frac{1}{p^2} + \frac{1}{q^2}\right)\left(\frac{1}{p^2} + \frac{n_2^2}{n_1^2}\frac{1}{q^2}\right), \quad (1.3.33)$$

where a prime indicates differentiation with respect to the argument.

For a given set of the parameters k_0, a, n_1, and n_2, the eigenvalue equation (1.3.33) can be solved numerically to determine the propagation constant β. In general, it may have multiple solutions for each integer value of m. It is customary to enumerate these solutions in descending numerical order and denote them by β_{mn} for a given m ($n = 1, 2, \ldots$). Each value β_{mn} corresponds to one possible mode of propagation of the optical field whose spatial distribution is obtained from Eqs. (1.3.27) through

(1.3.32). Since the field distribution does not change with propagation except for a phase factor and satisfies all boundary conditions, it is an optical mode of the fiber. In general, both E_z and H_z are nonzero (except for $m = 0$), in contrast to the planar waveguides, for which one of them can be taken to be zero. Fiber modes are therefore referred to as *hybrid modes* and are denoted by HE_{mn} or EH_{mn}, depending on whether H_z or E_z dominates. In the special case $m = 0$, HE_{0n} and EH_{0n} are also denoted by TE_{0n} and TM_{0n}, respectively, since they correspond to transverse-electric ($E_z = 0$) and transverse-magnetic ($H_z = 0$) modes of propagation. A different notation LP_{mn} is sometimes used for weakly guiding fibers [52] for which both E_z and H_z are nearly zero (LP stands for linearly polarized modes).

A mode is uniquely determined by its propagation constant β. It is useful to introduce a quantity $\bar{n} = \beta/k_0$, called the *mode index* or *effective index* and having the physical significance that each fiber mode propagates with an effective refractive index $\bar{n}$ whose value lies in the range $n_1 > \bar{n} > n_2$. A mode ceases to be guided when $\bar{n} \leq n_2$. This can be understood by noting that the optical field of guided modes decays exponentially inside the cladding layer since [51]

$$K_m(q\rho) = (\pi/2q\rho)^{1/2}\exp(-q\rho) \quad \text{for} \quad q\rho \gg 1. \tag{1.3.34}$$

When $\bar{n} \leq n_2$, $q^2 \leq 0$ from Eq. (1.3.26) and the exponential decay does not occur. The mode is said to reach cutoff when q becomes zero or when $\bar{n} = n_2$. From Eq. (1.3.25), $p = k_0(n_1^2 - n_2^2)^{1/2}$ when $q = 0$. A parameter that plays an important role in determining the *cutoff condition* is defined as

$$V = k_0 a(n_1^2 - n_2^2)^{1/2} \approx (2\pi/\lambda)an_1\sqrt{2\Delta}. \tag{1.3.35}$$

It is called the *normalized frequency* ($V \propto \omega$) or simply the V parameter. It is also useful to introduce a normalized propagation constant b as

$$b = \frac{\beta/k_0 - n_2}{n_1 - n_2} = \frac{\bar{n} - n_2}{n_1 - n_2}. \tag{1.3.36}$$

Figure 1.10 shows b as a function of V for several low-order fiber modes [53] obtained by solving the eigenvalue equation (1.3.33). A fiber with a large value of V supports many modes. A rough estimate of the number of modes for such a multimode fiber is given by $V^2/2$ [15]. For example, a typical multimode fiber with $a = 25$ μm and $\Delta = 5 \times 10^{-3}$ has $V \simeq 18$ at $\lambda = 1.3$ μm and would support about 162 modes. However, the number of modes decreases rapidly as V is reduced. As seen in Figure 1.10, a fiber with $V = 5$ supports seven modes. Below a certain value of V all modes except the HE_{11} mode reach cutoff. Such fibers support a single mode and are called single-mode fibers. The properties of single-mode fibers are described next.

1.3.3 Properties of Single-Mode Fibers

Single-mode fibers support only the HE_{11} mode, also known as the fundamental mode of the fiber. The fiber is designed such that all higher-order modes are cut off at the operating wavelength. As seen in Figure 1.10, the V parameter determines the number

1.3 Fiber Modes

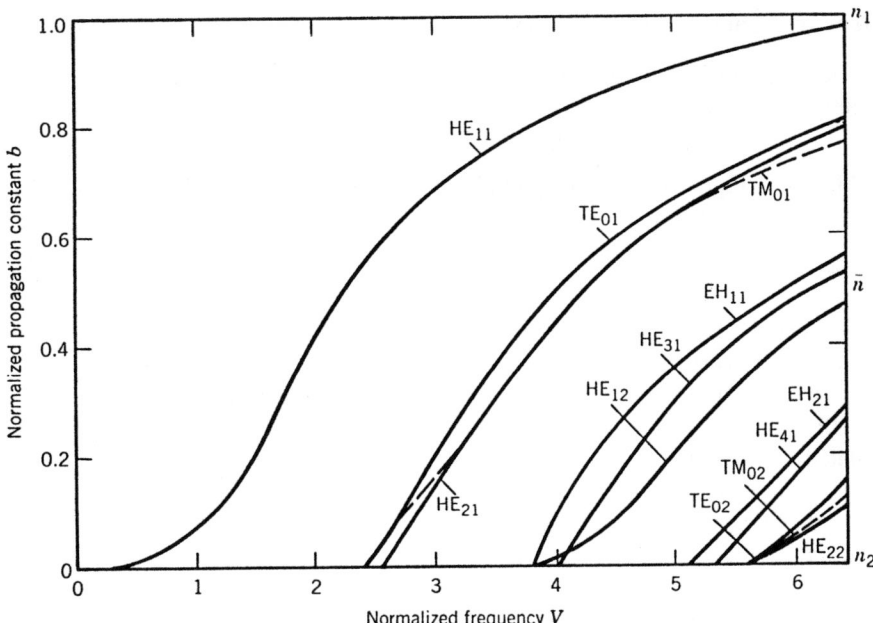

Figure 1.10: Normalized propagation constant b as a function of normalized frequency V for a few low-order fiber modes. The right scale shows the mode index $\bar{n}$. (After Ref. [53].)

of modes supported by a fiber. The cutoff condition of various modes is also determined by V. The fundamental mode has no cutoff and is always supported by a fiber.

The *single-mode condition* is determined by the value of V at which the TE_{01} and TM_{01} modes reach cutoff (see Figure 1.10). The eigenvalue equations for these two modes can be obtained by setting $m = 0$ in Eq. (1.3.33) and are given by

$$pJ_0(pa)K_0'(qa) + qJ_0'(pa)K_0(qa) = 0, \qquad (1.3.37)$$
$$pn_2^2 J_0(pa)K_0'(qa) + qn_1^2 J_0'(pa)K_0(qa) = 0. \qquad (1.3.38)$$

A mode reaches cutoff when $q = 0$. Since $pa = V$ when $q = 0$, the cutoff condition for both modes is simply given by $J_0(V) = 0$. The smallest value of V for which $J_0(V) = 0$ is 2.405. A fiber designed such that $V < 2.405$ supports only the fundamental HE_{11} mode. This is the single-mode condition.

We can use Eq. (1.3.35) to estimate the core radius of single-mode fibers used in lightwave systems. For the operating wavelength range 1.3–1.6 μm, the fiber is generally designed to become single mode for $\lambda > 1.2$ μm. By taking $\lambda = 1.2$ μm, $n_1 = 1.45$, and $\Delta = 5 \times 10^{-3}$, Eq. (1.3.35) shows that $V < 2.405$ for a core radius $a < 3.2$ μm. The required core radius can be increased to about 4 μm by decreasing Δ to 3×10^{-3}. Indeed, most telecommunication fibers are designed with $a \approx 4$ μm.

Mode Index and Field Distribution

The mode index $\bar{n}$ at the operating wavelength can be obtained by using Eq. (1.3.36), according to which

$$\bar{n} = n_2 + b(n_1 - n_2) \approx n_2(1 + b\Delta) \tag{1.3.39}$$

and by using Figure 1.10, which provides b as a function of V for the HE_{11} mode. An analytic approximation for b is [13]

$$b(V) \approx (1.1428 - 0.9960/V)^2 \tag{1.3.40}$$

and is accurate to within 0.2% for V in the range 1.5–2.5.

The field distribution of the fundamental mode is obtained by using Eqs. (1.3.27) through (1.3.32). The axial components E_z and H_z are quite small for $\Delta \ll 1$. Hence, the HE_{11} mode is approximately linearly polarized for weakly guiding fibers. It is also denoted as LP_{01}, following an alternative terminology in which all fiber modes are assumed to be linearly polarized [52]. One of the transverse components can be taken as zero for a linearly polarized mode. If we set $E_y = 0$, the E_x component of the electric field for the HE_{11} mode is given by [13]

$$E_x = E_0 \begin{cases} [J_0(p\rho)/J_0(pa)]\exp(i\beta z); & \rho \leq a, \\ [K_0(q\rho)/K_0(qa)]\exp(i\beta z); & \rho > a, \end{cases} \tag{1.3.41}$$

where E_0 is a constant related to the power carried by the mode. The dominant component of the corresponding magnetic field is given by $H_y = n_2(\varepsilon_0/\mu_0)^{1/2} E_x$. This mode is linearly polarized along the x axis. The same fiber supports another mode linearly polarized along the y axis. In this sense a single-mode fiber actually supports two orthogonally polarized modes that are degenerate and have the same mode index.

Fiber Birefringence

The degenerate nature of the orthogonally polarized modes holds only for an ideal single-mode fiber with a perfectly cylindrical core of uniform diameter. Real fibers exhibit considerable variation in the shape of their core along the fiber length. They may also experience nonuniform stress such that the cylindrical symmetry of the fiber is broken. Degeneracy between the orthogonally polarized fiber modes is removed because of these factors, and the fiber acquires birefringence. The degree of modal birefringence is defined by

$$B_m = |\bar{n}_x - \bar{n}_y|, \tag{1.3.42}$$

where $\bar{n}_x$ and $\bar{n}_y$ are the mode indices for the orthogonally polarized fiber modes. Birefringence leads to a periodic power exchange between the two polarization components. The period, referred to as the *beat length*, is given by

$$L_B = \lambda/B_m. \tag{1.3.43}$$

Typically, $B_m \sim 10^{-7}$, and $L_B \sim 10$ m for $\lambda \sim 1$ μm. From a physical viewpoint, linearly polarized light remains linearly polarized only when it is polarized along one of the principal axes. Otherwise, its state of polarization changes along the fiber length

1.3 Fiber Modes

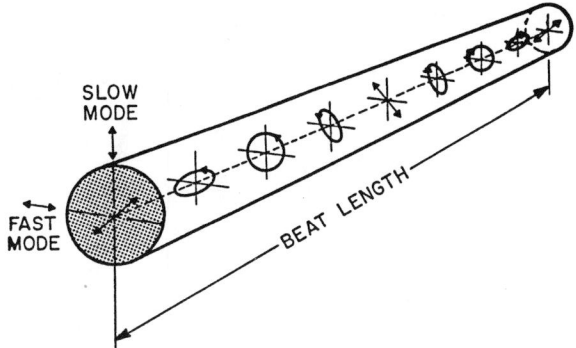

Figure 1.11: State of polarization in a birefringent fiber over one beat length. Input beam is linearly polarized at 45° with respect to the slow and fast axes.

from linear to elliptical, and then back to linear, in a periodic manner over the length L_B. Figure 1.11 shows schematically such a periodic change in the state of polarization for a fiber of constant birefringence B. The *fast axis* in this figure corresponds to the axis along which the mode index is smaller. The other axis is called the *slow axis*.

In conventional single-mode fibers, birefringence is not constant along the fiber but changes randomly, both in magnitude and direction, because of variations in the core shape (elliptical rather than circular) and the anisotropic stress acting on the core. As a result, light launched into the fiber with linear polarization quickly reaches a state of arbitrary polarization. Moreover, the orthogonally polarized components of the mode not only travel at different speeds but the speed difference changes randomly along the fiber length, resulting in pulse broadening. This phenomenon is called *polarization-mode dispersion* (PMD) and becomes a limiting factor for optical communication systems operating at high bit rates. It is possible to make fibers for which random fluctuations in the core shape and size are not the governing factor in determining the state of polarization. Such fibers are called *polarization-maintaining* fibers. A large amount of birefringence is introduced intentionally in these fibers through design modifications so that small random birefringence fluctuations do not affect the light polarization significantly. Typically, $B_m \sim 10^{-4}$ for such fibers.

Spot Size

Since the field distribution given by Eq. (1.3.41) is cumbersome to use in practice, it is often approximated by a *Gaussian distribution* of the form

$$E_x = A\exp(-\rho^2/w^2)\exp(i\beta z), \tag{1.3.44}$$

where w is the *field radius* and is referred to as the *spot size*. It is determined by fitting the exact distribution to the Gaussian function or by following a variational procedure [54]. Figure 1.12(a) shows the dependence of w/a on the V parameter. A comparison of the actual field distribution with the fitted Gaussian is shown in part (b) for $V = 2.4$. The quality of fit is generally quite good for values of V in the neigh-

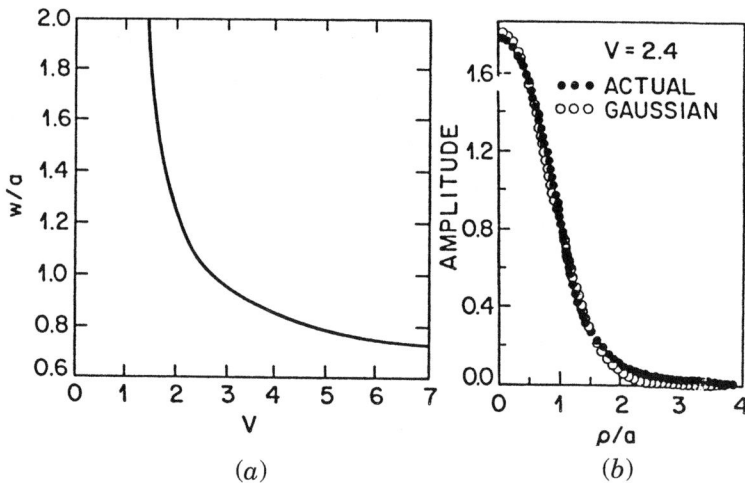

Figure 1.12: (a) Normalized spot size w/a as a function of the V parameter obtained by fitting the fundamental fiber mode to a Gaussian distribution; (b) quality of fit for $V = 2.4$. (After Ref. [54]; ©1978 OSA.)

borhood of 2. The spot size w can be determined from Figure 1.12(b). It can also be determined from an analytic approximation accurate to within 1% for $1.2 < V < 2.4$ and given by [54]

$$w/a \approx 0.65 + 1.619V^{-3/2} + 2.879V^{-6}. \qquad (1.3.45)$$

The effective core area, defined as $A_{\text{eff}} = \pi w^2$, is an important parameter for optical fibers as it determines how tightly light is confined to the core. It will be seen later that the nonlinear effects are stronger in fibers with smaller values of A_{eff}.

The fraction of the power contained in the core can be obtained by using Eq. (1.3.44) and is given by the *confinement factor*

$$\Gamma = \frac{P_{\text{core}}}{P_{\text{total}}} = \frac{\int_0^a |E_x|^2 \rho \, d\rho}{\int_0^\infty |E_x|^2 \rho \, d\rho} = 1 - \exp\left(-\frac{2a^2}{w^2}\right). \qquad (1.3.46)$$

Equations (1.3.45) and (1.3.46) determine the fraction of the mode power contained inside the core for a given value of V. Although nearly 75% of the mode power resides in the core for $V = 2$, this percentage drops down to 20% for $V = 1$. For this reason most single-mode fibers are designed to operate in the range $2 < V < 2.4$.

1.4 Fiber Losses

As mentioned earlier, fiber losses were the most limiting factor until low-loss silica fibers were developed during the 1970s. However, losses become of concern even for modern fibers whenever long lengths are required. This is the case for long-haul optical communication systems for which fiber length typically exceeds hundreds or

1.4 Fiber Losses

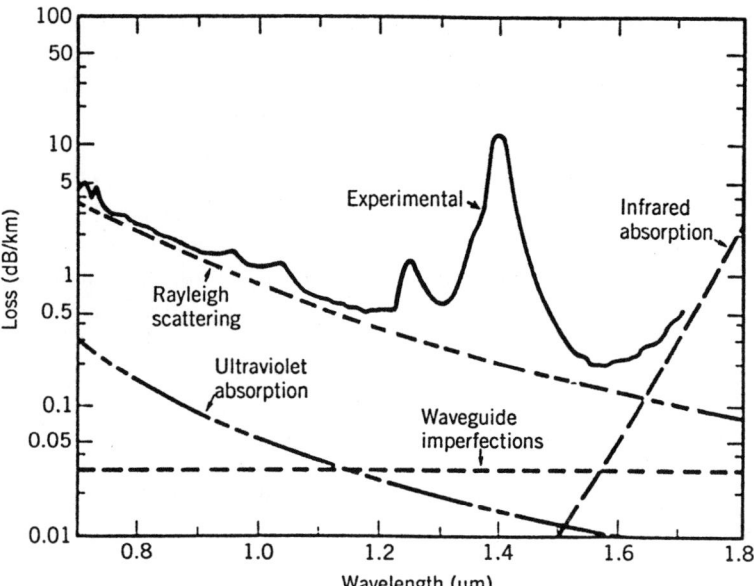

Figure 1.13: Loss spectrum of a single-mode fiber produced in 1979. Wavelength dependence of several fundamental loss mechanisms is also shown. (After Ref. [8]; ©1979 IEE.)

even thousands of kilometers. Losses can become an issue even for fiber-based devices and components whose length exceeds a few kilometers.

Under quite general conditions, changes in the average power of an optical beam propagating inside an optical fiber are governed by Beer's law:

$$P_{out} = P_{in} \exp(-\alpha L), \quad (1.4.1)$$

where P_{in} is the power launched at the input end of a fiber of length L and P_{out} is the output power. The parameter α is the loss parameter. Although denoted by the same symbol as the absorption coefficient in Eq. (1.3.12), α in Eq. (1.4.1) includes not only material absorption but also other sources of power attenuation. It is customary to express α in units of dB/km using the relation

$$\alpha\,(\text{dB/km}) = -\frac{10}{L} \log_{10}\left(\frac{P_{out}}{P_{in}}\right) \approx 4.343\alpha. \quad (1.4.2)$$

Fiber losses depend on the material used to make the fiber and are quite different for silica and polymer fibers. We consider the two types of fibers separately.

1.4.1 Loss Spectra of Silica Fibers

Figure 1.13 shows the loss spectrum $\alpha(\lambda)$ of a single-mode silica fiber made in 1979 with 9.4-μm core diameter, $\Delta = 1.9 \times 10^{-3}$, and 1.1-$\mu$m cutoff wavelength [8]. The fiber exhibited a loss of only about 0.2 dB/km in the wavelength region near 1.55 μm, the lowest value first realized in 1979. This value is close to the fundamental limit of about 0.16 dB/km for silica fibers. The loss spectrum exhibits a strong peak near 1.39 μm and several other smaller peaks, all of which are attributed to the presence of residual water vapors. A secondary minimum is found to occur near 1.3 μm, where the fiber loss is below 0.5 dB/km. Fiber losses are considerably higher at shorter wavelengths and exceed 5 dB/km in the visible region. Several factors contribute to overall losses; their relative contributions for silica fibers are also shown in Figure 1.13. We discuss the contribution of material absorption first. The contribution of Rayleigh scattering, a fundamental loss mechanism for all fibers, is discussed later in this section.

Material absorption can be divided into two categories. Intrinsic absorption losses correspond to absorption by fused silica (material used to make fibers), whereas extrinsic absorption is related to losses caused by impurities within silica. Any material absorbs at certain wavelengths corresponding to the electronic and vibrational resonances associated with specific molecules. For silica (SiO_2) molecules, electronic resonances occur in the ultraviolet region ($\lambda < 0.4$ μm), whereas vibrational resonances occur in the infrared region ($\lambda > 7$ μm). Because of the amorphous nature of fused silica, these resonances are in the form of absorption bands whose tails extend into the visible region. Figure 1.13 shows that intrinsic material absorption for silica in the wavelength range 0.8–1.6 μm is below 0.1 dB/km. In fact, it is less than 0.03 dB/km in the 1.3- to 1.6-μm wavelength window commonly used for lightwave systems.

Extrinsic absorption results from the presence of impurities. Transition-metal impurities such as Fe, Cu, Co, Ni, Mn, and Cr absorb strongly in the wavelength range 0.6–1.6 μm. Their amount should be reduced to below 1 part per billion to obtain a loss level below 1 dB/km. Such high-purity silica can be obtained by using modern techniques. The main source of extrinsic absorption in state-of-the-art silica fibers is the presence of water vapors. A vibrational resonance of the OH ion occurs near 2.73 μm. Its harmonic and combination tones with silica produce absorption at the 1.39-, 1.24-, and 0.95-μm wavelengths. The three spectral peaks seen in Figure 1.13 occur near these wavelengths and are due to the presence of residual water vapor in silica. Even a concentration of 1 part per million can cause a loss of about 50 dB/km at 1.39 μm. The OH ion concentration is reduced to below 10^{-8} in modern fibers to lower the 1.39-μm peak below 1 dB. In a new kind of fiber, known as the *dry fiber* (trade name AllWave fiber), the OH ion concentration can be reduced to such low levels that the 1.39-μm peak almost disappears [55].

1.4.2 Loss Spectra of Plastic Fibers

Figure 1.14 shows the loss spectra of several plastic fibers. A PMMA fiber exhibits low losses in the visible region but they typically exceed 100 dB even in this region. In contrast, losses of modern PFBVE fibers are close to 50 dB/km over a wide wavelength range extending from 800 to 1,300 nm and have the potential of being reduced to below 10 dB/km with further optimization.

Similar to the case of silica fibers, material absorption can be divided into intrinsic and extrinsic categories. Intrinsic absorption losses in plastic fibers result from the vibrational modes of various molecular bonds (C—C, C—O, C—H, and O—H, etc.) within the organic polymer used to make the fiber. Even though the vibrational fre-

1.4 Fiber Losses

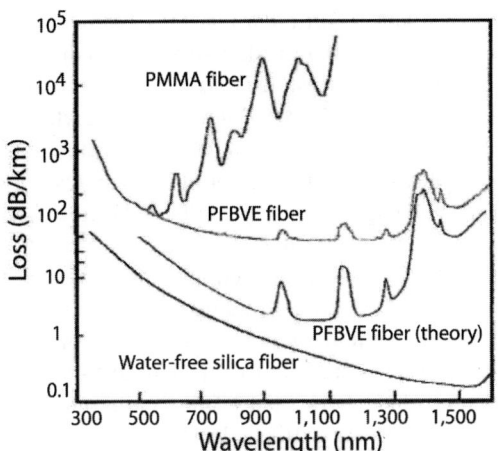

Figure 1.14: Loss spectra of several plastic optical fibers. The case of water-free silica fibers is also shown for comparison. (After Ref. [20]; ©2002 Elsevier.)

quencies of such modes lie in the wavelength region beyond 2 μm, their harmonics can introduce considerable losses for all plastic fibers even in the near-infrared and visible region. This is the main reason why the losses typically exceed 50 dB/km for plastic fibers. In the ultraviolet region losses increase quickly because of the electronic transitions.

Extrinsic absorption is related to the presence of impurities within the the fiber core. Transition-metal impurities such as Fe, Co, Ni, Mn, and Cr absorb strongly in the wavelength range 0.6–1.6 μm. Even a trace amount as small as a few parts per billion can add losses in excess of 10 dB/km. Similar to the case of silica fibers, any residual water vapor also leads to the presence of OH ions and a corresponding strong peak near 1,390 nm. This problem is less severe for fluorinated (PFBVE) fibers because fluropolymers do not absorb water easily. As seen in Figure 1.14, losses of such fibers can be reduced, in principle, to below 5 dB/km.

1.4.3 Rayleigh Scattering

Rayleigh scattering is a fundamental loss mechanism arising from local microscopic fluctuations in the material density. In any amorphous material, molecules move randomly in the molten state and freeze in place during fiber fabrication. Density fluctuations lead to random fluctuations of the refractive index on a scale smaller than the optical wavelength λ. Light scattering in such a medium is known as *Rayleigh scattering* [17], the same mechanism that is responsible for the blue color of the sky. As the Rayleigh-scattering cross section varies as λ^{-4}, the scattering is most dominant in the blue region and decreases rapidly as the wavelength increases. Thus, the intrinsic loss of optical fibers from Rayleigh scattering can be written as

$$\alpha_R = C/\lambda^4, \tag{1.4.3}$$

where C is a constant. Its value is different for silica and plastic fibers. In the case of silica fibers, C is in the range 0.7–0.9 (dB/km)-μm^4, depending on the constituents of the fiber core. These values of C correspond to $\alpha_R = 0.12$–0.16 dB/km at $\lambda = 1.55$ μm, indicating that fiber loss in Figure 1.13 is dominated by Rayleigh scattering near this wavelength. In the case of plastic fibers, even though C is close to 2 (dB/km)-μm^4, the resulting loss is still is a negligibly small factor of the total loss of plastic fibers.

The contribution of Rayleigh scattering can be reduced to below 0.01 dB/km for wavelengths longer than 3 μm. Existing optical fibers cannot be used in this wavelength region because of rather large infrared absorption at such wavelengths. Considerable effort has been directed toward finding other suitable materials with low absorption beyond 2 μm [56]. Fluorozirconate (ZrF$_4$) fibers have an intrinsic material absorption of about 0.01 dB/km near 2.55 μm and have the potential for exhibiting loss much smaller than that of silica fibers. State-of-the-art fluoride fibers, however, exhibit a loss of about 1 dB/km because of extrinsic losses. Chalcogenide and polycrystalline fibers exhibit minimum loss in the far-infrared region near 10 μm. The theoretically predicted minimum value of fiber loss for such fibers is below 10^{-3} dB/km because of reduced Rayleigh scattering. However, practical loss levels remain higher than those of silica fibers.

1.4.4 Waveguide Imperfections

An ideal single-mode fiber with a perfect cylindrical geometry guides the optical mode without energy leakage into the cladding layer. In practice, imperfections at the core–cladding interface (e.g., random core-radius variations) can lead to additional losses which contribute to the net fiber loss. The physical process behind such losses is *Mie scattering* [17], occurring because of index inhomogeneities on a scale longer than the optical wavelength. Care is generally taken to ensure that the core radius does not vary significantly along the fiber length during manufacture. Such variations can be kept below 1%, and the resulting scattering loss is typically below 0.03 dB/km.

Bends in the fiber constitute another source of scattering loss. The reason can be understood by using the ray picture. Normally, a guided ray hits the core–cladding interface at an angle greater than the critical angle to experience total internal reflection. However, the angle decreases near a bend and may become smaller than the critical angle for tight bends. The ray would then escape out of the fiber. In the mode description, a part of the mode energy is scattered into the cladding layer. The bending loss is proportional to $\exp(-R/R_c)$, where R is the radius of curvature of the fiber bend and $R_c = a/(n_1^2 - n_2^2)$. For single-mode fibers, $R_c = 0.2$–0.4 μm typically, and the bending loss is negligible (<0.01 dB/km) for bend radius $R > 5$ mm. Since most macroscopic bends exceed $R = 5$ mm, *macrobending losses* are negligible in practice.

A major source of fiber loss, particularly in cable form, is related to the random axial distortions that invariably occur during cabling when the fiber is pressed against a surface that is not perfectly smooth. Such losses are referred to as *microbending losses* and have been studied extensively [57]–[59]. Microbends cause an increase in the fiber loss for both multimode and single-mode fibers and can result in an excessively large loss ($\sim$100 dB/km) if precautions are not taken to minimize them. For single-mode fibers, microbending losses can be minimized by choosing the V parameter as close to the cutoff value of 2.405 as possible so that mode energy is confined primarily to the core. In practice, the fiber is designed to have V in the range 2.0–2.4 at the operating wavelength. Many other sources of optical loss exist in a fiber cable. These are related to splices and connectors used in forming the fiber link and are often treated as a part of the cable loss.

1.5 Fiber Dispersion

As discussed in Section 1.1, dispersion in multimode fibers leads to considerable broadening of optical pulses. In the geometrical-optics description, pulse broadening was attributed to different paths followed by different rays. In the wave-optics description, it is related to the different group velocities associated with different modes. For this reason, this source of dispersion is referred to as *modal dispersion*. Modal dispersion vanishes in single-mode fibers as the entire energy of the injected pulse is transported by a single mode. However, pulse broadening does not disappear altogether as the group velocity associated with the fundamental mode is frequency-dependent because of chromatic dispersion. As a result, different spectral components of the pulse travel at slightly different group velocities, a phenomenon called the *group-velocity dispersion* (GVD). This source of dispersion has two distinct origins referred to as material dispersion and waveguide dispersion. A third source of dispersion results from birefringence effects and is called the polarization-mode dispersion (PMD). This section considers all dispersion mechanisms in a systematic fashion.

1.5.1 Group-Velocity Dispersion

Consider a single-mode fiber of length L. A specific spectral component at the frequency ω would arrive at the output end of the fiber after a time delay $T = L/v_g$, where v_g is the *group velocity*, defined as [17]

$$v_g = (d\beta/d\omega)^{-1}. \quad (1.5.1)$$

By using $\beta = \bar{n}k_0 = \bar{n}\omega/c$ in Eq. (1.5.1), one can show that $v_g = c/\bar{n}_g$, where $\bar{n}_g$ is the *group index* given by

$$\bar{n}_g = \bar{n} + \omega(d\bar{n}/d\omega). \quad (1.5.2)$$

The frequency dependence of the group velocity leads to pulse broadening simply because different spectral components of the pulse disperse during propagation and do not arrive simultaneously at the fiber output. If $\Delta\omega$ is the spectral width of the pulse, the extent of pulse broadening for a fiber of length L is governed by

$$\Delta T = \frac{dT}{d\omega}\Delta\omega = \frac{d}{d\omega}\left(\frac{L}{v_g}\right)\Delta\omega = L\frac{d^2\beta}{d\omega^2}\Delta\omega = L\beta_2\Delta\omega, \quad (1.5.3)$$

where Eq. (1.5.1) was used. The parameter $\beta_2 = d^2\beta/d\omega^2$ is known as the GVD parameter. It determines how much an optical pulse would broaden on propagation inside the fiber.

In practice, it is customary to use wavelength spread $\Delta\lambda$ in place of $\Delta\omega$. By using $\omega = 2\pi c/\lambda$ and $\Delta\omega = (-2\pi c/\lambda^2)\Delta\lambda$, Eq. (1.5.3) can be written as

$$\Delta T = \frac{d}{d\lambda}\left(\frac{L}{v_g}\right)\Delta\lambda = DL\Delta\lambda, \tag{1.5.4}$$

where

$$D = \frac{d}{d\lambda}\left(\frac{1}{v_g}\right) = -\frac{2\pi c}{\lambda^2}\beta_2. \tag{1.5.5}$$

D is called the *dispersion parameter* and is expressed in units of ps/(km-nm).

The effect of dispersion on the bit rate B can be estimated by using the criterion $B\Delta T < 1$ in a manner similar to that used in Section 1.1. By using ΔT from Eq. (1.5.4), this condition becomes

$$BL|D|\Delta\lambda < 1. \tag{1.5.6}$$

Equation (1.5.6) provides an order-of-magnitude estimate of the BL product offered by single-mode fibers. For standard silica fibers, D is relatively small in the wavelength region near 1.3 μm [$D \sim 1$ ps/(km-nm)]. The BL product of single-mode fibers can exceed 1 (Tb/s)-km when optical sources with $\Delta\lambda$ below 1 nm are employed.

The wavelength dependence of D is governed by the frequency dependence of the mode index $\bar{n}$. From Eq. (1.5.5), D can be written as

$$D = \frac{d}{d\lambda}\left(\frac{\bar{n}_g}{c}\right) = \frac{1}{c}\frac{d}{d\lambda}\left(\bar{n} - \lambda\frac{d\bar{n}}{d\lambda}\right) = -\frac{\lambda}{c}\frac{d^2\bar{n}}{d\lambda^2}, \tag{1.5.7}$$

where Eq. (1.5.2) was used. If we substitute $\bar{n}$ from Eq. (1.3.39) and use Eq. (1.3.35), D can be written as the sum of two terms,

$$D = D_M + D_W, \tag{1.5.8}$$

where the *material dispersion* D_M and the *waveguide dispersion* D_W are given by

$$D_M = -\frac{\lambda}{c}\frac{d^2 n_2}{d\lambda^2}, \quad D_W = -\frac{n_2\Delta}{c\lambda}V\frac{d^2(Vb)}{dV^2}, \tag{1.5.9}$$

where the parameters V and b are given in Eqs. (1.3.35) and (1.3.36), respectively. In obtaining Eq. (1.5.9), the parameter Δ was assumed to be frequency-independent. A term known as differential material dispersion should be added to Eq. (1.5.8) when $d\Delta/d\omega \neq 0$. Its contribution is, however, negligible in practice.

1.5.2 Material Dispersion

Material dispersion occurs because the refractive index of the material used for fiber fabrication changes with the optical wavelength. On a fundamental level, the origin of material dispersion is related to the characteristic resonance frequencies at which the material absorbs the electromagnetic radiation. Far from any medium resonances, the refractive index $n(\omega)$ is well approximated by the *Sellmeier equation* [60]

$$n^2(\lambda) = 1 + \sum_{j=1}^{M}\frac{B_j\lambda^2}{\lambda^2 - \lambda_j^2}, \tag{1.5.10}$$

1.5 Fiber Dispersion

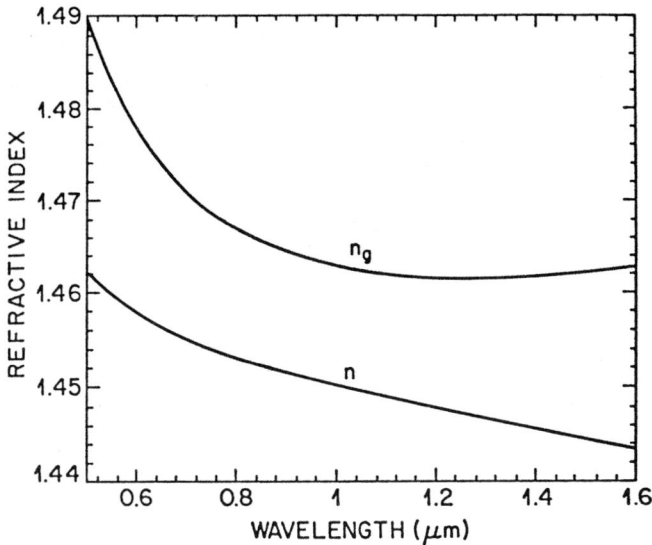

Figure 1.15: Variation of refractive index n and group index n_g with wavelength for fused silica.

where λ_j is the resonance wavelength and B_j is the oscillator strength. Here n stands for n_1 or n_2, depending on whether the dispersive properties of the core or the cladding are considered. The sum in Eq. (1.5.10) extends over all material resonances that contribute in the frequency range of interest. In the case of optical fibers, the parameters B_j and λ_j are obtained empirically by fitting the measured dispersion curves to Eq. (1.5.10) with $M = 3$. They depend on the amount of dopants and have been tabulated for several kinds of fibers [9]. For pure silica these parameters are found to be $B_1 = 0.6961663$, $B_2 = 0.4079426$, $B_3 = 0.8974794$, $\lambda_1 = 0.0684043$ μm, $\lambda_2 = 0.1162414$ μm, and $\lambda_3 = 9.896161$ μm, where $\lambda_j = 2\pi c/\omega_j$ with $j = 1$–3 [60]. The group index $n_g = n - \lambda(dn/d\lambda)$ can be obtained by using these parameter values.

Figure 1.15 shows the wavelength dependence of n and n_g in the range 0.5–1.6 μm for fused silica. Material dispersion D_M is related to the slope of n_g by the relation $D_M = c^{-1}(dn_g/d\lambda)$ [Eq. (1.5.9)]. It turns out that $dn_g/d\lambda = 0$ at $\lambda = 1.276$ μm. This wavelength is referred to as the *zero-dispersion wavelength* λ_{ZD}, since $D_M = 0$ at $\lambda = \lambda_{ZD}$. The dispersion parameter D_M is negative below λ_{ZD} and becomes positive above that. In the wavelength range 1.25–1.66 μm it can be approximated by an empirical relation

$$D_M \approx 122(1 - \lambda_{ZD}/\lambda). \tag{1.5.11}$$

It should be stressed that $\lambda_{ZD} = 1.276$ μm only for pure silica. It can vary in the range 1.27–1.29 μm for silica fibers whose core and cladding are generally doped to change their refractive index. The zero-dispersion wavelength of optical fibers also depends on the core radius a and the index step Δ through the waveguide contribution and can be made to vary in the range of 1.3–1.7 μm.

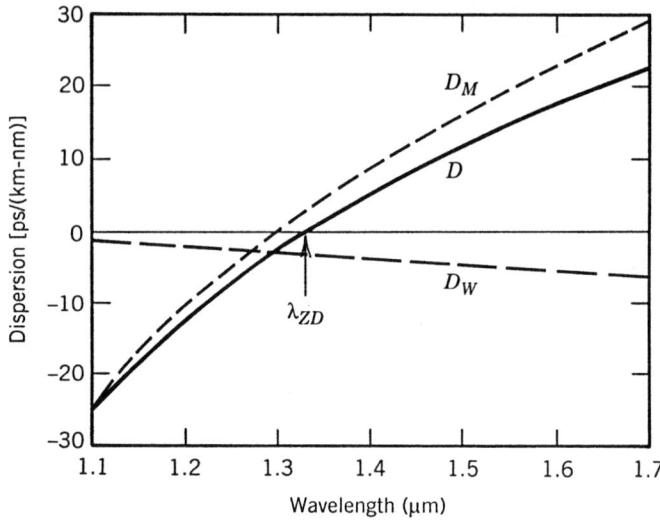

Figure 1.16: Total dispersion D and relative contributions of material dispersion D_M and waveguide dispersion D_W for a conventional single-mode fiber. The zero-dispersion wavelength shifts to a higher value because of the waveguide contribution.

1.5.3 Waveguide Dispersion

The contribution of waveguide dispersion D_W to the dispersion parameter D is given by Eq. (1.5.9) and depends on the V parameter of the fiber. It turns out that D_W is negative in general. On the other hand, D_M is negative for wavelengths below λ_{ZD} but becomes positive above that. Figure 1.16 shows D_M, D_W, and the sum $D = D_M + D_W$ for a typical single-mode silica fiber. The main effect of waveguide dispersion is to shift λ_{ZD} upward such that total dispersion is zero near 1.31 μm. It also reduces D from its material value D_M in the wavelength range 1.3–1.6 μm that is of interest for optical communication systems. Typical values of D are in the range of 15–18 ps/(km-nm) near 1.55 μm for standard silica fibers.

Since the waveguide contribution D_W depends on the core radius a and the index difference Δ, it is possible to design silica fibers such that λ_{ZD} is shifted into the vicinity of 1.55 μm [61], [62]. Such fibers are called *dispersion-shifted* fibers. It is also possible to tailor the waveguide contribution such that the total dispersion D is relatively small over a wide wavelength range extending from 1.3 to 1.6 μm [63]–[65]. Such fibers are called *dispersion-flattened* fibers. Figure 1.17 shows typical examples of the wavelength dependence of D for standard (conventional), dispersion-shifted, and dispersion-flattened fibers. The design of dispersion-modified fibers involves the use of multiple cladding layers and a tailoring of the refractive-index profile [61]–[67].

Waveguide dispersion can be used to produce new kind of fibers with interesting dispersive properties. For example, it has been used to make *dispersion-decreasing* fibers in which GVD decreases along the fiber length because of axial variations in the core radius. In another kind of fibers, known as the *dispersion-compensating* fibers,

1.5 Fiber Dispersion

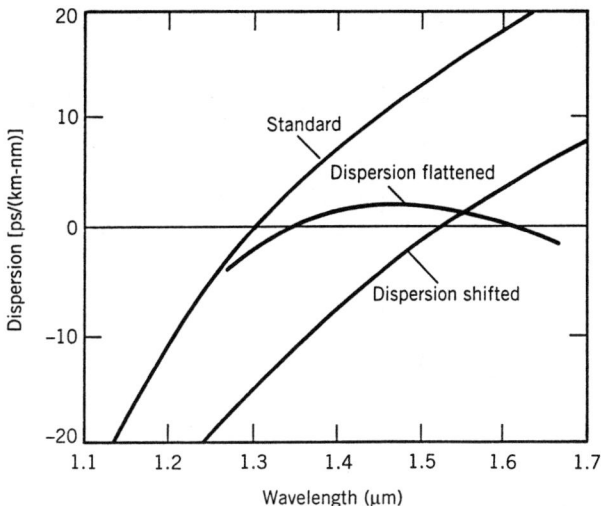

Figure 1.17: Typical wavelength dependence of the dispersion parameter D for standard, dispersion-shifted, and dispersion-flattened fibers.

GVD is made normal and large by shifting λ_{ZD} to beyond 1.6 μm. In contrast, it is possible to shift λ_{ZD} into the visible region near 0.8 μm in microstructure fibers using air holes within the cladding regions [43]. Figure 1.18(a) shows that because of air holes within the cladding region, the waveguide contribution D_W becomes positive and shifts the zero-dispersion wavelength toward 0.8 μm, in agreement with the experimental measurements (open circles). As seen in Figure 1.18(b), the total dispersion also depends on the core size of such fibers.

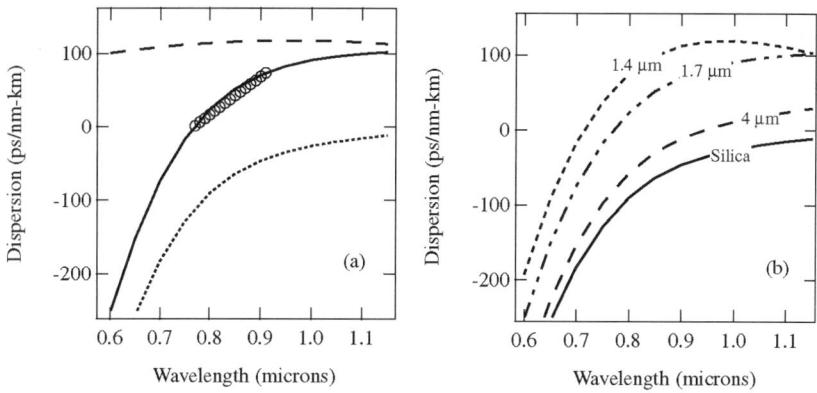

Figure 1.18: (a) Total dispersion (solid line) and contributions of material (dotted) and waveguide (dashed) dispersion for a microstructure fiber. Circles show the experimental data. (b) Dependence of dispersion parameter D on core size of the fiber. Solid line shows for comparison the case of bulk silica. (After Ref. [43]; ©2003 Springer.)

Table 1.1 Characteristics of several commercial fibers

Fiber Type and Trade Name	A_{eff} (μm^2)	λ_{ZD} (nm)	D (C band) [ps/(km-nm)]	Slope S [ps/(km-nm^2)]
Corning SMF-28	80	1,302–1,322	16 to 19	0.090
Lucent AllWave	80	1,300–1,322	17 to 20	0.088
Alcatel ColorLock	80	1,300–1,320	16 to 19	0.090
Corning Vascade	101	1,300–1,310	18 to 20	0.060
Lucent TrueWave-RS	50	1,470–1,490	2.6 to 6	0.050
Corning LEAF	72	1,490–1,500	2 to 6	0.060
Lucent TrueWave-XL	72	1,570–1,580	−1.4 to −4.6	0.112
Alcatel TeraLight	65	1,440–1,450	5.5 to 10	0.058

It appears from Eq. (1.5.6) that the *BL* product of a single-mode fiber can be increased indefinitely by operating at the zero-dispersion wavelength λ_{ZD} where $D = 0$. The dispersive effects, however, do not disappear completely at $\lambda = \lambda_{\text{ZD}}$. Optical pulses still experience broadening because of higher-order dispersive effects. This feature can be understood by noting that D cannot be made zero at all wavelengths contained within the pulse spectrum centered at λ_{ZD}. Clearly, the wavelength dependence of D will play a role in pulse broadening. Higher-order dispersive effects are governed by the *dispersion slope* $S = dD/d\lambda$. The parameter S is also called a *differential-dispersion* parameter. By using Eq. (1.5.5), it can be written as

$$S = (2\pi c/\lambda^2)^2 \beta_3 + (4\pi c/\lambda^3)\beta_2, \tag{1.5.12}$$

where $\beta_3 = d\beta_2/d\omega \equiv d^3\beta/d\omega^3$ is the third-order dispersion parameter. At $\lambda = \lambda_{\text{ZD}}$, $\beta_2 = 0$, and S is proportional to β_3.

The numerical value of the dispersion slope S plays an important role in the design of modern WDM systems. Since $S > 0$ for most fibers, different channels have slightly different GVD values. This feature makes it difficult to compensate dispersion for all channels simultaneously. To solve this problem, new kind of fibers have been developed for which S is either small (reduced-slope fibers) or negative (reverse-dispersion fibers). Table 1.1 lists dispersion characteristics for several commercially available fibers.

1.5.4 Polarization-Mode Dispersion

A potential source of pulse broadening is related to fiber birefringence. As discussed earlier, small departures from perfect cylindrical symmetry lead to birefringence because of different mode indices associated with the orthogonally polarized components of the fundamental fiber mode. If the input pulse excites both polarization components, it becomes broader as the two components disperse along the fiber because of their different group velocities. This phenomenon is called the PMD and has been studied extensively because it limits the performance of modern lightwave systems [68]–[77].

In fibers with constant birefringence (e.g., polarization-maintaining fibers), pulse broadening can be estimated from the time delay $\Delta \tau$ between the two polarization

1.5 Fiber Dispersion

components during propagation of the pulse. For a fiber of length L, $\Delta\tau$ is given by

$$\Delta\tau = \left| \frac{L}{v_{gx}} - \frac{L}{v_{gy}} \right| = L|\beta_{1x} - \beta_{1y}| = L(\Delta\beta_1), \tag{1.5.13}$$

where the subscripts x and y identify the two orthogonally polarized modes and $\Delta\beta_1$ is related to the difference in group velocities along the two principal states of polarization [68]. Equation (1.5.1) was used to relate the group velocity v_g to the propagation constant β. Similar to the case of modal dispersion discussed in Section 1.1, the quantity $\Delta\tau/L$ is a measure of PMD. For polarization-maintaining fibers, $\Delta\tau/L$ is quite large (~ 1 ns/km) when the two components are equally excited at the fiber input but can be reduced to zero by launching light along one of the principal axes.

The situation is somewhat different for conventional fibers in which birefringence varies along the fiber in a random fashion. It is intuitively clear that the polarization state of light propagating in fibers with randomly varying birefringence will generally be elliptical and would change randomly along the fiber during propagation. In the case of optical pulses, the polarization state will also be different for different spectral components of the pulse. The final polarization state is not of concern for most lightwave systems as photodetectors used inside optical receivers are insensitive to the state of polarization unless a coherent detection scheme is employed. What affects such systems is not the random polarization state but pulse broadening induced by random changes in the birefringence. This is referred to as *PMD-induced* pulse broadening.

The analytical treatment of PMD is quite complex in general because of its statistical nature. A simple model divides the fiber into a large number of segments. Both the degree of birefringence and the orientation of the principal axes remain constant in each section but change randomly from section to section. In effect, each fiber section can be treated as a phase plate using a Jones matrix [68]. Propagation of each frequency component associated with an optical pulse through the entire fiber length is then governed by a composite Jones matrix obtained by multiplying individual Jones matrices for each fiber section. The composite Jones matrix shows that two principal states of polarization exist for any fiber such that, when a pulse is polarized along them, the polarization state at fiber output is frequency-independent to first order, in spite of random changes in fiber birefringence. These states are analogous to the slow and fast axes associated with polarization-maintaining fibers. An optical pulse not polarized along these two principal states splits into two parts that travel at different speeds. The differential group delay $\Delta\tau$ is largest for the two principal states of polarization.

The principal states of polarization provide a convenient basis for calculating the moments of $\Delta\tau$. The PMD-induced pulse broadening is characterized by the root-mean-square (RMS) value of $\Delta\tau$, obtained after averaging over random birefringence changes. Several approaches have been used to calculate this average. The quantity $\langle(\Delta\tau)^2\rangle$ turns out to be the same in all cases and is given by [70]

$$\langle(\Delta\tau)^2\rangle = 2(\Delta\beta_1)^2 l_c^2 [\exp(-z/l_c) + z/l_c - 1], \tag{1.5.14}$$

where l_c is the correlation length defined as the length over which two polarization components remain correlated; its value can vary over a wide range from 1 m to 1 km for different fibers, typical values being ~ 10 m.

For short distances such that $z \ll l_c$, $\langle(\Delta\tau)^2\rangle^{1/2} = (\Delta\beta_1)z$ from Eq. (1.5.14), as expected for a polarization-maintaining fiber. For distances $z > 1$ km, a good estimate of pulse broadening is obtained using $z \gg l_c$. For a fiber of length L, $\langle(\Delta\tau)^2\rangle^{1/2}$ in this approximation becomes

$$\langle(\Delta\tau)^2\rangle^{1/2} \approx (\Delta\beta_1)\sqrt{2l_cL} \equiv D_p\sqrt{L}, \qquad (1.5.15)$$

where D_p is the PMD parameter. Measured values of D_p vary from fiber to fiber in the range $D_p = 0.01$–10 ps/$\sqrt{\text{km}}$. Fibers installed during the 1980s have relatively large PMD such that $D_p > 0.2$ ps/$\sqrt{\text{km}}$. In contrast, modern fibers are designed to have low PMD, and typically $D_p < 0.05$ ps/$\sqrt{\text{km}}$ for them. Because of the $\sqrt{L}$ dependence, PMD-induced pulse broadening is relatively small compared with the GVD effects. Indeed, $\langle(\Delta\tau)^2\rangle^{1/2} \sim 1$ ps for fiber lengths ~ 100 km and can be ignored for pulse widths >10 ps. However, PMD becomes a limiting factor for lightwave systems designed to operate over long distances at high bit rates [72]–[77]. Several schemes have been developed for compensating the PMD effects.

Several other factors need to be considered in practice. The derivation of Eq. (1.5.14) assumes that the fiber link has no elements exhibiting polarization-dependent loss or gain. The presence of polarization-dependent losses can induce additional broadening [73]. Also, the effects of second- and higher-order PMD become important at high bit rates (40 Gb/s or more) or for systems in which the first-order effects are eliminated using a PMD compensator [77].

1.5.5 Pulse-Propagation Equation

The discussion of pulse broadening in Section 1.5.1 is based on an intuitive phenomenological approach and provides only a rough estimate. In general, the extent of pulse broadening depends on the width and the shape of input pulses [14]. In this section we obtain the basic equation that governs propagation of pulsed optical fields in single-mode fibers.

For pulses whose spectral width is much smaller than their carrier frequency ω_0, or whose temporal width is much larger than the duration ω_0^{-1} of a single optical cycle, one can employ the slowly varying envelope approximation and write the optical field in the form

$$\mathbf{E}(\mathbf{r},t) = \text{Re}[\hat{\mathbf{e}}F(x,y)A(z,t)\exp(i\beta_0 z - i\omega_0 t)], \qquad (1.5.16)$$

where Re denotes the real part, $\hat{\mathbf{e}}$ is the polarization unit vector, $A(z,t)$ is the amplitude of the pulse envelope at a distance z inside the fiber, $\beta_0 \equiv \beta(\omega_0)$ is the propagation constant at the carrier frequency ω_0, and $F(x,y)$ is the spatial distribution of the fundamental fiber mode. To simplify the following analysis, we first assume that the birefringence effects can be ignored and treat $\hat{\mathbf{e}}$ as a constant.

The analysis of fiber modes in Section 1.3 showed that each frequency component of the optical field propagates in a single-mode fiber as a plane wave. For this reason, we introduce the Fourier transform of $A(z,t)$ as

$$A(z,t) = \frac{1}{2\pi}\int_{-\infty}^{\infty}\tilde{A}(z,\omega)\exp(-i\omega t)\,d\omega. \qquad (1.5.17)$$

1.5 Fiber Dispersion

Different spectral components $\tilde{A}(z,\omega)$ propagate inside an optical fiber according to the simple relation

$$\tilde{A}(z,\omega) = \tilde{A}(0,\omega)\exp[i\beta(\omega)z - i\beta_0 z], \qquad (1.5.18)$$

where $\tilde{A}(0,\omega)$ is the Fourier transform of the input amplitude $A(0,t)$. This equation can also be written as the following first-order differential equation:

$$\frac{d\tilde{A}}{dz} = i[\beta(\omega) - \beta_0]\tilde{A}(z,\omega). \qquad (1.5.19)$$

Pulse broadening results from the frequency dependence of β. Since the exact functional form of this dependence is not known in general, it is useful to expand $\beta(\omega)$ in a Taylor series around the carrier frequency ω_0. If we retain terms up to third order, we obtain

$$\beta(\omega) = \bar{n}(\omega)\frac{\omega}{c} \approx \beta_0 + \beta_1(\Delta\omega) + \frac{\beta_2}{2}(\Delta\omega)^2 + \frac{\beta_3}{6}(\Delta\omega)^3, \qquad (1.5.20)$$

where $\Delta\omega = \omega - \omega_0$ and $\beta_m = (d^m\beta/d\omega^m)_{\omega=\omega_0}$. From Eq. (1.5.1) $\beta_1 = 1/v_g$, where v_g is the group velocity. The GVD coefficient β_2 is related to the dispersion parameter D as in Eq. (1.5.5), and β_3 is related to the dispersion slope S through Eq. (1.5.12).

We substitute Eq. (1.5.20) in Eq. (1.5.17) and convert the resulting equation into the time domain after noting that $\Delta\omega$ is replaced by the differential operator $i(\partial/\partial t)$ in the time domain. The resulting equation can be written as [50]

$$\frac{\partial A}{\partial z} + \beta_1\frac{\partial A}{\partial t} + \frac{i\beta_2}{2}\frac{\partial^2 A}{\partial t^2} - \frac{\beta_3}{6}\frac{\partial^3 A}{\partial t^3} = 0. \qquad (1.5.21)$$

This is the basic propagation equation governing pulse evolution inside a single-mode fiber.

In the absence of dispersion ($\beta_2 = \beta_3 = 0$), the optical pulse propagates without change in its shape such that $A(z,t) = A(0, t - \beta_1 z)$. If we adopt a reference frame moving with the pulse and introduce the new variables

$$t' = t - \beta_1 z \qquad \text{and} \qquad z' = z, \qquad (1.5.22)$$

the β_1 term can be eliminated in Eq. (1.5.21) to yield

$$\frac{\partial A}{\partial z'} + \frac{i\beta_2}{2}\frac{\partial^2 A}{\partial t'^2} - \frac{\beta_3}{6}\frac{\partial^3 A}{\partial t'^3} = 0. \qquad (1.5.23)$$

This equation shows how the second- and third-order dispersion parameters affect an optical pulse as it propagates down the fiber. It is common to drop the primes over z' and t' whenever no confusion is likely to arise. If the third-order dispersive effects are negligible, Eq. (1.5.23) reduces to the following simple equation:

$$\frac{\partial A}{\partial z} + \frac{i\beta_2}{2}\frac{\partial^2 A}{\partial t^2} = 0. \qquad (1.5.24)$$

This equation is isomorphic to the paraxial equation governing diffraction of optical beams in one transverse dimension [17]. The analogy between dispersion in time and diffraction in space can often be exploited to advantage.

When the birefringence effects cannot be ignored, the polarization vector ê also changes with z. In that case, one can introduce the Jones vector, $|A\rangle = \hat{e}A$, as a two-dimensional column vector and write Eq. (1.5.19) as the following vector equation:

$$\frac{d|\tilde{A}(z,\omega)\rangle}{dz} = i\mathbf{b}(z,\omega)|\tilde{A}(z,\omega)\rangle, \quad (1.5.25)$$

where the birefringence matrix $\mathbf{b}(z,\omega)$ replaces the scalar quantity $\beta(\omega) - \beta_0$. The elements of this 2×2 matrix are not only frequency-dependent, but they can also vary along the fiber length in a random fashion to account for birefringence fluctuations responsible for the PMD effects. This stochastic differential equation is often solved in the Stokes space after introducing the Stokes parameters [77].

1.6 Fiber Nonlinearities

The response of any dielectric to light becomes nonlinear for intense electromagnetic fields, and optical fibers are no exception. Even though silica is intrinsically not a highly nonlinear material, the waveguide geometry that confines light to a small cross section over long fiber lengths makes nonlinear effects quite important in the design of modern optical communication systems [50]. We discuss in this section the nonlinear phenomena that are relevant for lightwave technology.

1.6.1 Self-Phase Modulation

The refractive index of silica was assumed to be power-independent in the discussion of fiber modes in Section 1.3. In reality, all materials behave nonlinearly at high intensities. The physical origin of this effect lies in the anharmonic response of electrons to optical fields. Mathematically, the induced electric polarization $\mathbf{P}$ appearing in the wave equation Eq. (1.3.8) has an additional nonlinear component related to the electric field $\mathbf{E}$ as

$$\mathbf{P}_{\text{NL}}(\mathbf{r},t) = \varepsilon_0 \chi^{(3)} \vdots \mathbf{E}(\mathbf{r},t)\mathbf{E}(\mathbf{r},t)\mathbf{E}(\mathbf{r},t), \quad (1.6.1)$$

where $\chi^{(3)}$ is the third-order nonlinear susceptibility [78], and the electronic response is assumed to be instantaneous. The tensorial nature of $\chi^{(3)}$ should be considered whenever the polarizations effects are of interest.

In general, $\chi^{(3)}$ leads to a large number of nonlinear effects depending on the incident field $\mathbf{E}(\mathbf{r},t)$. When $\mathbf{E}(\mathbf{r},t)$ can be written in the form of Eq. (1.5.16), the nonlinear effects can be included by modifying the core and cladding indices of the fiber as [50]

$$n'_j = n_j + \bar{n}_2(P/A_{\text{eff}}), \quad j = 1, 2, \quad (1.6.2)$$

where $\bar{n}_2 = (3/8n)\text{Re}(\chi^{(3)}_{xxxx})$ is the *nonlinear-index coefficient*, $P = |A|^2$ is the optical power, and A_{eff} is the effective core area defined as

$$A_{\text{eff}} = \frac{\left(\iint_{-\infty}^{\infty} |F(x,y)|^2 dx\,dy\right)^2}{\iint_{-\infty}^{\infty} |F(x,y)|^4 dx\,dy}. \quad (1.6.3)$$

1.6 Fiber Nonlinearities

The numerical value of $\bar{n}_2$ is about 2.6×10^{-20} m²/W for silica fibers and varies somewhat with the concentration of dopants used inside the core. Because of this relatively small value, the nonlinear part of the refractive index is quite small ($< 10^{-12}$ at a power level of 1 mW). Nevertheless, it affects modern lightwave systems considerably because of long fiber lengths.

If we use first-order perturbation theory to see how fiber modes are affected by the nonlinear term in Eq. (1.6.2), we find that the mode shape does not change but the propagation constant becomes power-dependent. It can be written as [50]

$$\beta' = \beta + k_0 \bar{n}_2 P / A_{\text{eff}} \equiv \beta + \gamma P, \tag{1.6.4}$$

where $\gamma = 2\pi \bar{n}_2 / (A_{\text{eff}} \lambda)$ is an important nonlinear parameter with values ranging from 1 to 5 W⁻¹/km depending on the values of A_{eff} and the wavelength. From a physical standpoint, the γ term produces a nonlinear phase shift given by

$$\phi_{\text{NL}} = \int_0^L (\beta' - \beta) dz = \int_0^L \gamma P(z) dz = \gamma P_{\text{in}} L_{\text{eff}}, \tag{1.6.5}$$

where $P(z) = P_{\text{in}} \exp(-\alpha z)$ accounts for fiber losses and L_{eff} is defined as

$$L_{\text{eff}} = [1 - \exp(-\alpha L)]/\alpha. \tag{1.6.6}$$

In deriving Eq. (1.6.5), P_{in} was assumed to be constant. In practice, the time dependence of P_{in} makes ϕ_{NL} vary with time. In fact, the optical phase changes with time in exactly the same fashion as the optical signal. Since this nonlinear phase modulation is self-induced, the nonlinear phenomenon responsible for it is called *self-phase modulation* (SPM). Noting that a time-dependent phase implies a frequency shift with the magnitude $\delta \omega(t) = -d\phi_{\text{NL}}/dt$, we conclude that SPM leads to chirping of optical pulses. Figure 1.19 shows how chirp varies with time for Gaussian ($m = 1$) and super-Gaussian pulses ($m = 3$) using

$$P_{\text{in}}(t) = P_0 \exp[-(t/T_0)^{2m}]. \tag{1.6.7}$$

The frequency chirp imposed by SPM manifests through broadening of the pulse spectrum, an effect that was first observed in 1978 for single-mode silica fibers [79].

The preceding discussion of SPM ignores the dispersive effects. In practice, the GVD and SPM act together on the optical pulse and should be considered simultaneously. Equation (1.5.24) can be generalized to include the SPM effects by using β' from Eq. (1.6.4) in Eq. (1.5.19) in place of β. The resulting equation is known as the nonlinear Schrödinger equation and has the form [50]

$$\frac{\partial A}{\partial z} + \frac{\alpha}{2} A + \frac{i\beta_2}{2} \frac{\partial^2 A}{\partial t^2} = i\gamma |A|^2 A, \tag{1.6.8}$$

where we added the term containing α to account for fiber losses. This equation should be used whenever fiber is long enough that the dispersive and nonlinear effects become important.

Since the nonlinear parameter γ depends inversely on the effective core area, the impact of fiber nonlinearities can be reduced considerably by enlarging A_{eff}. As seen in Table 1.1, A_{eff} is about 80 µm² for standard fibers but reduces to 50 µm² for dispersion-shifted fibers. A new kind of fiber known as large effective-area fiber (LEAF) has been developed for reducing the impact of fiber nonlinearities.

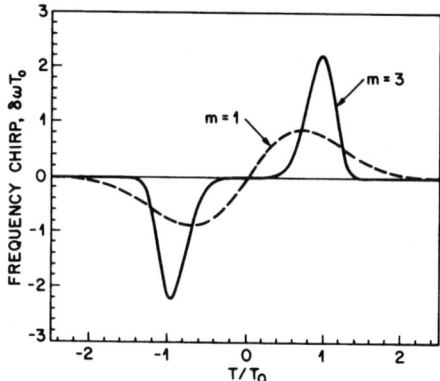

Figure 1.19: SPM-induced frequency chirp for Gaussian (dashed curve) and super-Gaussian (solid curve) pulses.

1.6.2 Cross-Phase Modulation

The intensity dependence of the refractive index in Eq. (1.6.2) can also lead to another nonlinear phenomenon known as *cross-phase modulation* (XPM). It occurs when two or more optical fields are transmitted simultaneously inside an optical fiber. In that case, the nonlinear phase shift for each field depends not only on the power of that field but also on the power of the other field.

Consider first, for simplicity, the case of two optical fields at the carrier frequencies ω_a and ω_b so that the total electric field is given by

$$\mathbf{E}(\mathbf{r},t) = \hat{e}\text{Re}[A_a(z,t)F_a(x,y)\exp(i\beta_{0a}z - i\omega_a t) + A_b(z,t)F_b(x,y)\exp(i\beta_{0b}z - i\omega_b t)], \quad (1.6.9)$$

where we assumed that both fields are polarized along the same direction which does not change during propagation. When this equation is substituted in Eq. (1.6.1) and the terms oscillating at frequencies ω_a and ω_b are collected, the propagation constants for the two fields are found to be modified as follows:

$$\beta_a' = \beta_a + \gamma_a(|A_a|^2 + 2|A_b|^2), \quad (1.6.10)$$
$$\beta_b' = \beta_b + \gamma_b(|A_b|^2 + 2|A_a|^2). \quad (1.6.11)$$

where $\gamma_j = (\omega_j/c)n_2/A_j^{\text{eff}}$ are generally different for $j = a$ and b.

The nonlinear phase shift for the two fields originates from the γ term. If we assume that fiber losses are nearly the same at two frequencies, this shift is found to be

$$\phi_a^{\text{NL}} = \gamma_a L_{\text{eff}}(P_a + 2P_b), \qquad \phi_b^{\text{NL}} = \gamma_b L_{\text{eff}}(P_b + 2P_a), \quad (1.6.12)$$

where P_a and P_b are the input power levels. The first term is due to SPM, whereas the second term represents the XPM contribution. The factor of 2 in this term indicates that XPM is twice as effective as SPM for the same amount of power. Similar to the SPM case, XPM leads to additional chirping and spectral broadening whenever the input fields are time-dependent.

1.6 Fiber Nonlinearities

In the case of a WDM lightwave system with multiple fields at distinct frequencies, the nonlinear phase shift for the jth channel is given by

$$\phi_j^{\text{NL}} = \gamma L_{\text{eff}} \left(P_j + 2 \sum_{m \neq j} P_m \right), \quad (1.6.13)$$

where the sum extends over the number of channels. The total phase shift depends on the powers in all channels and would vary from bit to bit depending on the bit pattern of the neighboring channels. It is difficult to estimate the impact of XPM on the performance of WDM lightwave systems. The reason is that the preceding discussion has ignored the dispersive effects. In practice, pulses in different channels travel at different speeds. The XPM-induced phase shift can occur only when two pulses overlap in time. For widely separated channels they overlap for such a short time that XPM effects are virtually negligible.

Equation (1.6.8) can be generalized to include the XPM effects by following the procedure discussed earlier. One still obtains a nonlinear Schrödinger equation for each field but these equations are coupled through XPM. In the case of two optical fields, the coupled equations take the form [50]

$$\frac{\partial A_a}{\partial z} + \beta_{1a} \frac{\partial A_a}{\partial t} + \frac{i\beta_{2a}}{2} \frac{\partial^2 A_a}{\partial t^2} = i\gamma_a (|A_a|^2 + 2|A_b|^2) A_a, \quad (1.6.14)$$

$$\frac{\partial A_b}{\partial z} + \beta_{1b} \frac{\partial A_b}{\partial t} + \frac{i\beta_{2b}}{2} \frac{\partial^2 A_b}{\partial t^2} = i\gamma_b (|A_b|^2 + 2|A_a|^2) A_b, \quad (1.6.15)$$

where β_{1j} and β_{2j} are the first- and second-order dispersion parameters at the frequency ω_j. Since $\beta_{1a} \neq \beta_{1b}$ in general, the two pulses propagate at different group velocities. It is this group-velocity mismatch that dictates how long the two pulses will continue to overlap.

1.6.3 Four-Wave Mixing

The nonlinear phenomenon of *four-wave mixing* (FWM) also originates from $\chi^{(3)}$. If three optical fields with carrier frequencies ω_1, ω_2, and ω_3 copropagate inside the fiber simultaneously, $\chi^{(3)}$ generates a fourth field whose frequency ω_4 is related to other frequencies by a relation $\omega_4 = \omega_1 \pm \omega_2 \pm \omega_3$. Several frequencies corresponding to different plus and minus sign combinations are possible in principle. In practice, most of these combinations do not build up because of a phase-matching requirement [50]. Frequency combinations of the form $\omega_4 = \omega_1 + \omega_2 - \omega_3$ are often troublesome for multichannel communication systems since they can become nearly phase-matched when channel wavelengths lie close to the zero-dispersion wavelength. In fact, the degenerate FWM process for which $\omega_1 = \omega_2$ is often the dominant process and impacts the system performance most.

On a fundamental level, a FWM process can be viewed as a scattering process in which two photons of energies $\hbar \omega_1$ and $\hbar \omega_2$ are destroyed, and their energy appears in the form of two new photons of energies $\hbar \omega_3$ and $\hbar \omega_4$. The *phase-matching condition* then stems from the requirement of momentum conservation. Since all four waves propagate in the same direction, the phase mismatch can be written as

$$\Delta = \beta(\omega_3) + \beta(\omega_4) - \beta(\omega_1) - \beta(\omega_2), \qquad (1.6.16)$$

where $\beta(\omega_j)$ is the propagation constant for an optical field at the frequency ω_j. In the degenerate case, $\omega_2 = \omega_1$, $\omega_3 = \omega_1 + \Omega$, and $\omega_3 = \omega_1 - \Omega$, where Ω represents the channel spacing. Using the Taylor expansion in Eq. (1.5.20), we find that the β_0 and β_1 terms cancel, and the phase mismatch is simply $\Delta = \beta_2 \Omega^2$. The FWM process is completely phase matched when $\beta_2 = 0$. When β_2 is small (<1 ps²/km) and channel spacing is also small ($\Omega < 100$ GHz), this process can still occur and transfer power from each channel to its nearest neighbors. Such a power transfer not only results in the power loss for the channel but also induces interchannel crosstalk that degrades the system performance severely. Modern WDM systems avoid FWM by using the technique of dispersion management in which GVD is kept locally high in each fiber section even though it is low on average. Commercial dispersion-shifted fibers are designed with a dispersion of about 4 ps/(km-nm), a value found large enough to suppress FWM.

As discussed in Chapters 8 and 9, FWM is used to advantage in several fiber-based (and semiconductor-based) devices. It is often used for optical amplification, resulting in amplifiers known as *parametric* amplifiers. It can also be used for wavelength conversion and for generating a spectrally inverted signal through the process of *optical phase conjugation*. Useful applications of FWM are discussed in several chapters in this book.

1.6.4 Stimulated Brillouin Scattering

Brillouin scattering can be understood as scattering of a photon to a lower-energy photon such that the energy difference appears in the form of an acoustic phonon. The physical process behind Brillouin scattering is the tendency of materials to become compressed in the presence of an electric field—a phenomenon termed electrostriction [78]. For an oscillating electric field at the pump frequency Ω_p, this process generates an acoustic wave at some frequency Ω. Spontaneous Brillouin scattering can be viewed as scattering of the pump wave from this acoustic wave, resulting in creation of a new wave at the pump frequency Ω_s. At high power levels, the nonlinear phenomena of stimulated Brillouin scattering (SBS) become important. The intensity of the scattered light grows exponentially once the incident power exceeds a threshold value [80]. SBS was first observed in optical fibers during the 1970s [81]–[83].

Any scattering process must conserve both the energy and the momentum. The energy conservation requires that the Stokes shift Ω equals $\omega_p - \omega_s$. The momentum conservation requires that the wave vectors satisfy $\mathbf{k}_A = \mathbf{k}_p - \mathbf{k}_s$. If we use the dispersion relation $|k_A| = \Omega/v_A$, where v_A is the acoustic velocity, this condition determines the acoustic frequency as [50]

$$\Omega = |k_A| v_A = 2 v_A |k_p| \sin(\theta/2), \qquad (1.6.17)$$

where $|k_p| \approx |k_s|$ was used and θ represents the angle between the pump and scattered waves. Note that Ω vanishes in the forward direction ($\theta = 0$) and is maximum in

1.6 Fiber Nonlinearities

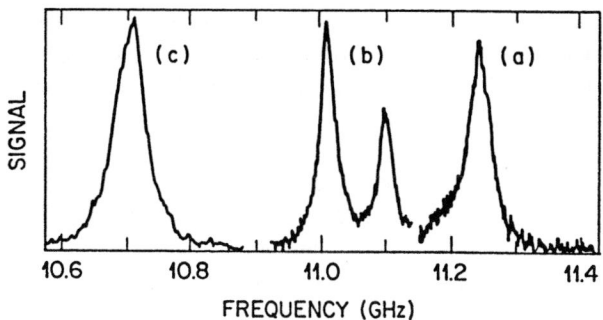

Figure 1.20: Brillouin-gain spectra measured using a 1.525-μm pump for three fibers with different germania doping: (a) silica-core fiber; (b) depressed-cladding fiber; (c) dispersion-shifted fiber. Vertical scale is arbitrary. (After Ref. [83]; ©1986 IEE.)

the backward direction ($\theta = \pi$). In a multimode fiber, SBS can be observed in the forward direction because it can propagate in the form of higher-order modes. In single-mode fibers, by contrast, SBS occurs in the backward direction with a frequency shift $\Omega_B = 2v_A|k_p|$. If we use $k_p = 2\pi\bar{n}/\lambda_p$, where λ_p is the pump wavelength, the *Brillouin shift* is given by

$$\nu_B = \Omega_B/2\pi = 2\bar{n}v_A/\lambda_p, \qquad (1.6.18)$$

where $\bar{n}$ is the mode index. Using $v_A = 5.96$ km/s and $\bar{n} = 1.45$ as typical values for silica fibers, $\nu_B = 11.1$ GHz at $\lambda_p = 1.55$ μm. Equation (1.6.18) shows that ν_B scales inversely with the pump wavelength.

Once the scattered wave is generated spontaneously, it beats with the pump and creates a frequency component at the beat frequency $\omega_p - \omega_s$, which is automatically equal to the acoustic frequency Ω. As a result, the beating term acts as source that increases the amplitude of the sound wave, which in turn increases the amplitude of the scattered wave, resulting in a positive feedback loop. SBS has its origin in this positive feedback, which ultimately can transfer all power from the pump to the scattered wave. The feedback process is governed by the following set of two coupled equations [78]:

$$\frac{dI_p}{dz} = -g_B I_p I_s - \alpha_p I_p, \qquad (1.6.19)$$

$$-\frac{dI_s}{dz} = +g_B I_p I_s - \alpha_s I_s, \qquad (1.6.20)$$

where I_p and I_s are the intensities of the pump and Stokes fields, g_B is the SBS gain, and α_p and α_p account for fiber losses.

The SBS gain g_B is frequency-dependent because of a finite damping time T_B of acoustic waves (the lifetime of acoustic phonons). If the acoustic waves decay as $\exp(-t/T_B)$, the Brillouin gain has a Lorentzian spectral profile given by [82]

$$g_B(\Omega) = \frac{g_B(\Omega_B)}{1 + (\Omega - \Omega_B)^2 T_B^2}. \qquad (1.6.21)$$

Figure 1.20 shows the Brillouin gain spectra at $\lambda_p = 1.525$ μm for three different kinds of single-mode silica fibers. Both the Brillouin shift v_B and the gain bandwidth Δv_B can vary from fiber to fiber because of the guided nature of light and the presence of dopants in the fiber core. The peak labeled (a) in Figure 1.20 is for a fiber with core of nearly pure silica (germania concentration of about 0.3% per mole). The measured Brillouin shift $v_B = 11.25$ GHz is in agreement with Eq. (1.6.18). The Brillouin shift is reduced for fibers (b) and (c) of a higher germania concentration in the fiber core. The double-peak structure for fiber (b) results from inhomogeneous distribution of germania within the core. The gain bandwidth in Figure 1.20 is larger than that expected for bulk silica ($\Delta v_B \approx 17$ MHz at $\lambda_p = 1.525$ μm). A part of the increase is due to the guided nature of acoustic modes in optical fibers. However, most of the increase in bandwidth can be attributed to variations in the core diameter along the fiber length. Because such variations are specific to each fiber, the SBS gain bandwidth is generally different for different fibers and can exceed 100 MHz; typical values are ~50 MHz for λ_p near 1.55 μm.

The peak value of the Brillouin gain in Eq. (1.6.21) occurs for $\Omega = \Omega_B$ and depends on various material parameters such as the density and the elasto-optic coefficient [78]. For silica fibers $g_B \approx 5 \times 10^{-11}$ m/W. The threshold power level for SBS can be estimated by solving Eqs. (1.6.19) and (1.6.20) and finding at what value of I_p, I_s grows from noise to a significant level. The threshold power $P_{th} = I_p A_{eff}$, where A_{eff} is the effective core area, satisfies the condition [80]

$$g_B P_{th} L_{eff}/A_{eff} \approx 21, \qquad (1.6.22)$$

where L_{eff} is the effective fiber length defined earlier in Eq. (1.6.6). For most lightwave systems L_{eff} can be approximated by $1/\alpha$ as $\alpha L \gg 1$ in practice. Using $A_{eff} = \pi w^2$, where w is the spot size, P_{th} can be as low as 1 mW depending on the values of w and α [82]. Once the power launched into an optical fiber exceeds the threshold level, most of the light is reflected backward through SBS. Clearly, SBS limits the launched power to a few milliwatts because of its low threshold.

The preceding estimate of P_{th} applies to a narrowband CW beam as it neglects the temporal and spectral characteristics of the incident light. In a lightwave system, the signal is in the form of a bit stream. For a single short pulse whose width is much smaller than the phonon lifetime, no SBS is expected to occur. However, for a high-speed bit stream, pulses come at such a fast rate that successive pulses build up the acoustic wave, similar to the case of a CW beam, although the SBS threshold increases. The exact value of the average threshold power depends on the modulation format (RZ versus NRZ) and is typically ~5 mW. It can be increased to 10 mW or more by increasing the bandwidth of the optical carrier to >200 MHz through phase modulation. SBS does not produce interchannel crosstalk in WDM systems because the 10-GHz frequency shift is much smaller than typical channel spacing.

1.6.5 Stimulated Raman Scattering

Spontaneous Raman scattering occurs in optical fibers when a pump wave is scattered by vibrating silica molecules. It can be understood using the energy-level diagram

1.6 Fiber Nonlinearities

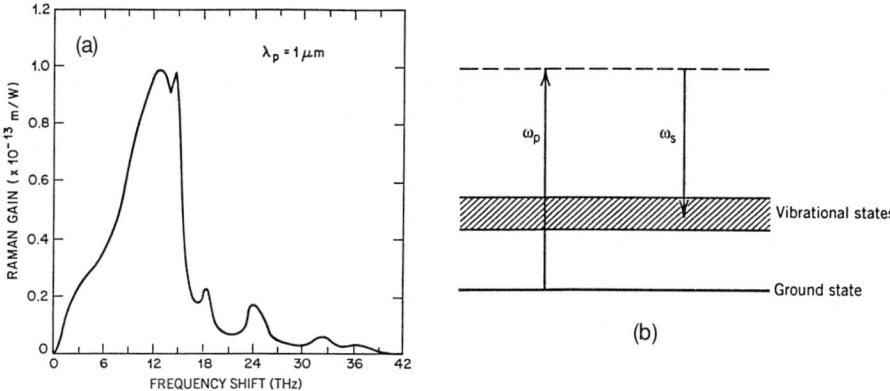

Figure 1.21: (a) Raman gain spectrum of fused silica at $\lambda_p = 1$ μm and (b) energy levels participating in the SRS process. (After Ref. [84]; ©1972 AIP.)

shown in Figure 1.21(b). Some pump photons give up their energy to create new photons of reduced energy at a lower frequency; the remaining energy is absorbed by silica molecules, which end up in an excited vibrational state. In this sense, SRS is quite similar to SBS, the only difference being that optical phonons participate in place of acoustic phonons. However, the different dispersion relations for acoustic and optical phonons lead to the following differences between the two from a practical standpoint: (i) SRS can occur in both the forward and backward directions even in single-mode fibers; (ii) the scattered light is shifted in frequency by about 13 THz for SRS; and (iii) the Raman gain spectrum is extremely broad as it extends over a 40-THz range. These differences stem from the fact that the vibrational energy levels of silica dictate the value of the Raman shift $\Omega_R = \omega_p - \omega_s$. As an acoustic wave is not involved, spontaneous Raman scattering is an isotropic process and occurs in all directions. SRS was first observed in optical fibers in a 1972 experiment [84].

Similar to the SBS case, the Raman scattering process becomes stimulated if the pump power exceeds a threshold value. SRS can occur in both the forward and backward directions in optical fibers. Physically speaking, the beating of the pump and with the scattered light in these two directions creates a frequency component at the beat frequency $\omega_p - \omega_s$, which acts as a source that derives molecular oscillations. Since the amplitude of the scattered wave increases in response to these oscillations, a positive feedback loop sets in. In the case of forward SRS, the feedback process is governed by the following set of two coupled equations [50]:

$$\frac{dI_p}{dz} = -g_R I_p I_s - \alpha_p I_p, \tag{1.6.23}$$

$$\frac{dI_s}{dz} = g_R I_p I_s - \alpha_s I_s, \tag{1.6.24}$$

where g_R is the SRS gain. In the case of backward SRS, a minus sign is added in front of the derivative in Eq. (1.6.24), and this set of equations becomes identical to the SBS case.

The spectrum of the Raman gain depends on the decay time associated with the excited vibrational state. In the case of a molecular gas or liquid, the decay time is relatively long (∼1 ns), resulting in a Raman-gain bandwidth of ∼1 GHz. In the case for optical fibers, the bandwidth exceeds 10 THz. Figure 1.21(a) shows the Raman-gain spectrum of silica fibers. The broadband and multipeak nature of the spectrum is due to the amorphous nature of glass. More specifically, vibrational energy levels of silica molecules merge together to form a band. As a result, the Stokes frequency ω_s can differ from the pump frequency ω_p over a wide range. The maximum gain occurs when the Raman shift $\Omega_R \equiv \omega_p - \omega_s$ is about 13 THz. Another major peak occurs near 15 THz, while minor peaks persist for values of Ω_R as large as 35 THz. The peak value of the Raman gain g_R is about 1×10^{-13} m/W at a wavelength of 1 μm. This value scales linearly with ω_p (or inversely with the pump wavelength λ_p), resulting in $g_R \approx 6 \times 10^{-13}$ m/W at 1.55 μm.

Similar to the case of SBS, the threshold power P_{th} is defined as the incident power at which half of the pump power is transferred to the Stokes field at the output end of a fiber of length L. It is estimated from [80]

$$g_R P_{th} L_{eff} / A_{eff} \approx 16, \quad (1.6.25)$$

where g_R is the peak value of the Raman gain. As before, L_{eff} can be approximated by $1/\alpha$. If we replace A_{eff} by πw^2, where w is the spot size, P_{th} for SRS is given by

$$P_{th} \approx 16\alpha(\pi w^2)/g_R. \quad (1.6.26)$$

If we use $\pi w^2 = 50$ μm^2 and $\alpha = 0.2$ dB/km as the representative values, P_{th} is about 570 mW near 1.55 μm. It is important to emphasize that Eq. (1.6.26) provides an order-of-magnitude estimate only as many approximations are made in its derivation.

As channel powers in optical communication systems are typically below 10 mW, SRS is not a limiting factor for single-channel lightwave systems. However, it affects the performance of WDM systems considerably through Raman-induced crosstalk. On the positive side, SRS can be used to amplify an optical signal when the fiber is pumped using a laser whose wavelength is shorter than the signal wavelength by an amount corresponding to the Raman shift. As discussed in Section 3.2, such fiber-based Raman amplifiers work in any wavelength region and can provide wide bandwidth (>100 nm) when multiple pump lasers are used at suitable wavelengths. Indeed, the Raman gain is used routinely for compensating fiber losses in modern lightwave systems [85].

Problems

1.1 A multimode fiber with a 50-μm core diameter is designed to limit the intermodal dispersion to 10 ns/km. What is the numerical aperture of this fiber? What is the limiting bit rate for transmission over 10 km at 0.88 μm? Use 1.45 for the refractive index of the cladding.

1.2 Use the ray equation in the paraxial approximation [Eq. (1.1.8)] to prove that intermodal dispersion is zero for a graded-index fiber with a quadratic index profile.

Problems

1.3 Use Maxwell's equations to express the field components E_ρ, E_ϕ, H_ρ, and H_ϕ in terms of E_z and H_z and obtain Eqs. (1.3.29) through (1.3.32).

1.4 Derive the eigenvalue equation (1.3.33) by matching the boundary conditions at the core–cladding interface of a step-index fiber.

1.5 A single-mode fiber has an index step $n_1 - n_2 = 0.005$. Calculate the core radius if the fiber has a cutoff wavelength of 1 μm. Estimate the spot size (full width at half maximum) of the fiber mode and the fraction of the mode power inside the core when this fiber is used at 1.3 μm. Use $n_1 = 1.45$.

1.6 Derive an expression for the confinement factor Γ of single-mode fibers defined as the fraction of the total mode power contained inside the core. Use the Gaussian approximation for the fundamental fiber mode. Estimate Γ for $V = 2$.

1.7 A single-mode fiber is measured to have $\lambda^2 (d^2 n / d\lambda^2) = 0.02$ at 0.8 μm. Calculate the dispersion parameters β_2 and D.

1.8 A 1.3-μm lightwave system uses a 50-km fiber link and requires at least 0.3 μW at the receiver. The fiber loss is 0.5 dB/km. Fiber is spliced every 5 km and has two connectors of 1-dB loss at both ends. Splice loss is only 0.2 dB. Determine the minimum power that must be launched into the fiber.

1.9 A 1.55-μm continuous-wave signal with 4-mW power is launched into a fiber with 50-μm^2 effective core area. After what fiber length would the nonlinear phase shift induced by SPM become 2π? Assume $\bar{n}_2 = 2.6 \times 10^{-20}$ m^2/W and neglect fiber losses.

1.10 Calculate the power launched into a 40-km-long single-mode fiber for which the SPM-induced nonlinear phase shift becomes 180°. Assume $\lambda = 1.55$ μm, $A_\text{eff} = 40$ μm^2, $\alpha = 0.2$ dB/km, and $\bar{n}_2 = 2.6 \times 10^{-20}$ m^2/W.

1.11 Find the maximum frequency shift occurring because of the SPM-induced chirp imposed on a Gaussian pulse of 20-ps width (full width at half-maximum) and 5-mW peak power after it has propagated 100 km. Use the fiber parameters of the preceding problem but assume $\alpha = 0$.

1.12 Starting from Eq. (1.6.16) prove that the phase mismatch for degenerate four-wave mixing ($\omega_1 = \omega_2$) is approximately given by $\Delta = \beta_2 \Omega^2$, where $\Omega = \omega_3 - \omega_1$ and β_2 is the dispersion parameter at the pump wavelength.

1.13 Use the law of momentum conservation to prove that stimulated Brillouin scattering can occur only in the backward direction in single-mode fibers.

1.14 Calculate the threshold power for stimulated Brillouin scattering for a 50-km fiber link operating at 1.3 μm and having a loss of 0.5 dB/km. How much does the threshold power change if the operating wavelength is changed to 1.55 μm, where the fiber loss is only 0.2 dB/km? Assume that $A_\text{eff} = 50$ μm^2 and $g_B = 5 \times 10^{-11}$ m/W at both wavelengths.

1.15 Repeat Problem 1.14 in the case of stimulated Raman scattering, assuming that $g_R = 6 \times 10^{-14}$ m/W at 1.55 μm and scales inversely with wavelength. Assume all other fiber parameters are the same.

References

[1] J. Hecht, *City of Light: Story of Fiber Optics*, Oxford University Press, New York, 1999.
[2] A. C. S. van Heel, *Nature* **173**, 39 (1954).
[3] B. I. Hirschowitz, L. E. Curtiss, C. W. Peters, and H. M. Pollard, *Gastro-enterology* **35**, 50 (1958).
[4] N. S. Kapany, *J. Opt. Soc. Am.* **49**, 779 (1959).
[5] N. S. Kapany, *Fiber Optics: Principles and Applications*, Academic Press, San Diego, CA, 1967.
[6] K. C. Kao and G. A. Hockham, *Proc. IEE* **113**, 1151 (1966); A. Werts, *Onde Electr.* **45**, 967 (1966).
[7] F. P. Kapron, D. B. Keck, and R. D. Maurer, *Appl. Phys. Lett.* **17**, 423 (1970).
[8] T. Miya, Y. Terunuma, T. Hosaka, and T. Miyoshita, *Electron. Lett.* **15**, 106 (1979).
[9] M. J. Adams, *An Introduction to Optical Waveguides*, Wiley, New York, 1981.
[10] T. Okoshi, *Optical Fibers*, Academic Press, San Diego, CA, 1982.
[11] A. W. Snyder and J. D. Love, *Optical Waveguide Theory*, Chapman & Hall, London, 1983.
[12] E. G. Neumann, *Single-Mode Fibers*, Springer, New York, 1988.
[13] L. B. Jeunhomme, *Single-Mode Fiber Optics*, Marcel Dekker, New York, 1990.
[14] D. Marcuse, *Theory of Dielectric Optical Waveguides*, 2nd ed., Academic Press, San Diego, CA, 1991.
[15] K. Iizuka, *Elements of Photonics*, Vol. II, Wiley, New York, 2002.
[16] J. A. Buck, *Fundamentals of Optical Fibers*, 2nd ed., Wiley, New York, 2003.
[17] M. Born and E. Wolf, *Principles of Optics*, 7th ed., Cambridge University Press, New York, 1999.
[18] T. Li, Ed., *Optical Fiber Communications*, Vol. 1, Academic Press, San Diego, CA, 1985, Chaps. 1–4.
[19] T. Izawa and S. Sudo, *Optical Fibers: Materials and Fabrication*, Kluwer Academic, Boston, 1987.
[20] D. J. DiGiovanni, S. K. Das, L. L. Blyler, W. White, R. K. Boncek, and S. E. Golowich, in *Optical Fiber Teleommunications*, Vol. 4A, I. Kaminow and T. Li, Eds., Academic Press, San Diego, CA, 2002, Chap. 2.
[21] M. Monerie, *IEEE J. Quantum Electron.* **18**, 535 (1982); *Electron. Lett.* **18**, 642 (1982).
[22] M. A. Saifi, S. J. Jang, L. G. Cohen, and J. Stone, *Opt. Lett.* **7**, 43 (1982).
[23] Y. W. Li, C. D. Hussey, and T. A. Birks, *J. Lightwave Technol.* **11**, 1812 (1993).
[24] R. Lundin, *Appl. Opt.* **32**, 3241 (1993); *Appl. Opt.* **33**, 1011 (1994).
[25] S. P. Survaiya and R. K. Shevgaonkar, *IEEE Photon. Technol. Lett.* **8**, 803 (1996).
[26] P. Geittner, H. J. Hagemann, J. Warnier, and H. Wilson, *J. Lightwave Technol.* **4**, 818 (1986).
[27] Y. Koike, T. Ishigure, and E. Nihei, *J. Lightwave Technol.* **13**, 1475 (1995).
[28] U. Fiedler, G. Reiner, P. Schnitzer, and K. J. Ebeling, *IEEE Photon. Technol. Lett.* **8**, 746 (1996).
[29] W. R. White, M. Dueser, W. A. Reed, and T. Onishi, *IEEE Photon. Technol. Lett.* **11**, 997 (1999).
[30] T. Ishigure, Y. Koike, and J. W. Fleming, *J. Lightwave Technol.* **18**, 178 (2000).
[31] A. Weinert, *Plastic Optical Fibers: Principles, Components, and Installation*, Wiley, New York, 2000.
[32] J. Zubia and J. Arrue, *Opt. Fiber Technol.* **7**, 101 (2001).

References

[33] W. Daum, J. Krauser, P. E. Zamzow, and O. Ziemann, *POF—Plastic Optical Fibers for Data Communication*, Springer, New York, 2002.

[34] J. C. Knight, T. A. Birks, P. S. J. Russell, and D. M. Atkin, *Opt. Lett.* **21**, 1547 (1996).

[35] T. A. Birks, J. C. Knight, and P. S. J. Russell, *Opt. Lett.* **22**, 961 (1997).

[36] J. C. Knight, T. A. Birks, P. S. J. Russell, and J. P. Sanders, *J. Opt. Soc. Am. A* **15**, 748 (1998).

[37] T. M. Monro, D. J. Richardson, N. G. R. Broderick, and P. J. Bennett, *J. Lightwave Technol.* **17**, 1093 (1999).

[38] J. Broeng, D. Mogilevstev, S. B. Barkou, and A. Bjarklev, *Opt. Fiber Technol.* **5**, 305 (1999).

[39] B. J. Eggleton, P. S. Westbrook, R. S. Windeler, S. Spälter, and T. A. Sreasser, *Opt. Lett.* **24**, 1460 (1999).

[40] R. F. Cregan, B. J. Mangan, J. C. Knight, T. A. Birks, P. S. J. Russell, P. J. Roberts, and D. C. Allan, *Science* **285**, 1537 (1999).

[41] T. M. Monro, P. J. Bennett, N. G. R. Broderick, and D. J. Richardson, *Opt. Lett.* **25**, 206 (2000).

[42] M. Ibanescu, Y. Fink, S. Fan, E. L. Thomas, J. D. Joannopoulos, *Science* **289**, 415 (2000).

[43] J. K. Ranka and A. L. Gaeta, in *Nonlinear Photonic Crystals*, R. E Slusher and B. J. Eggleton, Eds., Springer, New York, 2003.

[44] P. S. J. Russell, *Science* **299**, 358 (2003).

[45] H. Murata, *Handbook of Optical Fibers and Cables*, Marcel Dekker, New York, 1996.

[46] S. C. Mettler and C. M. Miller, in *Optical Fiber Telecommunications II*, S. E. Miller and I. P. Kaminow, Eds., Academic Press, San Diego, CA, 1988, Chap. 6.

[47] W. C. Young and D. R. Frey, in *Optical Fiber Telecommunications II*, S. E. Miller and I. P. Kaminow, Eds., Academic Press, San Diego, CA, 1988, Chap. 7.

[48] C. M. Miller, S. C. Mettler, and I. A. White, *Optical Fiber Splices and Connectors: Theory and Methods*, Marcel Dekker, New York, 1986.

[49] P. Diament, *Wave Transmission and Fiber Optics*, Macmillan, New York, 1990, Chap. 3.

[50] G. P. Agrawal, *Nonlinear Fiber Optics*, 3rd ed., Academic Press, San Diego, CA, 2001.

[51] M. Abramowitz and I. A. Stegun, Eds., *Handbook of Mathematical Functions*, Dover, New York, 1970, Chap. 9.

[52] D. Gloge, *Appl. Opt.* **10**, 2252 (1971); **10**, 2442 (1971).

[53] G. P. Agrawal, *Fiber-Optic Communication Systems*, 3rd ed., Wiley, New York, 2002.

[54] D. Marcuse, *J. Opt. Soc. Am.* **68**, 103 (1978).

[55] G. A. Thomas, B. L. Shraiman, P. F. Glodis, and M. J. Stephan, *Nature* **404**, 262 (2000).

[56] M. Saad and J. A. Harrington, *Infrared Optical Fibers and Their Applications*, SPIE Press, Bellingham, WA, 1999.

[57] W. B. Gardner, *Bell Syst. Tech. J.* **54**, 457 (1975).

[58] D. Marcuse, *Bell Syst. Tech. J.* **55**, 937 (1976).

[59] K. Petermann, *Electron. Lett.* **12**, 107 (1976); *Opt. Quantum Electron.* **9**, 167 (1977).

[60] I. H. Malitson, *J. Opt. Soc. Am.* **55**, 1205 (1965).

[61] L. G. Cohen, C. Lin, and W. G. French, *Electron. Lett.* **15**, 334 (1979).

[62] C. T. Chang, *Electron. Lett.* **15**, 765 (1979); *Appl. Opt.* **18**, 2516 (1979).

[63] L. G. Cohen, W. L. Mammel, and S. Lumish, *Opt. Lett.* **7**, 183 (1982).

[64] S. J. Jang, L. G. Cohen, W. L. Mammel, and M. A. Shaifi, *Bell Syst. Tech. J.* **61**, 385 (1982).

[65] V. A. Bhagavatula, M. S. Spotz, W. F. Love, and D. B. Keck, *Electron. Lett.* **19**, 317 (1983).

[66] P. Bachamann, D. Leers, H. Wehr, D. V. Wiechert, J. A. van Steenwijk, D. L. A. Tjaden, and E. R. Wehrhahn, *J. Lightwave Technol.* **4**, 858 (1986).
[67] B. J. Ainslie and C. R. Day, *J. Lightwave Technol.* **4**, 967 (1986).
[68] C. D. Poole and J. Nagel, in *Optical Fiber Telecommunications III*, Vol. A, I. P. Kaminow and T. L. Koch, Eds., Academic Press, San Diego, CA, 1997, Chap. 6.
[69] F. Bruyère, *Opt. Fiber Technol.* **2**, 269 (1996).
[70] P. K. A. Wai and C. R. Menyuk, *J. Lightwave Technol.* **14**, 148 (1996).
[71] M. Karlsson, *Opt. Lett.* **23**, 688 (1998).
[72] G. J. Foschini, R. M. Jopson, L. E. Nelson, and H. Kogelnik, *J. Lightwave Technol.* **17**, 1560 (1999).
[73] B. Huttner, C. Geiser, and N. Gisin, *IEEE J. Sel. Topics Quantum Electron.* **6**, 317 (2000).
[74] S. J. Savory and F. P. Payne, *J. Lightwave Technol.* **19**, 350 (2001).
[75] D. Wang and C. R. Menyuk, *J. Lightwave Technol.* **19** 487 (2001).
[76] J. P. Gordon and H. Kogelnik, *Proc. Natl. Acad. Sci. USA* **97**, 4541 (2000).
[77] H. Kogelnik, R. M. Jopson, and L. E. Nelson, in *Optical Fiber Teleommunications*, Vol. 4A, I. P. Kaminow and T. Li, Eds., Academic Press, San Diego, CA, 2002, Chap. 15.
[78] R. W. Boyd, *Nonlinear Optics*, 2nd ed., Academic Press, San Diego, CA, 2003.
[79] R. H. Stolen and C. Lin, *Phys. Rev. A* **17**, 1448 (1978).
[80] R. G. Smith, *Appl. Opt.* **11**, 2489 (1972).
[81] E. P. Ippen and R. H. Stolen, *Appl. Phys. Lett.* **21**, 539 (1972).
[82] D. Cotter, *Electron. Lett.* **18**, 495 (1982); *J. Opt. Commun.* **4**, 10 (1983).
[83] R. W. Tkach, A. R. Chraplyvy, and R. M. Derosier, *Electron. Lett.* **22**, 1011 (1986).
[84] R. H. Stolen, E. P. Ippen, and A. R. Tynes, *Appl. Phys. Lett.* **20**, 62 (1972).
[85] K. Rottwitt and A. J. Stentz, in *Optical Fiber Teleommunications*, Vol. 4A, I. P. Kaminow and T. Li, Eds., Academic Press, San Diego, CA, 2002, Chap. 5.

Chapter 2

Passive Fiber Components

Before optical fibers could be deployed in developing lightwave technology, it was necessary to develop a multitude of optical components that were compatible with the use of fibers. Such components can be divided into two main categories—active and passive. The active components, by definition, require an external energy source, whereas the passive ones manipulate the incident optical signal directly. This chapter focuses on passive components that are built using optical fibers and thus can be spliced directly into a fiber link. Fiber couplers, also known as *directional couplers*, are discussed first in Section 2.1 because they constitute an essential component of lightwave technology and are used routinely for splitting an optical field into two coherent but physically separated parts. Section 2.2 is devoted to fiber Bragg gratings. Even though fiber gratings became available commercially only during the 1990s, they constitute by now an integral part of lightwave technology. Section 2.3 focuses on several fiber-based interferometers whose operation requires coherent interference between two distinct optical fields. Section 2.4 is devoted to another class of important devices, such as isolators and circulators, whose operation requires one or more nonfiber components but the final device is packaged such that it can be used as any all-fiber device.

2.1 Directional Couplers

Fiber couplers are four-port devices (two input and two output ports) that are used routinely for a variety of applications related to fiber optics [1]–[4]. Their main function is to split coherently an optical field, incident on one of the input ports, and send the two parts to the two output ports. Since the output is directed in two different directions, such devices are referred to as *directional couplers*. They can also be fabricated using planar waveguides made of $LiNbO_3$ or other semiconductor materials (see Chapter 4). This section focuses on fiber couplers but most of the results apply to any directional coupler.

Several different techniques can be used to make fiber couplers [4]. Figure 2.1 shows schematically a fused fiber coupler in which the cores of two single-mode fibers are brought close together in a central region such that the spacing between the cores is

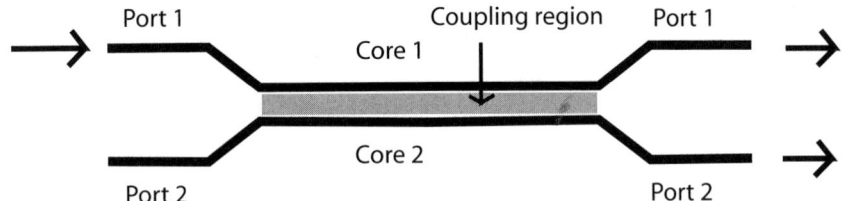

Figure 2.1: Schematic of a fiber coupler. Shaded area shows the region where the modes of two fibers overlap.

comparable to their diameters. A dual-core fiber, designed to have two cores close to each other throughout its length, can also act as a directional coupler. In both cases, the cores are close enough that the fundamental modes propagating in each core overlap partially in the cladding region between the two cores. Such evanescent-wave coupling between the tails of two modes leads to the transfer of optical power from one core to another under suitable conditions. Fiber couplers are called symmetric when their cores are identical in all respects. In general, the two cores need not be identical; such couplers are called asymmetric. In this section, we consider an asymmetric fiber coupler and discuss its properties using coupled-mode theory.

2.1.1 Coupled-Mode Equations

Coupled-mode theory is used commonly for treating coupling between two optical modes [5]–[10]. To derive the coupled-mode equations for a directional coupler, we begin with the Helmholtz equation (1.3.16) satisfied by a specific frequency component at the frequency ω and written as

$$\nabla^2 \tilde{\mathbf{E}} + \tilde{n}^2 k_0^2 \tilde{\mathbf{E}} = 0, \tag{2.1.1}$$

where $k_0 = \omega/c$ and $\tilde{\mathbf{E}}(\mathbf{r}, \omega)$ denotes the Fourier transform of the electric field $\mathbf{E}(\mathbf{r},t)$ with respect to time. The refractive index $\tilde{n}(x,y) = n_0$ everywhere in the x–y plane except in the region occupied by the two fiber cores, where it is larger by a constant amount.

The coupled-mode theory is based on the assumption that an approximate solution of Eq. (2.1.1) can be written as

$$\tilde{\mathbf{E}}(\mathbf{r},\omega) \approx \hat{e}[\tilde{A}_1(z,\omega)F_1(x,y) + \tilde{A}_2(z,\omega)F_2(x,y)]e^{i\beta z}, \tag{2.1.2}$$

where the propagation constant β is yet to be determined. The polarization direction $\hat{e}$ of the optical field is assumed to remain unchanged during propagation. The spatial distribution $F_m(x,y)$ with $m = 1$ or 2 corresponds to the fundamental mode supported by the mth core in the absence of the other core. It is obtained by solving Eq. (2.1.1) and satisfies the following equation:

$$\frac{\partial^2 F_m}{\partial x^2} + \frac{\partial^2 F_m}{\partial y^2} + [n_m^2(x,y)k_0^2 - \bar{\beta}_m^2]F_m = 0, \tag{2.1.3}$$

2.1 Directional Couplers

where $\bar{\beta}_m$ is the mode-propagation constant and $n_m(x,y) = n_0$ everywhere in the x–y plane except in the region occupied by the mth core, where it is larger by a constant amount. Equation (2.1.3) has been solved in Section 1.3.2 in terms of the Bessel functions. The same solution applies here.

The amplitudes A_1 and A_2 vary along the coupler length because of the overlap between the two modes. To find how they evolve with z, we substitute Eq. (2.1.2) in Eq. (2.1.1), multiply the resulting equation by F_1^* or F_2^*, use Eq. (2.1.3), and integrate over the entire x–y plane. This procedure leads to the following set of two coupled-mode equations in the frequency domain:

$$\frac{d\tilde{A}_1}{dz} = i(\bar{\beta}_1 - \beta)\tilde{A}_1 + i\kappa_{12}\tilde{A}_2, \tag{2.1.4}$$

$$\frac{d\tilde{A}_2}{dz} = i(\bar{\beta}_2 - \beta)\tilde{A}_2 + i\kappa_{21}\tilde{A}_1, \tag{2.1.5}$$

where the coupling coefficient κ_{mp} is defined as ($m, p = 1$ or 2)

$$\kappa_{mp} = \frac{k_0^2}{2\beta} \int\!\!\int_{-\infty}^{\infty} (\tilde{n}^2 - n_p^2) F_m^* F_p \, dx \, dy, \tag{2.1.6}$$

if we assume that the modes are normalized such that $\int\!\!\int_{-\infty}^{\infty} |F_m(x,y)|^2 \, dx \, dy = 1$.

The frequency-domain coupled-mode equations can be converted into time domain following the method used in Section 1.5.5. By expanding $\bar{\beta}_m(\omega)$ in a Taylor series around the carrier frequency ω_0 as

$$\bar{\beta}_m(\omega) = \beta_{0m} + (\omega - \omega_0)\beta_{1m} + \tfrac{1}{2}(\omega - \omega_0)^2 \beta_{2m} + \cdots, \tag{2.1.7}$$

retaining terms up to second order, and replacing $\omega - \omega_0$ by $i(\partial/\partial t)$ while taking the inverse Fourier transform, we obtain

$$\frac{\partial A_1}{\partial z} + \frac{1}{v_{g1}}\frac{\partial A_1}{\partial t} + \frac{i\beta_{21}}{2}\frac{\partial^2 A_1}{\partial t^2} = i\kappa_{12}A_2 + i\delta_a A_1, \tag{2.1.8}$$

$$\frac{\partial A_2}{\partial z} + \frac{1}{v_{g2}}\frac{\partial A_2}{\partial t} + \frac{i\beta_{22}}{2}\frac{\partial^2 A_2}{\partial t^2} = i\kappa_{21}A_1 - i\delta_a A_2, \tag{2.1.9}$$

where $v_{gm} \equiv 1/\beta_{1m}$ is the group velocity, β_{2m} is the group-velocity dispersion (GVD) parameter ($m = 1, 2$), and we have introduced the asymmetry parameter δ_a and the average propagation constant as

$$\delta_a = \tfrac{1}{2}(\beta_{01} - \beta_{02}), \qquad \beta = \tfrac{1}{2}(\beta_{01} + \beta_{02}). \tag{2.1.10}$$

Equations (2.1.8) and (2.1.9) are known as the coupled-mode equations and describe the behavior of most fiber couplers under quite general conditions. They do not include the nonlinear coupling induced by cross-phase modulation (XPM) because this coupling is relatively weak in practice. The self-phase modulation (SPM) term has also been ignored in view of the short length of most couplers. Both of these nonlinear effects should be included when input power levels are high and the coupler is relatively long [11].

For a symmetric coupler with two identical cores, $\delta_a = 0$ and $\kappa_{12} = \kappa_{21} \equiv \kappa$. The coupled-mode equations for symmetric couplers, after including both SPM and XPM, take the form:

$$\frac{\partial A_1}{\partial z} + \frac{1}{v_g}\frac{\partial A_1}{\partial t} + \frac{i\beta_2}{2}\frac{\partial^2 A_1}{\partial t^2} = i\kappa A_2 + i\gamma(|A_1|^2 + \sigma|A_2|^2)A_1, \quad (2.1.11)$$

$$\frac{\partial A_2}{\partial z} + \frac{1}{v_g}\frac{\partial A_2}{\partial t} + \frac{i\beta_2}{2}\frac{\partial^2 A_2}{\partial t^2} = i\kappa A_1 + i\gamma(|A_2|^2 + \sigma|A_1|^2)A_2, \quad (2.1.12)$$

where the subscript identifying cores has been dropped from the parameters v_g and β_2 since they have the same values for both cores. The nonlinear parameter γ was introduced in Section 1.6.1 as $\gamma = 2\pi n_2/(A_{\text{eff}}\lambda)$ for a fiber with the effective core area A_{eff}. The XPM parameter σ is quite small in practice because it is related to the the overlap between the fiber modes in the two cores [11]. We set $\gamma = 0$ in the following discussion as nonlinear effects can be neglected for most fiber couplers.

2.1.2 Power-Transfer Characteristics

Consider first the simplest case of a continuous-wave (CW) beam incident on one of the input ports of a fiber coupler. The time-dependent terms can then be set to zero in Eqs. (2.1.8) and (2.1.9) to obtain

$$\frac{dA_1}{dz} = i\kappa_{12}A_2 + i\delta_a A_1, \quad (2.1.13)$$

$$\frac{dA_2}{dz} = i\kappa_{21}A_1 - i\delta_a A_2. \quad (2.1.14)$$

By differentiating Eq. (2.1.13) and eliminating dA_2/dz using Eq. (2.1.14), we obtain the following equation for A_1:

$$\frac{d^2 A_1}{dz^2} + \kappa_e^2 A_1 = 0, \quad (2.1.15)$$

where the effective coupling coefficient κ_e is defined as

$$\kappa_e = \sqrt{\kappa^2 + \delta_a^2}, \qquad \kappa = \sqrt{\kappa_{12}\kappa_{21}}. \quad (2.1.16)$$

The same harmonic-oscillator-type equation is also satisfied by A_2.

By using the boundary condition that a single CW beam is incident on one of the input ports such that $A_1(0) = A_0$ and $A_2(0) = 0$, the solution of Eqs. (2.1.13) and (2.1.14) is given by

$$A_1(z) = A_0[\cos(\kappa_e z) + i(\delta_a/\kappa_e)\sin(\kappa_e z)], \quad (2.1.17)$$

$$A_2(z) = A_0(i\kappa_{21}/\kappa_e)\sin(\kappa_e z). \quad (2.1.18)$$

Thus, even though $A_2 = 0$ initially at $z = 0$, some power is transferred to the second core as light propagates inside the fiber coupler. Figure 2.2 shows the ratio $|A_2/A_0|^2$ as

2.1 Directional Couplers

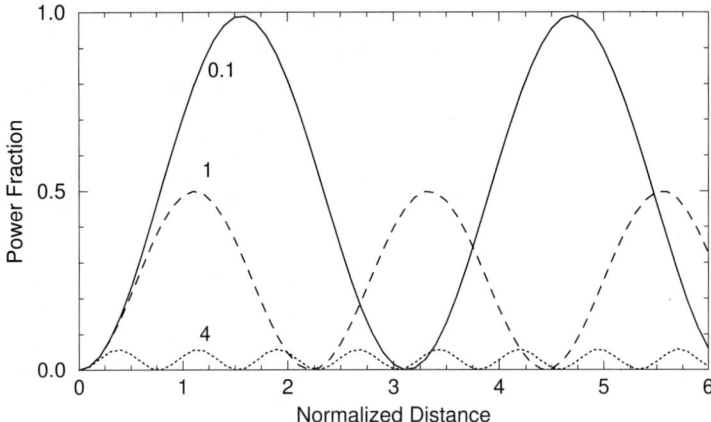

Figure 2.2: Fraction of power transferred to the second core plotted as a function of κz for three values of δ_a/κ when a CW beam is launched in one core at $z = 0$.

a function of z for several values of δ_a/κ. In all cases, power transfer to the second core occurs in a periodic fashion. The maximum power is transferred at distances such that $\kappa_e z = m\pi/2$, where m is an integer. The shortest distance at which maximum power is transferred to the second core for the first time is called the *coupling length* and is given by $L_c = \pi/(2\kappa_e)$.

The power coming out of the two output ports of a fiber coupler depends on the coupler length L and on the powers injected at the two input ends. For a symmetric coupler, the general solution of Eqs. (2.1.13) and (2.1.14) can be written in a matrix form as

$$\begin{pmatrix} A_1(L) \\ A_2(L) \end{pmatrix} = \begin{pmatrix} \cos(\kappa L) & i\sin(\kappa L) \\ i\sin(\kappa L) & \cos(\kappa L) \end{pmatrix} \begin{pmatrix} A_1(0) \\ A_2(0) \end{pmatrix}. \qquad (2.1.19)$$

Typically, only one beam is injected at the input end. The output powers, $P_1 = |A_1|^2$ and $P_2 = |A_2|^2$, are then obtained from Eq. (2.1.19) by setting $A_2(0) = 0$ and are given by

$$P_1(L) = P_0 \cos^2(\kappa L), \qquad P_2(L) = P_0 \sin^2(\kappa L), \qquad (2.1.20)$$

where $P_0 \equiv A_0^2$ is the incident power at the first input port. The coupler thus acts as a beam splitter, and the splitting ratio depends on the parameter κL.

2.1.3 Transfer Matrix of a Coupler

The concept of the transfer matrix, sometimes called the scattering matrix, is quite useful for couplers because a single matrix governs all properties of the coupler. If we introduce a quantity $\rho = P_1(L)/P_0 = \cos^2(\kappa L)$ that governs the fraction of the input power P_0 remaining in the same port of the coupler, this matrix can be written as

$$T_c = \begin{pmatrix} \sqrt{\rho} & i\sqrt{1-\rho} \\ i\sqrt{1-\rho} & \sqrt{\rho} \end{pmatrix}. \qquad (2.1.21)$$

The determinant of the transfer matrix is unity, as it should be for a lossless coupler. This matrix is also symmetric, a property required to ensure that the coupler behaves in the same way irrespective of the direction in which light propagates. In fact, it is easy to show [2] that a general symmetric unitary matrix has to have the form of Eq. (2.1.21). It is important to realize that a directional coupler introduces a relative phase shift of $\pi/2$ between the two output ports, as indicated by the factor i in the off-diagonal terms of the transfer matrix in Eq. (2.1.21). This phase shift plays an important role in the design of fiber interferometers (discussed in Section 2.3).

The simplest application of a fiber coupler is as an *optical tap*. If ρ is close to 1, most of the input signal remains in the same core but a small fraction, $1-\rho$, of the incident power is transferred to the other core and appears at the "cross port" of the coupler. Another application of fiber couplers consists of dividing the input power equally between the two output ports ($\rho = \frac{1}{2}$). For this purpose, the coupler length L is chosen such that $\kappa L = \pi/4$ or $L = L_c/2$. Such couplers are referred to as 3-dB couplers. Fiber couplers with $L = L_c$ transfer all of their input power to the cross port. In general, by choosing the coupler length appropriately, the power can be divided between the two output ports in an arbitrary manner.

The length of a coupler required to realize a given splitting ratio depends on the coupling coefficient κ, which in turn depends on the spacing d between the two cores. For a symmetric coupler, the integrals in Eq. (2.1.6) can be evaluated analytically [5]. The resulting expression is somewhat complicated as it involves the Bessel functions. The following empirical expression is useful in practice [12]:

$$\kappa = \frac{\pi V}{2k_0 n_0 a^2} \exp[-(c_0 + c_1 \bar{d} + c_2 \bar{d}^2)], \qquad (2.1.22)$$

where V is the fiber parameter introduced in Eq. (1.3.35), a is the core radius, and $\bar{d} \equiv d/a$ is the normalized center-to-center spacing between the two cores ($\bar{d} > 2$). The constants c_0, c_1, and c_2 depend on V as $c_0 = 5.2789 - 3.663V + 0.3841V^2$, $c_1 = -0.7769 + 1.2252V - 0.0152V^2$, and $c_2 = -0.0175 - 0.0064V - 0.0009V^2$. Equation (2.1.22) is accurate to within 1% for values of V and $\bar{d}$ in the range $1.5 \leq V \leq 2.5$ and $2 \leq \bar{d} \leq 4.5$. As an example, $\kappa \sim 1$ cm^{-1} for $\bar{d} = 3$, resulting in a coupling length of 1 cm or so. However, coupling length increases to 1 m or more when $\bar{d}$ exceeds 5.

2.1.4 Supermodes of a Coupler

One may ask whether the proximity of two cores always leads to periodic power transfer between the cores. In fact, the nature of power transfer depends on the launch conditions at the input end. The physics can be better understood by noting that, with a suitable choice of the propagation constant β in Eq. (2.1.2), the mode amplitudes $\tilde{A}_1$ and $\tilde{A}_2$ can be forced to become z-independent. From Eqs. (2.1.4) and Eq. (2.1.5), this can occur when the amplitude ratio $f = \tilde{A}_2/\tilde{A}_1$ is initially such that

$$f = \frac{\beta - \bar{\beta}_1}{\kappa_{12}} = \frac{\kappa_{21}}{\beta - \bar{\beta}_2}. \qquad (2.1.23)$$

2.1 Directional Couplers

This equation can be used to find the propagation constant β. Since β satisfies a quadratic equation, we find two values of β (called eigenvalues) such that

$$\beta_\pm = \tfrac{1}{2}(\bar{\beta}_1 + \bar{\beta}_2) \pm \sqrt{\delta_a^2 + \kappa^2}. \tag{2.1.24}$$

The spatial distribution corresponding to the two eigenvalues is given by

$$F_\pm(x,y) = (1 + f_\pm^2)^{-1/2}[F_1(x,y) + f_\pm F_2(x,y)], \tag{2.1.25}$$

where $f_\pm$ is obtained from Eq. (2.1.23) using $\beta = \beta_\pm$. These two specific linear combinations of F_1 and F_2 constitute the eigenmodes of a fiber coupler (also called supermodes), and the eigenvalues $\beta_\pm$ correspond to their propagation constants. In the case of a symmetric coupler, $f_\pm = \pm 1$ and the eigenmodes reduce to the even and odd combinations of F_1 and F_2. When the input conditions are such that an eigenmode of the coupler is excited, no power transfer occurs between the two cores.

The periodic power transfer between the two cores, occurring when light is incident on only one core, can be understood using the above modal description as follows. Under such launch conditions, both supermodes of the fiber coupler are excited simultaneously. Each supermode propagates with its own propagation constant. Since β_+ and β_- are not the same, the two supermodes develop a relative phase difference on propagation. This phase difference, $\psi(z) = (\beta_+ - \beta_-)z \equiv 2\kappa_e z$, is responsible for the periodic power transfer between two cores. The situation is analogous to that occurring in birefringent fibers when linearly polarized light is launched at an angle from a principal axis. In that case, the relative phase difference between the two orthogonally polarized eigenmodes leads to periodic evolution of the state of polarization, and the role of coupling length is played by the beat length. The analogy between fiber couplers and birefringent fibers turns out to be quite useful even when nonlinear effects are included [11].

2.1.5 Effects of Fiber Dispersion

In the case of optical pulses, the effects of fiber dispersion should be included. For symmetric couplers, the coupled-mode equations become:

$$\frac{\partial A_1}{\partial z} + \frac{i\beta_2}{2}\frac{\partial^2 A_1}{\partial T^2} = i\kappa A_2, \tag{2.1.26}$$

$$\frac{\partial A_2}{\partial z} + \frac{i\beta_2}{2}\frac{\partial^2 A_2}{\partial T^2} = i\kappa A_1, \tag{2.1.27}$$

where $T = t - z/v_g$ is the reduced time, β_2 accounts for the effects of GVD, and nonlinear effects are ignored ($\gamma = 0$).

We can introduce the dispersion length as $L_D = T_0^2/|\beta_2|$, where T_0 is related to the pulse width [13]. The GVD effects are negligible if the coupler length $L \ll L_D$. As L is comparable in practice to the coupling length ($L_c = \pi/2\kappa$), GVD has no effect on couplers for which $\kappa L_D \gg 1$. Since L_D exceeds 1 km for pulses with $T_0 > 1$ ps whereas $L_c < 1$ m typically, the GVD effects become important only for ultrashort pulses with $T_0 < 0.1$ ps. If we neglect the GVD term in Eqs. (2.1.26) and (2.1.27), the resulting equations become identical to those applicable for CW beams. Thus, picosecond optical pulses behave in the same way as CW beams. More specifically, their energy is transferred to the neighboring core periodically when such pulses are incident on one of the input ports of a fiber coupler.

The above conclusion is modified if the frequency dependence of the coupling coefficient κ cannot be ignored [14]. This feature can be included by expanding $\kappa(\omega)$ in a Taylor series around the carrier frequency ω_0 in a way similar to Eq. (2.1.7) so that

$$\kappa(\omega) \approx \kappa_0 + (\omega - \omega_0)\kappa_1 + \tfrac{1}{2}(\omega - \omega_0)^2 \kappa_2, \quad (2.1.28)$$

where $\kappa_m = d^m\kappa/d\omega^m$ is evaluated at $\omega = \omega_0$. When the frequency-domain coupled-mode equations are converted to time domain, two additional terms appear. With these terms included, Eqs. (2.1.26) and (2.1.27) become

$$\frac{\partial A_1}{\partial z} + \kappa_1 \frac{\partial A_2}{\partial T} + \frac{i\beta_2}{2}\frac{\partial^2 A_1}{\partial T^2} + \frac{i\kappa_2}{2}\frac{\partial^2 A_2}{\partial T^2} = i\kappa_0 A_2, \quad (2.1.29)$$

$$\frac{\partial A_2}{\partial z} + \kappa_1 \frac{\partial A_1}{\partial T} + \frac{i\beta_2}{2}\frac{\partial^2 A_2}{\partial T^2} + \frac{i\kappa_2}{2}\frac{\partial^2 A_1}{\partial T^2} = i\kappa_0 A_1. \quad (2.1.30)$$

In practice, the κ_2 term is negligible for pulses as short as 0.1 ps. The GVD term is also negligible if $\kappa L_D \gg 1$. Setting $\beta_2 = 0$ and $\kappa_2 = 0$, Eqs. (2.1.29) and (2.1.30) can be solved analytically to yield [14]:

$$A_1(z,T) = \tfrac{1}{2}\left[A_0(T - \kappa_1 z)e^{i\kappa_0 z} + A_0(T + \kappa_1 z)e^{-i\kappa_0 z}\right], \quad (2.1.31)$$

$$A_2(z,T) = \tfrac{1}{2}\left[A_0(T - \kappa_1 z)e^{i\kappa_0 z} - A_0(T + \kappa_1 z)e^{-i\kappa_0 z}\right], \quad (2.1.32)$$

where $A_0(T)$ represents the shape of the input pulse at $z = 0$. When $\kappa_1 = 0$, the solution reduces to

$$A_1(z,T) = A_0(T)\cos(\kappa_0 z), \qquad A_2(z,T) = iA_0(T)\sin(\kappa_0 z). \quad (2.1.33)$$

Equation (2.1.33) shows that the pulse switches back and forth between the two cores, while maintaining its shape, when the frequency dependence of the coupling coefficient can be neglected. However, when κ_1 is not negligible, Eq. (2.1.32) shows that the pulse will split into two subpulses after a few coupling lengths, and separation between the two would increase with propagation. This effect is referred to as intermodal dispersion and is similar in nature to polarization-mode dispersion occurring in birefringent fibers. Intermodal dispersion was observed in a 1997 experiment by launching short optical pulses (width about 1 ps) in one core of a dual-core fiber with the center-to-center spacing $d \approx 4a$ [15]. The autocorrelation traces showed the evidence of pulse splitting after 1.25 m, and the subpulses separated from each other at a rate of 1.13 ps/m. The coupling length was estimated to be about 4 mm. Intermodal dispersion in fiber couplers becomes of concern only when the coupler length $L \gg L_c$ and pulse widths are ~ 1 ps or shorter.

2.2 Fiber Gratings

Silica fibers can change their optical properties permanently when they are exposed to intense radiation from a laser operating in the blue or ultraviolet spectral region.

2.2 Fiber Gratings

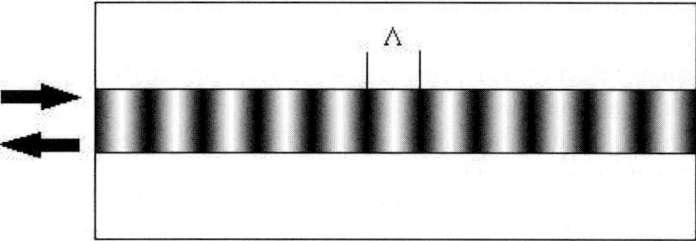

Figure 2.3: Schematic illustration of a fiber grating. Dark and light shaded regions within the fiber core show periodic variations of the refractive index.

This photosensitive effect can be used to induce periodic changes in the refractive index along the fiber length, resulting in the formation of an intracore Bragg grating. Fiber gratings can be designed to operate over a wide range of wavelengths although the wavelength region near 1.5 μm is of particular interest because of its relevance to fiber-optic communication systems. In this section, we discuss the phenomenon of Bragg diffraction, physical mechanism responsible for photosensitivity, and various techniques used to make fiber gratings. The coupled-mode theory is then used to introduce the concept of the photonic bandgap and discuss the role of fiber gratings as a narrowband optical filter.

2.2.1 Bragg Diffraction

The diffraction theory of gratings shows that when light is incident at an angle θ_i (measured with respect to the planes of constant refractive index), it is diffracted at an angle θ_r such that [16]

$$\sin\theta_i - \sin\theta_r = m\lambda/(\bar{n}\Lambda), \qquad (2.2.1)$$

where Λ is the grating period, $\lambda/\bar{n}$ is the wavelength of light inside the medium with an average refractive index $\bar{n}$, and m is the order of Bragg diffraction. This condition can be thought of as a phase-matching condition, similar to that occurring in the case of Brillouin scattering or four-wave mixing, and can be written as

$$\mathbf{k}_i - \mathbf{k}_d = m\mathbf{k}_g, \qquad (2.2.2)$$

where $\mathbf{k}_i$ and $\mathbf{k}_d$ are the wave vectors associated with the incident and diffracted light. The grating wave vector $\mathbf{k}_g$ has magnitude $2\pi/\Lambda$ and points in the direction in which the refractive index of the medium is changing in a periodic manner.

In the case of single-mode fibers, all three vectors lie along the fiber axis. As a result, $\mathbf{k}_d = -\mathbf{k}_i$ and the diffracted light propagates backward. Thus, as shown schematically in Figure 2.3, a fiber grating acts as a reflector for a specific wavelength of light for which the phase-matching condition is satisfied. In terms of the angles appearing in Eq. (2.2.1), $\theta_i = \pi/2$ and $\theta_r = -\pi/2$. If $m = 1$, the period of the grating Λ is related to the vacuum wavelength λ as $\lambda = 2\bar{n}\Lambda$. This condition is known as the *Bragg condition*, and gratings satisfying it are referred to as *Bragg gratings*. Physically, the Bragg condition ensures that weak reflections occurring throughout the grating add up in phase to produce a strong reflection. For a fiber grating reflecting light in the wavelength region near 1.5 μm, the grating period $\Lambda \approx 0.5$ μm.

Bragg gratings inside optical fibers were first formed in 1978 by irradiating a silica fiber for a few minutes with an intense argon-ion laser beam [17]. The grating period was fixed by the argon-ion laser wavelength, and the grating reflected light only within a narrow region around that wavelength. The mechanism behind grating formation can be understood as follows. The 4% reflection occurring at the two fiber–air interfaces creates a standing-wave pattern such that the laser light is absorbed only in the bright regions. If the glass structure changes in such a way that the refractive index increases permanently in the bright regions, an index grating is formed. Although this phenomenon attracted some attention during the next 10 years, it was not until 1989 that fiber gratings became a topic of intense investigation, fuelled partly by the observation of second-harmonic generation in photosensitive fibers. The impetus for this resurgence of interest was provided by a 1989 paper in which a side-exposed holographic technique was used to make fiber gratings with controllable period [18]. The holographic technique was quickly adopted to produce fiber gratings in the wavelength region near 1.55 μm [19]. Considerable work was done during the early 1990s to understand the physical mechanism behind photosensitivity of fibers and to develop techniques that were capable of making large changes in the refractive index [20], [21]. By 1995, fiber gratings were available commercially, and by 1997 they became a standard component of lightwave technology.

2.2.2 Photosensitivity

There is considerable evidence that photosensitivity of optical fibers is due to formation of defects inside the core of a Ge-doped silica fiber [22]–[24]. As discussed in Section 1.2.1, the core of silica fibers is often doped with germania to increase its refractive index and introduce an index step at the core–cladding interface. The Ge concentration is typically 3–5%.

The presence of Ge atoms in the fiber core leads to formation of oxygen-deficient bonds (such as Si–Ge, Si–Si, and Ge–Ge bonds), which act as defects in the silica matrix [20]. The most common defect is the GeO defect. It forms a defect band with an energy gap of about 5 eV (energy required to break the bond). Single-photon absorption of 244-nm radiation from an excimer laser (or two-photon absorption of 488-nm light from an argon-ion laser) breaks these defect bonds and creates GeE$'$ centers. Extra electrons associated with GeE$'$ centers are free to move within the glass matrix until they are trapped at hole-defect sites to form color centers known as Ge(1) and Ge(2). Such modifications in the glass structure change the absorption spectrum $\alpha(\omega)$. However, changes in the absorption also affect the refractive index since $\Delta\alpha$ and Δn are related through the Kramers–Kronig relation

$$\Delta n(\omega') = \frac{c}{\pi} \int_0^\infty \frac{\Delta\alpha(\omega)\,d\omega}{\omega^2 - \omega'^2}. \qquad (2.2.3)$$

2.2 Fiber Gratings

Even though absorption modifications occur mainly in the ultraviolet region, the refractive index can change even in the visible or infrared region. Moreover, since index changes occur only in the regions of fiber core where the ultraviolet light is absorbed, a periodic intensity pattern is transformed into an index grating. Typically, index change Δn is $\sim 10^{-4}$ in the 1.3- to 1.6-μm wavelength range, but can exceed 0.001 in fibers with high Ge concentration.

The presence of GeO defects is crucial for photosensitivity to occur in optical fibers. However, standard telecommunication fibers rarely have more than 3% of Ge atoms in their core, resulting in relatively small index changes. The use of other dopants such as phosphorus, boron, and aluminum can enhance the photosensitivity (and the amount of index change) to some extent, but these dopants also tend to increase fiber losses. It was discovered in the early 1990s that the amount of index change induced by ultraviolet absorption can be enhanced by two orders of magnitude ($\Delta n > 0.01$) by soaking the fiber in hydrogen gas at high pressures (200 atm) and room temperature [25]. The density of Ge–Si oxygen-deficient bonds increases in hydrogen-soaked fibers because hydrogen can recombine with oxygen atoms. Once hydrogenated, the fiber needs to be stored at low temperature to maintain its photosensitivity. However, gratings made in such fibers remain intact over long periods of time, indicating a nearly permanent nature of the resulting index changes [26]. Hydrogen soaking is commonly used for making fiber gratings.

It should be stressed that understanding of the exact physical mechanism behind photosensitivity is far from complete, and more than one mechanism may be involved [27]. Localized heating can also affect grating formation. For instance, in fibers with a strong grating (index change > 0.001), damage tracks were seen when the grating was examined under an optical microscope [28]; these tracks were due to localized heating to several thousand degrees of the core region where ultraviolet light was most strongly absorbed. At such high temperatures the local structure of amorphous silica can change considerably because of melting.

2.2.3 Fabrication Techniques

Fiber gratings can be made by using several different techniques, each having its own merits [20]. The original 1978 technique [17] in which a single argon-laser beam is launched into a germanium-doped silica fiber is rarely used in practice because it can be utilized only near the wavelength of the laser used to make the grating. It has been replaced in practice with a dual-beam holographic technique first used in 1989 [18].

The dual-beam holographic technique, shown schematically in Figure 2.4, makes use of an external interferometric scheme similar to that used for holography. Two optical beams, obtained from the same laser (operating in the ultraviolet region) and making an angle 2θ, are made to interfere at the exposed core of an optical fiber. A cylindrical lens is used to expand the beam along the fiber length. Similar to the single-beam scheme, the interference pattern creates an index grating. However, the grating period Λ is related to the ultraviolet laser wavelength λ_{uv} and the angle 2θ made by the two interfering beams through the simple relation $\Lambda = \lambda_{uv}/(2\sin\theta)$.

The most important feature of the holographic technique is that the grating period Λ can be varied over a wide range by simply adjusting the angle θ (see Figure 2.4).

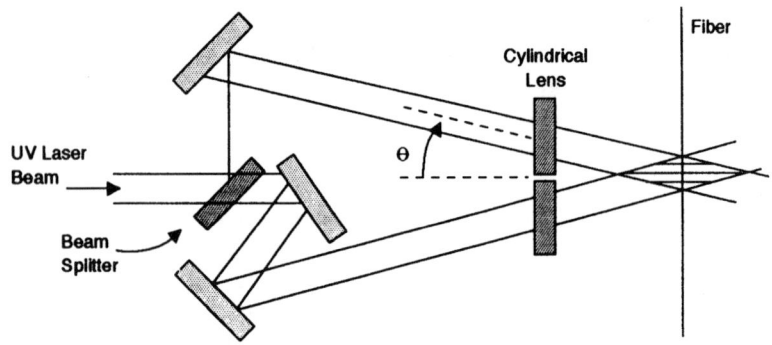

Figure 2.4: Schematic illustration of the dual-beam holographic technique.

The wavelength λ at which the grating will reflect light is related to Λ as $\lambda = 2\bar{n}\Lambda$. Since λ can be significantly larger than λ_{uv}, Bragg gratings operating in the visible or infrared region can be fabricated by the dual-beam holographic method even when λ_{uv} is in the ultraviolet region. In a 1989 experiment, Bragg gratings reflecting 580-nm light were made by exposing the 4.4-mm-long core region of a photosensitive fiber for 5 minutes with 244-nm ultraviolet radiation [18]. Reflectivity measurements indicated that the refractive index changes were $\sim 10^{-5}$ in the bright regions of the interference pattern. Bragg gratings formed by the dual-beam holographic technique were stable and remained unchanged even when the fiber was heated to 500°C.

Because of their practical importance, Bragg gratings operating in the 1.55-μm region were made in 1990 [19]. Since then, several variations of the basic technique have been used to make such gratings in a practical manner. An inherent problem for the dual-beam holographic technique is that it requires an ultraviolet laser with excellent temporal and spatial coherence. Excimer lasers commonly used for this purpose have relatively poor beam quality and require special care to maintain the interference pattern over the fiber core over a duration of several minutes.

It turns out that high-reflectivity fiber gratings can be written by using a single excimer laser pulse (with typical duration of 20 ns) if the pulse energy is large enough. Extensive measurements on gratings made by this technique indicate a threshold-like phenomenon near a pulse energy level of about 35 mJ [28]. For lower pulse energies, the grating is relatively weak since index changes are only about 10^{-5}. By contrast, index changes of about 10^{-3} are possible for pulse energies above 40 mJ. Bragg gratings with nearly 100% reflectivity have been made by using a single 40-mJ pulse at the 248-nm wavelength. The gratings remained stable at temperatures as high as 800°C. A short exposure time has an added advantage. The typical rate at which a fiber is drawn from a preform is about 1 m/s. Since the fiber moves only 20 nm in 20 ns, and since this displacement is a small fraction of the grating period Λ, a grating can be written during the drawing stage while the fiber is being pulled and before it is sleeved [29]. This feature makes the single-pulse holographic technique quite useful from a practical standpoint.

Commercial production of fiber gratings makes use of the phase-mask technique.

2.2 Fiber Gratings

The basic idea is to use a phase mask with a periodicity related to the grating period [30]. The phase mask acts as a master grating that is transferred to the fiber using a suitable method. In one realization of this technique [31], the phase mask was made on a quartz substrate on which a patterned layer of chromium was deposited using electron-beam lithography in combination with reactive ion etching. Phase variations induced in the 242-nm radiation passing through the phase mask translate into a periodic intensity pattern similar to that produced by the holographic technique. Photosensitivity of the fiber converts intensity variations into an index grating of the same periodicity as that of the phase mask.

The chief advantage of the phase-mask method is that the demands on the temporal and spatial coherence of the ultraviolet beam are much less stringent because of the noninterferometric nature of the technique. In fact, even a nonlaser source such as an ultraviolet lamp can be used. Furthermore, the phase-mask technique allows fabrication of fiber gratings with a variable period (chirped gratings) and can also be used to tailor the periodic index profile along the grating length. It is also possible to vary the Bragg wavelength over some range for a fixed mask periodicity by using a converging or diverging wavefront during the photolithographic process. On the other hand, the quality of fiber grating (length, uniformity, etc.) depends completely on the master phase mask, and all imperfections are reproduced precisely. Nonetheless, gratings with 5-mm length and 94% reflectivity were made in 1993, showing the potential of this technique [31].

The phase mask can also be used to form an interferometer. In this scheme, the ultraviolet laser beam falls normally on the phase mask and is diffracted into several beams in the Raman–Nath scattering regime. The zeroth-order beam (direct transmission) is blocked or cancelled by an appropriate technique. The two first-order diffracted beams interfere on the fiber surface and form a periodic intensity pattern. The grating period is exactly one-half of the phase mask period. In effect, the phase mask produces both the reference and object beams required for holographic recording.

There are several advantages to using a phase-mask interferometer. It is insensitive to the lateral translation of the incident laser beam and tolerant of any beam-pointing instability. Relatively long fiber gratings can be made by moving two side mirrors while maintaining their mutual separation. In fact, the two mirrors can be replaced by a single silica block that reflects the two beams internally through total internal reflection, resulting in a compact and stable interferometer [20]. The length of the grating formed inside the fiber core is limited by the size and optical quality of the silica block.

Long gratings can be formed by scanning the phase mask or by translating the optical fiber itself such that different parts of the optical fiber are exposed to the two interfering beams. In this way, multiple short gratings are formed in succession in the same fiber. Any discontinuity or overlap between the two neighboring gratings, resulting from positional inaccuracies, leads to the so-called stitching errors (also called phase errors) that can affect the quality of the whole grating substantially if left uncontrolled. Nevertheless, this technique was used in 1993 to produce a 5-cm-long grating [32]. Since then, gratings longer than 10 m have been made with success by employing techniques that minimize phase errors [33].

Another scheme, known as the point-by-point fabrication technique, bypasses the need of a master phase mask and fabricates the grating directly on the fiber, period by period, by exposing short sections of width w to a single high-energy pulse [34]. The fiber is translated by a distance $\Lambda - w$ before the next pulse arrives, resulting in a periodic index pattern such that only a fraction w/Λ in each period has a higher refractive index. The method is referred to as point-by-point fabrication since a grating is fabricated period by period even though the period Λ is typically below 1 μm. The technique works by focusing the spot size of the ultraviolet laser beam so tightly that only a short section of width w is exposed to it. Typically, w is chosen to be $\Lambda/2$ although it could be a different fraction if so desired. Only short fiber gratings (<1 cm) are typically produced because of the time-consuming nature of the point-to-point fabrication method. This method is more suitable for long-period gratings in which the grating period exceeds 10 μm and even can be longer than 100 μm, depending on the application [20]. Such gratings can be used for mode conversion and for flattening the gain profile of erbium-doped fiber amplifiers (see Chapter 3).

2.2.4 Grating Characteristics

Two different approaches have been used to study how a Bragg grating affects wave propagation in optical fibers. In one approach, Bloch formalism—used commonly for describing motion of electrons in semiconductors—is applied to Bragg gratings [35]. In another, the coupling between the forward- and backward-propagating waves is treated using coupled-mode theory. In this subsection we derive the coupled-mode equations and use them to discuss propagation of CW light through a Bragg grating. We also introduce the concept of photonic bandgap and use it to show how a Bragg grating introduces a large amount of dispersion.

Coupled-Mode Equations

Wave propagation in a linear periodic medium has been studied extensively using coupled-mode theory [10]. In the case of fiber gratings, the refractive index of the fiber mode can be written as

$$\tilde{n}(\omega,z) = \bar{n}(\omega) + \delta n_g(z), \qquad (2.2.4)$$

where $\delta n_g(z)$ accounts for periodic index variations inside the grating. The coupled-mode theory can be generalized to include the nonlinear effects [11]. However, the grating length is typically so short that nonlinear effects are negligible unless intensity levels approach a few GW/cm^2, and we ignore them in this section.

As in Section 2.1, the starting point consists of solving the Helmholtz equation (2.1.1) with the refractive index given in Eq. (2.2.4). Noting that $\tilde{n}$ is a periodic function of z, we expand $\delta n_g(z)$ in a Fourier series as

$$\delta n_g(z) = \sum_{m=-\infty}^{\infty} \delta n_m \exp[2\pi i m(z/\Lambda)]. \qquad (2.2.5)$$

Since both the forward- and backward-propagating waves should be included, $\tilde{E}$ in Eq. (2.1.1) is of the form

$$\tilde{E}(\mathbf{r},\omega) = F(x,y)[\tilde{A}_f(z,\omega)\exp(i\beta_B z) + \tilde{A}_b(z,\omega)\exp(-i\beta_B z)], \qquad (2.2.6)$$

2.2 Fiber Gratings

where $\beta_B = \pi/\Lambda$ is the Bragg wave number for a first-order grating. It is related to the Bragg wavelength through the Bragg condition $\lambda_B = 2\bar{n}\Lambda$ and can be used to define the Bragg frequency as $\omega_B = \pi c/(\bar{n}\Lambda)$. Transverse variations for the two counterpropagating waves are governed by the same modal distribution $F(x,y)$ in a single-mode fiber.

If we assume that $\tilde{A}_f$ and $\tilde{A}_b$ vary slowly with z and keep only the nearly phase-matched terms, the frequency-domain coupled-mode equations take the form [10]

$$\frac{\partial \tilde{A}_f}{\partial z} = i\delta(\omega)\tilde{A}_f + i\kappa \tilde{A}_b, \qquad (2.2.7)$$

$$-\frac{\partial \tilde{A}_b}{\partial z} = i\delta(\omega)\tilde{A}_b + i\kappa \tilde{A}_f, \qquad (2.2.8)$$

where δ is a measure of detuning from the Bragg frequency and is defined as

$$\delta(\omega) = (\bar{n}/c)(\omega - \omega_B) \equiv \beta(\omega) - \beta_B. \qquad (2.2.9)$$

The coupling coefficient κ governs the grating-induced coupling between the forward and backward waves. For a first-order grating, κ is given by

$$\kappa = \frac{k_0 \iint_{-\infty}^{\infty} \delta n_1 |F(x,y)|^2 \, dx \, dy}{\iint_{-\infty}^{\infty} |F(x,y)|^2 \, dx \, dy}. \qquad (2.2.10)$$

In this general form, κ can include transverse variations of δn_g occurring when the photoinduced index change is not uniform over the core area. For a transversely uniform grating $\kappa = 2\pi \delta n_1/\lambda$, as can be inferred from Eq. (2.2.10) by taking δn_1 as constant and using $k_0 = 2\pi/\lambda$. For a sinusoidal grating of the form $\delta n_g = n_a \cos(2\pi z/\Lambda)$, $\delta n_1 = n_a/2$ and the coupling coefficient is given by $\kappa = \pi n_a/\lambda$.

Equations (2.2.7) and (2.2.8) can be converted to time domain by following the procedure outlined earlier. We expand $\beta(\omega)$ in Eq. (2.2.9) in a Taylor series around the central frequency ω_0 as

$$\beta(\omega) = \beta_0 + (\omega - \omega_0)\beta_1 + \tfrac{1}{2}(\omega - \omega_0)^2 \beta_2 + \tfrac{1}{6}(\omega - \omega_0)^3 \beta_3 + \cdots, \qquad (2.2.11)$$

retain terms up to second order in $\omega - \omega_0$, and replace $\omega - \omega_0$ with the differential operator $i(\partial/\partial t)$. The resulting time-domain coupled-mode equations become

$$\frac{\partial A_f}{\partial z} + \frac{1}{v_g}\frac{\partial A_f}{\partial t} + \frac{i\beta_2}{2}\frac{\partial^2 A_f}{\partial t^2} = i\delta_0 A_f + i\kappa A_b, \qquad (2.2.12)$$

$$-\frac{\partial A_b}{\partial z} + \frac{1}{v_g}\frac{\partial A_b}{\partial t} + \frac{i\beta_2}{2}\frac{\partial^2 A_b}{\partial t^2} = i\delta_0 A_b + i\kappa A_f, \qquad (2.2.13)$$

where $\delta_0 = (\omega_0 - \omega_B)/v_g$. The other parameters have the same meaning as in Section 2.1. Specifically, v_g is the group velocity and β_2 is the GVD parameter.

The coupled-mode equations for a fiber grating should be compared with Eqs. (2.1.8) and (2.1.9) obtained for fiber couplers. The only major difference is the negative sign appearing in front of the $\partial A_b/\partial z$ term in Eq. (2.2.12) because of the backward propagation of A_b. However, this single difference changes the character of wave propagation profoundly.

Photonic Bandgap

In the case of a CW beam, we can use Eqs. (2.2.7) and (2.2.8). These frequency-domain coupled-mode equations have a general solution of the form

$$\tilde{A}_f(z) = A_1 \exp(iqz) + A_2 \exp(-iqz), \quad (2.2.14)$$
$$\tilde{A}_b(z) = B_1 \exp(iqz) + B_2 \exp(-iqz), \quad (2.2.15)$$

where q is to be determined. The constants A_1, A_2, B_1, and B_2 are interdependent and satisfy the following four relations:

$$(q-\delta)A_1 = \kappa B_1, \quad (q+\delta)B_1 = -\kappa A_1, \quad (2.2.16)$$
$$(q-\delta)B_2 = \kappa A_2, \quad (q+\delta)A_2 = -\kappa B_2. \quad (2.2.17)$$

These equations are satisfied for nonzero values of A_1, A_2, B_1, and B_2 if the possible values of q obey the dispersion relation

$$q = \pm\sqrt{\delta^2 - \kappa^2}. \quad (2.2.18)$$

This equation is of paramount importance for gratings. Its implications will become clear soon.

One can eliminate A_2 and B_1 by using Eqs. (2.2.14) through (2.2.17) and write the general solution in terms of an effective reflection coefficient $r(q)$ as

$$\tilde{A}_f(z) = A_1 \exp(iqz) + r(q)B_2 \exp(-iqz), \quad (2.2.19)$$
$$\tilde{A}_b(z) = B_2 \exp(-iqz) + r(q)A_1 \exp(iqz), \quad (2.2.20)$$

where

$$r(q) = \frac{q-\delta}{\kappa} = -\frac{\kappa}{q+\delta}. \quad (2.2.21)$$

The q dependence of r and the dispersion relation (2.2.18) indicate that both the magnitude and the phase of backward reflection depend on the frequency ω. The sign ambiguity in Eq. (2.2.18) can be resolved by choosing the sign of q such that $|r(q)| < 1$.

The dispersion relation of Bragg gratings exhibits an important property seen clearly in Figure 2.5, where Eq. (2.2.18) is plotted. If the frequency of the incident light is such that $-\kappa < \delta < \kappa$, q becomes purely imaginary. Since the grating does not support a propagating wave in this case, most of the incident field is reflected. The range $|\delta| \leq \kappa$ is referred to as the *photonic bandgap*, in analogy with the electronic energy bands occurring in crystals. It is also called the *stop band* since light stops transmitting through the grating when its frequency falls within the photonic bandgap.

To understand what happens when optical pulses propagate inside a fiber grating with their carrier frequency ω_0 outside the stop band but close to its edges, note that the effective propagation constant of the forward- and backward-propagating waves from Eqs. (2.2.6) and (2.2.14) is $\beta_e = \beta_B \pm q$, where q is given by Eq. (2.2.18) and is a function of optical frequency through δ. This frequency dependence of β_e indicates that a grating will exhibit dispersive effects even if it was fabricated in a nondispersive medium. In optical fibers, grating-induced dispersion adds to the material and

2.2 Fiber Gratings

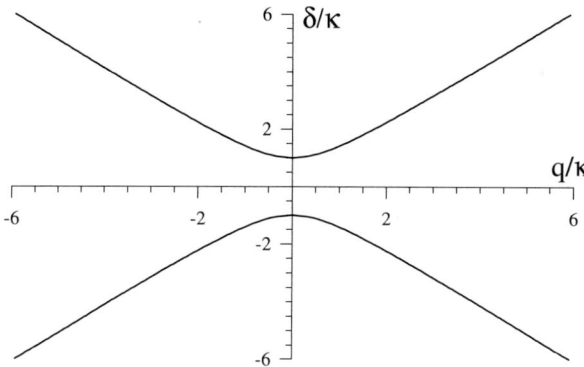

Figure 2.5: Dispersion curves showing variation of δ with q and the existence of the photonic bandgap for a fiber grating.

waveguide dispersion. In fact, the contribution of grating dominates among all sources responsible for dispersion. To see this more clearly, we expand β_e in a Taylor series in a way similar to Eq. (2.2.11) around the carrier frequency ω_0 of the pulse. The result is given by

$$\beta_e(\omega) = \beta_0^g + (\omega - \omega_0)\beta_1^g + \tfrac{1}{2}(\omega - \omega_0)^2 \beta_2^g + \tfrac{1}{6}(\omega - \omega_0)^3 \beta_3^g + \cdots, \qquad (2.2.22)$$

where β_m^g with $m = 1, 2, \ldots$ is defined as

$$\beta_m^g = \frac{d^m q}{d\omega^m} \approx \left(\frac{1}{v_g}\right)^m \frac{d^m q}{d\delta^m}, \qquad (2.2.23)$$

and derivatives are evaluated at $\omega = \omega_0$. The superscript g denotes that the dispersive effects have their origin in the grating. In Eq. (2.2.23), v_g is the group velocity of pulse in the absence of the grating ($\kappa = 0$).

Consider first the group velocity of the pulse inside the grating. If we use $V_G = 1/\beta_1^g$ and Eq. (2.2.23), it is given by

$$V_G = \pm v_g \sqrt{1 - \kappa^2/\delta^2}, \qquad (2.2.24)$$

where the choice of $\pm$ signs depends on whether the pulse is moving in the forward or the backward direction. Far from the band edges ($|\delta| \gg \kappa$), optical pulse is unaffected by the grating and travels at the group velocity expected in the absence of the grating. However, as $|\delta|$ approaches κ, the group velocity decreases and becomes zero at the two edges of the stop band where $|\delta| = \kappa$. Thus, close to the photonic bandgap, an optical pulse slows down inside a fiber grating. As an example, its speed is reduced by 50% when $|\delta|/\kappa \approx 1.18$.

Second- and third-order dispersive properties of the grating are governed by β_2^g and β_3^g, respectively. If we use Eq. (2.2.23) together with the dispersion relation, these

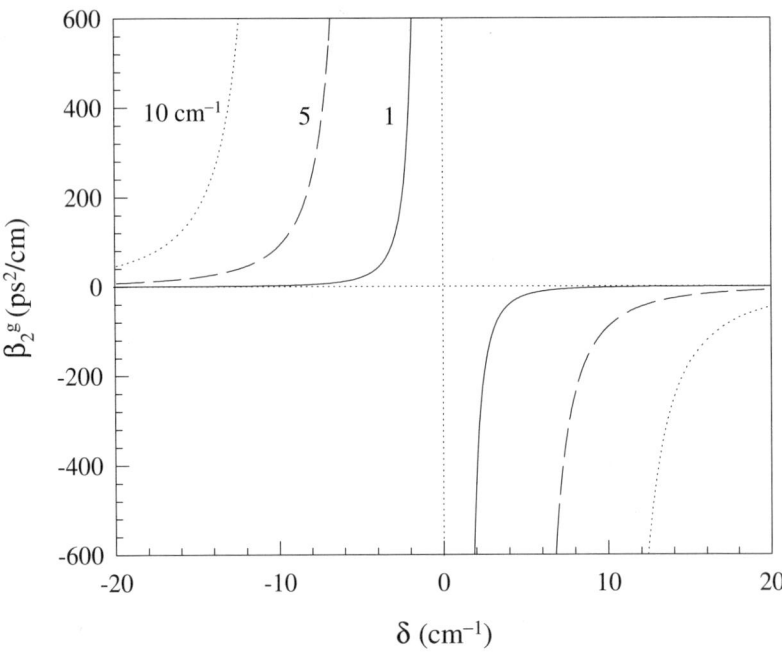

Figure 2.6: Grating-induced dispersion plotted as a function of frequency detuning δ for several values of the coupling coefficient κ.

parameters are given by

$$\beta_2^g = -\frac{\text{sgn}(\delta)\kappa^2/v_g^2}{(\delta^2 - \kappa^2)^{3/2}}, \qquad \beta_3^g = \frac{3|\delta|\kappa^2/v_g^3}{(\delta^2 - \kappa^2)^{5/2}}. \qquad (2.2.25)$$

The grating-induced GVD, governed by the parameter β_2^g, depends on the sign of detuning δ. The GVD is anomalous on the upper branch of the dispersion curve in Figure 2.5, where δ is positive and the carrier frequency exceeds the Bragg frequency. In contrast, GVD becomes normal ($\beta_2^g > 0$) on the lower branch of the dispersion curve, where δ is negative and the carrier frequency is smaller than the Bragg frequency. Notice that the third-order dispersion remains positive on both branches of the dispersion curve. Also note that both β_2^g and β_3^g become infinitely large at the two edges of the stop band.

The dispersive properties of a fiber grating are quite different than those of a uniform fiber. First, β_2^g changes sign on the two sides of the stop band centered at the Bragg wavelength, whose location is easily controlled and can be in any region of the optical spectrum. Second, β_2^g is anomalous on the shorter wavelength side of the stop band, whereas β_2 for fibers becomes anomalous for wavelengths longer than the zero-dispersion wavelength. Third, the magnitude of β_2^g exceeds that of β_2 by a large factor. Figure 2.6 shows how β_2^g varies with detuning δ for several values of κ. As seen there, $|\beta_2^g|$ can exceed 10^7 ps^2/km for a fiber grating.

2.2.5 Grating as an Optical Filter

What happens to optical pulses incident on a fiber grating depends very much on the location of the pulse spectrum with respect to the stop band associated with the grating. If the pulse spectrum falls entirely within the stop band, the entire pulse is reflected by the grating. On the other hand, if a part of the pulse spectrum is outside the stop band, that part will be transmitted through the grating. The shape of the reflected and transmitted pulses will be quite different than that of the incident pulse because of the splitting of the spectrum and the dispersive nature of a fiber grating. If the peak power of pulses is small enough that nonlinear effects are negligible, we can calculate the reflection and transmission coefficients for each spectral component and then integrate over the spectrum of the incident pulse. In the linear regime, a fiber grating acts an optical filter, and its reflectivity spectrum serves the role of the transfer function.

The reflection coefficient can be calculated by using Eqs. (2.2.19) and (2.2.20) with the appropriate boundary conditions. Consider a grating of length L and assume that light is incident only at the front end, located at $z = 0$. The reflection coefficient is then given by

$$r_g = \frac{\tilde{A}_b(0)}{\tilde{A}_f(0)} = \frac{B_2 + r(q)A_1}{A_1 + r(q)B_2}. \qquad (2.2.26)$$

If we use the boundary condition $\tilde{A}_b(L) = 0$ in Eq. (2.2.20),

$$B_2 = -r(q)A_1 \exp(2iqL). \qquad (2.2.27)$$

Using this value of B_2 and $r(q)$ from Eq. (2.2.21) in Eq. (2.2.26), we obtain

$$r_g(\delta) = \frac{i\kappa \sin(qL)}{q\cos(qL) - i\delta \sin(qL)}. \qquad (2.2.28)$$

The frequency dependence of $r_g(\delta)$ shows the filter characteristics associated with a fiber grating.

Figure 2.7 shows the reflectivity $|r_g|^2$ and the phase of r_g as a function of detuning δ for two values of κL. The grating reflectivity within the stop band approaches 100% for $\kappa L = 3$ or larger. Maximum reflectivity occurs at the center of the stop band and, by setting $\delta = 0$ in Eq. (2.2.28), we get for it:

$$R_{\max} = |r_g|^2 = \tanh^2(\kappa L). \qquad (2.2.29)$$

For $\kappa L = 2$, $R_{\max} = 0.93$. The condition $\kappa L > 2$ with $\kappa = 2\pi\delta n_1/\lambda$ can be used to estimate the grating length required for high reflectivity inside the stop band. For $\delta n_1 \approx 10^{-4}$ and $\lambda = 1.55$ μm, L should exceed 5 mm to yield $\kappa L > 2$. These requirements are easily met in practice. Indeed, reflectivity values in excess of 99% were achieved as early as 1993 for a 1.5-cm-long grating [28].

An undesirable feature of Figure 2.7(a) from a practical standpoint is the presence of multiple sidebands located on each side of the stop band. These sidebands originate from weak reflections occurring at the two grating ends where the refractive index changes suddenly compared to its value outside the grating region. Even though the change in refractive index is typically less than 1%, the reflections at the two grating

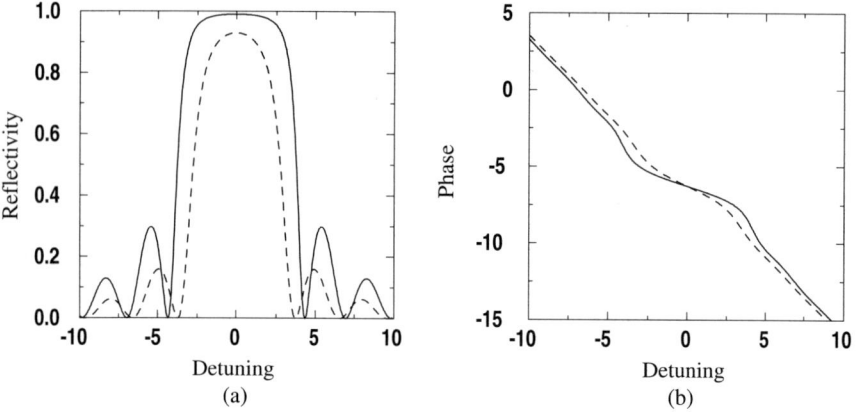

Figure 2.7: (a) The reflectivity $|r_g|^2$ and (b) the phase of r_g plotted as a function of detuning δ for $\kappa L = 2$ (dashed line) and $\kappa L = 3$ (solid line).

ends form a Fabry–Perot cavity with its own wavelength-dependent transmission. An apodization technique is commonly used to remove the sidebands seen in Figure 2.7(a) [20]. In this technique, the intensity of the ultraviolet laser beam used to form the grating is made nonuniform in such a way that the intensity drops to zero gradually near the two grating ends.

Figure 2.8(a) shows schematically the periodic index variation in an apodized fiber grating. In a transition region of width L_t near the grating ends, the value of the coupling coefficient κ increases from zero to its maximum value. These buffer zones can suppress the sidebands almost completely, resulting in fiber gratings with practically useful filter characteristics. Figure 2.8(b) shows the measured reflectivity spectrum for a 7.5-cm-long apodized fiber grating made by the scanning phase-mask technique. The

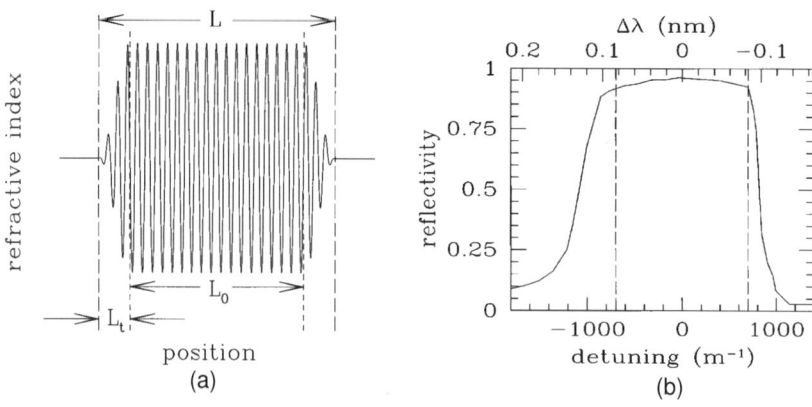

Figure 2.8: (a) Schematic variation of refractive index and (b) measured reflectivity spectrum for an apodized fiber grating. (After Ref. [37]); ©1999 OSA.)

2.2 Fiber Gratings

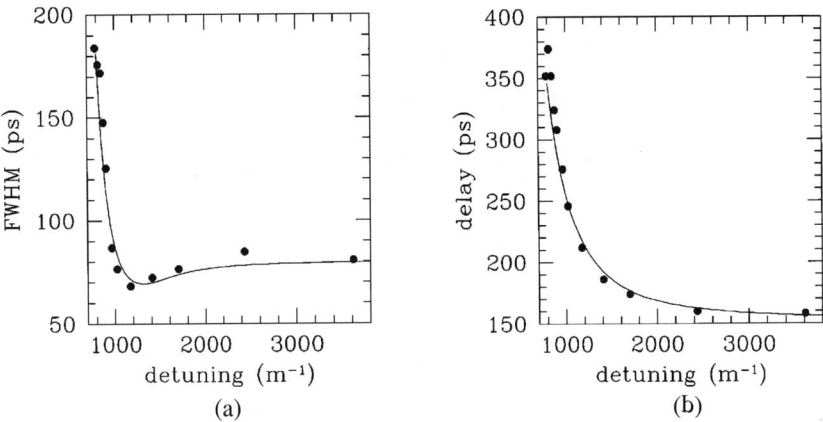

Figure 2.9: (a) Measured pulse width (FWHM) of 80-ps input pulses and (b) their arrival time as a function of detuning δ for an apodized 7.5-cm-long fiber grating. Solid lines show the prediction of the coupled-mode theory. (After Ref. [37]); ©1999 OSA.)

reflectivity exceeds 90% within the stop band, about 0.17-nm wide and centered at the Bragg wavelength of 1.053 μm, chosen to coincide with the wavelength of an Nd:YLF laser [37]. From the stop-band width, the coupling coefficient κ is estimated to be about 7 cm^{-1}. Note the sharp drop in reflectivity at both edges of the stop band and a complete absence of sidebands.

The same apodized fiber grating was used to investigate dispersive properties in the vicinity of a stop-band edge by transmitting 80-ps pulses (nearly Gaussian shape) through it [37]. Figure 2.9 shows the variation of the pulse width (a) and changes in the propagation delay during pulse transmission (b) as a function of the detuning δ from the Bragg wavelength on the upper branch of the dispersion curve. The most interesting feature is the increase in the arrival time observed as the laser is tuned close to the stop-band edge because of the reduced group velocity. Doubling of the arrival time for δ close to 900 m^{-1} shows that the pulse speed was only 50% of that expected in the absence of the grating. This result is in complete agreement with the prediction of coupled-mode theory.

Changes in the pulse width seen in Figure 2.9 can be attributed mostly to the grating-induced GVD effects in Eq. (2.2.25). The large broadening observed near the stop-band edge is due to an increase in $|\beta_2^g|$. Slight compression near $\delta = 1,200$ m^{-1} is due to a small amount of self-phase modulation that chirps the pulse. The nonlinear effects became quite significant at higher power levels.

2.2.6 Nonuniform Gratings

The properties of a Bragg grating can be considerably modified by introducing nonuniformities along their length such that the two grating parameters, κ and δ, become z-dependent. Examples of such nonuniform gratings include chirped gratings, phase-shifted gratings, and superstructure gratings. In a chirped grating, the optical period

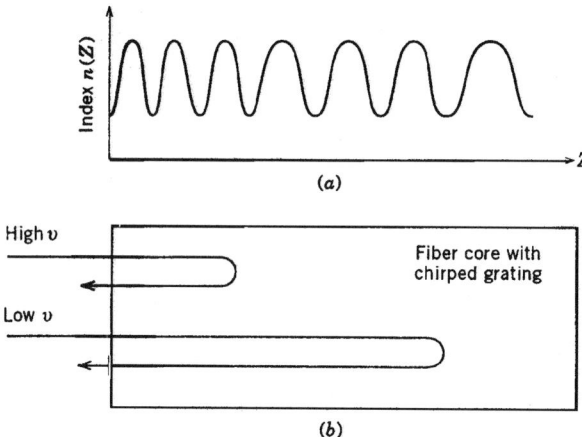

Figure 2.10: (a) Variations of refractive index in a chirped fiber grating. (b) Low- and high-frequency components of a pulse are reflected at different locations within the grating because of variations in the Bragg wavelength.

$\bar{n}\Lambda$ of the grating changes along the fiber length, as shown in Figure 2.10(a). Since the Bragg wavelength ($\lambda_B = 2\bar{n}\Lambda$) sets the frequency at which the stop band is centered, its axial variations translate into a shift of the stop band along the grating length. Mathematically, the parameter δ appearing in the nonlinear coupled-mode equations becomes z-dependent. Typically, Λ is designed to vary linearly along the grating, and $\delta(z) = \delta_0 + \delta_c z$, where δ_c is a chirp parameter. Such gratings are called *linearly chirped gratings*. From a device standpoint, chirped gratings typically have a much wider stop band compared with that of a constant-period grating. The reason is easy to understand. Since different parts of the grating have slightly different Bragg wavelengths, the stop band of a chirped grating results from a superposition of multiple stop bands of the same width but shifted continuously from one end to the other end of the grating.

Chirped fiber gratings have been fabricated using several different methods [20]. It is important to note that it is the optical period $\bar{n}\Lambda$ that needs to be varied along the grating (z axis). Thus, chirping can be induced either by varying the physical grating period Λ or by changing the effective mode index $\bar{n}$ along z. In the commonly used dual-beam holographic technique, the fringe spacing of the interference pattern is made nonuniform by using dissimilar curvatures for the interfering wavefronts, resulting in Λ variations. In practice, cylindrical lenses are used in one or both arms of the interferometer. Chirped fiber gratings can also be fabricated by tilting or stretching the fiber, by using strain or temperature gradients, or by stitching together multiple uniform sections.

Chirped Bragg gratings have several important practical applications. As seen in Figure 2.10(b), when a pulse—with its spectrum inside the stop band—is incident on a chirped grating, different spectral components of the pulse are reflected by different parts of the grating. As a result, even though the entire pulse is eventually reflected, it experiences a large amount of GVD whose nature (normal versus anomalous) and magnitude can be controlled by the chirp. For this reason, chirped gratings are commonly used for dispersion compensation [13]. Chirped gratings also exhibit interesting nonlinear effects when the incident pulse is sufficiently intense. In one experiment, 80-ps pulses were propagated through a 6-cm-long grating whose linear chirp could be varied over a considerable range through a temperature gradient established along its length [38]. The reflected pulses were split into a pair of pulses by the combination of SPM and XPM for peak intensities close to 10 GW/cm^2.

In a variation of the chirping idea, it is the coupling coefficient κ that is made nonuniform along the grating length. This occurs when the parameter δn_1 in Eq. (2.2.10) is a function of z. In practice, variations in the intensity of the ultraviolet laser beam used to make the grating translate into axial variations of κ. From a physical standpoint, since the width of the photonic bandgap is about 2κ, changes in κ translate into changes in the width of the stop band along the grating length. At a fixed wavelength of input light, such local variations in κ lead to axial variations of the group velocity V_G and the GVD parameter β_2^g, as seen from Eqs. (2.2.24) and (2.2.25). In effect, the dispersion provided by the grating becomes nonuniform and varies along its length.

In another class of gratings, grating parameters are designed to vary periodically along the length of a grating. Such devices have double periodicity and are called *sampled* or *superstructure gratings*. They were first used in the context of DFB semiconductor lasers [39]. Fiber-based superstructure gratings were made in 1994 [40]. Since then, their properties have attracted considerable attention [41]–[46]. A simple example of a superstructure grating is provided by a long grating with constant phase-shift regions occurring at periodic intervals. In practice, such a structure can be realized by placing multiple gratings next to each other with a small constant spacing among them or by blocking small regions during fabrication of a grating such that $\kappa = 0$ in the blocked regions. It can also be made by etching away parts of an existing grating. In all cases, $\kappa(z)$ varies periodically along z. It is this periodicity that modifies the stop band of a uniform grating. The period d of $\kappa(z)$ is typically about 1 mm. If the average index $\bar{n}$ also changes with the same period d, both δ and κ become periodic in the nonlinear coupled-mode equations. The most striking feature of a superstructure grating is the appearance of additional photonic bandgaps on both branches of the dispersion curve seen in Figure 2.5 for a uniform grating. Such bandgaps are referred to as *Rowland ghost gaps* [47].

2.3 Fiber Interferometers

The two fiber components covered in Sections 2.1 and 2.2 can be combined to form a variety of fiber-based optical devices. Four common ones among them are the fiber version of the well-known Fabry–Perot, Sagnac, Mach–Zehnder, and Michelson interferometers [16]. They exhibit interesting linear and nonlinear effects that are useful for lightwave technology. This section is devoted to such fiber-based interferometers after a brief discussion of fiber-ring resonators.

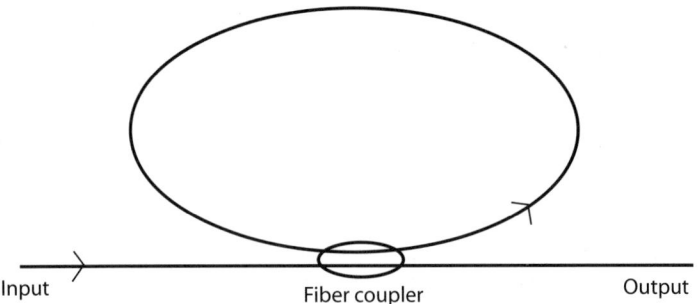

Figure 2.11: A fiber-ring resonators made by using a single directional coupler.

2.3.1 Fiber-Ring Resonators

A fiber-ring resonator can be made, as shown in Figure 2.11, by connecting the input and output ports associated with one core of a directional coupler through a piece of fiber to form a ring. The properties of such fiber-ring resonators were studied as early as 1982 and have continued to attract attention [48]–[53]. The most important feature of such a resonator is that it is a two-port device in which light propagates only in the forward direction. As we shall see shortly, this feature dictates that, under steady-state conditions and no losses, all input power much be transmitted through such a ring resonator irrespective of its frequency. For this reason, a fiber-ring resonator is sometimes referred to as an *all-pass* filter.

The transmission characteristics of a fiber-ring resonator can be obtained by using the transfer matrix of the coupler from Eq. (2.1.21). If A_i is the incident field at some frequency ω, A_t is the transmitted field, A_c is the field coupled to the ring, and A_f is the field fed back into the other input port of the coupler after one round trip inside the ring cavity, the four fields satisfy

$$\begin{pmatrix} A_f \\ A_i \end{pmatrix} = \begin{pmatrix} \sqrt{\rho} & i\sqrt{1-\rho} \\ i\sqrt{1-\rho} & \sqrt{\rho} \end{pmatrix} \begin{pmatrix} A_c \\ A_t \end{pmatrix}. \tag{2.3.1}$$

After one round trip inside the ring A_f and A_c are related as

$$A_f/A_c = \exp[-\alpha L/2 + i\beta(\omega)L] \equiv \sqrt{a}e^{i\phi}, \tag{2.3.2}$$

where $a = \exp(-\alpha L) \leq 1$ because of losses and $\phi(\omega) = \beta(\omega)L$ is the round-trip phase shift. Using this relation in Eq. (2.3.1), we find the transmission coefficient to be [52]

$$t_r(\omega) \equiv \sqrt{T_r}e^{i\phi_t} = \frac{A_t}{A_i} = \frac{\sqrt{a} - \sqrt{\rho}e^{-i\phi}}{1 - \sqrt{a\rho}e^{i\phi}}e^{i(\pi+\phi)}. \tag{2.3.3}$$

Consider first the lossless case by using $a = 1$. It is easy to verify that when $a = 1$, the transmissivity $T_r = 1$ for all values of ω. Thus, a lossless fiber-ring resonator acts as an all-pass filter and shows no frequency dependence as far as its transmitted power

2.3 Fiber Interferometers

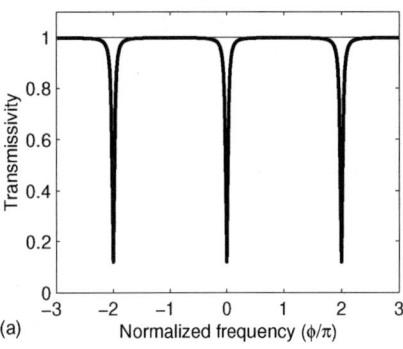 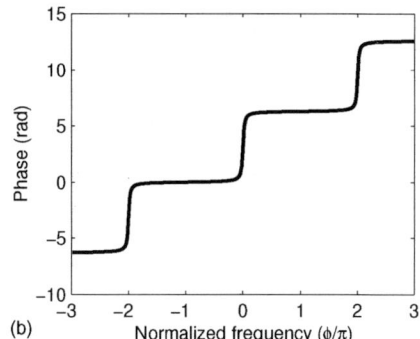

Figure 2.12: (a) Transmissivity and (b) phase shift at the output port of a fiber-ring resonator plotted as a function of frequency for $a = 0.95$ and $\rho = 0.9$. Thin solid line shows 100% transmission expected in the absence of losses.

is concerned. However, the phase ϕ_t of transmitted light does depend on frequency (or wavelength) and is given by

$$\phi_t(\omega) = \pi + \phi + 2\tan^{-1}\frac{\sqrt{\rho}\sin\phi}{1-\sqrt{\rho}\cos\phi}. \tag{2.3.4}$$

The three terms in this equation have the following origin. The π term results from the $\pi/2$ phase shift occurring as the light passes twice to the cross port inside the fiber coupler. The ϕ term is added during a round trip inside the ring. The third term represents an additional phase shift occurring whenever the optical frequency is close to a cavity resonance. Noting that $\phi = \beta(\omega)L = m\pi$, where m is an integer, is the resonance condition, one can identify resonances occurring for different values of m. Expanding $\beta(\omega)$ in a Taylor series as $\beta(\omega) = \beta_0 + \beta_1(\omega - \omega_0)$, where ω_0 is a resonance frequency, and introducing the round-trip time in the cavity as $\tau_r = \beta_1 L = L/v_g$, we can write ϕ in Eq. (2.3.3) as $\phi(\omega) = (\omega - \omega_0)\tau_r$ close to a cavity resonance at $\omega = \omega_0$. In this way, the variable ϕ can be viewed as a frequency detuning parameter.

Losses within the resonator modify the resonator behavior to a considerable extent. Figure 2.12 shows the (a) transmissivity T_r and the (b) phase shift ϕ_t as a function of frequency for $a = 0.95$ (5% loss per round trip) and $\rho = 0.9$. The transmissivity now becomes frequency-dependent and shows a dip at frequencies corresponding to cavity resonances. This behavior is easily understood by noting that close to each resonance, light is coupled strongly into the ring and experiences considerable losses over multiple round trips, resulting in reduced transmission. The phase exhibits a staircase-like structure seen in Figure 2.12 for all values of $a < 1$, although it jumps more gradually close to each resonance as losses increase.

The frequency dependence of transmitted phase can be used for many applications. For example, if an optical pulse is transmitted through the ring resonator, its different frequency components will be delayed by slightly different amounts whenever the pulse spectrum is in the vicinity of a cavity resonance. Thus, such a device introduces GVD in a manner similar to the case of a fiber grating. Noting that the group delay

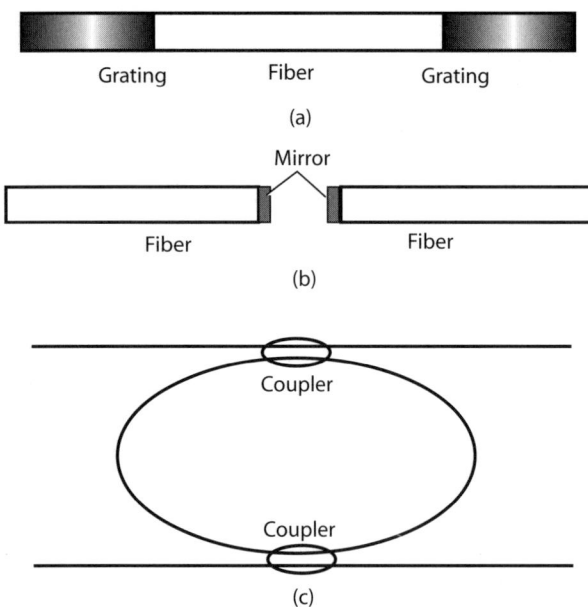

Figure 2.13: Three designs (a–c) for fiber-based Fabry–Perot resonators made by using Bragg gratings and directional couplers.

$\tau_d = d\phi_t/d\omega$, the effective GVD parameter for such a resonator of length L is given by

$$\beta_2 = \frac{1}{L}\frac{d^2\phi_t}{d\omega^2}. \tag{2.3.5}$$

Because of this inherent dispersion, a fiber-ring resonator can be used for dispersion compensation. If a single ring does not provide enough dispersion, one can cascade several ring resonators in series. Also, because of the periodic nature of the transmission spectrum, such a device can compensate for the dispersion of multiple WDM channels simultaneously.

2.3.2 Fabry–Perot Resonators

Fabry–Perot (FP) resonators are well-known devices used commonly for making lasers [54]. A fiber-based FP resonator can be constructed in several different ways. Figure 2.13 shows three designs. In a simple approach shown in Figure 2.13(a), a fiber Bragg grating is spliced at each end of a fiber of suitable length. In another design, facets of two optical fibers are made partially reflecting by depositing high-reflectivity coatings at the two ends [55]. As shown in Figure 2.13(b), the fibers with coated ends are separated by a gap filled with air or some other dielectric material [56]. The third configuration in Figure 2.13(c) shows an all-optical design. It represents a doubly-coupled ring resonator made by using two fiber couplers. All three devices behave identically as far as their operation is concerned.

2.3 Fiber Interferometers

Transmissivity of a FP resonator, formed using two mirrors (or Bragg gratings) of reflectivities R_1 and R_2, can be calculated by adding coherently the optical fields transmitted on successive round trips [16]. When an input CW field A_i at a frequency ω is incident at the left mirror, the transmitted field outside the resonator of length L can be written as [57]

$$A_t = A_i e^{i\pi}(1-R_1)^{1/2}(1-R_2)^{1/2}\left[1 + \sqrt{R_1 R_2}e^{i\phi_R} + R_1 R_2 e^{2i\phi_R} + \cdots \right], \quad (2.3.6)$$

where $\phi_R = 2\beta(\omega)L$ is the phase shift occurring over one round trip inside the resonator and $\beta(\omega)$ is the propagation constant inside the medium within the resonator. The geometrical series in Eq. (2.3.6) can be easily summed. When two mirrors (or gratings) are assumed to have equal reflectivity ($R_m = R_1 = R_2$), we obtain

$$A_t = \frac{(1-R_m)A_i e^{i\pi}}{1 - R_m \exp(i\tilde{\beta}L_R)}. \quad (2.3.7)$$

To keep the discussion simple, losses inside the resonator have been neglected as their presence does not introduce any new qualitative features.

In the case of a doubly-coupled ring resonator made using two fiber couplers as shown in Figure 2.13(c), directional couplers play the role of mirrors. This is evident by noting that only a fraction $(1-\rho)P_i$ of input power P_i is coupled into the ring, while the rest of the incident power passes through the bar port. Thus, even though a coupler does not reflect light, ρ plays the role of mirror reflectivity. For this reason, the bar port of the input coupler is equivalent to reflected signal, and the transmitted field A_t corresponds to the field leaking from the second coupler. In fact, one obtains Eq. (2.3.7) with $R_m = \rho$ when the transfer matrices of the two couplers are used to relate the transmitted field A_t to the incident field A_i.

Transmissivity of the FP resonator is obtained from Eq. (2.3.7) and is given by the well-known Airy formula [16]

$$T_R = \left|\frac{A_t}{A_i}\right|^2 = \frac{(1-R_m)^2}{(1-R_m)^2 + 4R_m \sin^2(\phi_R/2)}. \quad (2.3.8)$$

The round-trip phase shift can again be related to detuning from a resonance frequency ω_0 as $\phi_R = (\omega - \omega_0)\tau_r$, where τ_r is the round-trip time inside the resonator. Figure 2.14 shows transmissivity as a function of frequency detuning for several values of R_m. In contrast with the case of a ring resonator, transmissivity of a FP resonator is nearly zero except in the vicinity of cavity resonances. In fact, 100% of the incident light is transmitted ($T_R = 1$) whenever $\phi_R = 2m\pi$, m being an integer, irrespective of the exact value of mirror reflectivity. Frequencies that satisfy this resonance condition correspond to the longitudinal modes of a FP resonator.

The sharpness of the resonance peaks in Figure 2.14 does depend on mirror reflectivity and is quantified through the resonator finesse F_R, defined as

$$F_R = \frac{\Delta v_L}{\Delta v_R} = \frac{\pi \sqrt{R_m}}{1 - R_m}, \quad (2.3.9)$$

where Δv_R is the width of each resonance peak (at half-maximum) and Δv_L is the frequency spacing between two neighboring resonances, a quantity known as the *free*

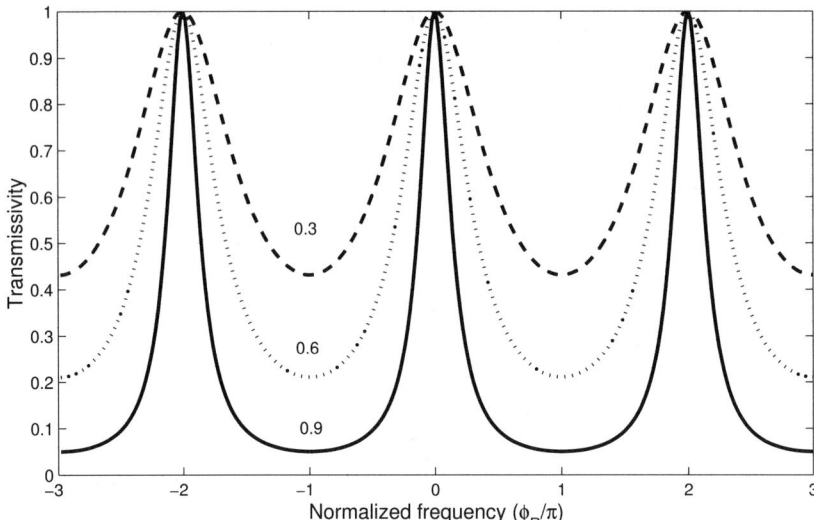

Figure 2.14: Transmissivity of a Fabry–Perot resonator as a function of ϕ_R/π for three values of mirror reflectivity R_m.

spectral range (also called longitudinal-mode spacing). The free spectral range can be obtained from the phase-matching condition

$$2[\beta(\omega + 2\pi\Delta\nu_L) - \beta(\omega)]L = 2\pi \qquad (2.3.10)$$

and is approximately given by $\Delta\nu_L = 1/\tau_r$, where $\tau_r = 2L/v_g$ is the round-trip time within the resonator. Because of group-velocity dispersion (GVD), the free spectral range of a fiber resonator is, in general, frequency-dependent. It can vary considerably in a FP resonator made by using Bragg gratings because of the large GVD associated with them [58].

FP resonators have found numerous applications, all stemming from the periodic nature of their strongly frequency-dependent transmission characteristics. In the field of lightwave technology, they are useful as an optical filter with periodic passbands whose spectral width can be tailored by adjusting the finesse. Moreover, it is often possible to tune the center frequencies of the passbands by changing the physical mirror spacing, or by modifying the refractive index of the intracavity medium electronically to affect the optical path length. It will be seen in Chapter 8 that such tunable optical filters are quite useful for WDM applications.

2.3.3 Sagnac Interferometers

Figure 2.15 shows schematically how a fiber coupler can be used to make a Sagnac interferometer. It is made by connecting a piece of long fiber to the two output ports of a fiber coupler to form a loop. It appears similar to a fiber-ring resonator (see Figure 2.11) but behaves quite differently because of two crucial differences. First, there

2.3 Fiber Interferometers

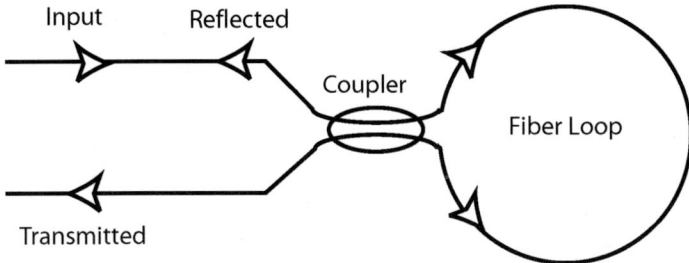

Figure 2.15: Schematic illustration of an all-fiber Sagnac interferometer acting as a nonlinear fiber-loop mirror.

is no feedback mechanism since all light entering from the input port exits from the resonator after a single round trip. Second, the entering optical field is split into two counterpropagating parts that share the same optical path and interfere at the coupler coherently. The relative phase difference between the counterpropagating beams determines whether an input beam is reflected or transmitted by the Sagnac interferometer. In fact, it turns out that when a 3-dB fiber coupler is used, any input is totally reflected, and the Sagnac loop acts as a perfect mirror. For this reason, such a device is also called the *fiber-loop mirror*.

Sagnac interferometers often exploit the nonlinear phenomena of self- and cross-phase modulation (SPM and XPM) occurring in optical fibers for realizing all-optical switching [59]–[61]. Such a *nonlinear* optical loop mirror can be designed to transmit a high-power signal while reflecting it at low power levels, thus acting as an all-optical switch. Such all-fiber devices have attracted considerable attention because they can be used for many applications such as mode locking, wavelength conversion, and channel demultiplexing (see Chapters 9 and 10).

Nonlinear Transmission

The physical mechanism behind nonlinear switching can be easily understood by considering a CW or a quasi-CW input beam. When such an optical signal is incident at one port of the fiber coupler, the transmissivity of a Sagnac interferometer depends on the power-splitting ratio of the coupler. The coupler splits the input into two beams traveling in opposite directions. If a fraction ρ of the input power P_0 remains in the clockwise direction, the transmissivity for a loop of length L is obtained by calculating the phase shifts acquired during a round trip by the two counterpropagating optical waves, and then recombining them interferometrically at the coupler. If we use the transfer matrix of a fiber coupler given in Eq. (2.1.21), the amplitudes of the clockwise and counterclockwise propagating fields are given by

$$A_f = \sqrt{\rho} A_0, \qquad A_b = i\sqrt{1-\rho} A_0, \qquad (2.3.11)$$

where the $\pi/2$ phase shift for A_b is introduced by the coupler. After one round trip, both fields acquire a linear phase shift as well as the SPM- and XPM-induced nonlinear

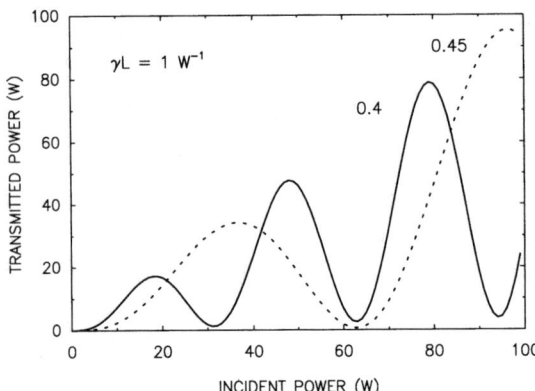

Figure 2.16: Transmitted power as a function of incident power for two values of ρ, showing the nonlinear response of an all-fiber Sagnac interferometer.

phase shifts. As a result, the two fields reaching the coupler take the following form:

$$A'_f = A_f \exp[i\phi_0 + i\gamma(|A_f|^2 + 2|A_b|^2)L], \quad (2.3.12)$$
$$A'_b = A_b \exp(i\phi_0 + i\gamma(|A_b|^2 + 2|A_f|^2)L], \quad (2.3.13)$$

where $\phi_0 \equiv \beta L$ is the linear phase shift for a loop of length L and β is the propagation constant within the loop.

The reflected and transmitted fields can now be obtained by using the transfer matrix of the fiber coupler a second time and are given by

$$\begin{pmatrix} A_t \\ A_r \end{pmatrix} = \begin{pmatrix} \sqrt{\rho} & i\sqrt{1-\rho} \\ i\sqrt{1-\rho} & \sqrt{\rho} \end{pmatrix} \begin{pmatrix} A'_f \\ A'_b \end{pmatrix}. \quad (2.3.14)$$

If we use Eqs. (2.3.11) through 2.3.14), the transmissivity $T_S \equiv |A_t|^2/|A_0|^2$ of the Sagnac loop is given by [61]

$$T_S = 1 - 2\rho(1-\rho)\{1 + \cos[(1-2\rho)\gamma P_0 L]\}, \quad (2.3.15)$$

where $P_0 = |A_0|^2$ is the input power. The linear phase shift does not appear in this equation because of its exact cancellation. For $\rho = 0.5$, T_S equals zero, and the loop reflectivity is 100% at all power levels. Physically, if the power is equally divided between the counterpropagating waves, the nonlinear phase shift is equal for both waves, resulting in no relative phase difference between them. However, if the power-splitting factor ρ is different than 0.5, the fiber-loop mirror exhibits different behavior at low and high powers and can act as an optical switch.

Figure 2.16 shows the transmitted power as a function of P_0 for two values of ρ. At low powers, little light is transmitted if ρ is close to 0.5 since $T_S \approx 1 - 4\rho(1-\rho)$. At high powers, the SPM-induced phase shift leads to 100% transmission of the input signal whenever

$$|1 - 2\rho|\gamma P_0 L = (2m-1)\pi, \quad (2.3.16)$$

2.3 Fiber Interferometers

where m is an integer. As seen in Figure 2.16, the device switches from low to high transmission periodically as input power increases. In practice, only the first transmission peak ($m = 1$) is likely to be used for switching because it requires the least power. The switching power for $m = 1$ can be estimated from Eq. (2.3.16) and is 31 W for a 100-m-long fiber loop when $\rho = 0.45$ and $\gamma = 10$ W^{-1}/km. It can be reduced by increasing the loop length, but one should then consider the effects of fiber loss and GVD that were neglected in deriving Eq. (2.3.15).

Nonlinear Switching

Nonlinear switching in all-fiber Sagnac interferometers was observed beginning in 1989 in several experiments [62]–[68]. Most experiments used short optical pulses with high peak powers. In this case, the power dependence of loop transmissivity in Eq. (2.3.15) can lead to considerable pulse distortion since only the central part of a pulse is intense enough to undergo switching. In a 1989 experiment, 180-ps pulses obtained from a Q-switched, mode-locked Nd:YAG laser were injected into a 25-m Sagnac loop [62]. Transmission increased from a few percent to 60% when peak power was increased beyond 30 W. Transmitted pulses were narrower than input pulses, as expected, because only the central part of the pulse was switched. The shape-induced deformation of optical pulses can be avoided in practice by using soliton effects since solitons have a uniform nonlinear phase across the entire pulse. Their use requires ultrashort pulses (width <10 ps) propagating in the anomalous-GVD regime of the fiber.

The switching threshold of a Sagnac interferometer is relatively large and requires intense pulses and long loop lengths. It can be reduced considerably by incorporating a fiber amplifier (see Chapter 3) within the loop [67]. If the amplifier is located close to the fiber coupler, its presence introduces an asymmetry whenever the counterpropagating pulses are not amplified simultaneously. Since such a Sagnac interferometer is unbalanced by the amplifier, even a 50:50 coupler ($\rho = 0.5$) can be used. The switching behavior in this case can be understood by noting that one wave is amplified at the entrance to the loop, while the counterpropagating wave experiences amplification just before exiting the loop. Since the intensities of the two waves differ by a large amount throughout the loop, the differential phase shift can be quite large. In fact, assuming that the clockwise wave is amplified first by a factor G, we can use Eq. (2.3.14) to calculate the transmissivity, provided that A_f in Eq. (2.3.12) is multiplied by $\sqrt{G}$. The result is given by

$$T_S = 1 - 2\rho(1-\rho)\{1 + \cos[(1 - \rho - G\rho)\gamma P_0 L]\}. \quad (2.3.17)$$

The condition for complete transmission is obtained from Eq. (2.3.16) by replacing $(1 - 2\rho)$ with $(1 - \rho - G\rho)$. For $\rho = 0.5$, the switching power is given by (if we use $m = 1$)

$$P_0 = 2\pi/[(G-1)\gamma L]. \quad (2.3.18)$$

Since the amplification factor G can be as large as 30 dB, the switching power is reduced by a factor of up to 1,000. Such a device, referred to as the *nonlinear amplifying-loop mirror*, can switch at peak power levels below 1 mW. Its implementation is relatively simple with the advent of fiber amplifiers (see Chapter 3). In a demonstration of the basic concept, 4.5 m of Nd-doped fiber was spliced within the 306-m fiber loop formed using a 3-dB coupler [67]. Quasi-CW-like switching was observed using 10-ns pulses. The switching power was about 0.9 W even when the amplifier provided only a 6-dB gain (a factor of 4). In a later experiment, the use of a semiconductor optical amplifier, providing different gains for counterpropagating waves inside a 17-m fiber loop, resulted in switching powers of less than 250 μW when 10-ns pulses were injected into the loop [68].

A Sagnac interferometer can also be unbalanced by using a fiber loop in which GVD is not constant but varies along the loop length [69]–[72]. The GVD can vary continuously as in a dispersion-decreasing fiber, or in a step-like fashion (using fibers with different dispersive properties connected in series). The simplest situation corresponds to the case in which the Sagnac loop is made with two types of fibers and is similar to a dispersion-management scheme used in lightwave systems for GVD compensation. Dispersion-varying fiber loops unbalance a Sagnac interferometer since the counterpropagating waves experience different GVD as they complete a round trip. The most noteworthy feature of such Sagnac loops is that they remain balanced for CW beams of any power levels since GVD does not affect them. However, evolution of optical pulses is affected both by GVD and SPM, resulting in a net relative phase shift between the counterpropagating waves. As a result, optical pulses can be switched to the output port while any CW background noise is reflected by dispersion-imbalanced Sagnac loops. An extinction ratio of 22 dB for the CW background was observed in an experiment [70] in which the 20-m loop was made using equal lengths of standard telecommunication fiber ($\beta_2 = -23$ ps^2/km) and dispersion-shifted fiber ($\beta_2 = -2.3$ ps^2/km).

An important class of applications is based on the XPM effects occurring when a control signal is injected into the Sagnac loop such that it propagates in only one direction and induces a nonlinear phase shift on one of the counterpropagating waves through XPM (see Chapter 10). In essence, the control signal is used to unbalance the Sagnac interferometer similar to the way an optical amplifier can be used to produce different SPM-induced phase shifts. As a result, the loop can be made using a 3-dB coupler so that a low-power CW beam is reflected in the absence of the control but transmitted when a control pulse is applied. Many experiments have shown the potential of XPM-induced switching [73]–[79]. As early as 1989, transmissivity of a 632-nm CW signal (obtained from a He–Ne laser) was switched from zero to close to 100% by using intense 532-nm picosecond pump pulses with peak powers of about 25 W [73].

When the signal and control wavelengths are far apart, one should consider the walk-off effects induced by the group-velocity mismatch. The walk-off problem can be solved by using a fiber whose zero-dispersion wavelength lies between the pump and signal wavelengths such that the two waves have the same group velocity. It can also be avoided by using an orthogonally polarized pump at the same wavelength as that of the signal [76]. There is still a group-velocity mismatch because of polarization-mode dispersion, but it is relatively small. Moreover, it can be used to advantage by constructing a Sagnac loop in which the slow and fast axes of polarization-maintaining fibers are interchanged in a periodic fashion. In one implementation of this idea [77], a 10.2-m loop consisted of 11 such sections. Two orthogonally polarized pump and

2.3 Fiber Interferometers

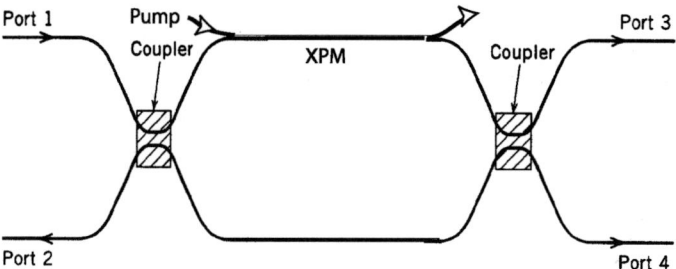

Figure 2.17: A Mach–Zehnder interferometer with two input and two output ports. An external pump is sometimes used to unbalance the device.

signal pulses (width about 230 fs) were injected into the loop and propagated as solitons. The pump pulse was polarized along the fast axis and delayed initially such that it overtook the signal pulse in the first section. In the second section, the signal pulse traveled faster because of the reversing of slow and fast axes and overtook the pump pulse. This process repeated in each section. As a result, the XPM-induced phase shift was enhanced considerably.

2.3.4 Mach–Zehnder Interferometers

An all-fiber Mach–Zehnder (MZ) interferometer is constructed by connecting two fiber couplers in series, as shown schematically in Figure 2.17. The first coupler splits the input signal into two parts, which acquire different phase shifts if arm lengths are different, before they interfere at the second coupler. Such a device has the same functionality as a Sagnac loop but has an added advantage that nothing is reflected back toward the input port. Moreover, a MZ interferometer can be unbalanced by simply using different lengths for its two arms since the two optical fields inside it take physically separated paths. However, the same feature also makes the interferometer susceptible to environmental fluctuations. MZ interferometers can also be made using planar waveguides and are used for a variety of applications. Nonlinear effects in MZ interferometers were considered starting in 1987 and have continued to be of interest [80]–[86].

The transmission characteristics of an MZ interferometer can be analyzed using a method similar to that used for Sagnac interferometers. The main difference is that the two fields produced at the output of the first fiber coupler take different physical paths, and thus may acquire different linear and nonlinear phase shifts. In general, two couplers need not be identical and can have different power-splitting fractions, ρ_1 and ρ_2. Two arms of the interferometer can also have different lengths and propagation constants. We consider such an asymmetric MZ interferometer and find the powers transmitted from the two output ports when a single CW beam with power P_0 is incident at one input port. If we use Eq. (2.3.11) at the first coupler and take into account both the linear and nonlinear phase shifts, the optical fields at the second coupler are given by

$$A_1 = \sqrt{\rho_1} A_0 \exp(i\beta_1 L_1 + i\rho_1 \gamma |A_0|^2 L_1), \qquad (2.3.19)$$

$$A_2 = i\sqrt{1-\rho_1} A_0 \exp[i\beta_2 L_2 + i(1-\rho_1)\gamma |A_0|^2 L_2], \qquad (2.3.20)$$

where L_1 and L_2 are the lengths and β_1 and β_2 are the propagation constants for the two arms of the MZ interferometer.

The optical fields exiting from the output ports of a MZ interferometer are obtained by using the transfer matrix of the second fiber coupler:

$$\begin{pmatrix} A_3 \\ A_4 \end{pmatrix} = \begin{pmatrix} \sqrt{\rho_2} & i\sqrt{1-\rho_2} \\ i\sqrt{1-\rho_2} & \sqrt{\rho_2} \end{pmatrix} \begin{pmatrix} A_1 \\ A_2 \end{pmatrix}. \qquad (2.3.21)$$

The fraction of power transmitted from the bar port of the MZ interferometer is obtained using $T_b = |A_3|^2/|A_0|^2$ and is given by

$$T_b = \rho_1\rho_2 + (1-\rho_1)(1-\rho_2) - 2[\rho_1\rho_2(1-\rho_1)(1-\rho_2)]^{1/2}\cos(\phi_L + \phi_{NL}), \qquad (2.3.22)$$

where the linear and nonlinear parts of the relative phase shift are given by

$$\phi_L = \beta_1 L_1 - \beta_2 L_2, \quad \phi_{NL} = \gamma P_0[\rho_1 L_1 - (1-\rho_1)L_2]. \qquad (2.3.23)$$

This equation simplifies considerably for a symmetric MZ interferometer made using two 3-dB couplers so that $\rho_1 = \rho_2 = \frac{1}{2}$. The nonlinear phase shift vanishes for such a coupler when $L_1 = L_2$, and the transmissivity of the bar port is given as $T_b = \sin^2(\phi_L/2)$. Since the linear phase shift ϕ_L is frequency-dependent, the output depends on the wavelength of light. Thus, an MZ interferometer acts as an optical filter. The spectral response can be improved by using a cascaded chain of such interferometers with relative path lengths adjusted suitably (see Chapter 8).

The nonlinear switching characteristics of an MZ interferometer are similar to those of a Sagnac loop in the sense that the output from one of the ports can be switched from low to high (or vice versa) by changing the input peak power of the incident signal. Figures 2.18(a) and (b) show the experimentally observed transmittance from the bar port (circles) and the cross port (crosses) as input peak power is varied over a range of 0 to 25 W for two values of ϕ_L [84]. Predictions of Eq. (2.3.22) are also shown for comparison using $\rho_1 = 0.34$ and $\rho_2 = 0.23$ for the power-splitting ratios of the two couplers. The arm lengths were identical in this experiment ($L_1 = L_2$) as the MZ interferometer was made using a dual-core fiber whose two identical cores were connected on each side to a fiber coupler. This configuration avoids temporal fluctuations occurring on a millisecond timescale. Such fluctuations occur invariably when two separate fiber pieces are used in each arm of the MZ interferometer and require an active stabilization scheme for controlling them [81].

Similar to the case of Sagnac interferometers, switching can also be accomplished by inducing XPM-induced phase shift in only one arm of the MZ interferometer through a control or pump propagating to that arm only (see Figure 2.17). The MZ interferometer is balanced in the absence of the pump, and the signal injected from port 1 appears at port 4. When the pump is intense enough to induce a π phase shift through XPM,

2.3 Fiber Interferometers

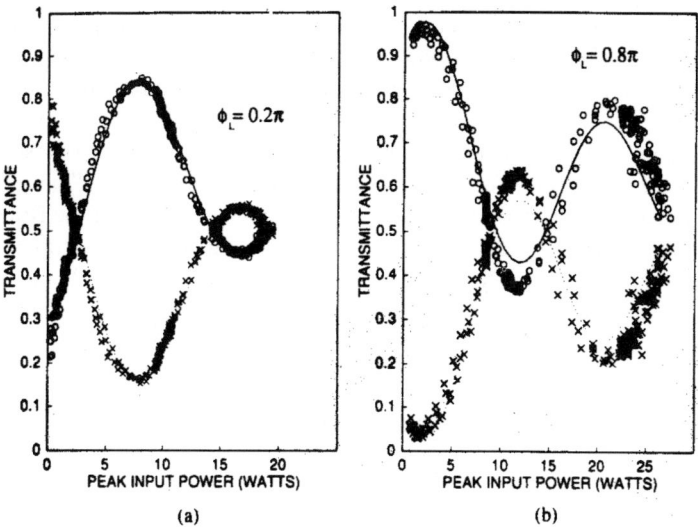

Figure 2.18: Nonlinear switching in a Mach–Zehnder interferometer for two values (a and b) of ϕ_L. Data from the bar and cross ports are shown by circles and crosses. Theoretical predictions are shown as solid and dotted curves. (After Ref. [84]); ©1991 OSA.)

the signal is directed toward port 3. Clearly, such a device can be used as a wavelength converter when the channel whose wavelength needs to be converted acts as a pump, while a CW beam at the desired wavelength is launched at an input port.

The pump power required for XPM-induced switching can be reduced to manageable levels by using several different techniques. The nonlinear parameter γ can be increased by reducing the effective core area A_{eff} of the fiber. In a 1988 experiment, an XPM-induced phase shift of 10° was measured at a pump power of about 15 mW by reducing A_{eff} to only 2 μm^2 in an MZ interferometer with 38 m of fiber in each arm [82]. The use of a ring resonator in one arm of the resonator can increase the XPM-induced phase shift by several orders of magnitude [52]. The pump power required for the π phase shift is reduced to under 10 mW for a 10-m-long fiber ring, although the switching speed is also reduced to below 1 GHz for such devices.

MZ interferometers can be used for a variety of applications. Most of them are based on the ability of an MZ interferometer to produce large changes in its output with small changes in the refractive index in one of its arms. As discussed in Chapter 6, MZ interferometers made with LiNbO$_3$ or semiconductor waveguides are used routinely as high-speed modulators since such electro-optic materials permit voltage-induced changes in the refractive index. Silica fibers do not have this property, but their refractive index can be changed either optically (through SPM and XPM) or through changes in the environment (such as temperature or pressure). The latter property is useful for making fiber sensors.

Another category of applications discussed in Chapter 8 employs MZ interferometers as optical filters [87]–[90]. The simplest scheme uses a series of interconnected fiber couplers, forming a chain of MZ interferometers. Such a device is sometimes referred to as a *resonant coupler* since it resonantly couples out a specific-wavelength channel from a WDM signal to one output port while the remaining channels appear at the other output port. Its performance can be optimized by controlling the power-splitting ratios of various directional couplers [87]. The wavelength selectivity of Bragg gratings can also be used to make add–drop filters [20]. In one scheme, two identical Bragg gratings are formed in the center of each arm of an MZ interferometer [88]. The channel whose wavelength λ_g falls within the stop band of the Bragg grating is totally reflected and appears at port 2. The remaining channels are not affected by the gratings and appear at port 4. The same device can add a channel at λ_g if the signal at that wavelength is injected from port 3. Stability of the MZ interferometer is of primary concern in these devices and requires active phase control in practice [89].

2.3.5 Michelson Interferometers

A Michelson interferometer is made by connecting two separate pieces of fibers to the output ports of a fiber coupler and attaching 100% reflecting mirrors or Bragg gratings at the other end of the fibers [20]. A Michelson interferometer functions much like an MZ interferometer, with the crucial difference that the light propagating in its two arms is forced to interfere at the same coupler where it was split. Because of this feature, a Michelson interferometer acts as a nonlinear mirror, similar to a Sagnac interferometer, with the important difference that the interfering optical fields do not share the same physical path. Nonlinear Michelson interferometers can also be made using bulk optics (beam splitters and mirrors) with a long piece of fiber in one arm acting as a nonlinear medium. Nonlinear effects in Michelson interferometers were first studied within the context of passive mode locking and have continued to remain of interest [91]–[94].

We can apply the analysis of Section 2.3.4 developed for an MZ interferometer to the case of a Michelson interferometer because of the similarity between the two. In both cases, an optical field is split into two parts at a fiber coupler, each part acquires a phase shift, and the two parts recombine interferometrically at the coupler. Since the same coupler is used for splitting and combining the optical fields in the case of a Michelson interferometer, we should set $\rho_1 = \rho_2 \equiv \rho$ in Eq. (2.3.22). For the same reason, transmission from the bar port of the coupler turns into reflection from the input port, and the reflectivity is given by

$$R_M = \rho^2 + (1-\rho)^2 - 2\rho(1-\rho)\cos(\phi_L + \phi_{NL}). \qquad (2.3.24)$$

The lengths L_1 and L_2 appearing in Eq. (2.3.23) should be interpreted as round-trip lengths in each arm of the Michelson interferometer. The transmissivity is, of course, given by $T_M = 1 - R_M$. The reflection and transmission characteristics of a Michelson interferometer are similar to those of a Sagnac loop with two major differences. First, the round-trip path lengths L_1 and L_2 can be different for a Michelson interferometer. Second, the reflectivity and transmissivity are reversed for the Sagnac loop. Indeed, Eq. (2.3.24) reduces to Eq. (2.3.15) if $\phi_L = 0$.

Because of the SPM-induced nonlinear phase shift, the reflectivity of a Michelson interferometer is power-dependent. As a result, such an interferometer tends to shorten an optical pulse and acts effectively as a fast-responding saturable absorber [91]. The pulse-shortening mechanism can be understood as follows. When the relative linear phases are set appropriately, the nonlinear phase shift may lead to constructive interference near the peak of the pulse, while the wings of the pulse experience destructive interference. The pulse-shortening capability of Michelson interferometers can be exploited for passive mode locking of lasers. This technique is commonly referred to as *additive-pulse mode locking* since it is the interferometric addition of an optical pulse at the coupler that is responsible for mode locking [95]. This topic is discussed in Chapter 4 within the context of mode-locked fiber lasers.

2.4 Isolators and Circulators

So far the focus in this chapter has been on all-fiber optical components. However, lightwave systems often need passive components that cannot be built using only optical fibers. An example is provided by optical isolators and optical circulators, both of which fall into the category of nonreciprocal devices in the sense that they break the time-reversal symmetry inherent in optics and require that the device behave differently if the direction of light propagation is reversed. A magnetic field can break this symmetry but it requires the use of magnetic materials. In this section we first discuss the Faraday effect, the basic physical mechanism responsible for nonreciprocal behavior, and then consider the design of optical isolators and circulators [96]. All such devices use a magnetic material but are packaged in such a way that the final device is fully compatible with an all-fiber design concept.

2.4.1 Faraday Effect

The Faraday effect is a consequence of different refractive indices seen by the right- and left-circularly polarized components of an optical beam inside a magneto-optic material in the presence of a magnetic field. It manifests as a change in the state of polarization as the beam propagates through the medium. On a more fundamental level, the Faraday effect has its origin in the motion of an electron in the presence of an applied magnetic field [96]. The nonreciprocal nature of the Faraday effect results from the fact that changes in the state of polarization depend on the direction of the magnetic field but not on the direction in which light is traveling.

Mathematically, the propagation constant β inside the medium becomes different for the two circularly polarized components of an optical field at frequency ω in the presence of a magnetic field and is given by

$$\beta^{\pm} = n^{\pm}(\omega/c), \qquad (2.4.1)$$

where the $+$ and $-$ signs correspond to right and left circular polarization, respectively. The induced circular birefringence is directly proportional to the magnitude of the applied magnetic field and can be written as $\delta n = n^+ - n^- = K_F H_{\text{dc}}$, where K_F is a constant and the magnetic field is oriented along the direction of propagation. In a medium of length l_M, the relative phase shift between the right- and left-circularly polarized components of an optical beam is thus given by

$$\delta\phi = (\omega/c)K_F H_{\text{dc}} l_M = V_c H_{\text{dc}} l_M, \qquad (2.4.2)$$

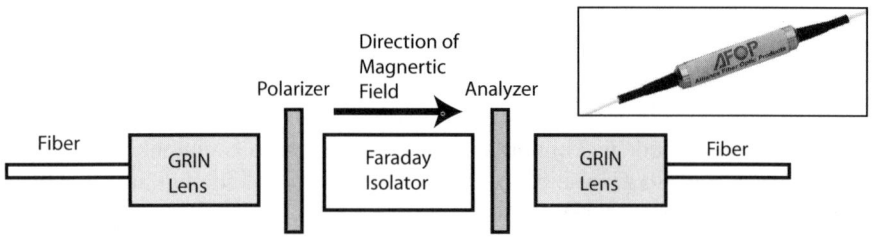

Figure 2.19: Components used to make an optical isolator. The inset shows a packaged device.

where $V_c = (\omega/c)K_F$ is known as the Verdet constant. As a result of this phase shift, the plane of polarization of the incident light is rotated by the Faraday angle $\theta_F = \frac{1}{2}\delta\phi$.

The Verdet constant for most materials is extremely small. For example, its value for silica is $\sim 10^{-5}$ rad/(Oe-cm). The largest values of the Verdet constant occur for materials containing paramagnetic ions such as terbium. The most commonly used materials for practical devices are terbium-doped glasses and a crystal known as terbium gallium garnet. The Verdet constant of a terbium gallium garnet is ~ 0.1 rad/(Oe-cm).

Nonreciprocal changes in the state of polarization induced by the Faraday effect are useful for making a device known as the Faraday rotator. In this device, the magnetic field and the garnet length are chosen such that the direction of a linearly polarized light is rotated by 45° from its original direction. As discussed next, optical isolators are made using a Faraday rotator.

2.4.2 Optical Isolators

An isolator is the optical analog of a rectifying diode in the sense that ir permits light propagation in one direction only. Figure 2.19 shows the design of a fiber-pigtailed optical isolator [3]. It uses a Faraday rotator sandwiched between two polarizers. Incoming light from the fiber on the left is first collimated using a gradient-index (GRIN) lens before it falls on the left polarizer, which passes only the vertically polarized component. The Faraday rotator is designed to rotate the plane of polarization by 45° (say, clockwise). The second polarizer, acting as an analyzer, passes the light completely because it is tilted at 45° compared with the first polarizer. The second GRIN lens focuses the light passed through the analyzer and couples it into the fiber on the right. Sometimes a wave plate is inserted after the analyzer such that it rotates the polarization back in the vertical direction. Such a device transmits a vertically polarized signal without any change but would block the horizontally polarized one.

Now consider what happens to the light traveling backward. The analyzer would pass it unchanged if its polarization direction is tilted 45° from the vertical and coincides with the polarizer axis. The Faraday rotator still rotates the plane of polarization by 45° in the clockwise direction. As a result, light becomes horizontally polarized by the time it arrives at the left polarizer and is completely blocked by it. Thus, the device shown in Figure 2.19 transmits light in only a forward direction and acts as an optical isolator.

2.4 Isolators and Circulators

Real isolators do not provide complete isolation because of imperfections inherent in any practical device. For example, if polarizers are not oriented perfectly or if the Faraday rotator does not rotate the plane of polarization by exactly 45°, some light will pass in the backward direction. However, the state of the art is such that commercial isolators can routinely provide better than 30-dB isolation, and the isolation level can exceed 40 dB for suitably designed devices.

Another practical issue is related to the insertion loss, defined as the loss in power as the signal is transmitted through an isolator. It is evident that the insertion loss of the isolator shown in Figure 2.19 depends on the state of polarization of the incident light. It passes the vertically polarized light with negligible losses but will introduce some losses for any other state of polarization. Such polarization-dependent losses are not desirable in any lightwave system because optical fibers generally do not maintain the state of polarization of passing light unless they are specifically designed to do so. A simple technique for making a polarization-independent isolator splits the incoming beam into its orthogonally polarized components (using a polarization beam splitter, for example), processes the two components separately, and combines them before coupling the light into the output fiber. Although simple in concept, this scheme increases the cost because of the large number of optical components employed. Nevertheless, such isolators are available commercially with insertion losses of about 0.5 dB while providing >40 dB isolation. They are reasonably compact in size (typically 4 cm long and 5 mm wide) and have fiber pigtails at both ends so that they virtually act as an all-fiber device.

2.4.3 Optical Circulators

Optical circulators extend the basic idea behind an optical isolator and add more functionality to the device. A circulator does not discard the backward propagating light, as an isolator does, but directs it to another port, thus resulting in a three-port device in the simplest configuration. More ports can be added if one wants to redirect light coming from the third port to a fourth port. Figure 2.20(a) shows a device with six ports. Such devices are called circulators because they direct light to different ports in a circular fashion depending on which port light enters [96]. The last port is generally not coupled back to the first port to ensure the isolation of all ports.

As one may expect, the design of optical circulators becomes increasingly complex as the number of ports increases. A second layer of complexity is added for polarization-independent circulators because they must split the incoming light from any port into its orthogonally polarized components and process each component separately. In general, a circulator requires a large number of parts, including those shown in Figure 2.19. An important new component is the beam displacer, made from a strongly birefringent medium such that it displaces the orthogonally polarized components spatially by different amounts. In spite of their complexity, optical circulators are available commercially in a relatively compact size (about 6 cm long and 5 mm wide) with fiber pigtails at each end. Insertion losses of such devices are also quite acceptable (close to 0.5 dB).

Optical circulators have found many applications in designing lightwave systems. Figure 2.20(b) shows, as an example, how a three-port circulator can be used with

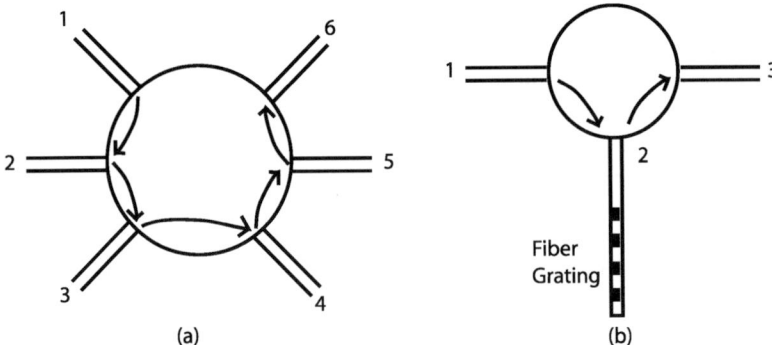

Figure 2.20: (a) Schematic of a six-port optical circulator. Arrows indicate the direction of output port. (b) A three-port circulator used in combination with a fiber grating.

a fiber grating to realize a narrowband bandpass filter working in transmission. As discussed in Section 2.2, a fiber grating acts as a bandpass filter but it reflects the filtered light. The circulator converts it into a transmission filter for all practical purposes. The input light entering from port 1 is reflected by the grating located at port 2, but the circulator directs it to port 3 where the output fiber is connected. Without a circulator, one would need to use a 3-dB fiber coupler but that will also introduce 3-dB losses. The use of circulator allows to realize the the same goal with less than 1-dB insertion losses.

Problems

2.1 Explain in physical terms why the proximity of two cores in a fiber coupler leads to power exchange between the two cores.

2.2 Starting from the wave equation, derive the coupled-mode equations for a fiber coupler in the frequency domain.

2.3 Convert Eqs. (2.1.4) and (2.1.5) into time-domain equations, treating both the propagation constants and the coupling coefficients as frequency-dependent. Assume a symmetric coupler to simplify the algebra.

2.4 Evaluate the integral in Eq. (2.1.6) to find the coupling coefficient for a symmetric fiber coupler whose core centers are separated by a distance d. Assume that the fundamental mode in each waveguide has a Gaussian shape with width (FWHM) w_0.

2.5 Discuss how κ obtained in previous problem depends on the ratio d/w_0 by plotting it. Find the coupling length when $d/w_0 = 3$.

2.6 Derive the coupled-mode equations (2.2.12) and (2.2.13) for fiber gratings starting from the Maxwell equations.

2.7 What is meant by the stop band of a grating? Starting from the linear coupled-mode equations (2.2.7) and (2.2.8), find the dispersion relation and the width of the stop band.

2.8 An optical pulse is transmitted through a fiber grating with its spectrum located close to but outside the stop band. Its energy is small enough that nonlinear effects are negligible. Derive an expression for the group velocity of the pulse.

2.9 For the previous problem, derive expressions for the second- and third-order dispersion induced by the grating. You can neglect the material and waveguide dispersion of silica fibers.

2.10 Derive an expression for the reflectivity of a fiber grating by solving the coupled-mode equations (2.2.7) and (2.2.8). Plot it as a function of δ/κ using $\kappa L = 3$.

2.11 Derive an expression for the transmittivity of a fiber-ring resonator of length L formed using a fiber coupler with bar-state transmission of ρ. Include both losses and phase shifts incurred over each round trip. Prove that the transmittivity does not depend on frequency of light in a lossless ring resonator.

2.12 Derive Eq. (2.3.8) by considering multiple round trips inside a Fabry–Perot resonator.

2.13 Derive Eq. (2.3.15) by considering the phase shifts experienced by the counterpropagating waves inside a Sagnac loop. Use it to estimate the minimum switching power required when $\rho = 0.4$ and $\gamma L = 0.1$ W^{-1}.

2.14 Use Eq. (2.3.15) for a Gaussian pulse for which $P_i(t) = P_0 \exp[-(t/T_0)^2]$. Plot the shape of the transmitted pulse using $T_0 = 1$ ps, $\rho = 0.45$, and $\gamma P_0 L = 1, 2,$ and 4. Estimate the compression factor in each case.

2.15 Derive an expression for the transmittivity of a Sagnac loop containing an optical amplifier next to the fiber coupler. Assume G is the amplifier gain, ρ is the bar-state transmission of the coupler, and a CW beam with power P_0 is injected into the loop.

2.16 Use the expression derived in the previous problem to find the switching power when a 3-dB coupler is used ($\rho = 0.5$) to make the Sagnac loop. Estimate its numerical value for a 100-m loop when $G = 30$ dB. Use $\gamma = 2$ W^{-1}/km.

References

[1] V. J. Tekippe, *Fiber Integ. Opt.* **9**, 97 (1990).
[2] P. E. Green, Jr., *Fiber-Optic Networks*, Prentice-Hall, Upper Saddle River, NJ, 1993, Chap. 3.
[3] J. Hecht, *Understanding Fiber Optics*, Prentice-Hall, Upper Saddle River, NJ, 1999, Chap. 15.
[4] A. K. Ghatak and K. Thyagarajan, *Introduction to Fiber Optics*, Cambridge University Press, New York, 1999, Chap. 17.
[5] A. W. Snyder and J. D. Love, *Optical Waveguide Theory*, Chapman and Hall, London, 1983.

[6] H. A. Haus, *Waves and Fields in Optoelectronics*, Prentice-Hall, Englewood Cliffs, NJ, 1984.
[7] D. Marcuse, *Theory of Dielectric Optical Waveguides*, Academic Press, San Diego, CA, 1991, Chap. 6.
[8] H. A. Haus and W. P. Huang, *Proc. IEEE* **79**, 1505 (1991).
[9] G. P. Agrawal and N. K. Dutta, *Semiconductor Lasers*, 2nd ed., Van Nostrand Reinhold, New York, 1993.
[10] A. Yariv, *Optical Electronics in Modern Communications*, 5th ed., Oxford University Press, New York, 1997.
[11] G. P. Agrawal, *Applications of Nonlinear Fiber Optics*, Academic Press, San Diego, CA, 2001.
[12] R. Tewari and K. Thyagarajan, *J. Lightwave Technol.* **4**, 386 (1986).
[13] G. P. Agrawal, *Fiber-Optic Communication Systems*, 3rd ed., Wiley, New York, 2002.
[14] K. S. Chiang, *Opt. Lett.* **20**, 997 (1995); *IEEE J. Quantum Electron.* **33**, 950 (1997).
[15] K. S. Chiang, Y. T. Chow, D. J. Richardson, D. Taverner, L. Dong, L. Reekie, and K. M. Lo, *Opt. Commun.* **143**, 189 (1997).
[16] M. Born and E. Wolf, *Principles of Optics*, 7th ed., Cambridge University Press, New York, 1999, Section 8.6.
[17] K. O. Hill, Y. Fujii, D. C. Johnson, and B. S. Kawasaki, *Appl. Phys. Lett.* **32**, 647 (1978).
[18] G. Meltz, W. W. Morey, and W. H. Glen, *Opt. Lett.* **14**, 823 (1989).
[19] R. Kashyap, J. R. Armitage, R. Wyatt, S. T. Davey, and D. L. Williams, *Electron. Lett.* **26**, 730 (1990).
[20] R. Kashyap, *Fiber Bragg Gratings*, Academic Press, San Diego, CA, 1999.
[21] A. Othonos and K. Kalli, *Fiber Bragg Gratings*, Artec House, Boston, 1999.
[22] P. S. J. Russell, L. J. Poyntz-Wright, and D. P. Hand, *Proc. SPIE* **1373**, 126 (1991).
[23] G. Meltz and W. W. Morey, *Proc. SPIE* **1516**, 185 (1991).
[24] T. E. Tsai, C. G. Askins, and E. J. Friebele, *Appl. Phys. Lett.* **61**, 390 (1992).
[25] P. J. Lemaire, R. M. Atkins, V. Mizrahi, and W. A. Reed, *Electron. Lett.* **29**, 1191 (1993).
[26] T. Erdogan, V. Mizrahi, P. J. Lemaire, and D. Monroe, *J. Appl. Phys.* **76**, 73 (1994).
[27] J. Canning, *Opt. Fiber Technol.* **6**, 275 (2000).
[28] J. L. Archambault, L. Reekie, and P. S. J. Russell, *Electron. Lett.* **29**, 28 (1993); *Electron. Lett.* **29**, 453 (1993).
[29] L. Long, J. L. Archambault, L. Reekie, P. S. J. Russell, and D. N. Payne, *Opt. Lett.* **18**, 861 (1993).
[30] K. O. Hill, B. Malo, F. Bilodeau, D. C. Johnson, and J. Albert, *Appl. Phys. Lett.* **62**, 1035 (1993).
[31] D. Z. Anderson, V. Mizrahi, T. Erdogan, and A. E. White, *Electron. Lett.* **29**, 566 (1993).
[32] K. C. Byron, K. Sugden, T. Bricheno, and I. Bennion, *Electron. Lett.* **29**, 1659 (1993).
[33] J. Brennan III and D. LaBrake, in Proc. Bragg Gratings, Photosensitivity, and Poling, Optical Society of America, Washington, DC, 1999, pp. 35–37.
[34] K. O. Hill, B. Malo, K. A. Vineberg, F. Bilodeau, D. C. Johnson, and I. Skinner, *Electron. Lett.* **26**, 1270 (1990).
[35] P. S. J. Russell, *J. Mod. Opt.* **38**, 1599 (1991).
[36] N. M. Litchinitser, B. J. Eggleton, and D. B. Patterson, *J. Lightwave Technol.* **15**, 1303 (1997).
[37] B. J. Eggleton, C. M. de Sterke, and R. E. Slusher, *J. Opt. Soc. Am. B* **16**, 587 (1999).
[38] R. E. Slusher, B. J. Eggleton, T. A. Strasser, and C. M. de Sterke, *Opt. Eng.* **3**, 465 (1998).

[39] V. Jayaraman, D. Cohen, amd L. Coldren, *Appl. Phys. Lett.* **60**, 2321 (1992).
[40] B. J. Eggleton, P. A. Krug, L. Poladian, and F. Ouellette, *Electron. Lett.* **30**, 1620 (1994).
[41] N. G. R. Broderick, C. M. de Sterke, and B. J. Eggleton, *Phys. Rev. E* **55**, R5788 (1995).
[42] C. M. de Sterke and N. G. R. Broderick, *Opt. Lett.* **20**, 2039 (1995).
[43] B. J. Eggleton, C. M. de Sterke, and R. E. Slusher, *Opt. Lett.* **21**, 1223 (1996).
[44] N. G. R. Broderick and C. M. de Sterke, *Phys. Rev. E* **55**, 3232 (1997).
[45] C. M. de Sterke, B. J. Eggleton, and P. A. Krug, *J. Lightwave Technol.* **15**, 1494 (1997).
[46] P. Petropoulos, M. Isben, M. N. Zervas, and D. J. Richardson, *Opt. Lett.* **25**, 521 (2000).
[47] P. S. J. Russell, *Phys. Rev. Lett.* **56**, 596 (1986).
[48] L. F. Stokes, M. Chodorow, and H. J. Shaw, *Opt. Lett.* **7**, 288 (1982).
[49] C. Y. Yue, J. D. Peng, Y. B. Liao, and B. K. Zhou, *Electron. Lett.* **24**, 622 (1988).
[50] K. P. Poo and A. D. Kersey, *IEEE Photon. Technol. Lett.* **7**, 209 (1005).
[51] C. K. Madsen and L. H. Zhao, *Optical Filter Design and Analysis: A Signal Processing Approach*, Wiley, New York, 1999.
[52] J. E. Heebner and R. W. Boyd, *Opt. Lett.* **24**, 847 (1999).
[53] L. Wang, *Opt. Commun.* **213**, 27 (2002).
[54] O. Svelto, *Principles of Lasers*, 4th ed., Plenum Press, New York, 1998.
[55] J. Stone and L. W. Stulz, *Electron. Lett.* **23**, 781 (1987); *Electron. Lett.* **26**, 1290 (1990).
[56] I. P. Kaminow, *IEEE J. Sel. Areas Commun.* **8**, 1005 (1990).
[57] K. IIzuka, *Elements of Photonics*, Vol. II, Wiley, New York, 2002.
[58] S. Legoubin, M. Douay, P. Bernage, and P. Niay, *J. Opt. Soc. Am. A* **12**, 1687 (1995).
[59] K. Otsuka, *Opt. Lett.* **8**, 471 (1983).
[60] D. B. Mortimore, *J. Lightwave Technol.* **6**, 1217 (1988).
[61] N. J. Doran and D. Wood, *Opt. Lett.* **13**, 56 (1988).
[62] N. J. Doran, D. S. Forrester, and B. K. Nayar, *Electron. Lett.* **25**, 267 (1989).
[63] K. J. Blow, N. J. Doran, and B. K. Nayar, *Opt. Lett.* **14**, 754 (1989).
[64] M. N. Islam, E. R. Sunderman, R. H. Stolen, W. Pleibel, and J. R. Simpson, *Opt. Lett.* **14**, 811 (1989).
[65] N. Takato, T. Kaminato, A. Sugita, K. Jinguji, H. Toba, and M. Kawachi, *IEEE J. Sel. Areas Commun.* **8**, 1120 (1990).
[66] K. J. Blow, N. J. Doran, and B. P. Nelson, *Electron. Lett.* **26**, 962 (1990).
[67] M. E. Fermann, F. Haberl, M. Hofer, and H. Hochstrasser, *Opt. Lett.* **15**, 752 (1990).
[68] A. W. O'Neill and R. P. Webb, *Electron. Lett.* **26**, 2008 (1990).
[69] A. L. Steele and J. P. Hemingway, *Opt. Commun.* **123**, 487 (1996).
[70] W. S. Wong, S. Namiki, M. Margalit, H. A. Haus, and I. P. Ippen, *Opt. Lett.* **22**, 1150 (1997).
[71] I. Y. Khrushchev, I. D. Philips, A. D. Ellis, R. J. Manning, D. Nesset, D. G. Moodie, R. V. Penty, and I. H. White, *Electron. Lett.* **35**, 1183 (1999).
[72] Y. J. Chai, I. Y. Khrushchev, and I. H. White, *Electron. Lett.* **36**, 1565 (2000).
[73] M. C. Farries and D. N. Payne, *Appl. Phys. Lett.* **55**, 25 (1989).
[74] K. J. Blow, N. J. Doran, B. K. Nayar, and B. P. Nelson, *Opt. Lett.* **15**, 248 (1990).
[75] M. Jinno and T. Matsumoto, *IEEE Photon. Technol. Lett.* **2**, 349 (1990); *Electron. Lett.* **27**, 75 (1991).
[76] H. Avramopoulos, P. M. W. French, M. C. Gabriel, H. H. Houh, N. A. Whitaker, and T. Morse, *IEEE Photon. Technol. Lett.* **3**, 235 (1991).

[77] J. D. Moores, K. Bergman, H. A. Haus, and E. P. Ippen, *Opt. Lett.* **16**, 138 (1991); *J. Opt. Soc. Am. B* **8**, 594 (1991).

[78] M. Jinno, *J. Lightwave Technol.* **10**, 1167 (1992); *Opt. Lett.* **18**, 726 (1993); *Opt. Lett.* **18**, 1409 (1993).

[79] H. Bülow and G. Veith, *Electron. Lett.* **29**, 588 (1993).

[80] N. J. Doran and D. Wood, *J. Opt. Soc. Am. B* **4**, 1843 (1987).

[81] N. Imoto, S. Watkins, and Y. Sasaki, *Opt. Commun.* **61**, 159 (1987).

[82] I. H. White, R. V. Penty, and R. E. Epworth, *Electron. Lett.* **24**, 340 (1988).

[83] M. N. Islam, S. P. Dijaili, and J. P. Gordon, *Opt. Lett.* **13**, 518 (1988).

[84] B. K. Nayar, N. Finlayson, N. J. Doran, S. T. Davey, W. L. Williams, and J. W. Arkwright, *Opt. Lett.* **16**, 408 (1991).

[85] K. I. Kang, T. G. Chang, I. Glesk, and P. R. Prucnal, *Appl. Opt.* **35**, 1485 (1996).

[86] P. Elango, J. W. Arkwright, P. L. Chu, and G. R. Atkins, *IEEE Photon. Technol. Lett.* **8**, 1032 (1996).

[87] M. Kuznetsov, *J. Lightwave Technol.* **12**, 226 (1994).

[88] T. J. Cullen, H. N. Rourke, C. P. Chew, S. R. Baker, T. Bircheno, K. Byron, and A. Fielding, *Electron. Lett.* **30**, 2160 (1994).

[89] G. Nykolak, M. R. X. de Barros, T. N. Nielsen, and L. Eskildsen, *IEEE Photon. Technol. Lett.* **9**, 605 (1997).

[90] T. Mizuochi, T. Kitayama, K. Shimizu, and K. Ito, *J. Lightwave Technol.* **16**, 265 (1998).

[91] F. Ouellette and M. Piché, *Opt. Commun.* **60**, 99 (1986).

[92] E. M. Dianov and O. G. Okhotnikov, *IEEE Photon. Technol. Lett.* **3**, 499 (1991).

[93] C. Spielmann, F. Krausz, T. Brabec, E. Wintner, and A. J. Schmidt, *Appl. Phys. Lett.* **58**, 2470 (1991).

[94] C. X. Shi, *Opt. Lett.* **18**, 1195 (1993).

[95] H. A. Haus, J. G. Fujimoto, and E. P. Ippen, *J. Opt. Soc. Am. B* **7**, 2068 (1991).

[96] E. Collett, *Polarized Light in Fiber Optics*, Polawave, Lincroft, NJ, 2003, Chap. 10.

Chapter 3

Active Fiber Components

Optical fibers can be used to make active devices even though silica glass is not an electrically conducting material. The only requirement is that all fiber-based active devices must be pumped optically. This feature limits such devices to fall into just two categories, namely optical amplifiers and lasers. As seen in Chapter 1, transmission distance of any long-haul lightwave system is eventually limited by fiber losses. A commonly used loss-management technique makes use of fiber-based optical amplifiers. Several kinds of optical amplifiers have been developed during the last 20 years or so. In Section 3.1 we consider doped-fiber amplifiers in which a rare-earth element doped within the fiber core provides the optical gain. Section 3.2 focuses on Raman amplifiers in which the nonlinear phenomenon of stimulated Raman scattering is employed for realizing optical gain. Section 3.3 is devoted to parametric amplifiers in which the nonlinear phenomenon of four-wave mixing is exploited for this purpose. All three types of amplifiers can be used to make fiber lasers. We focus on erbium-doped fiber lasers in Section 3.4.

3.1 Doped-Fiber Amplifiers

An important class of fiber amplifiers makes use of a *rare-earth* element such as erbium for providing gain by doping the fiber core with it during the manufacturing process (see Section 1.2). Although doped-fiber amplifiers were studied as early as 1964 [1], their use became practical only 25 years later, after the fabrication and characterization techniques were perfected [2]. Amplifier properties such as the operating wavelength and the gain bandwidth are determined by the rare-earth element rather than the optical fiber, which plays the role of a host. Many different rare-earth elements, such as erbium, holmium, neodymium, samarium, thulium, and ytterbium, can be used to realize fiber amplifiers operating at different wavelengths in the range 0.5–3.5 μm. Erbium-doped fiber amplifiers (EDFAs) have attracted the most attention because they operate in the wavelength region near 1.55 μm [3]–[8]. Their deployment in WDM lightwave systems after 1995 revolutionized the field of fiber-optic communications and led by 2000 to the commercialization of lightwave systems with capacities exceeding 1 Tb/s.

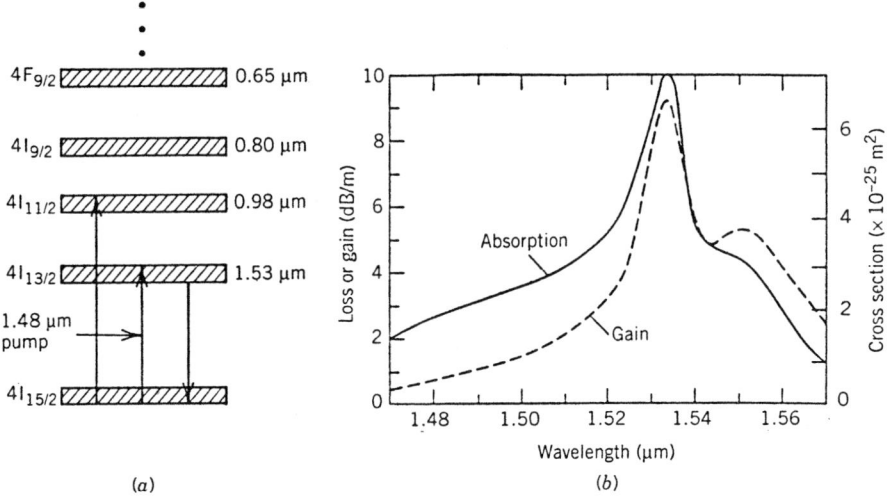

Figure 3.1: (a) Energy bands (shaded regions) of erbium ions in silica fibers. Numbers on right show the wavelengths corresponding to energy gaps. (b) Absorption and gain spectra of an EDFA whose core was codoped with germania.

3.1.1 Pumping and Amplifier Gain

All doped-fiber amplifiers amplify incident light through stimulated emission, the same mechanism that generates coherent light in lasers. Indeed, an optical amplifier is nothing but a laser without feedback. Its main ingredient is the optical gain realized when the amplifier is pumped to achieve *population inversion*. In the case of EDFAs, fiber core contains erbium ions (Er^{3+}), and pumping at a suitable wavelength provides gain through population inversion. Figure 3.1(a) shows a few energy levels of Er^{3+} in silica glasses. The amorphous nature of silica broadens all energy levels of Er^{3+} into bands. The lowest two bands are used for providing amplification, while any of the higher energy bands can be used for pumping an EDFA. Early experiments used the visible radiation emitted from argon-ion, Nd:YAG, or dye lasers even though such pumping schemes are relatively inefficient. From a practical standpoint, the use of semiconductor lasers is preferred (see Chapter 5).

Efficient EDFA pumping is possible using semiconductor lasers operating near 0.98- and 1.48-μm wavelengths. Indeed, the development of such pump lasers was fueled with the advent of EDFAs. It is possible to realize 30-dB gain with only 10–15 mW of absorbed pump power. Efficiencies as high as 11 dB/mW were achieved by 1990 with 0.98-μm pumping [9]. The pumping transition $^4I_{15/2} \to {}^4I_{9/2}$ can use high-power GaAs lasers, although pumping efficiency around 800 nm is relatively low (about 1 dB/mW). The required pump power can be reduced by codoping the silica fiber with aluminum and phosphorus or by using fluorophosphate fibers [10]. With the availability of visible semiconductor lasers, EDFAs have been pumped in the wavelength range of 0.6–0.7 μm. In a 1993 experiment [11], a 33-dB gain was realized at 27 mW of pump power obtained from an AlGaInP laser operating at 670 nm. The

3.1 Doped-Fiber Amplifiers

pumping efficiency was as high as 3 dB/mW at lower pump powers. Most EDFAs use 980-nm pump lasers, as such lasers are commercially available and can provide more than 100 mW of pump power. Pumping at 1,480 nm requires longer fibers and higher powers because it uses the tail of the absorption band shown in Figure 3.1(b).

The optical gain, in general, depends not only on the pump power P_p but also on the frequency ω (or wavelength) and the local power $P(z)$ of the signal at any point inside the amplifier. When the gain medium can be modeled as a homogeneously broadened two-level system, we can write the *gain coefficient* as [13]

$$g(\omega) = \frac{g_0(P_p)}{1 + (\omega - \omega_0)^2 T_2^2 + P/P_s}, \tag{3.1.1}$$

where the peak value g_0 of the gain depends on the pump power and ω_0 is the atomic transition frequency. The saturation power P_s depends on atomic parameters such as the fluorescence time T_1 and the transition cross section associated with the dopant. The parameter T_2 in Eq. (3.1.1), known as the *dipole relaxation time*, is typically quite small (<1 ps). The fluorescence time T_1 can vary from 1 μs–10 ms, depending on the rare-earth element used to make the amplifier; its value is close to 10 ms for EDFAs.

The amplification of a continuous-wave (CW) signal is governed by

$$\frac{dP}{dz} = g(\omega)P, \tag{3.1.2}$$

where $P(z)$ is the optical power at a distance z from the input end of the amplifier of length L located at $z = 0$. In the small-signal regime in which $P/P_s \ll 1$ throughout the amplifier, a straightforward integration of Eq. (3.1.2) shows that the signal power grows exponentially as $P(z) = P(0)\exp(gz)$. The amplifier gain G or the *amplification factor* can now be found using

$$G(\omega) = P_{\text{out}}/P_{\text{in}} = P(L)/P(0) = \exp[g(\omega)L], \tag{3.1.3}$$

where the frequency dependence of both G and g is shown explicitly. This equation shows that the gain of an EDFA depends on the signal wavelength. For this reason, the uniformity of the gain spectrum plays an important role. The exponential growth of signal ceases to occur when input power is large enough that $P(z)$ becomes comparable to or exceeds P_s. Amplifier gain G can be obtained from Eqs. (3.1.1) and (3.1.2) and is found to be significantly reduced in the saturation regime.

EDFAs can be designed to operate in such a way that the pump and signal beams propagate in opposite directions, a configuration referred to as backward pumping. The performance is nearly the same in the forward and backward pumping configurations when the signal power is small enough for the amplifier to remain unsaturated. In the saturation regime, the efficiency is generally better in the backward-pumping configuration [12], mainly because of the important role played by the amplified spontaneous emission (ASE). In the bidirectional pumping configuration, the amplifier is pumped in both directions simultaneously by using two semiconductor lasers located at the two fiber ends. This configuration requires two pump lasers but has the advantage that the population inversion, and hence the gain coefficient g, is relatively uniform along the entire amplifier length.

3.1.2 Gain Spectrum

If we neglect the term P/P_s in Eq. (3.1.1), the frequency dependence of gain coefficient is of the form

$$g(\omega) = \frac{g_0}{1+(\omega-\omega_0)^2 T_2^2}. \tag{3.1.4}$$

The gain is maximum when the incident frequency ω coincides with the atomic transition frequency ω_0. The gain reduction for $\omega \neq \omega_0$ is governed by a Lorentzian profile that is a characteristic of homogeneously broadened two-level systems [13]. The gain bandwidth is defined as the full width at half-maximum (FWHM) of the gain spectrum $g(\omega)$. For the Lorentzian spectrum, the gain bandwidth is given by $\Delta\omega_g = 2/T_2$, or by

$$\Delta v_g = \frac{\Delta\omega_g}{2\pi} = \frac{1}{\pi T_2}. \tag{3.1.5}$$

The concept of *amplifier bandwidth* is commonly used in place of the gain bandwidth. The difference is clear from Eq. (3.1.3). Both the amplifier gain $G(\omega)$ and the gain coefficient $g(\omega)$ are maximum when $\omega = \omega_0$ and decrease with the signal detuning $\omega - \omega_0$. However, $G(\omega)$ decreases much faster than $g(\omega)$. The amplifier bandwidth Δv_A is defined as the FWHM of $G(\omega)$ and is related to the gain bandwidth Δv_g as

$$\Delta v_A = \Delta v_g \left[\frac{\ln 2}{\ln(G_0/2)}\right]^{1/2}, \tag{3.1.6}$$

where $G_0 = \exp(g_0 L)$. The amplifier bandwidth is smaller than the gain bandwidth, and the difference depends on the amplifier gain itself. Notice that this distinction disappears if G is expressed in decibels since $G(\text{dB}) \approx 4.343 gL$.

The gain spectrum of EDFAs deviates considerably from the simple Lorentzian profile obtained from Eq. (3.1.4). Figure 3.1(b) shows the absorption and gain spectra of a silica fiber whose core was doped with both erbium and germania. The flatness of the gain spectrum is important for an EDFA, as it determines the uniformity of individual channel gains when a WDM signal is amplified. The shape of the gain spectrum is affected considerably by the amorphous nature of silica and by the presence of other codopants within the fiber core such as germania and alumina [5]. The gain spectrum of erbium ions in isolation is homogeneously broadened and follows the Lorentzian shape dictated by Eq. (3.1.4). However, it is considerably broadened in the presence of randomly located silica molecules. Structural disorders lead to inhomogeneous broadening of the gain spectrum [4]. Moreover, *Stark splitting* of various energy levels provides additional homogeneous broadening. Mathematically, the gain in Eq. (3.1.4) should be averaged over the distribution of atomic transition frequencies ω_0 such that the effective gain is given by

$$g_{\text{eff}}(\omega) = \int_{-\infty}^{\infty} g(\omega,\omega_0) f(\omega_0) \, d\omega_0, \tag{3.1.7}$$

where $f(\omega_0)$ is the distribution function whose form also depends on the presence of other dopants within the fiber core.

3.1 Doped-Fiber Amplifiers

The gain spectrum seen in Figure 3.1(b) is quite broad and has a double-peak structure. The actual shape can be modified to some extent by using suitable dopants. For example, the addition of alumina to the fiber core broadens the gain spectrum even more. Attempts have been made to isolate the contributions of homogeneous and inhomogeneous broadening through measurements of *spectral hole burning*. In germania-doped EDFAs, the contribution of inhomogeneous broadening is relatively small [5]. In contrast, the gain spectrum of aluminosilicate glasses has roughly equal contributions from homogeneous and inhomogeneous broadening mechanisms. The gain bandwidth of such EDFAs typically exceeds 35 nm.

The gain spectrum of EDFAs can vary from amplifier to amplifier even when core composition is the same because it also depends on the amplifier length. The reason is that the gain depends on both the absorption and emission cross sections having different spectral characteristics. The local inversion or local gain varies along the fiber length because of pump power variations. The total gain is obtained by integrating over the amplifier length. This feature can be used to realize EDFAs that provide amplification in the L band covering the spectral region 1,570–1,610 nm. This issue is discussed later in this section.

3.1.3 Simple Theory

The gain of an EDFA depends on a large number of device parameters such as erbium-ion concentration, amplifier length, core radius, and pump power [14]–[16]. A three-level rate-equation model commonly used for lasers [13] can be adapted for EDFAs. It is sometimes necessary to add a fourth level to include the *excited-state absorption*. In general, the resulting equations must be solved numerically.

Considerable insight can be gained by using a simple two-level model that is valid when ASE and excited-state absorption are negligible. The model assumes that the top level of the three-level system remains nearly empty because of a rapid transfer of the pumped population to the excited state. It is, however, important to take into account the different emission and absorption cross sections for the pump and signal fields. The population densities of the two states, N_1 and N_2, satisfy the following two rate equations [5]:

$$\frac{\partial N_2}{\partial t} = (\sigma_p^a N_1 - \sigma_p^e N_2)\phi_p + (\sigma_s^a N_1 - \sigma_s^e N_2)\phi_s - \frac{N_2}{T_1}, \quad (3.1.8)$$

$$\frac{\partial N_1}{\partial t} = (\sigma_p^e N_2 - \sigma_p^a N_1)\phi_p + (\sigma_s^e N_2 - \sigma_s^a N_1)\phi_s + \frac{N_2}{T_1}, \quad (3.1.9)$$

where σ_j^a and σ_j^e are the absorption and emission cross sections at the frequency ω_j with $j = p, s$. The quantities ϕ_p and ϕ_s represent the photon flux for the pump and signal waves, defined such that $\phi_j = P_j/(a_j h \nu_j)$, where P_j is the optical power, σ_j is the transition cross section at the frequency ν_j, and a_j is the cross-sectional area of the fiber mode for $j = p, s$.

The pump and signal powers vary along the amplifier length because of absorption, stimulated emission, and spontaneous emission. If the contribution of spontaneous emission is neglected, P_s and P_p satisfy the simple equations

$$\frac{\partial P_s}{\partial z} = \Gamma_s(\sigma_s^e N_2 - \sigma_s^a N_1)P_s - \alpha P_s, \tag{3.1.10}$$

$$s\frac{\partial P_p}{\partial z} = \Gamma_p(\sigma_p^e N_2 - \sigma_p^a N_1)P_p - \alpha' P_p, \tag{3.1.11}$$

where α and α' take into account fiber losses at the signal and pump wavelengths, respectively, both of which can be neglected for typical amplifier lengths of 10–20 m. However, they must be included in the case of distributed amplification over lengths >1 km. The confinement factors Γ_s and Γ_p account for the fact that the doped region within the core provides the gain for the entire fiber mode. The parameter $s = \pm 1$ in Eq. (3.1.11), depending on the direction of pump propagation; $s = -1$ in the case of a backward-propagating pump.

Equations (3.1.8) through (3.1.11) can be solved analytically, in spite of their complexity, after some justifiable approximations [14]. For lumped amplifiers, the fiber length is short enough that both α and α' can be set to zero. If we note that $N_1 + N_2 = N_t$ where N_t is the total ion density, only one equation, say, Eq. (3.1.8) for N_2, need be solved. Noting again that the absorption and stimulated-emission terms in the field and population equations are related, we can write the steady-state solution of Eq. (3.1.8), obtained by setting the time derivative to zero, as

$$N_2(z) = -\frac{T_1}{a_d h v_s}\frac{\partial P_s}{\partial z} - \frac{sT_1}{a_d h v_p}\frac{\partial P_p}{\partial z}, \tag{3.1.12}$$

where $a_d = \Gamma_s a_s = \Gamma_p a_p$ is the cross-sectional area of the doped portion of the fiber core. Substituting this solution into Eqs. (3.1.10) and (3.1.11) and integrating them over the fiber length, we can obtain the powers P_s and P_p at the fiber output in an analytical form. This model can be extended to include the ASE propagation in both the forward and backward directions.

The total amplifier gain G for an EDFA of length L is obtained using

$$G = \Gamma_s \exp\left[\int_0^L (\sigma_s^e N_2 - \sigma_s^a N_1)\,dz\right], \tag{3.1.13}$$

where $N_1 = N_t - N_2$ and N_2 is given by Eq. (3.1.12). Figure 3.2 shows the calculated gain at 1.55 μm as a function of pump power and amplifier length using typical parameter values. For a given amplifier length L, the amplifier gain initially increases exponentially with the pump power, but the increase becomes much smaller when the pump power exceeds a certain value corresponding to the "knee" in Figure 3.2(a). For a given pump power, the amplifier gain becomes maximum at an optimum value of L and drops sharply when L exceeds this optimum value. The reason is that the latter portion of the amplifier remains unpumped and absorbs the amplified signal.

Since the optimum value of L depends on the pump power P_p, it is necessary to choose both L and P_p appropriately. Figure 3.2(b) shows that a 35-dB gain can be realized at a pump power of 5 mW for $L = 30$ m and 1.48-μm pumping. It is possible to design amplifiers such that high gain is obtained for amplifier lengths as short as a

3.1 Doped-Fiber Amplifiers

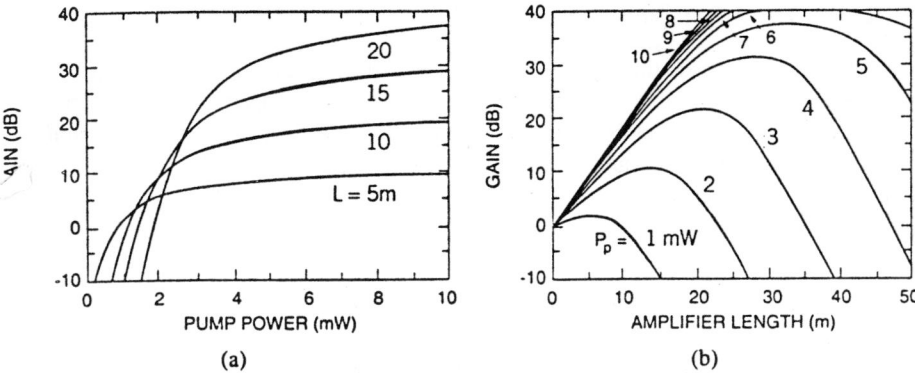

Figure 3.2: Calculated gain as a function of (a) pump power and (b) amplifier length for an EDFA pumped at 1.48 μm. (After Ref. [15]; ©1991 IEEE.)

few meters. The qualitative features seen in Figures 3.2(a) and (b) are observed in all EDFAs, and the agreement between theory and experiment is generally quite good [16]. In general, the output saturation power can vary over a wide range, depending on the EDFA design, with typical values $\sim$10 mW. For this reason the output power levels of EDFAs are generally limited to below 100 mW, although powers as high as 250 mW can be obtained with a proper design.

The foregoing analysis assumes that both pump and signal waves exist in the form of CW beams. In practice, EDFAs are pumped by using CW semiconductor lasers, but the signal is in the form of a pulse train (containing a random sequence of 1 and 0 bits), and the duration of individual pulses is inversely related to the bit rate. The question is whether all pulses experience the same gain or not. Fortunately, the gain remains constant with time in an EDFA for even microsecond-long pulses. The reason is related to a relatively large value of the fluorescence time associated with erbium ions ($T_1 \sim 10$ ms). When the time scale of signal-power variations is much shorter than T_1, erbium ions are unable to follow such fast variations. As single-pulse energies typically fall much below the saturation energy ($\sim$10 μJ), EDFAs respond to the average power. As a result, gain saturation is governed by the average signal power, and amplifier gain does not vary from pulse to pulse even for a WDM signal.

In some applications such as packet-switched networks, signal power may vary on a time scale comparable to T_1. Amplifier gain in that case is likely to become time-dependent, an undesirable feature from the standpoint of system performance. A gain-control mechanism that keeps the amplifier gain pinned at a constant value consists of forcing the EDFA to oscillate at a controlled wavelength outside the range of interest (typically below 1.5 μm). Since the gain remains clamped at the threshold value for a laser, the signal is amplified by the same factor despite variations in the signal power. In one implementation of this scheme, an EDFA was forced to oscillate at 1.48 μm by fabricating two fiber Bragg gratings acting as high-reflectivity mirrors at the two ends of the amplifier [17].

3.1.4 Amplifier Noise

All amplifiers degrade the signal-to-noise ratio (SNR) of the amplified signal because of spontaneous emission that adds noise to the signal during its amplification. The SNR degradation is quantified through a parameter F_n, called the *amplifier noise figure* in analogy with the electronic amplifiers and defined as

$$F_n = \frac{(\text{SNR})_{\text{in}}}{(\text{SNR})_{\text{out}}}, \qquad (3.1.14)$$

where SNR refers to the electric power generated when the optical signal is converted into an electric current. In general, F_n depends on several detector parameters that govern thermal noise associated with the detector (see Chapter 7). A simple expression for F_n can be obtained by considering an ideal detector whose performance is limited by shot noise only [18].

Consider an amplifier with the gain G such that the output and input powers are related by $P_{\text{out}} = GP_{\text{in}}$. The SNR of the input signal is given by

$$(\text{SNR})_{\text{in}} = \frac{\langle I \rangle^2}{\sigma_s^2} = \frac{(R_d P_{\text{in}})^2}{\sigma_s^2}, \qquad (3.1.15)$$

where $\langle I \rangle = R_d P_{\text{in}}$ is the average photocurrent, R_d is the responsivity of the photodetector (see Section 7.1), and

$$\sigma_s^2 = 2q(R_d P_{\text{in}})\Delta f \qquad (3.1.16)$$

is the shot noise for a detector of bandwidth Δf.

To evaluate the SNR of the amplified signal, we need to add the contribution of spontaneous emission to the receiver noise. It turns out that the dominant contribution to the receiver noise comes from the beating of spontaneous emission with the signal [18]. For an amplifier of gain G, the SNR is found to be

$$(\text{SNR})_{\text{out}} = \frac{(R_d G P_{\text{in}})^2}{G\sigma_s^2 + \sigma_b^2}, \qquad (3.1.17)$$

where σ_b^2 is the beat noise induced by spontaneous emission. If we use the definition in Eq. (3.1.14), the noise figure is given by

$$F_n = \frac{1}{G}\left(1 + \frac{\sigma_b^2}{G\sigma_s^2}\right). \qquad (3.1.18)$$

To calculate σ_b^2, we use the spectral density of spontaneous-emission noise in the form [18]

$$S_{\text{sp}}(v) = n_{\text{sp}}(G-1)hv, \qquad (3.1.19)$$

where hv is the average photon energy. The parameter n_{sp} is called the *spontaneous-emission factor* (or the population-inversion factor) and is given by

$$n_{\text{sp}} = N_2/(N_2 - N_1). \qquad (3.1.20)$$

3.1 Doped-Fiber Amplifiers

The spontaneously emitted radiation mixes with the amplified signal and produces the current $I = R_d|\sqrt{G}E_{\text{in}} + E_{\text{sp}}|^2$. Noting that the fields E_{in} and E_{sp} oscillate at different frequencies, we write them in the following form:

$$E_{\text{in}} = \sqrt{P_{\text{in}}}\exp(i\phi_s - i\omega_s t), \quad E_{\text{sp}} = \int \sqrt{S_{\text{sp}}}\exp(i\phi_n - i\omega_n t)\,\omega_n, \qquad (3.1.21)$$

where ω_s is the signal frequency, ω_n is the noise frequency, and ϕ_s and ϕ_n are the corresponding phases. The frequency integral in E_{sp} is over the entire bandwidth of the amplifier or over the bandwidth of the optical filter if such a filter is used.

It is easy to see that the beating of spontaneous emission with the signal produces a noise current $\Delta I = 2R_d \int (GP_{\text{in}}S_{\text{sp}})^{1/2}\cos\theta_n\,d\omega_n$, where $\theta_n = (\omega_n - \omega_s)t + (\phi_n - \phi_s)$ is a random phase difference. If we average over this random phase, the variance of the photocurrent is found to be [19]

$$\sigma_b^2 \approx 4(R_d G P_{\text{in}})(R_d S_{\text{sp}})\Delta f. \qquad (3.1.22)$$

Using this result in Eq. (3.1.17), we obtain the simple result

$$F_n = 2n_{\text{sp}}(1 - 1/G) + 1/G \approx 2n_{\text{sp}}, \qquad (3.1.23)$$

where the last approximation is valid for $G \gg 1$. This equation shows that the SNR of the amplified signal is degraded by 3 dB even for an ideal amplifier for which $n_{\text{sp}} = 1$. For most practical amplifiers, F_n exceeds 3 dB. For its application in optical communication systems, an optical amplifier should have F_n as low as possible.

The calculation of amplifier noise figure is somewhat complicated for EDFAs since n_{sp} depends on the relative populations N_1 and N_2 of the ground and excited states, both of which vary along the amplifier length [20]–[23]. It can be calculated by using the rate-equation model discussed earlier and averaging n_{sp} over the amplifier length. As a result, the noise figure depends both on the amplifier length L and the pump power P_p, just as the amplifier gain does. Figure 3.3(a) shows the variation of F_n with the amplifier length for several values of $P_p' = P_p/P_p^{\text{sat}}$ when a 1.53-μm signal is amplified with an input power of 1 mW. The amplifier gain under the same conditions is also shown in Figure 3.3(b). The results show that a noise figure close to 3 dB can be obtained for a high-gain amplifier pumped such that $P_p \gg P_p^{\text{sat}}$ [20].

The experimental results confirm that F_n close to 3 dB is possible in EDFAs. A noise figure of 3.2 dB was measured in a 30-m-long EDFA pumped at 0.98 μm with 11 mW of power [21]. A similar value was found for another EDFA pumped with only 5.8 mW of pump power at 0.98 μm [22]. In general, it is difficult to achieve high gain, low noise, and high pumping efficiency simultaneously. The main limitation is imposed by the ASE traveling backward toward the pump and depleting the pump power. Incorporation of an internal isolator alleviates this problem to a large extent. In one implementation, 51-dB gain was realized with a 3.1-dB noise figure at a pump power of only 48 mW [24].

The measured values of F_n are generally larger for EDFAs pumped at 1.48 μm. A noise figure of 4.1 dB was obtained for a 60-m-long EDFA when pumped at 1.48 μm with 24 mW of pump power [21]. The reason for a larger noise figure for 1.48-μm pumped EDFAs can be understood from Figure 3.1(a), which shows that the pump level

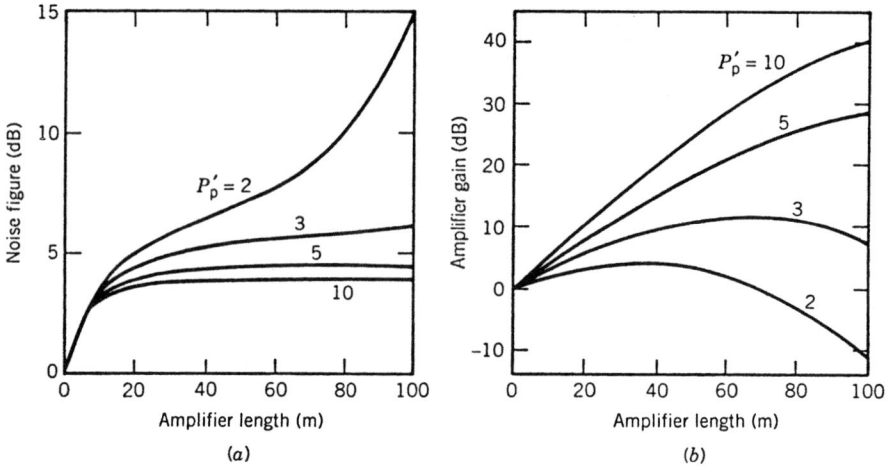

Figure 3.3: (a) Noise figure and (b) amplifier gain as a function of the length for several pumping levels. (After Ref. [23]; ©1990 IEE.)

and the excited level lie within the same band for 1.48-μm pumping. It is difficult to achieve complete population inversion ($N_1 \approx 0$) under such conditions. It is nonetheless possible to realize $F_n < 3.5$ dB for pumping wavelengths near 1.46 μm.

Relatively low noise levels of EDFAs make them an ideal choice for WDM lightwave systems. In spite of low noise, the performance of long-haul fiber-optic communication systems employing multiple EDFAs is often limited by amplifier noise. The noise problem is particularly severe when the system operates in the anomalous-dispersion region of the fiber because a nonlinear phenomenon known as the modulation instability [25] enhances amplifier noise [26] and degrades the signal spectrum [27]. Amplifier noise also introduces a timing jitter and limits the performance of all lightwave systems [19].

3.1.5 Amplifier Design

The bandwidth of EDFAs is large enough that it can cover the entire C band extending from 1,525 to 1,565 nm for WDM applications. The main practical limitation of an EDFA stems from the spectral nonuniformity of the amplifier gain. Even though the gain spectrum of an EDFA is relatively broad, as seen in Figure 3.1, the gain is far from uniform (or flat) over a wide wavelength range. This problem becomes quite severe for long-haul systems employing a cascaded chain of EDFAs. The reason is that even a 0.2-dB gain difference grows to 20 dB over a chain of 100 in-line amplifiers, an unacceptable power variation range in practice. To amplify all channels by nearly the same amount, the double-peak nature of the EDFA gain spectrum forces one to pack all channels near the gain peak.

The entire bandwidth of 35–40 nm can be used if the gain spectrum is flattened by introducing wavelength-selective losses through an optical filter. The basic idea behind gain flattening is quite simple. If an optical filter whose transmission losses mimic the

3.1 Doped-Fiber Amplifiers

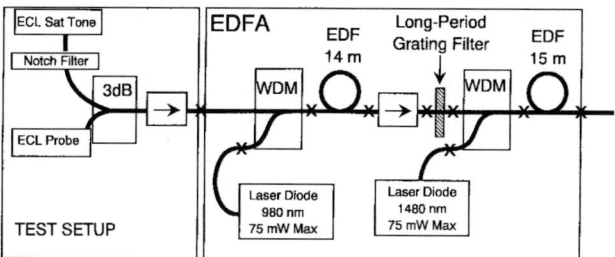

Figure 3.4: Schematic of an EDFA designed to provide uniform gain over the C band using an optical filter containing several long-period fiber gratings. The two-stage design helps to reduce the noise level. (After Ref. [29]; ©1997 IEEE.)

gain profile (high in the high-gain region and low in the low-gain region) is inserted after the doped fiber, the output power will become constant for all channels. Although the fabrication of such a filter is not simple, several gain-flattening techniques have been developed [5]. In a commonly used technique, long-period fiber gratings are used for flattening the gain profile and equalizing channel gains [28].

The gain-flattening techniques can be divided into active and passive categories. Most filter-based methods are passive in the sense that channel gains cannot be adjusted in a dynamic fashion. The location of the optical filter itself requires some thought because of high losses associated with it. Placing it before the amplifier increases the noise, while placing it after the amplifier reduces the output power. Often a two-stage configuration, as shown in Figure 3.4, is used. The second stage acts as a power amplifier while the noise figure is mostly determined by the first stage whose noise is relatively low because of its low gain. A combination of several long-period fiber gratings acting as the optical filter in the middle of two stages resulted by 1997 in an EDFA whose gain was flat to within 1 dB over the 40-nm bandwidth in the wavelength range of 1,530–1,570 nm [29].

Ideally, an optical amplifier should provide the same gain for all channels under all possible operating conditions. This is not the case in general. For instance, if the number of channels being transmitted changes, the gain of each channel will change since it depends on the total signal power (because of gain saturation). The active control of channel gains is thus desirable for WDM applications. Many techniques have been developed for this purpose. The most commonly used technique stabilizes the gain dynamically by incorporating within the amplifier a laser that operates outside the used bandwidth. Such devices are called gain-clamped EDFAs as their gain is clamped by a built-in laser [30]–[35].

Modern WDM lightwave systems use the so-called C and L bands simultaneously and need uniform amplifier gain over a bandwidth exceeding 70 nm. The use of the L band requires optical amplifiers capable of providing gain in the wavelength range 1,570–1,610 nm. It turns out that EDFAs can provide gain over this wavelength range if designed appropriately. An L-band EDFA requires long fiber lengths (>100 m) to keep the inversion level relatively low. Figure 3.5 shows a two-stage design for an L-band amplifier [36]. The first stage is pumped at 980 nm and acts as a traditional

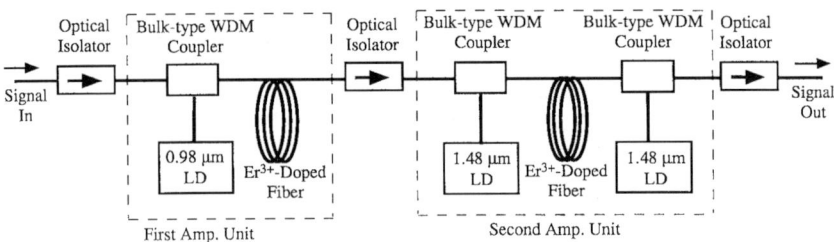

Figure 3.5: Schematic of an L-band EDFA providing uniform gain over the L band with a two-stage design. (After Ref. [36]; ©1999 IEEE.)

EDFA (fiber length 20–30 m). In contrast, the second stage has 200-m-long doped fiber and is pumped bidirectionally using 1,480-nm lasers. An optical isolator between the two stages passes the ASE from the first stage to the second stage (necessary for pumping the second stage), but blocks the backward-propagating ASE from entering the first stage. Such cascaded, two-stage amplifiers can provide flat gain over a wide bandwidth while maintaining a relatively low noise level. As early as 1996, flat gain to within 0.5 dB was realized over the wavelength range of 1,544–1,561 nm [37]. The second EDFA was codoped with ytterbium and phosphorus and was optimized such that it acted as a power amplifier. Since then, EDFAs providing flat gain over the entire C and L bands have been made [5]. As discussed in Section 3.2, Raman amplification can be used in any wavelength range. If Raman amplification is combined with one or two EDFAs, uniform gain can be realized over a 75-nm bandwidth covering the C and L bands [38].

A parallel configuration has also been developed for EDFAs capable of amplifying over the C and L bands simultaneously [39]. In this approach, the incoming WDM signal is split into two branches, which amplify the C-band and L-band signals separately using an optimized EDFA in each branch. The two-arm design has produced a relatively uniform gain of 24 dB over a bandwidth as large as 80 nm when pumped with 980-nm semiconductor lasers while maintaining a noise figure of about 6 dB [5]. The two-arm or two-stage amplifiers are complex devices and contain multiple components, such as optical filters and isolators, within them for optimizing the amplifier performance. An alternative approach to broadband EDFAs uses a *fluoride* fiber in place of silica fibers as the host medium in which erbium ions are doped. Gain flatness over a 76-nm bandwidth has been realized by doping a tellurite fiber with erbium ions [40]. Although such EDFAs are simpler in design compared with multistage amplifiers, they suffer from the splicing difficulties because of the use of nonsilica glasses.

The S band extending from 1,470 to 1,530 nm may also be used for some advanced lightwave systems. It turns out that erbium ions can provide gain even in this spectral band with suitable design modifications. However, the gain is relatively low even at pump-power levels approaching 100 mW. The gain is also quite nonuniform over the S band. In a 2003 experiment [41], it was possible to obtain nearly uniform 20-dB gain over a 30-nm bandwidth by employing 9 amplification stages, pumped using 9 lasers with power levels in the range 72–160 mW, but the noise figure exceeded 6 dB. For this reason, thulium-doped fibers are often used for S-band amplification [42]. Raman

3.2 Raman Amplifiers

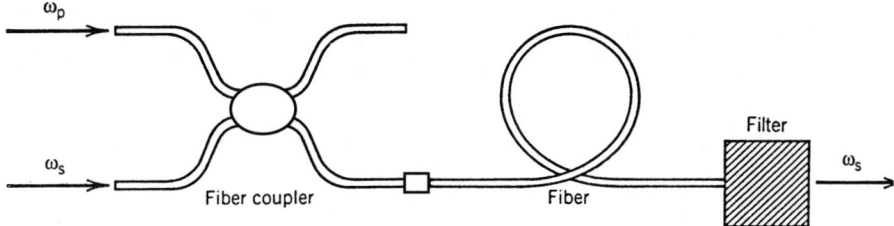

Figure 3.6: Schematic of a fiber-based Raman amplifier in the forward-pumping configuration.

amplifiers provide another alternative because they can work in any wavelength region. We focus on them in the next section.

3.2 Raman Amplifiers

A fiber-based Raman amplifier uses stimulated Raman scattering (SRS) occurring in silica fibers when an intense pump beam propagates through it [25]. As discussed in Section 1.6.4, SRS is normally considered a harmful process because it introduces crosstalk in WDM lightwave systems. However, the same process is useful for making Raman amplifiers, which are capable of providing large gain over a wide bandwidth in any spectral region. They require much longer fiber lengths (>1 km) compared with EDFAs. However, even this feature can be turned into an advantage by using the transmission fiber itself as a Raman-gain medium. Since fiber length then typically exceeds 50 km, this scheme is referred to as *distributed Raman amplification*.

3.2.1 Raman Gain and Bandwidth

SRS differs from the stimulated-emission process used for EDFAs in one fundamental aspect. Whereas incident signal photons in an EDFA stimulate emission of identical photons without losing their energy, in the case of SRS an incident pump photon gives up its energy to create another photon of reduced energy at a lower frequency; the remaining energy is absorbed by the medium in the form of molecular vibrations (see Figure 1.21). Similar to EDFAs, Raman amplifiers must be pumped optically to provide gain. Figure 3.6 shows how a fiber can be used as a Raman amplifier. The pump and signal beams at frequencies ω_p and ω_s are injected into the fiber through a fiber coupler. The energy is transferred from the pump to the signal through SRS as the two beams copropagate inside the fiber. The pump and signal counterpropagate in the backward-pumping configuration often used in practice.

The Raman-gain spectrum, $g_R(\Omega)$, where $\Omega = \omega_p - \omega_s$ is the Raman shift, has been measured for silica fibers [43]–[48]. In general, g_R depends on composition of the fiber core and can vary significantly with the use of different dopants. Figure 1.21 shows g_R for fused silica as a function of the Raman shift Ω at a pump wavelength $\lambda_p = 1$ μm. For other pump wavelengths, g_R can be obtained from the inverse dependence of g_R on λ_p (as gain is proportional to the energy of pump photons). The most significant

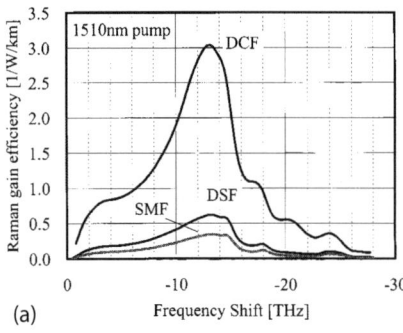 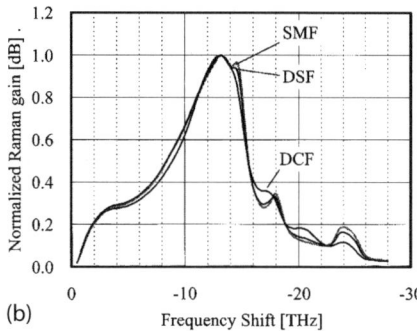

Figure 3.7: (a) Raman-gain efficiency (ratio g_R/A_{eff}) and (b) normalized gain profiles for standard (SMF), dispersion-shifted (DSF) and dispersion-compensating (DCF) fibers. (After Ref. [49]; ©2001 IEEE.)

feature of the Raman gain in silica fibers is that $g_R(\Omega)$ extends over a large frequency range (up to 40 THz) with a broad peak located near 13.2 THz. This behavior is due to the noncrystalline nature of silica glasses. In amorphous materials such as fused silica, molecular vibrational frequencies spread out into bands that overlap and create a continuum. As a result, in contrast to most molecular media for which the Raman gain occurs at specific well-defined frequencies, it extends continuously over a broad range in silica fibers. Optical fibers act as broadband Raman amplifiers because of this feature.

The amplification process is governed by Eqs. (1.6.23) and (1.6.24). The Raman-gain coefficient g_R can be related to the optical gain $g(z)$ as $g(z) = g_R I_p(z)$, where I_p is the pump intensity. In terms of the pump power P_p, the gain can be written as

$$g(\omega) = g_R(\omega)(P_p/A_{\text{eff}}), \qquad (3.2.1)$$

where A_{eff} is the effective core area of the pump beam inside the fiber. Since A_{eff} can vary considerably for different types of fibers, the ratio g_R/A_{eff} is a measure of the Raman-gain efficiency [49]. This ratio is plotted in Figure 3.7(a) for three different fibers. A dispersion-compensating fiber (DCF) can be 8 times more efficient than a standard silica fiber (SMF) because of its smaller core diameter. The frequency dependence of the Raman gain is almost the same for the three kinds of fibers as evident from the normalized gain spectra shown in Figure 3.7(b). The gain bandwidth Δv_g is about 6 THz if we define it as the FWHM of the dominant peak in Figure 3.7(b).

It is evident from Figure 3.7(b) that, when a pump beam is launched into the fiber together with a weak signal beam, it will be amplified because of the Raman gain as long as the frequency difference $\Omega = \omega_p - \omega_s$ lies within the bandwidth of the Raman-gain spectrum. The signal gain depends considerably on the frequency difference Ω and is maximum when the signal beam is downshifted from the pump frequency by 13.2 THz (about 100 nm in the 1.55-μm region). However, it is important to note that the Raman gain exists in all spectral regions, that is, optical fibers can be used to amplify any signal, provided an appropriate pump source is used. This remarkable feature of Raman amplifiers is quite different from EDFAs, which can amplify only

3.2 Raman Amplifiers

signals whose wavelength is close to the atomic transition wavelength occurring near 1.53 μm.

Similar to the case of EDFAs, the nonuniform nature of the Raman-gain spectrum in Figure 3.7(b) is of concern for WDM lightwave systems because different channels will be amplified by different amounts. However, this problem is easily solved in practice by using multiple pumps at slightly different wavelengths. Each pump provides nonuniform gain but the gain spectra associated with different pumps overlap partially. With a suitable choice of wavelengths and powers for each pump laser, it is possible to realize a nearly flat gain profile over a considerably wide wavelength range. We discuss the single-pump scheme first, as it allows us to introduce the basic ideas in a simple manner, and then focus on the multiple-pump configuration of Raman amplifiers.

3.2.2 Single-Pump Raman Amplification

Consider the simplest situation in which a single CW pump beam is launched into an optical fiber used to amplify a CW signal. Even in this case, the Raman-amplification process is governed by the following set of two coupled nonlinear equations:

$$\frac{dP_s}{dz} = \frac{g_R}{A_{\text{eff}}} P_p P_s - \alpha_s P_s, \tag{3.2.2}$$

$$\eta \frac{dP_p}{dz} = -\frac{\omega_p}{\omega_s} \frac{g_R}{A_{\text{eff}}} P_p P_s - \alpha_p P_p, \tag{3.2.3}$$

where α_s and α_p account for fiber losses at the signal and pump wavelengths, respectively. The parameter η takes values ± 1 depending on the pumping configuration; the minus sign should be used in the backward-pumping case. Equations (3.2.2) and (3.2.3) can be derived rigorously from Maxwell's equations. They can also be written phenomenologically by considering the processes through which photons appear in or disappear from each beam. The frequency ratio ω_p/ω_s appears in Eq. (3.2.3) because the pump and signal photons have different energies.

Equations (3.2.2) and (3.2.3) are not easy to solve analytically because of their nonlinear nature. In many practical situations, pump power is so large compared with the signal power that pump depletion can be neglected by setting $g_R = 0$ in Eq. (3.2.3), which is then easily solved. As an example, $P_p(z) = P_0 \exp(-\alpha_p z)$ in the forward-pumping case, where P_0 is the input pump power at $z = 0$. If we substitute this solution in Eq. (3.2.2), we obtain

$$\frac{dP_s}{dz} = \frac{g_R}{A_{\text{eff}}} P_0 \exp(-\alpha_p z) P_s - \alpha_s P_s. \tag{3.2.4}$$

This equation can be easily integrated, and the result is given as

$$P_s(L) = P_s(0) \exp(g_R P_0 L_{\text{eff}}/A_{\text{eff}} - \alpha_s L) \equiv G(L) P_s(0), \tag{3.2.5}$$

where G is the amplification factor, L is the amplifier length, and L_{eff} is the effective fiber length defined as

$$L_{\text{eff}} = [1 - \exp(-\alpha_p L)]/\alpha_p. \tag{3.2.6}$$

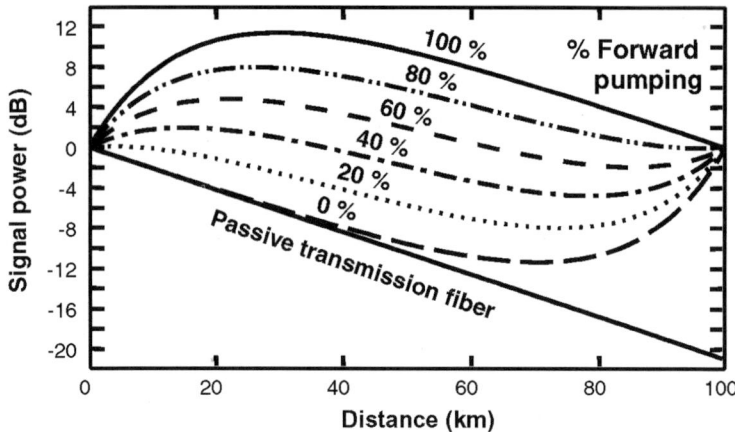

Figure 3.8: Evolution of signal power in a 100-km-long, bidirectionally pumped, Raman amplifier as the contribution of forward pumping is varied from 0 to 100%. The straight line shows for comparison the case of no Raman gain. (After Ref. [50]; ©2003 Springer.)

The solution (3.2.5) shows that, because of pump losses, the effective amplification length is reduced from L to L_{eff}.

The backward-pumping case can be considered in a similar fashion. In this case, Eq. (3.2.3) should be solved with $g_R = 0$ using the boundary condition $P_p(L) = P_0$, resulting in the solution $P_p(z) = P_0 \exp[-\alpha_p(L-z)]$. The integration of Eq. (3.2.2) yields the same solution given in Eq. (3.2.5), indicating that the amplified signal power remains the same in the backward-pumping case.

The case of bidirectional pumping is slightly more complicated because we now have two pump lasers at the two fiber ends. However, Eq. (3.2.2) can still be integrated to obtain

$$P_s(z) \equiv G(z)P_s(0) = P_s(0)\exp\left(\frac{g_R}{A_{\text{eff}}} \int_0^z P_p(z)\,dz - \alpha_s L\right), \quad (3.2.7)$$

where $P_p(z)$ is the total pump power at a distance z. For two pump lasers located at the left and right ends of the amplifier, $P_p(z)$ is given by

$$P_p(z) = P_0\{f_p \exp(-\alpha_p z) + (1-f_p)\exp[-\alpha_p(L-z)]\}, \quad (3.2.8)$$

where P_0 is the total power and f_p is the fraction of it used in the forward direction. Figure 3.8 shows how the signal power changes along a 100-km-long Raman amplifier as f_p varies in the range 0 to 1. In all cases, $g_R/A_{\text{eff}} = 0.7$ W^{-1}/km, $\alpha_s = 0.2$ dB/km, $\alpha_p = 0.25$ dB/km, and the total pump power is chosen such that the Raman gain is just sufficient to compensate for fiber losses, that is, $G(L) = 1$.

One may ask which pumping configuration is the best from a system standpoint. The answer is not so simple as it depends on many factors. Forward pumping is superior from the noise viewpoint. However, for a long-haul system limited by fiber nonlinearities, backward pumping may offer better performance because the signal power is the smallest throughout the link length in this case. The total accumulated nonlinear

3.2 Raman Amplifiers

phase shift induced by self-phase modulation (SPM) can be obtained from [25]

$$\phi_{NL} = \gamma \int_0^L P_s(z)\,dz = \gamma P_s(0) \int_0^L G(z)\,dz, \tag{3.2.9}$$

where $\gamma = 2\pi n_2/(\lambda_s A_{\text{eff}})$ is the nonlinear parameter responsible for SPM. The increase in the nonlinear phase shift occurring because of Raman amplification can be quantified through the ratio [50]

$$R_{NL} = \frac{\phi_{NL}(\text{pump on})}{\phi_{NL}(\text{pump off})} = L_{\text{eff}}^{-1} \int_0^L G(z)\,dz. \tag{3.2.10}$$

It turns out that this ratio is the smallest and the nonlinear effects are the least in the case of backward pumping and become enhanced by more than 10 dB when forward pumping is used.

The quantity $G(L)$ represents the net signal gain and can be even <1 (net loss) if the Raman gain is not sufficient to overcome fiber losses. It is useful to the introduce the concept of the on–off Raman gain using the definition

$$G_A = \frac{P_s(L) \text{ with pump on}}{P_s(L) \text{ with pump off}} = \exp(g_R P_0 L_{\text{eff}}/A_{\text{eff}}). \tag{3.2.11}$$

Clearly, G_A represents the total amplifier gain distributed over a length L_{eff}. If we use a typical value of $g_R/A_{\text{eff}} = 3$ W^{-1}/km for a DCF from Figure 3.7 together with $L_{\text{eff}} = 1$ km, the signal can be amplified by 20 dB for $P_0 \approx 1.5$ W. The required pump power can be reduced by using longer fibers.

The main drawback of Raman amplifiers from the standpoint of lightwave system applications is that they require a high-power CW laser for pumping. Most experiments performed in the 1980s in the 1.55-μm spectral region used tunable color-center lasers as a pump; such lasers are too bulky for telecommunication applications. For this reason, with the advent of EDFAs around 1989, Raman amplifiers were rarely used in the 1.55-μm wavelength region. The situation changed with the availability of compact high-power semiconductor and fiber lasers. Indeed, the development of Raman amplifiers underwent a virtual renaissance during the 1990s [51]–[61].

3.2.3 Multiple-Pump Raman Amplification

Starting in 1998, the use of multiple pumps for Raman amplification was pursued for developing broadband optical amplifiers required for WDM lightwave systems operating in the 1.55-μm region [55]–[61]. Massive WDM systems (100 or more channels) typically require optical amplifiers capable of providing uniform gain over a 70–80-nm wavelength range. In a simple approach, hybrid amplifiers made by combining erbium doping with Raman gain were used. In one implementation of this idea [38], nearly 80-nm bandwidth was realized by combining an EDFA with two Raman amplifiers, pumped simultaneously at three different wavelengths (1,471, 1,495, and 1,503 nm) using four pump modules, each module launching more than 150 mW of power into the fiber. The combined gain of 30 dB was nearly uniform over the wavelength range of 1.53–1.61 μm.

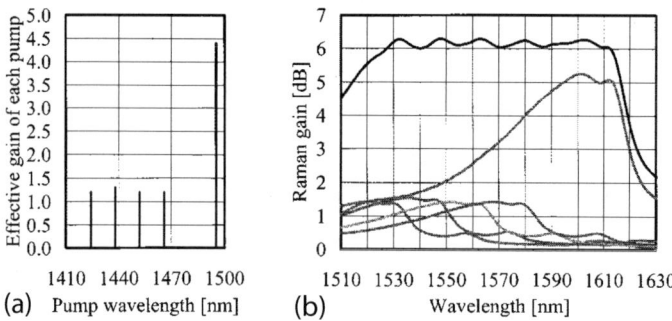

Figure 3.9: Composite Raman gain (b) for a Raman amplifier pumped with five lasers with different wavelengths and relative powers (a) chosen to provide nearly uniform gain over an 80-nm bandwidth. Other curves show the Raman gain provided by individual pumps. (After Ref. [49]; ©2001 IEEE.)

Broadband amplification over 80 nm or more can also be realized by using a pure Raman-amplification scheme [49]. In this case, a relatively long span (typically >5 km) of a fiber with a relatively narrow core (such as a DCF) is pumped using multiple pump lasers. Alternatively, one can use the transmission fiber itself as the Raman-gain medium. In the latter scheme, the entire long-haul fiber link is divided into multiple segments (60 to 100 km long), each one pumped backward using a pump module consisting of multiple pump lasers. The Raman gain accumulated over the entire segment length compensates for fiber losses of that segment in a distributed manner.

The basic idea behind multipump Raman amplifiers makes use of the property that the Raman gain exists at any wavelength as long as the pump wavelength is suitably chosen. Thus, even though the gain spectrum of a single pump is not very wide and is flat only over a few nanometers (see Figure 3.7), it can be broadened and flattened considerably by using several pumps of different wavelengths. Each pump creates the gain that mimics the spectrum shown in Figure 3.7(b), and superposition of several such spectra can create relatively flat gain over a wide spectral region when pump wavelengths and power levels are chosen judiciously. Figure 3.9(a) shows an example for the case in which five pump lasers operating at wavelengths of 1,420, 1,435, 1,450, 1,465, and 1,495 nm are used for pumping the Raman amplifier [49]. The individual pump powers are chosen to provide different gain spectra shown in part (b) such that the total Raman gain is nearly uniform over a 80-nm bandwidth (top trace).

Raman gain spectra with a bandwidth of more than 100 nm have been realized using multiple pump lasers [56]–[61]. In a 2000 demonstration of this technique, 100 WDM channels with 25-GHz channel spacing, each operating at a bit rate of 10 Gb/s, were transmitted over 320 km [59]. All channels were amplified simultaneously by pumping each 80-km fiber span in the backward direction using four semiconductor lasers. Such a distributed Raman amplifier provided 15-dB gain at a pump power of 450 mW. An undesirable feature of SRS is that the Raman gain is polarization-sensitive. In general, the gain is maximum when the signal and pump are polarized along the same direction

3.2 Raman Amplifiers

but is nearly zero when they are orthogonally polarized. The polarization problem can be solved by pumping a Raman amplifier such that two orthogonally polarized lasers are used at each pump wavelength or by depolarizing the output of each pump laser. It should be stressed that the state of polarization of the pump and signal fields changes randomly in any realistic fiber because of birefringence variations along the fiber length. This issue is discussed later in this section.

The design of broadband Raman amplifiers suitable for WDM applications requires consideration of several factors. The most important among them is the inclusion of pump–pump interactions. In general, multiple pump beams are also affected by the Raman gain, and some power from each short-wavelength pump is invariably transferred to each longer-wavelength pump. An appropriate model that includes pump interactions, Rayleigh backscattering, and spontaneous Raman scattering considers each frequency component separately and requires the solution of the following set of coupled equations [58]:

$$\frac{dP_f(\nu)}{dz} = \int_{\mu>\nu} g'(\mu,\nu)[P_f(\mu)+P_b(\mu)][P_f(\nu)+2h\nu n_{sp}(\mu-\nu)]d\mu$$
$$- \int_{\mu<\nu} g'(\nu,\mu)[P_f(\mu)+P_b(\mu)][P_f(\nu)+4h\nu n_{sp}(\nu-\mu)]d\mu,$$
$$- \alpha(\nu)P_f(\nu) + r_s P_b(\nu), \qquad (3.2.12)$$

where $g' = g_R/A_{\text{eff}}$, μ and ν denote optical frequencies, and the subscripts f and b denote forward- and backward-propagating waves, respectively. A similar equation holds for the backward propagating waves. In all equations, the parameter n_{sp} accounts for spontaneous Raman scattering that acts as a noise to the amplified signal and is defined as

$$n_{sp}(\Omega) = [1 - \exp(-\hbar\Omega/k_BT)]^{-1}, \qquad (3.2.13)$$

where $\Omega = |\mu - \nu|$ is the Raman shift and T denotes absolute temperature of the amplifier. In Eq. (3.2.12), the first and second terms account for the Raman-induced power transfer into and out of each frequency band. The factor of 4 in the second term results from the polarization-independent nature of spontaneous Raman scattering [49]. Fiber losses and Rayleigh backscattering are included through the last two terms and are governed by the parameters α and r_s, respectively. As discussed later in this section, double Rayleigh backscattering can be a limiting factor for Raman amplifiers.

To design broadband Raman amplifiers, the entire set of such equations is solved numerically to find the channel gains, and input pump powers are adjusted until the gain is nearly the same for all channels. Figure 3.10 shows an example of the gain spectrum measured for a Raman amplifier made by pumping a 25-km-long dispersion-shifted fiber with 12 diode lasers. The frequencies and powers of the pump lasers, required to achieve a nearly flat gain profile, are also shown. Notice that all power levels are under 100 mW. The amplifier provides about 10.5 dB gain over an 80-nm bandwidth with a ripple of less than 0.1 dB. Such an amplifier is suitable for dense WDM systems covering both the C and L bands. Several experiments have used broadband Raman amplifiers to demonstrate transmission over long distances at high bit rates. In one 3-Tb/s experiment, 77 channels, each operating at 42.7 Gb/s, were transmitted over 1,200 km by using the C and L bands simultaneously [60].

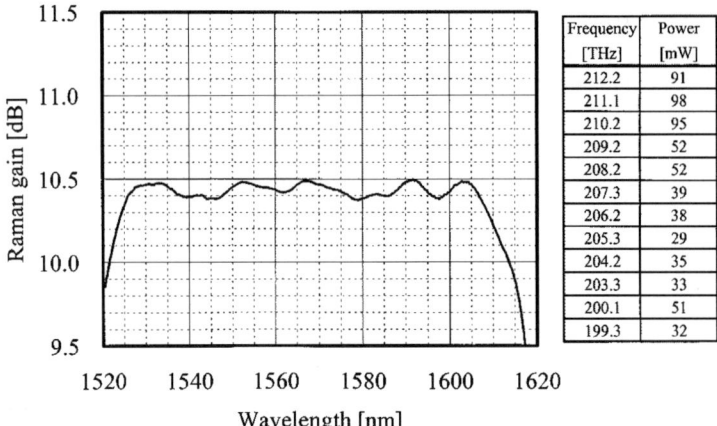

Figure 3.10: Measured Raman gain as a function of signal wavelength for a Raman amplifier pumped with 12 lasers. Pump frequencies and power levels needed to realize nearly flat gain over a 80-nm bandwidth used are shown in the table on the right. (After Ref. [49]; ©2001 IEEE.)

3.2.4 Performance Limiting Factors

The performance of modern Raman amplifiers is affected by several factors that need to be controlled. A few among them are spontaneous Raman scattering, double Rayleigh backscattering, polarization-mode dispersion (PMD), and pump-noise transfer. We briefly discuss their importance in this subsection.

Spontaneous Raman Scattering

Spontaneous Raman scattering adds to the amplified signal and appears as a noise because of random phases associated with all spontaneously generated photons. This noise mechanism is similar to the spontaneous emission that affects the performance of EDFAs except that, in the Raman case, it depends on the phonon population in the vibrational state, which in turn depends on temperature of the Raman amplifier. On a more fundamental level, one should consider the evolution of signal with the noise added by spontaneous Raman scattering by solving

$$\frac{dA_s}{dz} = \frac{g_R}{2A_{\text{eff}}} P_p(z) A_s - \frac{\alpha_s}{2} A_s + f_n(z,t), \quad (3.2.14)$$

where A_s is the signal field defined such that $P_s = |A_s|^2$, P_p is the pump power, and the Langevin noise source $f_n(z,t)$ takes into account the noise added through spontaneous Raman scattering. Since each scattering event is independent of others, this noise can be modeled as a Gaussian stochastic process for which $\langle f_n(z,t) \rangle = 0$ but the second moment is given by

$$\langle f_n(z,t) f_n(z',t') \rangle = n_{\text{sp}} h \nu_0 g_R P_p(z) \delta(z-z') \delta(t-t'), \quad (3.2.15)$$

3.2 Raman Amplifiers

where n_{sp} is the spontaneous-scattering factor given in Eq. (3.2.12) and $h\nu_0$ is the average photon energy. The two delta functions ensure that all spontaneous events are independent of each other.

Equation (3.2.14) can be easily integrated to obtain $A_s(L) = \sqrt{G(L)}A_s(0) + A_{sp}$, where $G(L)$ is the amplification factor defined earlier in Eq. (3.2.5) and the contribution of spontaneous Raman scattering is given by

$$A_{sp} = \sqrt{G(L)} \int_0^L \frac{f_n(z,t)}{\sqrt{G(z)}} dz, \quad G(z) = \exp\left(\int_0^z [g_R P_p(z') - \alpha_s] dz'\right). \quad (3.2.16)$$

It is clear from this equation that the spontaneous contribution added to the signal depends on the distributed nature of the Raman gain. We can calculate the power of this contribution using $P_{sp} = \langle|A_{sp}|^2\rangle$ together with Eq. (3.2.15). The result is found to be

$$P_{sp} = n_{sp} h\nu_0 g_R B_{opt} G(L) \int_0^L \frac{P_p(z)}{G(z)} dz, \quad (3.2.17)$$

where B_{opt} is the bandwidth of the Raman amplifier or of the optical filter, if such a filter is used to reduce the noise. The total noise power is higher by factor of 2 when both polarization components are considered.

The noise figure can now be calculated following the procedure used for EDFAs and is found to be

$$F_n = \frac{P_{sp}}{Gh\nu_0 \Delta f} = n_{sp} g_R \frac{B_{opt}}{\Delta f} \int_0^L \frac{P_p(z)}{G(z)} dz. \quad (3.2.18)$$

This equation shows that the noise figure of a Raman amplifier depends not only on the optical and electrical bandwidths but also on the pumping scheme. It is quite useful for lumped Raman amplifiers for which fiber length is ~ 1 km and the net signal gain exceeds 10 dB. In the case of distributed amplification, the length of the fiber section typically exceeds 50 km, and pumping is such that $G(z) < 1$ throughout the fiber length. In this case, F_n predicted by Eq. (3.2.18) can be very large. It is common to introduce the concept of an *effective noise figure* using the definition $F_{eff} = F_n \exp(-\alpha_s L)$. Physically, this concept can be understood by noting that a passive fiber reduces the SNR of an optical signal by a factor of $\exp(-\alpha_s L)$ because of the reduction in signal power [61]. To find the effective noise figure of a Raman amplifier, one should remove the contribution of the passive fiber.

It should be stressed that F_{eff} can be less than 1 (or negative on the decibel scale). In fact, it is this feature of distributed Raman amplification that makes it so attractive for long-haul WDM lightwave systems. Physically speaking, the distributed gain counteracts fiber losses in the transmission fiber itself and results in a SNR that is improved, compared with the case in which losses are compensated at the end of fiber using lumped amplifiers. The forward pumping results in even less noise because Raman gain is then concentrated toward the input end of the fiber.

Rayleigh Backscattering

The phenomenon that limits the performance of distributed Raman amplifiers most turns out to be Rayleigh backscattering [62]–[67]. As discussed in Section 1.4.3,

Rayleigh scattering occurs in all fibers and is the fundamental loss mechanism for them. Although most of the scattered light escapes through the cladding, a part of backscattered light can couple into the core mode supported by a single-mode fiber. Normally, such Rayleigh backscattered noise is negligible because its power level is $\sim 10^{-4}$ or less compared with the forward propagating signal. However, it can be amplified over long lengths in fibers with distributed Raman gain.

Rayleigh backscattering affects the performance of Raman amplifiers in two ways. First, a part of backward propagating spontaneous noise can appear in the forward direction, enhancing the overall noise. This is a relatively minor problem for Raman amplifiers. Second, *double Rayleigh backscattering* of the signal can create a crosstalk component in the forward direction that has nearly the same spectral range as the signal (in-band crosstalk). It is this Rayleigh-induced noise, amplified by the distributed Raman gain, that becomes the major source of power penalty in Raman-amplified lightwave systems. Physically speaking, density fluctuations inherent in any optical glass reflect a small portion of the signal through Rayleigh backscattering. This backward propagating field is reflected a second time by density fluctuations at a different location, and thus ends up propagating in the same direction as the signal. The location of the two reflections can vary over the entire length of the fiber, and the total noise field is obtained by summing over all possible paths. For this reason, the Rayleigh noise is also said to originate from *multiple-path interference* [50].

Even though the statistical properties of Rayleigh-induced noise require a complicated calculation, it is relatively easy to calculate the fraction of *average* signal power that ends up propagating in the forward direction after double Rayleigh backscattering. In a simple approach, one supplements the signal equation (3.2.4) with two more equations of the form

$$-\frac{dP_1}{dz} = [g_R P_p(z) - \alpha_s]P_1 + r_s P_s, \qquad (3.2.19)$$

$$\frac{dP_2}{dz} = [g_R P_p(z) - \alpha_s]P_2 + r_s P_1, \qquad (3.2.20)$$

where P_1 and P_2 represent the average power levels of the noise components created through single and double Rayleigh backscattering, respectively. The three equations can be integrated over the fiber length to find the fraction f_{DRS} that ends up coming out with the signal at the output end at $z = L$. The result is found to be [64]

$$f_{\text{DRS}} = \frac{P_2(L)}{P_s(L)} = r_s^2 \int_0^L dz\, G^{-2}(z) \int_z^L G^2(z')\, dz'. \qquad (3.2.21)$$

Figure 3.11 shows how f_{DRS} increases with the on–off Raman gain G_A using $r_s^* = 10^{-4}$, $g_R = 0.7$ W^{-1}/km, $\alpha_s = 0.2$ dB/km, $\alpha_p = 0.25$ dB/km, and backward pumping for a 100-km-long Raman amplifier. The crosstalk begins to exceed the -35-dB level for the 20-dB Raman gain needed for compensating fiber losses. Since this crosstalk accumulates over multiple amplifiers, it can lead to large power penalties in long-haul lightwave systems.

3.2 Raman Amplifiers

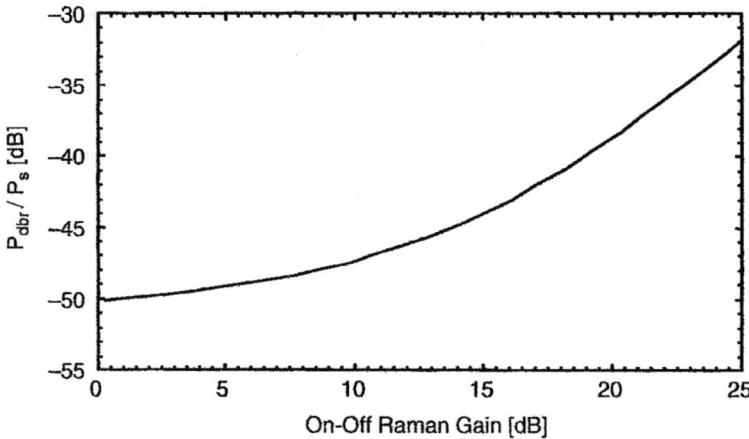

Figure 3.11: Fraction of signal power that is converted to noise by double Rayleigh backscattering plotted as a function of Raman gain for a 100-km-long backward pumped distributed Raman amplifier. (After Ref. [61]; ©2002 Elsevier.)

Polarization Effects

As mentioned earlier, the Raman-gain coefficient g_R is maximum when pump and signal are copolarized and nearly vanishes when they are orthogonally polarized, in contrast with EDFAs that provide the same gain for any state of polarization (SOP). This feature creates a problem for most systems in which the signal polarization does not remain fixed. The problem can be solved to a large extent by using orthogonally polarized pumps at each wavelength so that each pump amplifies the signal in such a way that the total gain is independent of the signal SOP.

However, this simple solution neglects the PMD effects in fibers [68]–[70]. In practice, birefringence fluctuations randomize the SOP of all fields propagating inside the fiber. Of even more concern is the fact that the PMD level can fluctuates with time, resulting in gain fluctuations. Indeed, the effects of PMD on Raman amplification have been observed in several experiments [71], [72]. A vector theory of Raman amplification capable of including the PMD-induced random evolution of the pump and signal SOPs has been developed to study the PMD effects [73]. This theory has been used to find the statistics of polarization-dependent gain (PDG) and its relationship with the operating parameters of Raman amplifiers [74].

The concept of PDG is important because the input SOP of the signal is unknown in any long-haul lightwave system when the signal arrives at the Raman amplifier. The PDG is defined as the difference between the maximum and minimum values of G while the SOP of the input signal is varied over the entire range of possible values. The gain difference $\Delta = G_{\max} - G_{\min}$ is itself random because both $G_{\max}$ and $G_{\min}$ are random. It is useful to know the statistics of Δ and its relationship to the operating parameters of a Raman amplifier because they can identify the conditions under which PDG may be reduced to acceptable low levels.

The probability density function of PDG, $p(\Delta)$, has been calculated [74] and is

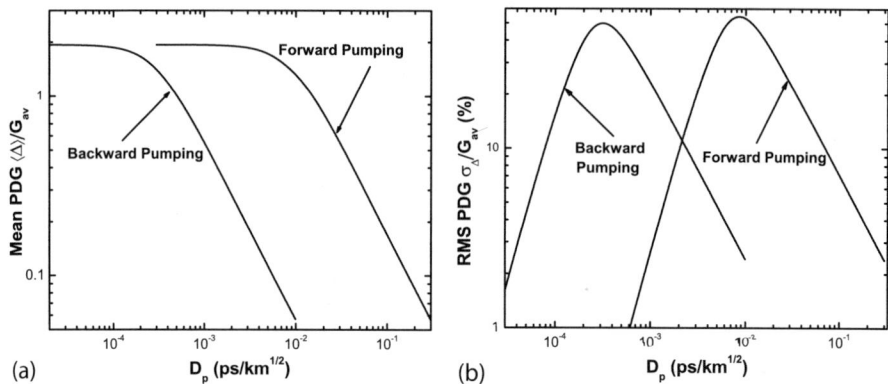

Figure 3.12: (a) Mean PDG and (b) standard deviation σ_Δ (both normalized to the average gain) as a function of PMD parameter under forward and backward pumping conditions.

found to depend on the PMD parameter D_p of the fiber used to make the Raman amplifier. In the limit $D_p \to 0$, $p(\Delta)$ becomes a delta function located at the maximum gain difference (almost $2G_{av}$) as little gain exists for the orthogonally polarized signal. As D_p increases, $p(\Delta)$ broadens quickly because PMD changes the signal SOP randomly. If D_p is relatively small, the diffusion length L_d is larger than or comparable to the effective fiber length L_{eff}, and $p(\Delta)$ remains centered at almost the same location but broadens because of large fluctuations. Its shape mimics a Gaussian distribution. When D_p is large enough that $L_{eff} \gg L_d$, $p(\Delta)$ becomes Maxwellian and its peak shifts to smaller values. This behavior has been observed experimentally [71].

The mean value of PDG, $\langle\Delta\rangle$, and the variance of PDG fluctuations, $\sigma_\Delta^2 = \langle\Delta^2\rangle - \langle\Delta\rangle^2$, can be calculated using the PDG distribution. Figures 3.12(a) and (b) shows how these two quantities vary with the PMD parameter for a Raman amplifier. Both $\langle\Delta\rangle$ and σ_Δ are normalized to the average gain G_{av}. As expected, the mean PDG decreases monotonically as D_p increases. Note, however, that $\langle\Delta\rangle$ can be as large as 30% of the average gain for $D_p = 0.05$ ps/$\sqrt{\text{km}}$ in the case of forward pumping and it decreases slowly with D_p after that, reaching a value of 8% for $D_p = 0.2$ ps/$\sqrt{\text{km}}$. In the case of backward pumping, the behavior is similar but the curve shifts to a value of D_p smaller by about a factor of 30. As a result, $\langle\Delta\rangle$ is quite small for typical PMD values associated with modern silica fibers.

As seen in Figure 3.12(b), the level of PDG fluctuations increases rapidly as D_p becomes nonzero, peaks to a value close to 56% of G_{av} for D_p near 0.01 ps/$\sqrt{\text{km}}$ in the case of forward pumping, and then begins to decrease [71]. A similar behavior occurs for backward pumping. The D_p^{-1} dependence of $\langle\Delta\rangle$ and σ_Δ can be deduced analytically in the limit $L_{eff} \gg L_d$. In this limit, the PDG distribution $p(\Delta)$ becomes approximately Maxwellian, and the average and standard deviation of PDG in the case of forward pumping are given by [74]

$$\langle\Delta\rangle \approx \frac{9.8 g_R P_0}{D_p |\Omega_R|}\sqrt{L_{eff}(1-\alpha_p L_{eff}/2)}, \qquad \sigma_\Delta = \sqrt{(3\pi/8 - 1)}\langle\Delta\rangle. \qquad (3.2.22)$$

3.2 Raman Amplifiers

The same equations hold in the case of backward pumping except that $|\Omega_R| = \omega_p - \omega_s$ should be replaced with $\omega_p + \omega_s$. It is this feature that makes PDG negligible for a Raman amplifier when they are pumped backward. Physically, the relative SOP of pump and signal changes much more rapidly when the two counterpropagate, thereby producing an averaging effect.

The PMD effects can be reduced in practice by employing the technique of polarization scrambling [49]. In this technique, the pump SOP is changed randomly so that the signal experiences different local gain in different parts of the fiber, resulting effectively in an average gain that is independent of the signal SOP. The vector theory of Raman amplification shows that Eq. (3.2.22) remains valid, provided we replace the pump power P_0 with $d_p P_0$, where d_p is the degree of polarization of the pump. As a result, both $\langle \Delta \rangle$ and σ_Δ are reduced by d_p and can be made negligible by reducing the degree of polarization to very low levels.

Pump-Noise Transfer

As discussed in Chapter 5, all semiconductor lasers exhibit power fluctuations. As seen from Eq. (3.2.11), the gain of a Raman amplifier depends on the pump power exponentially. It is intuitively expected from this equation that any fluctuation in the pump power would be magnified and result in even larger fluctuations in the amplified signal power. This is indeed the case, as also observed experimentally, although details of the noise transfer depend on many factors such as the pumping scheme and the dispersion characteristics of the fiber used for making the Raman amplifier [75].

Power fluctuations of a semiconductor laser are quantified through a frequency-dependent quantity called the relative intensity noise (RIN). In essence, RIN is a measure of the noise-power spectrum. When the RIN of the amplified signal is calculated, it is found that it is enhanced by a factor r_n in a certain frequency range when compared with the RIN of the pump laser. Both the enhancement factor and the frequency range depend on the pumping scheme and the dispersion parameter D of the fiber. In general, the enhancement is negligible for the backward pumping configuration and for large values of D [75]. This feature can be understood physically as follows. The distributed Raman gain builds up as the signal propagates inside the fiber. In the case of forward pumping and low dispersion, pump and signal travel at nearly the same speed. As a result, any fluctuation in the pump power stays in the same temporal window of the signal. In contrast, when dispersion is large, the signal moves out of the temporal window associated with the fluctuation, and sees a somewhat averaged gain. The averaging is much stronger in the case of backward pumping because the relative speed is extremely large (twice that of the signal group velocity). In this configuration, the effects of pump-power fluctuations are smoothed out so much that almost no RIN enhancement occurs.

Figure 3.13 shows, as an example, the measured and calculated values of r_n for an 81-km-long Raman amplifier made using a fiber with $D = 2.3$ ps/(km-nm). In the case of forward pumping, RIN is enhanced almost by a factor of 10 in the frequency range 0–10 MHz because of the relatively low value of D. However, when backward pumping is used, RIN is actually reduced at all frequencies except for frequencies below 2 kHz. The backward pumping is often used in practice because it helps with

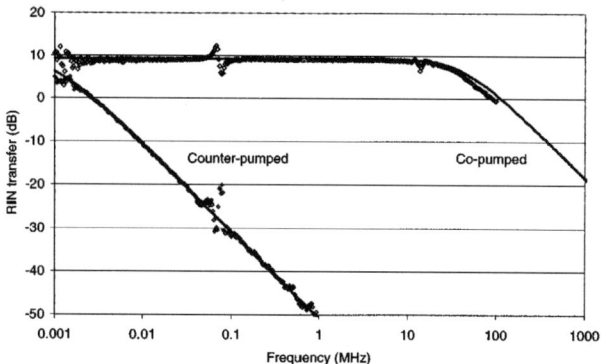

Figure 3.13: Measured (symbols) and calculated (solid lines) values of the RIN-enhancement factor as a function of noise frequency in the two pumping configurations for a 81-km-long distributed Raman amplifier. (After Ref. [75]; ©IEEE 2001.)

both the PMD and RIN problems. Forward pumping can be employed for Raman amplifiers when fiber dispersion is made large and lasers with low RIN are used to reduce the RIN enhancement and polarization scrambling is used to avoid the PMD-induced degradation.

3.3 Parametric Amplifiers

Similar to the cases of Raman gain induced by stimulated Raman scattering, the nonlinear phenomenon of four-wave mixing (FWM) leads to parametric gain in optical fibers, which can be exploited for making parametric amplifiers, phase conjugators, and wavelength converters. Such devices have attracted considerable attention in recent years because of their potential applications in lightwave systems [76]–[84]. This section describes the main characteristics of fiber-based parametric amplifiers.

3.3.1 FWM Interaction

As discussed in Section 1.6.3, FWM requires phase-matched interaction among four copropagating optical fields inside optical fibers. It is useful to write the total field as

$$\tilde{\mathbf{E}}(\mathbf{r},t) = \text{Re}[\hat{e}_j \sum_{j=1}^{4} A_j(z,t) F_j(x,y) \exp(i\beta_j z - i\omega_j t)], \qquad (3.3.1)$$

where ω_j is the carrier frequency, β_j is the propagation constant, $\hat{e}_j$ is the polarization direction, $F_j(x,y)$ is the modal distribution, and $A_j(z,t)$ is the slowly varying envelope amplitude for the four waves whose frequencies satisfy the FWM condition $\omega_1 + \omega_2 = \omega_3 + \omega_4$. We can use Eq. (3.3.1) in the Maxwell equations to study the FWM process but we need to make several approximations to simplify the analysis. First, we ignore the polarization effects and assume that all four field are linearly polarized along the

3.3 Parametric Amplifiers

same direction that does not change during propagation. Second, we assume that four waves have nearly the same frequencies so that $F_j(x,y)$ can be replaced with a single quantity $F(x,y)$. Third, we assume that the pump waves 1 and 2 are so intense that they remain nearly undepleted at the output end of the fiber. Fourth, we focus on the CW case and assume that A_j is time-independent for $j = 1\text{--}4$. With these simplifications, the FWM process is governed by the following set of two equations satisfied by the signal field A_3 and the idler field A_4 [25]:

$$\frac{dA_3}{dz} = 2i\gamma[(P_1 + P_2)A_3 + \sqrt{P_1 P_2}e^{-i\theta}A_4^*], \tag{3.3.2}$$

$$\frac{dA_4^*}{dz} = -2i\gamma[(P_1 + P_2)A_4^* + \sqrt{P_1 P_2}e^{i\theta}A_3], \tag{3.3.3}$$

where $P_1 = |A_1|^2$ and $P_2 = |A_2|^2$ are the pump powers, and

$$\theta = [\Delta\beta - 3\gamma(P_1 + P_2)]z \tag{3.3.4}$$

is the phase-mismatch parameter with

$$\Delta\beta = \beta_3 + \beta_4 - \beta_1 - \beta_2, \tag{3.3.5}$$

and $\beta_j = \tilde{n}_j \omega_j / c$. The effective mode indices $\tilde{n}_j$ for the four waves ($j = 1\text{--}4$) are generally different because of fiber dispersion. The nonlinear parameter γ is responsible not only for SPM and XPM but also leads to FWM.

To solve Eqs. (3.3.2) and (3.3.3), we eliminate first the XPM terms through the transformation

$$B_j = A_j \exp[-2i\gamma(P_1 + P_2)z], \quad (j = 3, 4), \tag{3.3.6}$$

and obtain a simpler set of two equations:

$$\frac{dB_3}{dz} = 2i\gamma\sqrt{P_1 P_2}\exp(-i\kappa z)B_4^*, \tag{3.3.7}$$

$$\frac{dB_4^*}{dz} = -2i\gamma\sqrt{P_1 P_2}\exp(i\kappa z)B_3, \tag{3.3.8}$$

where the net phase mismatch is given by

$$\kappa = \Delta\beta + \gamma(P_1 + P_2). \tag{3.3.9}$$

The general solution of these equations is of the form [85]

$$B_3(z) = (a_3 e^{gz} + b_3 e^{-gz})\exp(-i\kappa z/2), \tag{3.3.10}$$
$$B_4^*(z) = (a_4 e^{gz} + b_4 e^{-gz})\exp(i\kappa z/2), \tag{3.3.11}$$

where a_3, b_3, a_4, and b_4 are determined from the boundary conditions. The parametric gain g depends on the pump power and is defined as

$$g = \sqrt{(\gamma P_0 r)^2 - (\kappa/2)^2}, \tag{3.3.12}$$

where we have introduced the parameters r and P_0 as

$$r = 2(P_1 P_2)^{1/2}/P_0, \qquad P_0 = P_1 + P_2. \tag{3.3.13}$$

The derivation of the parametric gain has assumed that the two pump waves are distinct. If the pump fields were indistinguishable on the basis of their frequency or polarization, the FWM process is said to be degenerate because a single pump provides both photons for creating a pair of signal and idler photons. The parametric gain in the degenerate case is still given by Eq. (3.3.12) if we choose $P_1 = P_2 = P_0$ and set $r = 1$. The maximum gain, $g_{\max} = \gamma P_0$, occurs when $\kappa = 0$, or at $\Delta\beta = -2\gamma P_0$. The shift of the gain peak from the phase-matching condition $\Delta\beta = 0$ is due to the contribution of SPM and XPM to the phase mismatch.

It is interesting to compare the peak value of the parametric gain with that of the Raman gain seen in Figure 1.21. From Eq. (3.3.12) the maximum parametric gain is given by (with $r = 1$)

$$g_{\max} = \gamma P_0 = g_P(P_0/A_{\text{eff}}), \tag{3.3.14}$$

where g_P is defined as $g_P = 2\pi n_2'/\lambda_1$ at the pump wavelength λ_1. Using $\lambda_1 = 1$ μm and $n_2 \approx 2.6 \times 10^{-20}$ m^2/W, we obtain $g_P \approx 2 \times 10^{-13}$ m/W. This value is larger by about a factor of 2 compared with g_R. As a result, the pump power needed for a fixed amplifier length is expected to be lower for a parametric amplifier than that of a Raman amplifier if phase matching is achieved. In practice, however, Raman amplifiers use less pump power because they employ long fiber lengths ($\sim$10 km). Such long lengths cannot be used for parametric amplifiers because it is difficult to maintain phase matching. It is useful to define a length scale, known as the coherence length, as $L_{\text{coh}} = 2\pi/|\Delta\kappa|$, where $\Delta\kappa$ is the maximum value of the wave-vector mismatch that can be tolerated. Significant FWM occurs if $L < L_{\text{coh}}$. In practice, it is common to keep fiber lengths below 1 km and still maintain pump power levels below 1 W by using the so-called highly nonlinear fibers for which γ exceeds 10 W^{-1}/km.

3.3.2 Single-Pump Parametric Amplifiers

We first focus on the degenerate case in which a single pump at frequency ω_p is used to pump a parametric amplifier. Both the amplifier gain and the gain bandwidth can be deduced from the approximate analytic solution given in Eqs. (3.3.10) and (3.3.11). The constants a_3, b_3, a_4, and b_4 in these equations are determined from the boundary conditions. If we assume that idler field $A_4(0) = 0$ at the input end located at $z = 0$, the signal and idler powers at the fiber output ($z = L$) are given by [85]

$$P_3(L) = P_3(0)[1 + (1 + \kappa^2/4g^2)\sinh^2(gL)], \tag{3.3.15}$$
$$P_4(L) = P_3(0)(1 + \kappa^2/4g^2)\sinh^2(gL), \tag{3.3.16}$$

where the parametric gain g is given by Eq. (3.3.12). Parametric amplifiers differ from any other amplifier in one important aspect: As the signal at the frequency ω_s is amplified, at the same time, a new field wave is generated at the idler $2\omega_p - \omega_s$. Moreover, it follows from Eq. (3.3.11) that the amplitude of this new field is related to the complex

3.3 Parametric Amplifiers

conjugate of the signal field or, equivalently, its spectrum is inverted with respect to the signal spectrum. This property is useful for dispersion compensation.

The amplification factor is obtained from Eq. (3.3.15) and can be written using Eq. (3.3.12) in the form

$$G_p = \frac{P_3(L)}{P_3(0)} = 1 + \frac{\sinh^2(gL)}{(gL_{NL})^2}, \qquad (3.3.17)$$

where the nonlinear length is defined as $L_{NL} = (\gamma P_0)^{-1}$, P_0 is the pump power, and g is given by Eq. (3.3.12) with $r = 1$. The parametric gain depends on the phase mismatch κ and can become quite small if the phase-matching condition is not satisfied. On the other hand, when phase matching is perfect ($\kappa = 0$) and $gL \gg 1$, the amplifier gain increases exponentially with P_0 as

$$G_p \approx \tfrac{1}{4} \exp(2\gamma P_0 L). \qquad (3.3.18)$$

In terms of the nonlinear length, the gain increases as $\exp(2L/L_{NL})$ for a parametric amplifier as long as gain saturation and pump depletion remain negligible. Notice that the amplifier length must be longer than the nonlinear length L_{NL} for realizing significant gain and that G_p exceeds 28 dB for $L = 4L_{NL}$. Since $L_{NL} = 100$ m for a fiber with $\gamma = 10$ W^{-1}/km and pumped with $P_0 = 1$ W, a 0.5-km-long parametric amplifier can provide more than 30-dB gain.

The gain spectrum can be obtained from Eq. (3.3.17) by plotting G_p as a function of pump-signal detuning defined as $\Omega_s = \omega_s - \omega_p$. Assuming that the pump is located in the anomalous-dispersion regime such that its wavelength is close to the zero-dispersion wavelength of the fiber, and expanding $\beta(\omega)$ in a Taylor series around the pump frequency ω_p, we can easily show that $\Delta\beta \approx |\beta_2|\Omega_s^2$, where β_2 is the GVD parameter at the pump frequency. Figure 3.14 shows the gain as a function of signal detuning from the pump wavelength at three pump-power levels for a 500-m-long parametric amplifier designed using a fiber with $\gamma = 10$ W^{-1}/km and $\beta_2 = -0.5$ ps^2/km. As seen there, both the peak gain and the amplifier bandwidth increase as pump power is increased. The peak value of the gain is close to 38 dB at a 1-W pump level and occurs when the signal is detuned by 1 THz (about 8 nm) from the pump wavelength. The gain is smaller for smaller detunings because of the phase mismatch. The shape of the gain spectrum is related to the nonlinear phenomenon of modulation instability [25], which generates spectral side bands on each side of the pump at a location such that the phase-matching condition is exactly satisfied ($\kappa = 0$).

The amplifier bandwidth $\Delta\Omega_A$ can be estimated from Eq. (3.3.17) and depends on both the fiber length L and the pump power P_0. A convenient definition of $\Delta\Omega_A$ corresponds to the frequency at which the mismatch $\kappa = 2\pi/L$ [85], where κ is given in Eq. (3.3.9). If we use $\Delta\beta \approx |\beta_2|\Omega_s^2$ in Eq. (3.3.17), the amplifier bandwidth is found to be [85]

$$\Delta\Omega_A = \frac{1}{|\beta_2|\Omega_s} \left[\left(\frac{\pi}{L}\right)^2 + (\gamma P_0)^2 \right]^{1/2}. \qquad (3.3.19)$$

At high pump powers the bandwidth can be approximated as

$$\Delta\Omega_A \approx \gamma P_0/|\beta_2\Omega_s| \equiv (|\beta_2|\Omega_s L_{NL})^{-1}, \qquad (3.3.20)$$

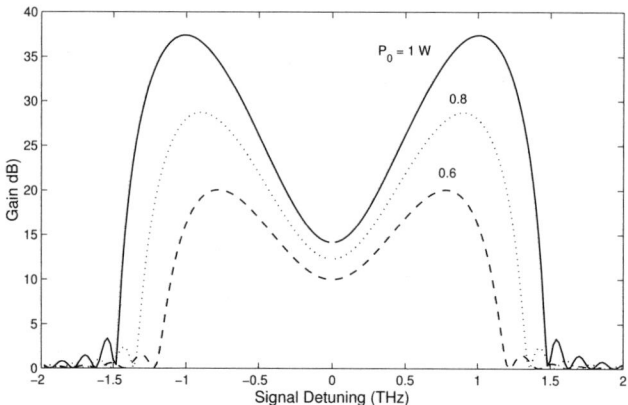

Figure 3.14: Theoretical gain spectra of a 0.5-km-long parametric amplifier at several pump power levels.

As a rough estimate, the bandwidth is ~ 1 THz for $|\beta_2| = 1$ ps^2/km, $\Omega_s/2\pi = 1$ THz, and $L_{\rm NL} = 100$ m. The bandwidth of a single-pump parametric amplifier is typically smaller than that of Raman amplifiers. It can be increased considerably by locating the pump close to the zero-dispersion wavelength of the fiber so that $|\beta_2|$ is reduced. In the limit β_2 approaches zero, the bandwidth is determined by the fourth-order dispersion parameter β_4 and can exceed 5 THz or 40 nm.

Single-pump parametric amplifiers are indeed designed with their pump wavelength λ_p selected close to the zero-dispersion wavelength λ_0 of the fiber. In one implementation [77], a DFB semiconductor laser operating near $\lambda_p \approx 1.54$ μm was used for pumping while a tunable external-cavity semiconductor laser provided the signal. The length of the dispersion-shifted fiber ($\lambda_0 = 1.5393$) used for parametric amplification was 200 m. The amplifier bandwidth changed considerably as pump wavelength was varied in the vicinity of λ_0 and was larger when pump wavelength was detuned such that $\lambda_p - \lambda_0 = 0.8$ nm. These results are readily understood when the effects of fourth-order dispersion are included in the analysis. The gain was relatively low in this experiment. However, the use of dispersion management and highly nonlinear fibers provides higher gain and larger bandwidths for parametric amplifiers [79].

The experimental results on single-pump parametric amplifiers agree with the preceding simple FWM theory. Figure 3.15 shows the measured gain spectra at three power levels for a parameter amplifier made using 500 m of highly nonlinear fiber, for which γ was measured to be 11 W^{-1}/km. Sold curves show in each case the theoretical prediction based on Eq. (3.3.17). A distributed feedback (DFB) laser whose wavelength exceeded the zero-dispersion wavelength by 2 nm was used for pumping. Two EDFAs were used to boost the pump power to a level as high as 2 W. It was necessary to broaden the pump spectrum from 10 MHz to >1 GHz to suppress the onset of stimulated Brillouin scattering. Both the peak gain and the amplifier bandwidth increased as pump power was increased. At a pump power level of 32.2 dBm, the maximum net gain was measured to be 39 dB, while the bandwidth exceeded 35 nm. The theoretical

3.3 Parametric Amplifiers

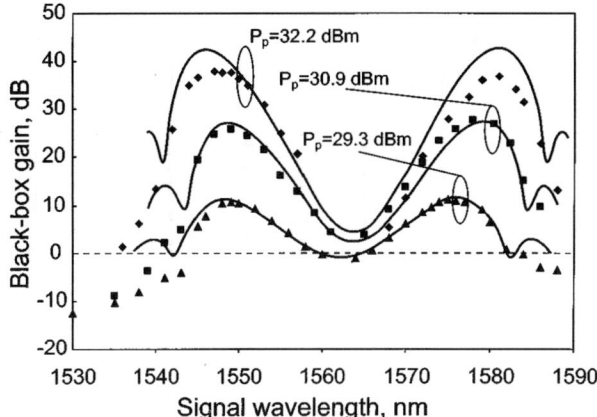

Figure 3.15: Measured gain spectra (symbols) and theoretical predictions (solid lines) at three pump power levels. A 0.5-km-long highly nonlinear fiber with $\gamma = 11$ W^{-1}/km was used for the experiment. (After Ref. [80]; ©IEEE 2002.)

predictions are in excellent agreement with the measured data except in the vicinity of the gain peak. This discrepancy is attributed to pump depletion that is not included in the simple FWM theory used here.

3.3.3 Dual-Pump Parametric Amplifiers

A fundamental shortcoming of single-pump parametric amplifiers is obvious from Figure 3.15. As seen there, the gain is far from being uniform over the entire bandwidth. In fact, only a small portion of the gain spectrum near the two gain peaks can be used in practice. This problem can be solved by using two pumps whose wavelengths are suitably chosen to produce a flat gain profile over a wide spectral range.

A dual-pump parametric amplifier is fundamentally different from the conventional single-pump device [82]. Its operating principle can be understood by considering two pumps positioned in the anomalous and normal dispersion regimes, as shown in Figure 3.16. When either pump is used in isolation, parametric gain is narrowband and relatively small. More specifically, the pump in the anomalous region produces spectral features similar to those seen in Figure 3.14, whereas the pump in the normal region exhibits almost no gain. However, when both pumps are turned on simultaneously, the gain is not only larger but the spectral range over which gain is nearly uniform is also considerably wider. It is this central flat-gain region that turns parametric amplifiers into devices that are useful for lightwave systems.

Dual-pump parametric amplifiers provide uniform gain over a wide bandwidth by balancing three distinct processes that can produce parametric gain in multiple spectral regions [82]–[84]. Consider, as shown in Figure 3.16, two pumps at frequencies ω_1 and ω_2, and a signal at the frequency ω_{1+}. First, the ω_1 pump generates an idler at ω_{1-} through a degenerate FWM process (or modulation instability) as

$$\omega_1 + \omega_1 \rightarrow \omega_{1+} + \omega_{1-}, \quad (3.3.21)$$

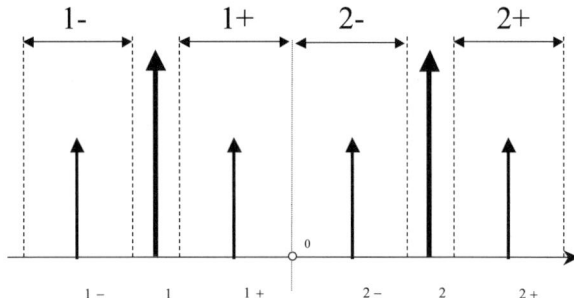

Figure 3.16: Spectral bands of a parametric amplifier pumped at two frequencies ω_1 and ω_2 located symmetrically around the zero-dispersion frequency ω_0. A signal at ω_{1+} produces three dominant idlers through three distinct FWM processes.

whenever the pump wavelength lies in the anomalous-dispersion region of the fiber. The second pump can also produce a new idler through $\omega_2 + \omega_2 \rightarrow \omega_{1+} + \omega_i$, but this process is expected to be relatively weak when the separation between the second pump and signal is large. Second, the nondegenerate FWM produces its own parametric gain through the process

$$\omega_1 + \omega_2 \rightarrow \omega_{1+} + \omega_{2-}. \tag{3.3.22}$$

Third, a Bragg-scattering process produces additional gain through the combinations denoted as [82]

$$\omega_1 + \omega_{1+} \rightarrow \omega_2 + \omega_{2-}, \qquad \omega_2 + \omega_{1+} \rightarrow \omega_1 + \omega_{2+}. \tag{3.3.23}$$

Several other weak idler waves are also created but the three dominant idlers identified in Figure 3.16 must be considered for any dual-pump parametric amplifier. The relative strength of the three processes depends on the pump powers as well as on the detuning of the two pumps from the zero-dispersion wavelength. This feature allows one to produce a desired parametric response by simply choosing the pump wavelengths and powers appropriately.

The theory of dual-pump parametric amplifiers is somewhat involved because one must consider at least six fields at the frequencies shown in Figure 3.16 simultaneously for an accurate analysis. If we neglect pump depletion, Eqs. (3.3.2) and (3.3.3) found in the single-pump case are replaced by a set of four coupled equations [82]. It turns out that the gain spectrum depends on a larger number of parameters such as pump wavelengths, zero-dispersion wavelength, and the third- and fourth-order dispersion parameters of the fiber. In one configuration that can provide flat gain over a wide bandwidth, the two pump wavelengths are chosen almost symmetrically on the opposite sides of the zero-dispersion wavelength λ_0. Figure 3.17 shows examples of the gain spectra in this case at three pump-power levels. A 500-m-long highly nonlinear fiber with $\gamma = 10$ W^{-1}/km and $\lambda_0 = 1570$ nm is assumed to be pumped using two lasers at wavelengths of 1,525 and 1,618 nm. The third- and fourth-order dispersion parameters of the fiber are $\beta_3 = 0.038$ ps^3/km and $\beta_4 = 1 \times 10^{-4}$ ps^3/km, respectively. When each pump provides 500 mW of power, a relatively large and uniform gain (38 dB) occurs

3.3 Parametric Amplifiers

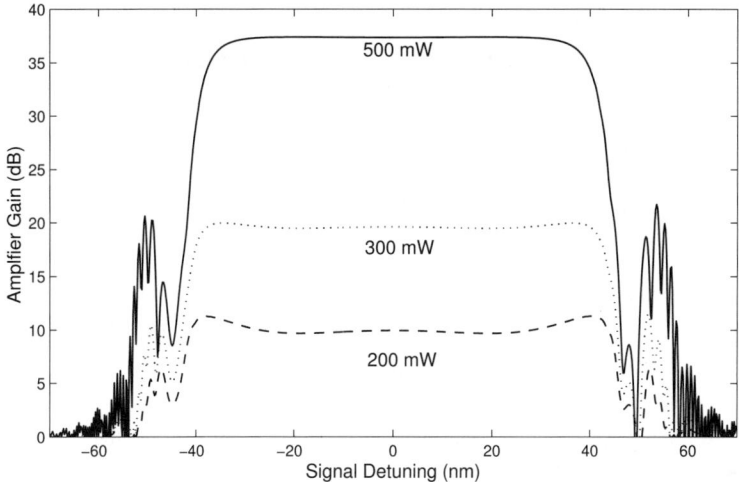

Figure 3.17: Theoretical gain spectra for a dual-pump parametric amplifier for three pump powers. Two pumps have equal powers and are located 35 nm away on each side of the zero-dispersion wavelength. Signal detuning is specified with respect to this wavelength.

over a bandwidth of more than 70 nm. The central flat gain region is mostly due to the nondegenerate FWM process. The other two FWM processes only affect the spectral wings and lead to oscillations seen in Figure 3.17.

Several experiments have shown that parametric amplifiers can provide flat gain over a wide bandwidth when pumped using the configuration shown in Figure 3.16. In one 2003 experiment [86], a gain of more than 40 dB was obtained over a bandwidth of 34 nm when a 1-km-long highly nonlinear fiber ($A_{\text{eff}} = 11$ μm^2) with its zero-dispersion wavelength at 1,583.5 nm was pumped at 1,559 and 1,610 nm simultaneously. Figure 3.18 shows the measured gain (symbols) as a function of signal detuning when the pump powers were 600 and 200 mW at 1,559 and 1,610 nm, respectively. Unequal input pump powers were used because the Raman gain transfers power from the shorter-wavelength pump to the other pump throughout the fiber length. The onset of SBS was avoided by modulating the pump phases at 10 GHz using a pseudorandom bit pattern. The theoretical fit in Figure 3.18 was obtained using a general theory that included the Raman-induced transfer of powers between the pumps, signal, and idlers with the parameter values $\gamma = 9.18$ W^{-1}/km, $\beta_3 = 0.0315$ ps^3/km, and $\beta_4 = 1.9 \times 10^{-4}$ ps^3/km.

3.3.4 Performance Limiting Factors

Similar to the case of Raman amplifiers, the performance of parametric amplifiers is also affected by factors such as spontaneous parametric fluorescence, double Rayleigh backscattering, PMD, and pump-noise transfer. The Rayleigh backscattering problem is of less concern for parametric amplifiers because of the relatively short fiber length used to make them. On the other hand, the RIN-enhancement problem is much more

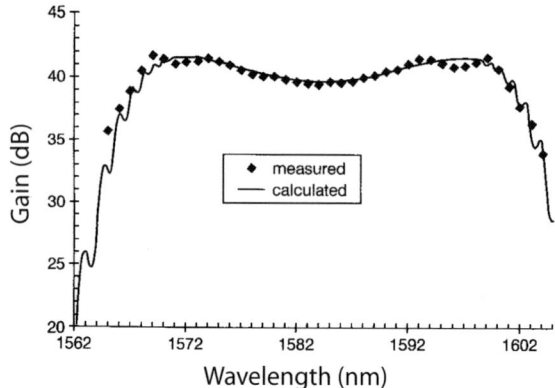

Figure 3.18: Measured (symbols) and theoretical (solid curve) gain plotted as a function of signal wavelength for a parametric amplifier pumped with 600 and 200 mW of powers at 1,559 and 1,610 nm, respectively. The fiber was 1 km long and its zero-dispersion wavelength was 1583.5 nm. (After Ref. [86]; ©IEE 2003.)

serious for parametric amplifiers because of the copropagating nature of the pumps and signal [81]. It can only be solved by using pump lasers with low RIN and by using a short fiber length.

Spontaneous parametric fluorescence is the source of noise in parametric amplifiers and enhances the noise figure above the minimum value of 3 dB expected for an ideal amplifier. Recent measurements indicate that the noise figure remains close to 4 dB [87]. Such noise levels are low enough that they are not likely to limit the usefulness of parametric amplifiers. It should be mentioned that, in principle, noise can be reduced to below the 3-dB quantum limit using the phenomenon of squeezing, whereby noise is reduced in a certain frequency range in one quadrature even though it is enhanced considerably in the other quadrature [88]. Indeed, a noise figure of 1.8 dB was measured in a 1999 experiment for a phase-sensitive parametric amplifier providing 16-dB gain [89]. However, this technique is unlikely to be used in practical lightwave systems because of the difficulty in controlling the signal phase.

Similar to the case of Raman amplifiers, the parametric gain is extremely polarization-dependent because it is maximum when the pumps and signal are copolarized but considerably smaller when they are orthogonally polarized. Since the signal SOP in most lightwave systems does not remain fixed, the PDG of parametric amplifiers is not acceptable in practice. The problem can be solved for dual-pump parametric amplifiers by using orthogonally polarized pumps with equal powers. It is common to employ linearly polarized pumps for this purpose. The signal in any SOP can be decomposed into its two linearly polarized components, each of which is amplified by the two pumps such that the total gain is independent of the signal SOP. Although this simple idea works in practice [84], it turns out that the amplifier gain is reduced drastically from the case in which both pumps and signals are copolarized along the same direction.

Much higher values of the gain can be realized if the two pumps are left and right

3.3 Parametric Amplifiers

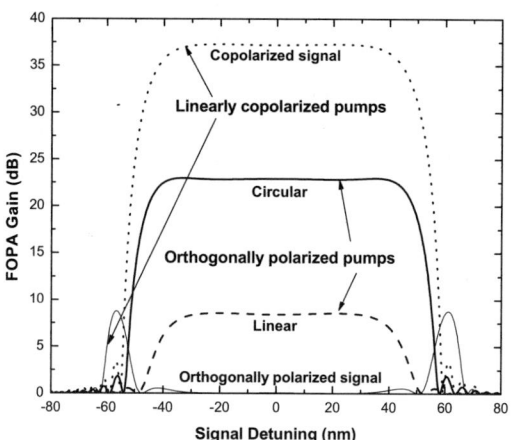

Figure 3.19: Parametric gain as a function of signal detuning for a 500-m-long amplifier pumped using two 0.5-W pumps in several different polarization configurations. The thin and dotted lines show that dependence of gain on signal polarization in the case of linearly copolarized pumps.

circularly polarized rather then being linearly polarized in the orthogonal directions [90]. Figure 3.19 shows the improvement realized by this simple change for a 500-m-long amplifier designed using a highly nonlinear fiber with $\gamma = 10 \text{ W}^{-1}/\text{km}$. The zero-dispersion wavelength of the fiber is assumed to be $\lambda_0 = 1,580$ nm, while $\beta_3 = 0.04 \text{ ps}^3/\text{km}$ and $\beta_4 = 1.0 \times 10^{-4} \text{ ps}^4/\text{km}$. The two pumps with the wavelengths 1,535 and 1,628 nm are launched with equal powers of 0.5 W.

The dotted curve shows the ideal case in which both pumps and the signal are linearly copolarized signal, resulting in a fairly flat gain of 37 dB over a wide wavelength range. However, this gain is highly polarization-dependent. When the signal becomes orthogonally polarized to the two pumps, the amplifier gain is reduced to almost zero in the central part (thin solid curve) because of a large reduction in the FWM coupling efficiency and the phase mismatch induced by cross-phase modulation. When the pumps are linearly polarized along two orthogonal directions, the gain becomes independent of the signal SOP (dashed curve) but it is reduced dramatically to a quite small value of around 8.5 dB. However, as shown by the solid curve, the gain can be increased from 8.5 to 23 dB over a wider spectral region if the two pumps are orthogonally but circularly polarized. Thus, the amplifier gain is enhanced by a factor of 25 by simply changing linear polarization to circular polarization for both pumps while maintaining orthogonality.

The polarization effects seen in Figure 3.19 can be understood physically using a basis in which ↑ and ↓ denote left circular polarization (spin up) and right circular polarization (spin down), respectively. Because of the angular momentum conservation requirement, the nondegenerate FWM process must satisfy certain selection rules. If both pump photons are in the ↑ or ↓ state, the signal and idler photons must also be in the same state. In contrast, when the two pump photons are orthogonally circularly polarized (one ↑ and the other ↓), the signal and idler photons must also be in orthogonal states to conserve the zero angular momentum. However, this is possible through two combinations, $\uparrow_s + \downarrow_i$ and $\downarrow_s + \uparrow_i$, both of which are equally probable and thus lead to higher polarization-independent signal gain. Note also from the above argument that single-pump parametric amplifiers will always be polarization-dependent because the two pump photons have identical polarizations.

The results shown in Figure 3.19 have ignored the PMD effects. In contrast with the case of Raman amplifiers, the PMD is of less concern for parametric amplifiers because of much smaller fiber lengths used in practice. Nonetheless, it affects the amplifier performance whenever the diffusion length becomes comparable to the fiber length [84]. Numerical simulations based on the vector FWM equations show that not only the average value of the gain is reduced considerably, but the gain also becomes polarization-dependent even for two orthogonally polarized pumps [91]. Another source of degradation is related to fluctuations in the zero-dispersion wavelength of the fiber along its length. Such fluctuations reduce the parametric gain and make it relatively nonuniform. This problem becomes less severe when the two pumps are moved closer to the average value of zero-dispersion wavelength.

3.4 Fiber Lasers

Any fiber amplifier can be converted into a laser by placing it inside a cavity designed to provide optical feedback. Such lasers are called *fiber lasers*. This section focuses on fiber lasers made using EDFAs and called erbium-doped fiber lasers (EDFLs). Such lasers have found applications in the domain of lightwave technology as optical sensors and sources of ultrashort pulses.

3.4.1 Cavity Design

Fiber lasers can be designed with a variety of choices for the laser cavity [7]. The most common type of laser cavity is the Fabry–Perot cavity, made by placing the gain medium between two high-reflecting mirrors. In the case of fiber lasers, mirrors are often butt-coupled to the fiber ends to avoid diffraction losses. Dielectric mirrors, made by depositing multiple thin layers of different dielectrics, are often used for this purpose because they can be designed to be highly reflective at the laser wavelength while transmitting the pump wavelength. Alignment of such a cavity is not easy since cavity losses increase rapidly with a tilt of the fiber end or the mirror. This problem can be solved by depositing dielectric mirrors directly onto the polished ends of a doped fiber [92]. However, end-coated mirrors are quite sensitive to imperfections at the fiber tip. Furthermore, since pump light passes through the same mirrors, dielectric coatings can be easily damaged when high-power pump light is coupled into the fiber.

Several alternatives exist to avoid passing the pump light through dielectric mirrors. For example, one can take advantage of fiber couplers. It is possible to design a fiber coupler such that most of the pump power comes out of the port that is a part of

3.4 Fiber Lasers

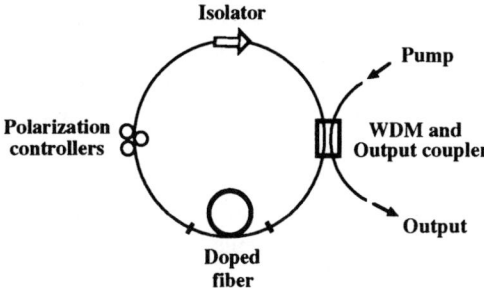

Figure 3.20: Schematic of a unidirectional ring cavity used for fiber lasers.

the laser cavity. Such couplers are called WDM couplers. Another solution is to use fiber gratings as mirrors [28]. As discussed in Section 2.2, a fiber grating can act as a high-reflectivity mirror for the laser wavelength while being transparent to pump radiation. The use of two such gratings results in an all-fiber Fabry–Perot cavity [93]. An added advantage of Bragg gratings is that the laser can be forced to operate in a single longitudinal mode. A third approach makes use of a Sagnac interferometer discussed in Section 2.3.3 (also known as fiber-loop mirrors).

Ring cavities are often used to realize the unidirectional operation of a laser. In the case of fiber lasers, an additional advantage is that a ring cavity can be made without using mirrors, resulting in an all-fiber cavity. In the simplest design, two ports of a WDM coupler are connected together to form a ring cavity containing the doped fiber, as shown in Figure 3.20. An isolator is inserted within the loop for unidirectional operation. A polarization controller is also needed for conventional fibers that do not preserve polarization.

Figure 3.21 shows a ring-cavity design used for mode-locked fiber lasers. This configuration is referred to as the *figure-8 cavity* because of its appearance. The ring cavity on the right acts as a nonlinear amplifying-loop mirror, whose switching characteristics were discussed in Section 2.3.3. The nonlinear effects play an important role in the operation of figure-8 lasers. At low powers, loop transmittivity is relatively small, resulting in relatively large cavity losses for CW operation. The Sagnac loop becomes

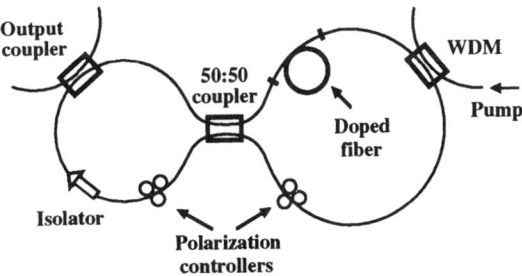

Figure 3.21: Schematic of a figure-8 cavity useful for mode-locked fiber lasers.

fully transmissive for pulses whose peak power attains a critical value [see Eq. (3.2.8)]. For this reason, a figure-8 cavity favors mode locking. An isolator in the left cavity ensures unidirectional operation. The laser output is taken through a fiber coupler with low transmission (<10%) to minimize cavity losses. An interesting property of the figure-8 cavity is that it permits passive mode locking without any active mode-locking elements.

Many other cavity designs are possible. For example, one can use two coupled Fabry–Perot cavities. In the simplest scheme, one mirror is separated from the fiber end by a controlled amount. The 4% reflectivity of the fiber–air interface acts as a low-reflectivity mirror that couples the fiber cavity with the empty air-filled cavity. Such a compound resonator has been used to reduce the line width of an EDFL. Three fiber gratings in series also produce two coupled Fabry–Perot cavities.

3.4.2 Continuously Operating Fiber Lasers

This subsection focuses on CW fiber lasers. As early as 1989, a 0.98-μm-pumped EDFL exhibited a slope efficiency of 58% against absorbed pump power [94], a figure that is close to the quantum limit of 63% obtained by taking the ratio of signal-to-pump photon energies. EDFLs pumped at 1.48 μm also exhibit good performance. In fact, the choice between 0.98 and 1.48 μm is not always clear since each pumping wavelength has its own merits. Both pump wavelengths have been used for developing practical EDFLs with excellent performance characteristics.

An important property of continuously operating EDFLs from a practical standpoint is their ability to provide output that is tunable over a wide wavelength range. In practice, it is important to keep the laser line width relatively small as its wavelength is tuned. Many techniques can be used for this purpose [95]. In a 1989 experiment, an intracavity étalon formed between the bare fiber end and the output mirror led to a 620-MHz line width even though the fiber was 13 m long [96]. The laser wavelength can also be tuned by using an external grating in combination with an étalon. Figure 3.22(a) shows the experimental setup together with the tuning curves (b) obtained for two different fiber lengths. This laser was tunable over a 70-nm range. The output power was more than 250 mW in the wavelength range from 1.52 to 1.57 μm.

Ring cavities can also be used to make tunable EDFLs. A common technique uses a fiber-based étalon that can be tuned electrically inside the ring cavity. Such EDFLs have shown low threshold (absorbed pump power of 2.9 mW) with 15% slope efficiency [97]. They can be tuned over 60 nm and provide a line width of $\sim$1.4 kHz. In an optimized EDFL, 15.6-mW of output power was obtained with 48% slope efficiency (68% with respect to absorbed pump power) while the tuning range (at the 3-dB point) was 42 nm [98].

Many other tuning techniques have been used for fiber lasers. In one experiment, a fiber laser was tuned over 33 nm through strain-induced birefringence [99]. In another, a fiber laser could be tuned over 39 nm by using a reflection Mach–Zehnder interferometer that acts as a wavelength-selective loss element within the ring-laser cavity [100]. The wavelength for which cavity losses are minimum is changed by controlling the optical path length in one of the interferometer arms either electro-optically or by applying stress.

3.4 Fiber Lasers

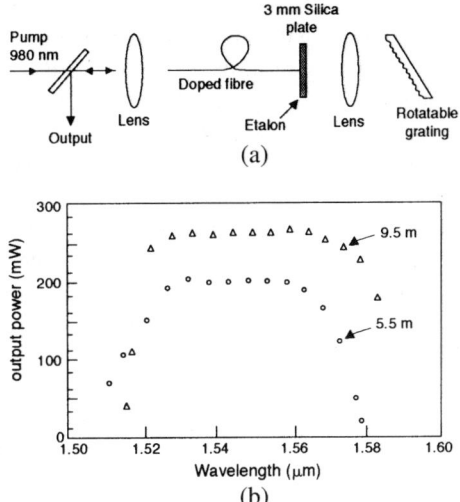

Figure 3.22: (a) Experimental setup for a broadly tunable EDFL. (b) Tuning curves for two different fiber lengths at 540 mW of launched power. (After Ref. [96], ©IEE.)

Fiber gratings can also be used to improve the performance of EDFLs [28]. As early as 1990, a Bragg grating was used to realize a line width of about 1 GHz. Since then, fiber gratings have been used inside EDFAs for a variety of reasons [101]–[109]. The simplest configuration splices a Bragg grating at each end of an erbium-doped fiber, forming a Fabry–Perot cavity. Such devices are called *distributed Bragg reflector* (DBR) lasers, following the terminology used for semiconductor lasers (see Chapter 5). DBR fiber lasers can be tuned continuously while exhibiting a narrow line width. They can also be made to oscillate in a single longitudinal mode by decreasing the fiber length. In a 1993 experiment, an EDFL was made to oscillate at two distinct wavelengths, with a narrow line width at each wavelength, by fabricating two different gratings or by using a single grating with dual-peak reflectivity [101].

Multiple fiber gratings can be used to make coupled-cavity fiber lasers. Such lasers have operated at two different wavelengths (0.5 nm apart) simultaneously such that each spectral line was stable to within 3 MHz and had a line width of only 16 kHz [102]. Fiber gratings have been used to make efficient, low-noise EDFLs. In one such laser, up to 7.6 mW of output power was obtained while the relative intensity-noise level was below −145 dB/Hz at frequencies above 10 MHz [103]. Even higher powers can be obtained by using the master oscillator/power amplifier (MOPA) configuration in which a fiber laser acting as a master oscillator is coupled to a fiber amplifier through an intracore Bragg grating. Output powers of up to 62 mW were obtained in 1994 by using such a configuration through active feedback while maintaining intensity-noise levels below −110 dB/Hz at all frequencies [104].

Another approach consists of making a distributed-feedback (DFB) fiber laser. Analogous to DFB semiconductor lasers (see Chapter 5), a Bragg grating is formed directly into the erbium-doped fiber that provides gain [105]. Phase-shifted DFB lasers have also been made by leaving a small region of the doped fiber in the middle without a grating [106]. Multiple gratings with slightly different Bragg wavelengths can also be formed into the same doped fiber, resulting in several DFB lasers cascaded together.

Multiwavelength optical sources, capable of emitting light at several well-defined wavelengths simultaneously, are useful for WDM applications (see Chapter 8). Fiber lasers can be used for this purpose, and several schemes have been developed [110]–[118]. A dual-frequency fiber laser was demonstrated in 1993 using a coupled-cavity configuration [101]. In 1997, simultaneous operation of a fiber laser at up to 29 wavelengths was realized by cooling the doped fiber to 77 K using liquid nitrogen [115]. The cavity length was quite small ($\sim$1 mm or so) in this laser because an 1-mm cavity length corresponds to a 100-GHz wavelength spacing commonly used in WDM lightwave systems. Cooling of the doped fiber helps to reduce homogeneous broadening of the gain spectrum to below 0.5 nm. The gain spectrum is then predominantly inhomogeneously broadened, resulting in multimode operation through spectral hole burning. Long cavities with several meters of doped fibers can also be used. Wavelength selection is then made using an intracavity comb filter such as a Fabry–Perot interferometer [110]. In a dual-filter approach, a tunable comb filter in combination with a set of fiber gratings provided a multiwavelength source that was switchable on a microsecond timescale to precise preselected wavelengths [116]. In a 2002 experiment, a fiber-ring laser provided output at 52 channels, designed to be 50 GHz apart, with nearly equal power levels [118].

3.4.3 Mode-Locked Fiber Lasers

Although fiber lasers can be mode-locked using an amplitude or phase modulator, passive mode locking produces ultrashort optical pulses without requiring any active component inside the laser cavity. It makes use of a nonlinear device whose response to an entering optical pulse is intensity-dependent such that the exiting pulse is narrower than the input pulse. Several implementations of this basic idea have been used to make passively mode-locked fiber lasers.

Saturable Absorbers

Saturable absorbers are commonly used for passive mode locking of solid-state lasers. The basic mechanism behind mode locking is easily understood by considering a fast saturable absorber whose absorption can change on a timescale of the pulse width. When an optical pulse propagates through such an absorber, its wings experience more loss than the central part, which is intense enough to saturate the absorber. The net result is that the pulse is shortened during its passage through the absorber. Pulse shortening provides a mechanism through which a laser can minimize cavity losses by generating intense pulses if the CW radiation is unable to saturate the absorber. The pulse continues to shorten until it becomes so short that its spectral width is comparable to the gain bandwidth. The reduced gain in spectral wings then provides the broadening mechanism that stabilizes the pulse width to a specific value. In the case of fiber lasers, GVD and SPM also play an important role in the evolution of mode-locked pulses and should be included.

3.4 Fiber Lasers

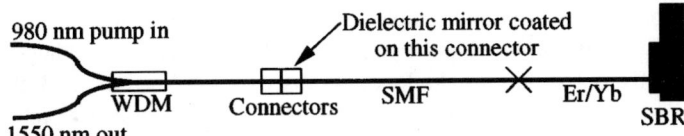

Figure 3.23: Schematic of a fiber laser that was mode-locked using a saturable Bragg reflector. (After Ref. [120]; ©1998 OSA.)

It is not easy to find a fast saturable absorber responding at time scales of 1 ps or shorter. The most suitable material for fiber lasers is a semiconductor absorbing layer attached to one of the cavity mirrors. The saturable absorber can be made using either a single or a stack of quantum-well layers. In the latter case, it forms a periodic structure called the *superlattice*. Each period of the superlattice consists of alternating absorbing and transparent layers. In some cases, the mirror attached to the saturable absorber is also made using a periodic arrangement of thin layers that form a grating and reflect light through Bragg diffraction. Such a device is referred to as a *saturable Bragg reflector* to emphasize the use of a Bragg grating. A superlattice saturable absorber integrated with a Bragg reflector requires the growth of hundreds of thin layers using molecular-beam epitaxy (see Chapter 4).

Fiber lasers that are mode-locked using a saturable Bragg reflector inside a short Fabry–Perot cavity have quite interesting properties. Figure 3.23 shows the cavity design schematically. A short piece (with a length of 15 cm) of doped fiber is butt-coupled to the saturable Bragg reflector. Its other end is sliced to a section of standard fiber (length about 30 cm) and terminated with a connector on which a high-reflectivity dielectric mirror has been coated. In a series of experiments, total cavity length was changed from 0.5 to 2.5 m, and the average GVD was varied over a wide range (normal to anomalous) by using the dispersion-compensating fiber [119]. A mode-locked pulse train could be formed even in the case of normal GVD, but the pulse width was close to 16 ps at a repetition rate of about 40 MHz. Much shorter pulses were observed when the average GVD was anomalous. Pulse widths below 0.5 ps formed over a wide range of average GVD ($\beta_2 = -2$ to -14 ps^2/km) although they were chirped. For short laser cavities (under 50 cm), harmonic mode locking was found to occur. A 45-cm-long laser produced transform-limited, 300-fs pulses at a repetition rate of 2.6 GHz, and the spacing between pulses was uniform to within 4% of the expected value [120].

Nonlinear Fiber-Loop Mirrors

An undesirable aspect of semiconductor-based saturable absorbers is that fiber lasers using them lose their all-fiber nature. A solution is provided by the nonlinear fiber-loop mirrors (Sagnac interferometers) whose power-dependent transmission can shorten an optical pulse just as saturable absorbers do (see Section 2.3.3). The figure-8 fiber laser shown in Figure 3.21 falls into this category. The physical mechanism responsible for mode locking in such lasers is known as additive-pulse mode locking.

The operation of a figure-8 laser can be understood as follows. The central 3-dB coupler in Figure 3.21 splits the entering radiation into two equal counterpropagating parts. The doped fiber providing amplification is placed close to the central coupler such that one wave is amplified at the entrance to the loop while the other experiences amplification just before exiting the loop. As discussed in Section 2.3.3, the counter-propagating waves acquire different nonlinear phase shifts while completing a round trip inside such a nonlinear amplifying-loop mirror (NALM). Moreover, the phase difference is not constant but varies along the pulse profile. If the NALM is adjusted such that the phase shift is close to π for the central intense part, this part of the pulse is transmitted, while pulse wings are reflected because of their lower power levels and smaller phase shifts. The net result is that the pulse exiting from the NALM is narrower compared with that entering it. Because of this property, a NALM behaves like a fast saturable absorber but responds at femtosecond timescales because of the ultrafast origin of fiber nonlinearity.

NALMs were first used in 1991 for mode locking a fiber laser and produced pulses as short as 290 fs from an EDFL pumped at 1.48-μm [121]. The threshold for mode-locked operation was only 50 mW. Once mode locking was initiated, pump power could be decreased to as low as 10 mW. It is difficult to produce pulses shorter than 100 fs from figure-8 lasers. However, pulses as short as 30 fs were obtained by amplifying the laser output and then compressing the amplified pulse in a dispersion-shifted fiber [122]. Pulses as short as 98 fs were generated directly from a figure-8 laser by using a polarization-sensitive isolator and a short piece of normal-GVD fiber for chirp compensation [123].

Passively mode-locked fiber lasers suffer from a major drawback that has limited their usefulness. It was observed in several experiments that the repetition rate of mode-locked pulses was essentially uncontrollable and could vary over a wide range. Typically, several pulses circulate simultaneously inside the laser cavity, and the spacing among them is not necessarily uniform. In contrast with the case of active mode locking, nothing in the cavity determines the relative location of pulses. The key to stabilizing a figure-8 laser consists of implementing a scheme that can adjust the repetition rate of the laser. The use of normal GVD also helps but the resulting mode-locked pulses are considerably wider. In a 1997 experiment ($\beta_2 > 0$), pulses inside the cavity were stretched considerably during amplification inside the doped fiber [124]. This permitted energy levels as high as 0.5 nJ. Pulses could be compressed down to 125 fs by using dispersion-shifted fiber inside the cavity. Both the central wavelength and the spectral width of mode-locked pulses were tunable by adjusting the polarization controllers within the cavity.

Nonlinear Polarization Rotation

Fiber lasers can also be mode-locked by using intensity-dependent changes in the state of polarization (occurring because of SPM and XPM) when the orthogonally polarized components of a single pulse propagate inside an optical fiber (see Section 1.6). The physical mechanism behind mode locking makes use of the nonlinear birefringence. From a conceptual point of view, the mode-locking mechanism is identical to that used for figure-8 lasers except that the orthogonally polarized components of the same pulse are used in place of counterpropagating waves. From a practical standpoint, passive mode locking can be accomplished by using a cavity with a single fiber ring.

3.4 Fiber Lasers

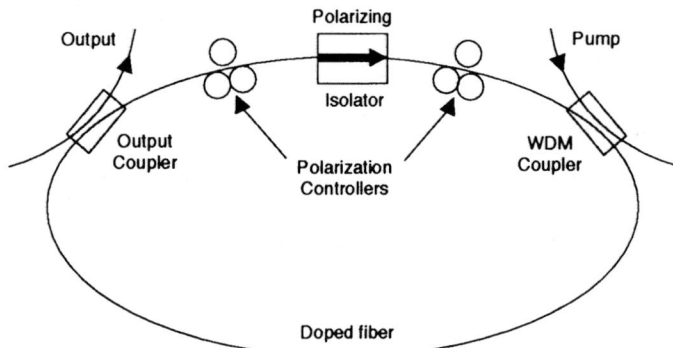

Figure 3.24: Schematic design of a fiber laser passively mode-locked via nonlinear polarization-rotation.

The mode-locking process can be understood using the ring cavity shown in Figure 3.24. A polarizing isolator placed between two polarization controllers acts as the mode-locking element. It plays the double role of an isolator and a polarizer such that light leaving the isolator is linearly polarized. Consider a linearly polarized pulse just after the isolator. The polarization controller placed after the isolator changes the polarization state to elliptical. The polarization state evolves nonlinearly during propagation of the pulse because of SPM- and XPM-induced phase shifts imposed on the orthogonally polarized components. The state of polarization is nonuniform across the pulse because of the intensity dependence of the nonlinear phase shift. The second polarization controller (one before the isolator) is adjusted such that it forces the polarization to be linear in the central part of the pulse. The polarizing isolator lets the central intense part of the pulse pass but blocks (absorbs) the low-intensity pulse wings. The net result is that the pulse is slightly shortened after one round trip inside the ring cavity, an effect identical to that produced by a fast saturable absorber.

The technique of nonlinear polarization rotation was first used in 1992 for passive mode locking of fiber lasers. In one experiment, stable, self-starting pulse trains of 450-fs width were generated at a 42-MHz repetition rate using this technique [125]. Further improvements occurred when it was realized that the presence of anomalous GVD within the laser cavity is not necessarily beneficial since it limits both the width and the energy of mode-locked pulses. In a 1993 experiment, 76-fs pulses with 90-pJ energy and 1 kW of peak power were generated by using a ring cavity in which the average GVD was normal [126]. Mode-locked pulses as short as 63 fs have been generated with proper optimization [127].

For practical applications, environmental stability is often an important issue. The main source of environmental instability is the relatively long length of the fiber inside the laser cavity required to produce a large enough nonlinear phase shift. Temperature and stress variations can lead to birefringence fluctuations that affect the mode-locking process. The problem can be solved to a large extent by reducing the fiber length to under 10 m and using a fiber with high built-in birefringence (polarization-maintaining fiber) so that linear birefringence is not affected by environmental changes. In one scheme, a Fabry–Perot cavity in which one of the mirrors acts as a Faraday rotator has been used to realize environmentally stable operation [128].

Hybrid and Harmonic Mode Locking

Hybrid mode locking combines active and passive techniques to improve the laser performance. The most obvious combination incorporates an amplitude or phase modulator inside a passively mode-locked fiber laser. The modulator provides periodic timing slots to produce a regular pulse train, while passive mode locking shortens the pulse compared to that expected from active mode locking alone. An added benefit is that the modulator can be operated at a frequency that is a high multiple of the round-trip frequency, resulting in a well-defined repetition rate that can exceed 10 GHz or more while the mode spacing remains close to 10 MHz. This is referred to as harmonic mode locking, a technique that is useful for operating fiber lasers at high repetition rates.

Harmonic mode locking of fiber lasers is accomplished using electro-optic modulators. As discussed in Chapter 6, $LiNbO_3$ modulators are relatively compact and can be modulated at speeds as high as 40 GHz. They can also be integrated within the fiber cavity with relatively small coupling losses. For these reasons, $LiNbO_3$ modulators are commonly used for mode-locking fiber lasers. As early as 1990, the pulse-repetition rate was extended to 30 GHz by using a high-speed $LiNbO_3$ modulator, although pulses were relatively wide [129]. A ring cavity of 30-m length was used with an intracavity isolator for unidirectional operation. In a 1992 experiment, a fiber laser provided transform-limited pulses as short as 3.5 ps at repetition rates of up to 20 GHz [130]. Polarization-maintaining fibers were used in a 1993 experiment to produce 6-ps pulses at repetition rates of up to 40 GHz and at wavelengths tunable over a wide range from 40 to 50 nm [131]. By 2000, fiber lasers operating at 40 GHz were routinely used to produce harmonically mode-locked pulse trains and pulses as short as 0.85 ps were obtained using this approach [132]. The laser wavelength was also tunable over 15 nm or so.

Problems

3.1 The Lorentzian gain profile of an optical amplifier has a FWHM of 1 THz. Calculate the amplifier bandwidths when it is operated to provide 20- and 30-dB gain. Neglect gain saturation.

3.2 An optical amplifier can amplify a 1-μW signal to the 1-mW level. What is the output power when a 1-mW signal is incident on the same amplifier? Assume that the saturation power is 10 mW.

3.3 Explain the concept of noise figure for an optical amplifier. Why does the SNR of the amplified signal degrade by 3 dB even for an ideal amplifier?

3.4 Explain the gain mechanism in EDFAs. Use Eqs. (3.1.8) and (3.1.9) to derive an expression for the small-signal gain in the steady state.

3.5 Discuss how EDFAs can be used to provide gain in the L band. How can you use them to provide amplification over both the C and L bands?

3.6 A Raman amplifier is pumped in the backward direction using 1 W of power. Find the output power when a 1-μW signal is injected into the 5-km-long amplifier. Assume losses of 0.2 and 0.25 dB/km at the signal and pump wavelengths, respectively, $A_{\text{eff}} = 50$ μm^2, and $g_R = 6 \times 10^{-14}$ m/W. Neglect gain saturation.

3.7 Extend Eqs. (3.2.2) and (3.2.3) to the case of a bidirectionally pumped Raman amplifier. Solve them to obtain an expression for the amplifier gain in terms of pump powers at the two ends, assuming that both pumps remain undepleted.

3.8 Solve Eqs. (3.2.2) and (3.2.3) analytically after assuming that $\alpha_s = \alpha_p$ and derive an expression for the saturated gain in terms of the input signal and pump powers at $z = 0$.

3.9 Using Eqs. (3.2.19) and (3.2.20), derive Eq. (3.2.21) for the fraction of power at the output resulting from double Rayleigh backscattering.

3.10 Starting from Eqs. (3.3.2) and (3.3.3), derive an expression for the amplification factor for a parametric amplifier of length L.

3.11 How would you design a fiber laser without using actual mirrors? Sketch two such designs and explain their operation.

3.12 Derive an expression for the output power when N longitudinal modes with identical powers become mode-locked such that phase difference between two neighboring modes is constant. Estimate the pulse width for $N = 10^4$, assuming a fiber-ring cavity of 5-m perimeter and a mode index of 1.5.

3.13 Explain how cross-phase modulation can be used to induce mode locking in fiber lasers. Use diagrams as necessary.

3.14 Explain the mode-locking process in a figure-8 fiber laser. What limits the pulse width in such lasers?

3.15 How can nonlinear birefringence be used to advantage for the passive mode locking of fiber lasers? Draw the laser cavity schematically and explain the purpose of each component.

References

[1] C. J. Koester and E. Snitzer, *Appl. Opt.* **3**, 1182 (1964).
[2] S. B. Poole, D. N. Payne, R. J. Mears, M. E. Fermann, and R. E. Laming, *J. Lightwave Technol.* **4**, 870 (1986).
[3] A. Bjarklev, *Optical Fiber Amplifiers: Design and System Applications*, Artech House, Boston, 1993.
[4] E. Desurvire, *Erbium-Doped Fiber Amplifiers: Principles and Applications*, Wiley, New York, 1994.
[5] P. C. Becker, N. A. Olsson, and J. R. Simpson, *Erbium-Doped Fiber Amplifiers: Fundamentals and Technology*, Academic Press, San Diego, CA, 1999.
[6] G. P. Agrawal, *Applications of Nonlinear Fiber Optics*, Academic Press, San Diego, CA, 2001, Chap. 4.
[7] M. J. F. Digonnet, Ed., *Rare-Earth Doped Fiber Lasers and Amplifiers*, 2nd ed., Marcel Dekker, New York, 2001.

[8] E. Desurvire, D. Bayart, B. Desthieux, and S. Bigo, *Erbium-Doped Fiber Amplifiers: Devices and System Developments*, Wiley, New York, 2002.
[9] M. Shimizu, M. Yamada, H. Horiguchi, T. Takeshita, and M. Okayasu, *Electron. Lett.* **26**, 1641 (1990).
[10] B. Pederson, A. Bjarklev, H. Vendeltorp-Pommer, and J. H. Povlesen, *Opt. Commun.* **81**, 23 (1991).
[11] M. Horiguchi, K. Yoshino, M. Shimizu, and M. Yamada, *Electron. Lett.* **29**, 593 (1993).
[12] R. I. Laming, J. E. Townsend, D. N. Payne, F. Meli, G. Grasso, and E. J. Tarbox, *IEEE Photon. Technol. Lett.* **3**, 253 (1991).
[13] A. E. Siegman, *Lasers*, University Science Books, Mill Valley, CA, 1986.
[14] A. A. M. Saleh, R. M. Jopson, J. D. Evankow, and J. Aspell, *IEEE Photon. Technol. Lett.* **2**, 714 (1990).
[15] C. R. Giles and E. Desurvire, *J. Lightwave Technol.* **9**, 271 (1991).
[16] K. Nakagawa, S. Nishi, K. Aida, and E. Yoneda, *J. Lightwave Technol.* **9**, 198 (1991).
[17] E. Delevaque, T. Georges, J. F. Bayon, M. Monerie, P. Niay, and P. Benarge, *Electron. Lett.* **29**, 1112 (1993).
[18] A. Yariv, *Opt. Lett.* **15**, 1064 (1990); H. Kogelnik and A. Yariv, *Proc. IEEE* **52**, 165 (1964).
[19] G. P. Agrawal, *Fiber-Optic Communications Systems*, 3rd ed., Wiley New York, 2002, Chap. 6.
[20] R. Olshansky, *Electron. Lett.* **24**, 1363 (1988).
[21] M. Yamada, M. Shimizu, M. Okayasu, T. Takeshita, M. Horiguchi, Y. Tachikawa, and E. Sugita, *IEEE Photon. Technol. Lett.* **2**, 205 (1990).
[22] R. I. Laming and D. N. Payne, *IEEE Photon. Technol. Lett.* **2**, 418 (1990).
[23] K. Kikuchi, *Electron. Lett.* **26**, 1851 (1990).
[24] R. I. Laming, M. N. Zervas, and D. N. Payne, *IEEE Photon. Technol. Lett.* **4**, 1345 (1992).
[25] G. P. Agrawal, *Nonlinear Fiber Optics*, 3rd ed., Academic Press, San Diego, CA, 2001.
[26] K. Kikuchi, *IEEE Photon. Technol. Lett.* **5**, 221 (1993).
[27] M. Murakami and S. Saito, *IEEE Photon. Technol. Lett.* **4**, 1269 (1992).
[28] R. Kashyap, *Fiber Bragg Gratings*, Academic Press, San Diego, CA, 1999.
[29] P. F. Wysocki, J. B. Judkins, R. P. Espindola, M. Andrejco, A. M. Vengsarkar, and K. Walker, *IEEE Photon. Technol. Lett.* **9**, 1343 (1997).
[30] X. Y. Zhao, J. Bryce, and R. Minasian, *IEEE J. Sel. Topics Quantum Electron.* **3**, 1008 (1997).
[31] R. H. Richards, J. L. Jackel, and M. A. Ali, *IEEE J. Sel. Topics Quantum Electron.* **3**, 1027 (1997).
[32] G. Luo, J. L. Zyskind, J. A Nagel, and M. A. Ali, *J. Lightwave Technol.* **16**, 527 (1998).
[33] M. Karasek and J. A. Valles, *J. Lightwave Technol.* **16**, 1795 (1998).
[34] A. Bononi and L. Barbieri, *J. Lightwave Technol.* **17**, 1229 (1999).
[35] M. Karasek, M. Menif, and R. A. Rusch, *J. Lightwave Technol.* **19**, 933 (2001).
[36] H. Ono, M. Yamada, T. Kanamori, S. Sudo, and Y. Ohishi, *J. Lightwave Technol.* **17**, 490 (1999).
[37] P. F. Wysocki, N. Park, and D. DiGiovanni, *Opt. Lett.* **21**, 1744 (1996).
[38] M. Masuda and S. Kawai, *IEEE Photon. Technol. Lett.* **11**, 647 (1999).
[39] M. Yamada, H. Ono, T. Kanamori, S. Sudo, and Y. Ohishi, *Electron. Lett.* **33**, 710 (1997).
[40] M. Yamada, A. Mori, K. Kobayashi, H. Ono, T. Kanamori, K. Oikawa, Y. Nishida, and Y. Ohishi, *IEEE Photon. Technol. Lett.* **10**, 1244 (1998).

[41] H. Ono, Y. Yamada, and M. Shimizu, *J. Lightwave Technol.* **21**, 2240 (2003).
[42] T. Kasamatsu, Y. Yano, and T. Ono, *J. Lightwave Technol.* **20**, 1826 (2002).
[43] R. H. Stolen, E. P. Ippen, and A. R. Tynes, *Appl. Phys. Lett.* **20**, 62 (1972); R. H. Stolen and E. P. Ippen, *Appl. Phys. Lett.* **22**, 276 (1973).
[44] F. L. Galeener, Phys. Rev. B**19**, 4292 (1979).
[45] N. Shibata, M. Horigudhi, and T. Edahiro, *J. Noncrys. Solids* **45**, 115 (1981).
[46] R. H. Stolen, J. P. Gordon, W. J. Tomlinson, and H. A. Haus, *J. Opt. Soc. Am. B* **6**, 1159 (1989).
[47] D. J. Dougherty, F. X. Kartner, H. A. Haus, and E. P. Ippen, *Opt. Lett.* **20**, 31 (1995).
[48] D. Mahgerefteh, D. L. Butler, J. Goldhar, B. Rosenberg, and G. L. Burdge, *Opt. Lett.* **21**, 2026 (1996).
[49] S. Namiki and Y. Emori, *IEEE J. Sel. Topics Quantum Electron.* **7**, 3 (2001).
[50] J. Bromage, P. J. Winzer, and R.-J. Essiambre, in *Raman Amplifiers for Telecommunications*, M. N. Islam, Ed., Springer, New York, 2003, Chap. 15.
[51] S. V. Chernikov, Y. Zhu, R. Kashyap, and J. R. Taylor, *Electron. Lett.* **31**, 472 (1995).
[52] P. B. Hansen, L. Eskilden, S. G. Grubb, A. J. Stentz, T. A. Strasser, J. Judkins, J. J. DeMarco, J. R. Pedrazzani, and D. J. DiGiovanni, *IEEE Photon. Technol. Lett.* **9**, 262 (1997).
[53] A. Bertoni and G. C. Reali, *Appl. Phys. B* **67**, 5 (1998).
[54] D. V. Gapontsev, S. V. Chernikov, and J. R. Taylor, *Opt. Commun.* **166**, 85 (1999).
[55] H. Masuda, S. Kawai, K. Suzuki, and K. Aida, *IEEE Photon. Technol. Lett.* **10**, 516 (1998).
[56] Y. Emori, K. Tanaka, and S. Namiki, *Electron. Lett.* **35**, 1355 (1999).
[57] S. A. E. Lewis, S. V. Chernikov, and J. R. Taylor, *Electron. Lett.* **35**, 1761 (1999).
[58] H. D. Kidorf, K. Rottwitt, M. Nissov, M. X. Ma, and E. Rabarijaona, *IEEE Photon. Technol. Lett.* **12**, 530 (1999).
[59] H. Suzuki, J. Kani, H. Masuda, N. Takachio, K. Iwatsuki, Y. Tada, and M. Sumida, *IEEE Photon. Technol. Lett.* **12**, 903 (2000).
[60] B. Zhu, L. Leng, L. E. Nelson, Y. Qian, L. Cowsar, S. Stulz, C. Doerr, L. Stulz, S. Chandrasekhar, et al., *Electron. Lett.* **37**, 844 (2001).
[61] K. Rottwitt and A. J. Stentz, in *Optical Fiber Telecommunications*, Vol. 4A, I. Kaminow and T. Li, Eds., Academic Press, San Diego, 2002, Chap. 5.
[62] P. Wan and J. Conradi, *J. Lightwave Technol.* **14**, 288 (1996).
[63] P. B. Hansen, L. Eskilden, A. J. Stentz, T. A. Strasser, J. Judkins, J. J. DeMarco, R. Pedrazzani, and D. J. DiGiovanni, *IEEE Photon. Technol. Lett.* **10**, 159 (1998).
[64] M. Nissov, K. Rottwitt, H. D. Kidorf, and M. X. Ma, *Electron. Lett.* **35**, 997 (1999).
[65] S. R. Chinn, *IEEE Photon. Technol. Lett.* **11**, 1632 (1999).
[66] S. A. E. Lewis, S. V. Chernikov, and J. R. Taylor, *IEEE Photon. Technol. Lett.* **12**, 528 (2000).
[67] C. H. Kim, J. Bromage, and R. M. Jopson, *IEEE Photon. Technol. Lett.* **14**, 573 (2002).
[68] J. P. Gordon and H. Kogelnik, *Proc. Natl. Acad. Sci. USA* **97**, 4541 (2000).
[69] D. Wang and C. R. Menyuk, *J. Lightwave Technol.* **19** 487-494 (2001).
[70] H. Kogelnik, R. M. Jopson, and L. E. Nelson, in *Optical Fiber Telecommunications*, Vol. 4B, I. P. Kaminow and T. Li, Eds., Academic, San Diego, 2002, Chap. 15.
[71] P. Ebrahimi, M. C. Hauer, Q. Yu, R. Khosravani, D. Gurkan, D. W. kim, D. W. Lee, and A. E. Willner, Proc. Conf. on Lasers and Electro-optics (Optical Society of America, Washington, D.C., 2001), pp. 143-144.

[72] S. Popov, E. Vanin, and G. Jacobsen, *Opt. Lett.* **27**, 848 (2002).
[73] Q. Lin and G. P. Agrawal, *Opt. Lett.* **27**, 2194 (2002); *Opt. Lett.* **28**, 227 (2003);
[74] Q. Lin and G. P. Agrawal, *J. Opt. Soc. Am. B* **32**, 1616 (2003).
[75] C. R. S. Fludger, V. Handerek, and R. J. Mears, *J. Lightwave Technol.* **19**, 1140 (2001).
[76] K. Inoue, *J. Lightwave Technol.* **12**, 1916 (1994).
[77] M. E. Marhic, N. Kagi, T. K. Chiang, and L. G. Kazovsky, *Opt. Lett.* **21**, 573 (1996).
[78] M. Onishi, T. Okuno, T. Kashiwada, S. Ishikawa, N. Akasaka, and M. Nishimura, *Opt. Fiber Technol.* **4**, 204 (1998).
[79] M. E. Marhic, F. S. Yang, M. C. Ho, and L. G. Kazovsky, *J. Lightwave Technol.* **17**, 210 (1999).
[80] J. Hansryd, P. A. Andrekson, M. Westlund, J. Li, and P. O. Hedekvist, *IEEE J. Sel. Topics Quantum Electron.* **8**, 506 (2002).
[81] M. N. Islam and Ö. Boyraz, *IEEE J. Sel. Topics Quantum Electron.* **8**, 527 (2002).
[82] C. J. McKinstrie, S. Radic, and A. Chraplyvy, *IEEE J. Sel. Topics Quantum Electron.* **8**, 538 (2002).
[83] S. Radic, C. J. McKinstrie, A. R. Chraplyvy, G. Raybon, J. C. Centanni, C. G. Jorgensen, K. Brar, and C. Headley, *IEEE Photon. Technol. Lett.* **14**, 1406 (2002).
[84] S. Radic and C. J. McKinstrie, *Opt. Fiber Technol.* **9**, 7 (2003).
[85] R. H. Stolen and J. E. Bjorkholm, *IEEE J. Quantum Electron.* **QE-18**, 1062 (1982).
[86] S. Radic, C. J. McKinstrie, R. M. Jopson, J. C. Centanni, Q. Lin, and G. P. Agrawal, *Electron. Lett.* **39**, 838 (2003).
[87] J. L. Blows and S. E. French, *Opt. Lett.* **27**, 491 (2003).
[88] S. K. Choi, R. D. Li, C. Kim, and P. Kumar, *J. Opt. Soc. Am. B* **14**, 1564 (1997).
[89] W. Imajuku, A. Takad, and Y. Yamabayashi, *Electron. Lett.* **35**, 1954 (1999).
[90] M. E. Marhic, K. K. Y. Wong, and L. G. Kazovsky, *Electron. Lett.* **39**, 350 (2003).
[91] F. Yaman, Q. Lin and G. P. Agrawal, *IEEE Photon. Technol. Lett.* **16**, 431 (2004).
[92] M. Shimizu, H. Suda, and M. Horiguchi, *Electron. Lett.* **23**, 768 (1987).
[93] G. A. Ball, W. W. Morey, and W. H. Glenn, *IEEE Photon. Technol. Lett.* **3**, 613 (1991).
[94] W. J. Barnes, S. B. Poole, J. E. Townsend, L. Reekie, D. J. Taylor, and D. N. Payne, *J. Lightwave Technol.* **7**, 1461 (1989).
[95] P. R. Morkel, in *Rare Earth Doped Fiber Lasers and Amplifiers*, M. J. F. Digonnet, Ed., Marcel Dekker, New York, 2001, Chap. 6.
[96] R. Wyatt, *Electron. Lett.* **26**, 1498 (1989).
[97] J. L. Zyskind, J. W. Sulhoff, J. W. Stone, D. J. DiGiovanni, L. W. Stultz, H. M. Presby, A. Piccirilli, and P. E. Pramayon, *Electron. Lett.* **27**, 2148 (1991).
[98] C. V. Poulsen and M. Sejka, *IEEE Photon. Technol. Lett.* **5**, 646 (1993).
[99] P. D. Humphrey and J. E. Bowers, *IEEE Photon. Technol. Lett.* **5**, 32 (1993).
[100] Y. T. Chieng and R. A. Minasian, *IEEE Photon. Technol. Lett.* **6**, 153 (1994).
[101] S. V. Chernikov, R. Kashyap, P. F. McKee, and J. R. Taylor, *Electron. Lett.* **29**, 1089 (1993).
[102] S. V. Chernikov, J. R. Taylor, and R. Kashyap, *Opt. Lett.* **18**, 2024 (1993); *Electron. Lett.* **29**, 1788 (1993).
[103] J. T. Kringlebton, P. R. Morkel, L. Reekie, J. L. Archambault, and D. N. Payne, *IEEE Photon. Technol. Lett.* **5**, 1162 (1993).
[104] G. A. Ball, C. E. Holton, G. Hull-Allen, and W. W. Morey, *IEEE Photon. Technol. Lett.* **6**, 192 (1994).
[105] W. H. Loh and R. I. Lamming, *Electron. Lett.* **31**, 1440 (1995).

References

[106] M. Sejka, P. Varming, J. Hübner, and M. Kristensen, *Electron. Lett.* **31**, 1445 (1995).
[107] J. L. Archambault and S. G. Grubb, *J. Lightwave Technol.* **15**, 1378 (1997).
[108] K. Hsu, W. H. Loh, L. Dong, and C. M. Miller, *J. Lightwave Technol.* **15**, 1438 (1997).
[109] W. H. Loh, B. N. Samson, L. Dong, G. J. Cowle, and K. Hsu, *J. Lightwave Technol.* **16**, 114 (1998).
[110] J. Chow, G. Town, B. Eggleton, M. Isben, K. Sugden, and I. Bennion, *IEEE Photon. Technol. Lett.* **8**, 60 (1996).
[111] O. Graydon, W. H. Loh, R. I. Laming, and L. Dong, *IEEE Photon. Technol. Lett.* **8**, 63 (1996).
[112] S. Yamashita and K. Hotate, *Electron. Lett.* **32**, 1298 (1996).
[113] N. Park and P. F. Wysocki, *IEEE Photon. Technol. Lett.* **8**, 1459 (1996).
[114] J. Hübner, P. Varming, and M. Kristensen, *Electron. Lett.* **33**, 139 (1997).
[115] S. Yamashita, K. Hsu, and W. H. Loh, *IEEE J. Sel. Topics Quantum Electron.* **3**, 1058 (1997).
[116] N. J. C. Libatique and R. K. Jain, *IEEE Photon. Technol. Lett.* **11**, 1584 (1999).
[117] J. Sun, J. Qiu, and D. Huang, *Opt. Commun.* **182**, 193 (2000).
[118] N. Pleros, C. Bintjas, M. Kalyvas, G. Theophilopoulos, K. Yiannopoulos, S. Sygletos, and H. Avramopoulos, *IEEE Photon. Technol. Lett.* **14**, 693 (2002).
[119] B. C. Collings, K. Bergman, S. T. Cundiff, S. Tsuda, J. N. Kutz, J. E. Cunningham, W. Y. Jan, M. Koch, and W. H. Knox, *IEEE J. Sel. Topics Quantum Electron.* **3**, 1065 (1997).
[120] B. C. Collings, K. Bergman, and W. H. Knox, *Opt. Lett.* **23**, 123, (1998).
[121] M. Nakazawa, E. Yoshida, and Y. Kimura, *Appl. Phys. Lett.* **59**, 2073 (1991).
[122] D. J. Richardson, A. B. Grudinin, and D. N. Payne, *Electron. Lett.* **28**, 778 (1992).
[123] M. Nakazawa, E. Yoshida, and Y. Kimura, *Electron. Lett.* **29**, 63 (1993).
[124] T. O. Tsun, M. K. Islam, and P. L. Chu, *Opt. Commun.* **141**, 65 (1997).
[125] K. Tamura, H. A. Haus, and E. P. Ippen, *Electron. Lett.* **28**, 2226 (1992).
[126] K. Tamura, E. P. Ippen, H. A. Haus, and L. E. Nelson, *Opt. Lett.* **18**, 1080 (1993).
[127] K. Tamura, E. P. Ippen, and H. A. Haus, *Appl. Phys. Lett.* **67**, 158 (1995).
[128] M. E. Fermann, L. M. Yang, M. L. Stock, and M. J. Andrejco, *Opt. Lett.* **19**, 43 (1994).
[129] A. Takada and H. Miyazawa, *Electron. Lett.* **26**, 216 (1990).
[130] H. Takara, S. Kawanishi, M. Saruwatari, and K. Noguchi, *Electron. Lett.* **28**, 2095 (1992).
[131] T. Pfeiffer and G. Veith, *Electron. Lett.* **29**, 1849 (1993).
[132] M. Nakazawa and E. Yoshida, *IEEE Photon. Technol. Lett.* **12**, 1613 (2000).

Chapter 4

Planar Waveguides

As optical fibers neither conduct electricity nor respond to an external voltage, they cannot be used for many functions that are needed for lightwave technology. Planar waveguides can be made using several different types of materials and have thus found a variety of important applications. Fiber-optic communication systems need optical sources such as light-emitting diodes and semiconductor lasers that make use of planar waveguides. Electro-optic materials are employed for fabricating planar waveguides suitable for modulators. Even silica and organic polymers can be used to form planar waveguides that are useful for a wide variety of applications. This chapter focuses on the physics and technologies behind the fabrication of planar waveguides that are useful for making a multitude of active and passive components for lightwave technology. After introducing the concept of waveguide modes in Section 4.1 through a three-layer planar structure, we consider rectangular waveguides in Section 4.2. In Section 4.3 we discuss the materials and the techniques used for fabricating different kinds of waveguides. Simple passive components made with planar waveguides are described in Section 4.4, whereas Section 4.5 focuses on more advanced devices built in the form of planar lightwave circuits.

4.1 Basic Concepts

A planar waveguide differs from an optical fiber in its design but not in the basic physics. The main design difference is that a planar waveguide consists of a central core film (thickness <0.1 mm) sandwiched between two layers whose refractive indices must be lower than that of the core layer but need not be identical. Figure 4.1 shows schematically such a three-layer asymmetric structure and their different refractive indices. In practice, the bottom layer (refractive index n_s) is often a substrate over which the core layer is deposited. The top cladding layer is sometimes called the cover layer (refractive index n_c) as it covers the waveguide core. In some cases, the top layer is absent, and air acts as a cover ($n_c = 1$). Of course, one can design a planar waveguide such that both cladding layers have the same refractive index ($n_c = n_s$). Such waveguides are referred to as being *symmetric*.

4.1 Basic Concepts

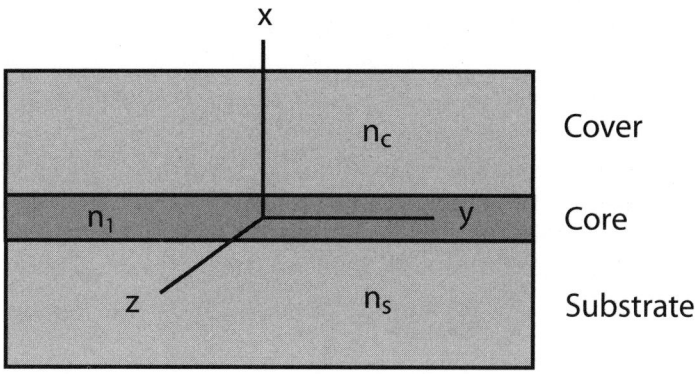

Figure 4.1: Cross section of a planar waveguide consisting of three layers with different refractive indices such that both n_s and n_c are smaller than n_1.

Similar to the case of optical fibers, the phenomenon of total internal reflection is responsible for the confinement of light to the central layer in Figure 4.1. A ray picture based on geometrical optics can be used to discuss how light is totally reflected at both interfaces whenever its angle of incidence exceeds the largest of two critical angles defined through $\sin\phi_c = n_c/n_1$ and $\sin\phi_s = n_s/n_1$. However, the geometrical-optics description is valid only when the waveguide thickness $2d$ is much larger than the light wavelength λ. In practice, the two are often comparable to ensure the single-mode operation. For this reason, we first focus on the waveguide modes found by solving the Maxwell equations [1]–[3].

4.1.1 Modes of Planar Waveguides

An *optical mode* refers to a specific solution of the Maxwell equations, given in Section 1.3.1 as Eqs. (1.3.1) through (1.3.4), such that it satisfies all boundary conditions and has the property that its spatial distribution does not change with propagation. A simple way to obtain optical modes is to assume that the electric and magnetic fields vary with time as $\exp(-i\omega t)$, where ω is the frequency of the mode. If we use $\mathbf{M} = 0$ and $\mathbf{P} = \tilde{\chi}\mathbf{E}$ in Eqs. (1.3.5) and (1.3.6), the six electromagnetic components are related to each other through

$$\nabla \times \mathbf{E} = i\omega\mu_0 \mathbf{H}, \qquad (4.1.1)$$

$$\nabla \times \mathbf{H} = -i\omega\varepsilon_0 n^2 \mathbf{E}, \qquad (4.1.2)$$

where the refractive index n is defined in Eq. (1.3.13). We have neglected losses as they are small in practice and can be included in a simple manner if necessary.

We now use the geometry of planar waveguides to simplify the two curl equations. Assuming that the x axis is normal to the waveguide plane and that the waveguide is infinitely wide along y axis, we note that both $\mathbf{E}$ and $\mathbf{H}$ are y-independent. Moreover, both of them vary with z for an optical mode propagating along the z axis as $\exp(i\beta z)$, where β is the propagation constant of the mode. We can thus remove the y and z

derivatives from Eqs. (4.1.1) and (4.1.2) using

$$\frac{\partial \mathbf{E}}{\partial y} = 0, \quad \frac{\partial \mathbf{H}}{\partial y} = 0, \quad \frac{\partial \mathbf{E}}{\partial z} = i\beta \mathbf{E}, \quad \frac{\partial \mathbf{H}}{\partial z} = i\beta \mathbf{H}. \quad (4.1.3)$$

The resulting set of six equations, when written in Cartesian coordinates, has two distinct sets of linearly polarized solutions, known as the transverse electric (TE) and transverse magnetic (TM modes), depending on whether we choose $E_z = 0$ or $H_z = 0$.

In the case of TE modes, $E_x = E_z = 0$, while E_y satisfies

$$\frac{d^2 E_y}{dx^2} + (n^2 k_0^2 - \beta^2) E_y = 0, \quad (4.1.4)$$

where $k_0 = \omega\sqrt{\varepsilon_0 \mu_0} = \omega/c$. The magnetic filed components are related to E_y as

$$H_x = -\frac{\beta}{\omega\mu_0} E_y, \quad H_y = 0, \quad H_z = -\frac{i}{\omega\mu_0}\frac{dE_y}{dx}. \quad (4.1.5)$$

In the case of TM modes, $H_x = H_z = 0$, while H_y satisfies

$$\frac{d^2 H_y}{dx^2} + (n^2 k_0^2 - \beta^2) H_y = 0. \quad (4.1.6)$$

The electric filed components are now related to H_y as

$$E_x = \frac{\beta}{\omega\varepsilon_0 n^2} H_y, \quad E_y = 0, \quad E_z = \frac{i}{\omega\varepsilon_0 n^2}\frac{dH_y}{dx}. \quad (4.1.7)$$

Consider the TE modes first. Equation (4.1.4) can be solved within each layer separately using $n = n_c$, n_1, and n_s for the three layers seen in Figure 4.1. Since the refractive index is constant in each layer, the general solution can be written in terms of sinusoidal and exponential functions as [2]

$$E_y(x) = \begin{cases} B_c \exp[-q_1(x-d)]; & x > d, \\ A \cos(px - \phi) & ; \quad |x| \leq d, \\ B_s \exp[q_2(x+d)] & ; \quad x < -d, \end{cases} \quad (4.1.8)$$

where we discarded the exponentially growing terms in the cover and substrate layers because each guided mode is confined to the core. The constants p, q_1, and q_2 are defined as

$$p^2 = n_1^2 k_0^2 - \beta^2, \quad q_1^2 = \beta^2 - n_c^2 k_0^2, \quad q_2^2 = \beta^2 - n_s^2 k_0^2. \quad (4.1.9)$$

The four constants B_c, B_s, A, and ϕ need to be determined from the boundary conditions at the two interfaces requiring that the tangential components of **E** and **H** be continuous across them. In the case of TE modes, these boundary conditions are satisfied if E_y and H_z are continuous at $x = \pm d$. From Eq. (4.1.8), E_y is continuous at $x = \pm d$, provided

$$B_c = A\cos(pd - \phi); \quad B_s = A\cos(pd + \phi). \quad (4.1.10)$$

4.1 Basic Concepts

Since H_z is related to the derivative dE_y/dx, the continuity of H_z is ensured if this derivative is continuous at $x = \pm d$. This requirement leads to the following two relations between A and ϕ:

$$pA\sin(pd - \phi) = q_1 A\cos(pd - \phi), \qquad (4.1.11)$$

$$pA\sin(pd + \phi) = q_2 A\cos(pd + \phi). \qquad (4.1.12)$$

If we eliminate A from these equations, ϕ must satisfy simultaneously

$$\tan(pd - \phi) = q_1/p, \qquad \tan(pd + \phi) = q_2/p. \qquad (4.1.13)$$

These equations can be solved to obtain the following expression for ϕ:

$$2\phi = m\pi - \tan^{-1}(q_1/p) + \tan^{-1}(q_2/p), \qquad (4.1.14)$$

where $m = 0, 1, 2, \ldots$ is an integer.

The propagation constant β can also be determined from Eq. (4.1.13). Eliminating ϕ, we obtain a single relation, known as the *eigenvalue equation*, among p, q_1, and q_2 in the form

$$2pd = m\pi + \tan^{-1}(q_1/p) + \tan^{-1}(q_2/p), \qquad (4.1.15)$$

where $m = 0, 1, 2, \ldots$ is an integer. From Eq. (4.1.9), this eigenvalue equation determines β for the TE modes in terms of the waveguide parameters. Since a different solution exists for each value of m, the modes are labeled as TE$_m$ modes.

In the case of TM modes, we need to solve Eq. (4.1.6). Noting that this equation is identical to Eq. (4.1.4), we see that the solution for H_y is also given by Eq. (4.1.8). However, the boundary condition demanding the continuity of E_z requires that the quantity $n^{-2}(dH_y/dx)$ be continuous at $x = \pm d$. Since n is different on the two sides of each interface, the eigenvalue equation is modified slightly and takes the form

$$2pd = m\pi + \tan^{-1}\left(\frac{n_1^2 q_1}{n_c^2 p}\right) + \tan^{-1}\left(\frac{n_1^2 q_2}{n_s^2 p}\right). \qquad (4.1.16)$$

Again multiple solutions are possible for different values of m and are labeled as TM$_m$ modes.

4.1.2 TE and TM Modes of Symmetric Waveguides

The preceding eigenvalue equations for TE and TM modes are simplified considerably for symmetric waveguides for which the two cladding layers surrounding the core layer have the same refractive index ($n_c = n_s$). Noticing from Eq. (4.1.9) that $q_1 = q_2 \equiv q$, we can write the eigenvalue equation for TE modes, Eq. (4.1.15), as

$$q = p\tan(pd - m\pi/2). \qquad (4.1.17)$$

Similar to the fiber case, we introduce a normalized parameter,

$$V = d\sqrt{p^2 + q^2} = k_0 d\sqrt{n_1^2 - n_s^2}, \qquad (4.1.18)$$

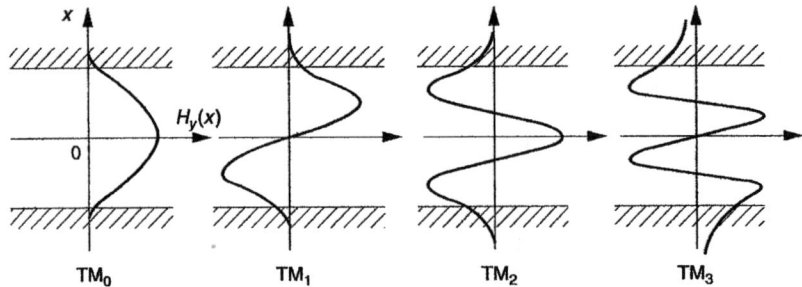

Figure 4.2: Spatial profiles associated with the first four TM modes of a symmetric planar waveguide. Shaded regions mark core–cladding interfaces. (After Ref. [3]; ©2002 Wiley.)

which governs the properties of all waveguide modes. If we use $u = pd$, the eigenvalue equation can be written in terms of V as

$$\sqrt{V^2 - u^2} = u\tan(u - m\pi/2). \tag{4.1.19}$$

For given value of V and m, this equation is solved to find $p = u/d$. The propagation constant for that mode is then found from Eq. (4.1.9) using $\beta = \bar{n}k_0 = (n_1^2 k_0^2 - p^2)^{1/2}$, where $\bar{n}$ is the effective mode index. The spatial distribution of the mode is found from Eq. (4.1.8). For a symmetric waveguide, $\phi = m\pi/2$, and E_y can be written as

$$E_y(x) = \begin{cases} B_\pm \exp[-q(|x|-d)]; & |x| > d, \\ A\cos(px - m\pi/2) \; ; & |x| \le d, \end{cases} \tag{4.1.20}$$

where $B_\pm = A\cos(pd \mp m\pi/2)$ and the lower sign is chosen for $x < 0$. In general, the modes with even values of m are symmetric around $x = 0$ and are called *even* modes. In contrast, the modes with odd values of m are asymmetric and are called *odd* modes.

The TM modes can be determined following a similar procedure. Using $q_1 = q_2 \equiv q$ in Eq. (4.1.16), we can write the eigenvalue equation for TM modes as

$$(n_1/n_s)^2 q = p\tan(pd - m\pi/2). \tag{4.1.21}$$

The mode index for TM modes is slightly lower because of the factor $(n_1/n_s)^2$ appearing in the preceding equation. The field distribution for each mode is completely specified in terms of $H_y(x)$, which can be written in the form of Eq. (4.1.20). Figure 4.2 shows, as an example, the mode profiles for the first four TM$_m$ modes ($m = 0, 1, 2$, and 3) for a symmetric planar waveguide. Similar to the case of TE modes, TM modes can also be divided into even and odd categories depending on whether m is even or odd. The value of m denotes the number of nodes within the core layer at which the field amplitude vanishes. The TE$_0$ and TM$_0$ modes for which $m = 0$ have no nodes within the core are called the fundamental modes of a planar waveguide.

The number of modes supported by a waveguide depends on the V parameter. A mode ceases to exist when $q = 0$ since, from Eq. (4.1.20), the mode is then no longer confined to the core. From Eqs. (4.1.17) and (4.1.21), this occurs for both the TE and TM modes when $V = V_m = m\pi/2$. Thus, for a given value of V, the number of modes

4.1 Basic Concepts

equals the largest value of m for which V_m first exceeds V. For example, a waveguide with $V = 10$ supports 7 TE modes and 7 TM modes since $V_6 = 9.42$ but V_7 exceeds 10. A waveguide will support a single TE and a single TM mode when its width is chosen such that $V < \pi/2$. This is the *single-mode condition* for planar waveguides.

4.1.3 TE and TM Modes of Asymmetric Waveguides

In the case of an asymmetric waveguide with $n_c \ne n_s$, one must solve the eigenvalue equations (4.1.15) and (4.1.16) to find the effective mode index $\bar{n}$ for the TE and TM modes supported by the waveguide. We assume for definiteness that $n_1 > n_s > n_c$. The guided modes exist as long as $n_1 > \bar{n} > n_s$. It is useful to introduce two normalized parameters as

$$b = \frac{\bar{n}^2 - n_s^2}{n_1^2 - n_s^2}, \qquad \delta = \frac{n_s^2 - n_c^2}{n_1^2 - n_s^2}. \tag{4.1.22}$$

The b parameter lies in the range $0 < b < 1$ and is called the normalized propagation constant as it is related to β. The parameter δ is a measure of waveguide asymmetry and it vanishes for a symmetric waveguide with $n_c = n_s$.

In terms of the normalized parameters V, b, and δ, the eigenvalue equation (4.1.15) for the TE modes can be written as

$$2V\sqrt{1-b} = m\pi + \tan^{-1}\sqrt{\frac{b}{1-b}} + \tan^{-1}\sqrt{\frac{b+\delta}{1-b}}. \tag{4.1.23}$$

The solutions of this equation provide the dispersion curves for the TE modes. Figure 4.3 shows such dispersion curves for a few low-order modes by plotting b as a function of V for $\delta = 5$. The symmetric case ($\delta = 0$) is also shown for comparison. For a given value of V, the effective index of each mode decreases as δ increases.

The cutoff condition for each TE mode can be obtained by finding the value of V for which the mode ceases to decay exponentially in one or both of the cladding layers. It is obtained by setting $b = 0$ in Eq. (4.1.23) and is given by

$$V_m(\text{TE}) = \frac{m\pi}{2} + \frac{1}{2}\tan^{-1}\sqrt{\delta}. \tag{4.1.24}$$

The eigenvalue equation for the TM modes can be obtained by following a similar procedure and is found to be

$$2V\sqrt{1-b} = m\pi + \tan^{-1}\left(\frac{n_1^2}{n_s^2}\sqrt{\frac{b}{1-b}}\right) + \tan^{-1}\left(\frac{n_1^2}{n_c^2}\sqrt{\frac{b+\delta}{1-b}}\right). \tag{4.1.25}$$

The cutoff condition is again found by setting $b = 0$ and is given by

$$V_m(\text{TM}) = \frac{m\pi}{2} + \frac{1}{2}\tan^{-1}\left(\frac{n_1^2}{n_c^2}\sqrt{\delta}\right). \tag{4.1.26}$$

For a symmetric waveguide ($\delta = 0$), the two cutoff conditions reduce to a single condition, $V_m = m\pi/2$, as they should. In general, any asymmetry in the waveguide raises

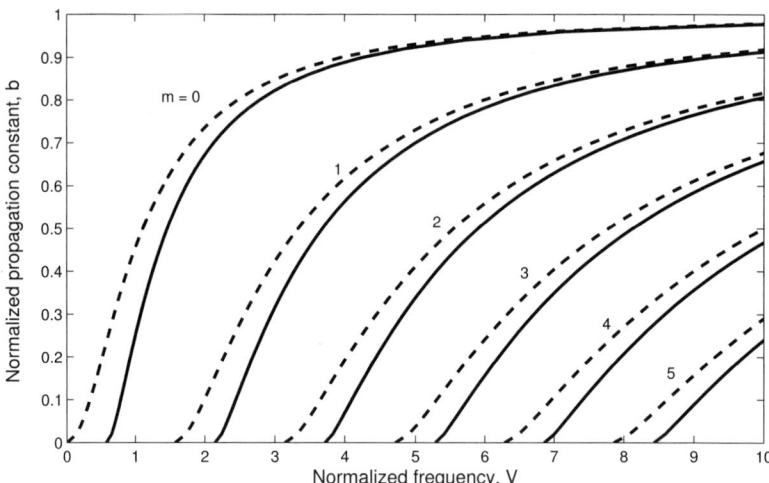

Figure 4.3: Dispersion curves (solid curves) for TE modes of an asymmetric planar waveguide with $\delta = 5$. Dashed curves show the symmetric case ($\delta = 0$) for comparison.

the value of V at which the mode ceases to exist. This has an interesting consequence for thin waveguides. Whereas the fundamental mode has no cutoff and thus always exists no matter how thin a symmetric waveguide is, this is not the case for asymmetric waveguides. For example, a waveguide for which $2V < \tan^{-1}\sqrt{\delta}$ does not support any bounded mode.

The spatial distribution of each TE mode is governed by Eq. (4.1.8) with B_c and B_s as given in Eq. (4.1.10). The constant A in Eq. (4.1.8) is related to the total power P launched into the waveguide through the relation [2]

$$P = \frac{\beta}{2\omega\mu_0} \int_{-\infty}^{\infty} |E_y(x)|^2 dx. \tag{4.1.27}$$

The integral can be performed using E_y from Eq. (4.1.8) and leads to

$$P = \frac{\beta A^2}{4\omega\mu_0}\left(2d + \frac{1}{q_1} + \frac{1}{q_2}\right). \tag{4.1.28}$$

It is also possible to calculate the confinement factor Γ representing the fraction of the power propagating inside the waveguide layer. It is found to be

$$\Gamma = \frac{\int_{-d}^{d} |E_y(x)|^2 dx}{\int_{-\infty}^{\infty} |E_y(x)|^2 dx} = \frac{2d + \sin^2(pd-\phi)/q_1 + \sin^2(pd+\phi)/q_2}{2d + 1/q_1 + 1/q_2}. \tag{4.1.29}$$

Similar to the case of optical fibers, Γ is close to 1 for the fundamental mode until V becomes significantly smaller than $\pi/2$.

4.2 Rectangular Waveguides

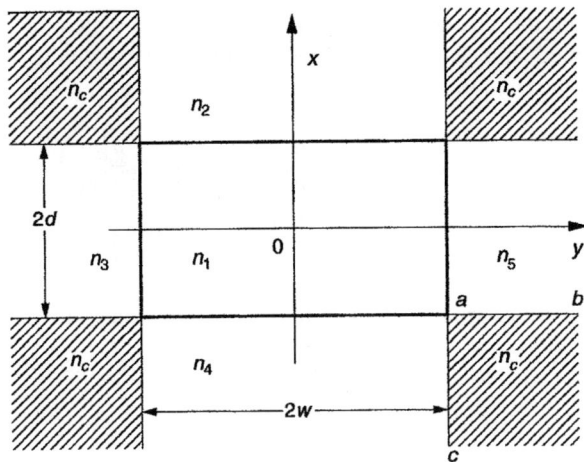

Figure 4.4: Schematic of a rectangular waveguide with different lower-index materials on its four sides. All shaded regions have the same refractive index.

4.2 Rectangular Waveguides

A planar waveguide of the form shown in Figure 4.1 guides light only in the vertical x direction. In practice, it is necessary to confine the light in both transverse dimensions by using a rectangular waveguide. This is realized by restricting the high-index region in the middle core layer to a finite width $2w$ and surrounding it on all sides by lower-index materials. In this section we first discuss the modes associated with rectangular waveguides and then focus on the design of such waveguides [4]–[6].

4.2.1 Modes of Rectangular Waveguides

In general, the refractive index can be different on the four sides of a rectangular waveguide. Figure 4.4 shows the geometry in the most general case in which a rectangular core of width $2w$, thickness $2d$, and index n_1 is surrounded by materials with refractive indices n_2 to n_5. To simplify the analysis, all shaded cladding regions are assumed to have the same refractive index n_c. Even then, it is not easy to analyze such a waveguide, and a numerical approach must be used whenever one needs the exact solution of Maxwell's equations. However, considerable physical insight is gained by making two simplifications. First, the boundary conditions associated with the hatched regions are ignored. Second, it is assumed that the core-cladding index differences are relatively small on all sides. One can then reduce the problem to two planar-waveguide problems in the x and y directions, both of which can be solved using the method discussed in Section 4.1.

Optical modes of a rectangular waveguide with the preceding two approximations were first found in Ref. [4]. Even though pure TE and TM modes do not exist, one can find TE- and TM-like modes for which either E_z, or H_z is nearly negligible compared to other field components. They are denoted as E^x_{mn} and E^y_{mn}, where the subscripts take integer values and the superscript denotes the direction in which the mode is linearly polarized. The mode indices for the E^x_{mn} modes are found by solving the following two coupled eigenvalue equations:

$$2p_x d = m\pi + \tan^{-1}\left(\frac{n_1^2 q_2}{n_2^2 p_x}\right) + \tan^{-1}\left(\frac{n_1^2 q_4}{n_4^2 p_x}\right), \quad (4.2.1)$$

$$2p_y w = n\pi + \tan^{-1}\left(\frac{q_3}{p_y}\right) + \tan^{-1}\left(\frac{q_5}{p_y}\right), \quad (4.2.2)$$

where the six quantities are related to various refractive indices as

$$p_x^2 = n_1^2 k_0^2 - \beta^2 - p_y^2, \quad p_y^2 = n_1^2 k_0^2 - \beta^2 - p_x^2, \quad (4.2.3)$$

$$q_2^2 = \beta^2 + p_y^2 - n_2^2 k_0^2, \quad q_4^2 = \beta^2 + p_y^2 - n_4^2 k_0^2, \quad (4.2.4)$$

$$q_3^2 = \beta^2 + p_x^2 - n_3^2 k_0^2, \quad q_5^2 = \beta^2 + p_x^2 - n_5^2 k_0^2, \quad (4.2.5)$$

and β is the mode propagation constant. These equations can be solved numerically to find the mode index $\bar{n} = \beta/k_0$.

In many practical cases one can employ a technique known as the *effective-index method* [6]. This method makes use of the typical practical situation in which the thickness of a rectangular waveguide is much smaller than its width ($d \ll w$). The planar waveguide problem in the x direction is solved first to obtain the effective mode index n_e, which is a function of y because of the finite width of the waveguide. The problem in the y direction is then solved by considering a planar waveguide of width $2w$ such that $n_y = n_e$ if $|y| < w$ but takes values n_3 and n_5 outside of this region. The single-mode condition in this case is given by

$$V_x = k_0 d \sqrt{n_1^2 - n_4^2} < \pi/2, \quad V_y = k_0 w \sqrt{n_e^2 - n_5^2} < \pi/2, \quad (4.2.6)$$

where we assumed $n_4 > n_2$ and $n_5 > n_3$.

4.2.2 Design of Rectangular Waveguides

The design of rectangular waveguides can vary considerable depending on the practical considerations and the fabrication technique used. Figure 4.5 shows several designs known by names that reflect the shape of the core of the waveguide.

For the ridge waveguide shown in Figure 4.5(a), the core is in the form of a ridge on top of a substrate that acts as the bottom cladding layer. If air is used for the top cladding, a relatively large index difference confines the mode to the ridge in the lateral direction. In the case of a rib waveguide shown in Figure 4.5(b), the thickness of the core layer is made larger over a central region. It is this thickness difference that confines the mode in the lateral direction. In both cases, the core layer is surrounded by air, and the core–air interface should be smooth to avoid scattering losses. In the strip-loaded waveguide shown in Figure 4.5(c), the top cladding layer is in the form of a strip, which provides mode confinement in the lateral direction. Physically, the effective index of the mode is larger in the strip region compared with that occurring

4.2 Rectangular Waveguides

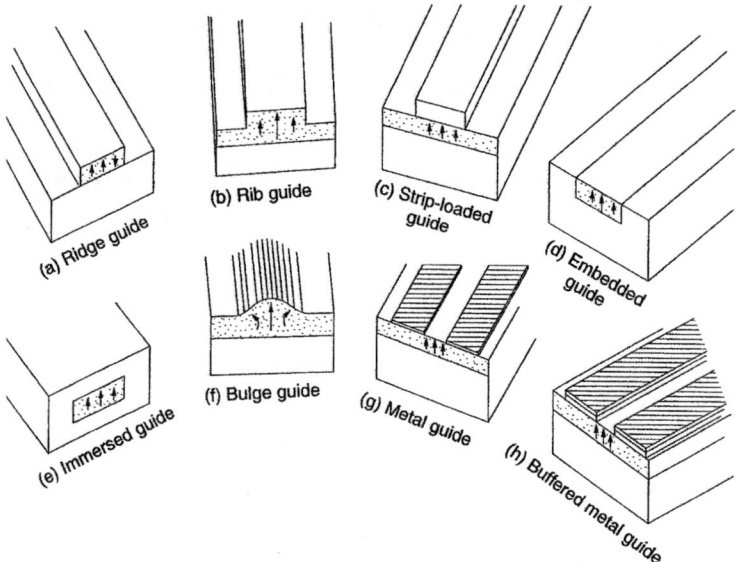

Figure 4.5: Schematic of several designs (a–h) used for rectangular waveguides. Arrows illustrate electric-field lines within the core, while shaded regions show metal stripes. (After Ref. [3]; ©2002 Wiley.)

in the region outside the strip. The resulting index difference confines the mode to the core region below the strip.

Figure 4.5(d) shows an embedded or channel waveguide in which the rectangular core region is embedded within the cladding material while air acts as the top cladding. Since only one surface of the core is exposed to air, scattering losses are reduced in this structure compared with the ridge and rib waveguides. The top of a channel waveguide can be used to place electrodes and apply an electric field directly over the core region. In the case of an immersed or buried waveguide shown in Figure 4.5(e), the rectangular core region is totally immersed within a lower-index material. This design provides the smallest scattering losses but such waveguides are also the most difficult to fabricate. The bulge waveguide shown in Figure 4.5(f) is similar to a rib waveguide in the sense that the core thickness is not constant in the lateral direction. Often, the exact shape of the bulge is not critical for waveguide operation while the smooth nature of the bulge can reduce scattering losses.

In the structure shown in Figure 4.5(g), the core layer is covered with two metal stripes that confine the mode in the lateral direction to the region between them. Physically, metal stripes have a refractive index lower than the surrounding air and thus reduce the effective mode index in the region under them. In this sense, this structure is equivalent to the strip-loaded design shown in Figure 4.5(c), except that air itself acts as a higher-index material. Of course, metal stripes introduce some loss, which can be reduced by placing a thin buffer layer before the metal stripes are deposited. This results in the buffered metal-waveguide structure shown in Figure 4.5(h).

4.3 Materials and Waveguide Fabrication

Planar waveguides can be fabricated using a variety of materials including semiconductors such as GaAs and InP, insulators such as silica (SiO_2) and organic polymers, and electro-optic crystals such as lithium niobate ($LiNbO_3$). In this section we discuss five categories of planar waveguides that have found a multitude of applications in the domain of lightwave technology.

4.3.1 Semiconductor Waveguides

Semiconductor materials are often used to make planar waveguides because they allow the fabrication of electrically active devices. Compound semiconductors belonging to the III–V Group, so named because they use two distinct elements from columns 3 and 5 of the Periodic Table, are used almost exclusively for making optical sources and photodetectors suitable for lightwave applications. We focus on such semiconductors in this section.

When a semiconductor material is used to make a waveguide, one needs at least two semiconductors with different refractive indices so that they can be used for the core and cladding layers. Since the refractive index depends inversely on the bandgap, two semiconductors must have different bandgaps. Typically, one uses GaAs or InP as a substrate on which one or more thin layers are deposited to form a waveguide. The performance of such waveguides depends on the quality of the interface between two semiconductors of different bandgaps. To reduce the formation of lattice defects, the lattice constant of the two materials should match to a degree better than 0.1%. Nature does not provide semiconductors whose lattice constants match to such precision. However, they can be fabricated artificially by forming ternary or quaternary compounds in which a fraction of the lattice sites in a naturally occurring binary semiconductor (e.g., GaAs) is replaced by other elements. In the case of GaAs, a ternary compound $Al_xGa_{1-x}As$ can be made by replacing a fraction x of Ga atoms by Al atoms. The resulting semiconductor has nearly the same lattice constant but its bandgap increases. The bandgap depends on the fraction x and can be approximated by a simple linear relation [7]

$$E_g(x) = 1.424 + 1.247x \qquad (0 < x < 0.45), \qquad (4.3.1)$$

where E_g is expressed in electron-volt (eV) units.

Figure 4.6 shows the interrelationship between the bandgap E_g and the lattice constant a for several ternary and quaternary compounds. Solid dots represent the binary semiconductors, and lines connecting them correspond to ternary compounds. The dashed portion of the line indicates that the resulting ternary compound has an indirect bandgap. The area of a closed polygon corresponds to quaternary compounds. The bandgap is not necessarily direct for such semiconductors. The shaded area in Figure 4.6 represents the ternary and quaternary compounds with a direct bandgap formed by using the elements indium (In), gallium (Ga), arsenic (As), and phosphorus (P).

The horizontal line connecting GaAs and AlAs corresponds to the ternary compound $Al_xGa_{1-x}As$, whose bandgap is direct for values of x up to about 0.45 and is

4.3 Materials and Waveguide Fabrication

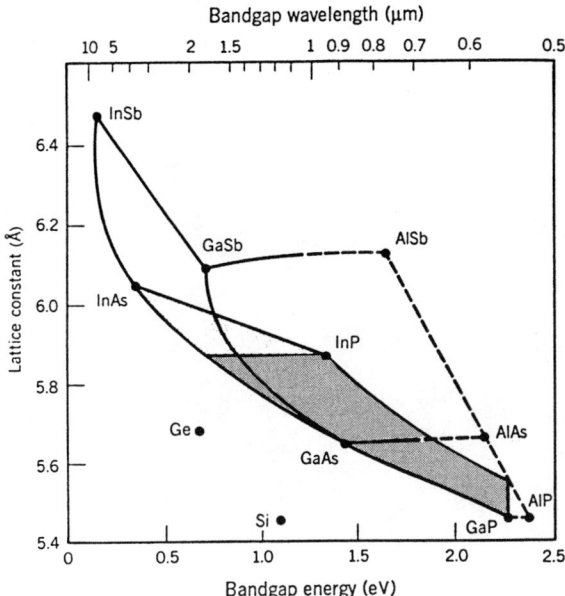

Figure 4.6: Lattice constants and bandgap energies of ternary and quaternary compounds formed by using nine Group III–V semiconductors. Shaded area corresponds to possible InGaAsP and AlGaAs structures. Horizontal lines passing through InP and GaAs show the lattice-matched designs.

given by Eq. (4.3.1). The core and cladding layers are formed such that x is larger for the cladding layers compared with the value of x for the core layer. When such a structure is used to make a laser, the wavelength λ of the emitted light is determined by the bandgap of the core layer since photon energy approximately equals the bandgap. By using $E_g \approx h\nu = hc/\lambda$, one finds that $\lambda \approx 0.87$ μm for a core layer made of GaAs ($E_g = 1.424$ eV). The wavelength can be reduced to near 0.8 μm by using an active layer with $x = 0.1$. Lasers based on GaAs operate in the wavelength range of 0.78–0.87 μm and are used in many commercial products such as compact-disk players and laser printers.

As discussed in Chapter 1, the wavelength range of 1.3–1.6 μm is of special importance for lightwave technology because both the dispersion and losses in silica fibers are considerably reduced compared with the values occurring in the 0.85-μm region. InP is the base material for optical sources emitting light in this wavelength region. As seen in Figure 4.6 by the horizontal line passing through InP, the bandgap of InP can be reduced considerably by making the quaternary compound $In_{1-x}Ga_xAs_yP_{1-y}$ while the lattice constant remains matched to InP. The fractions x and y cannot be chosen arbitrarily but are related by $x/y = 0.45$ to ensure matching of the lattice constant. The bandgap of the quaternary compound can be expressed in terms of y only and is well approximated by [7]

$$E_g(y) = 1.35 - 0.72y + 0.12y^2, \qquad (4.3.2)$$

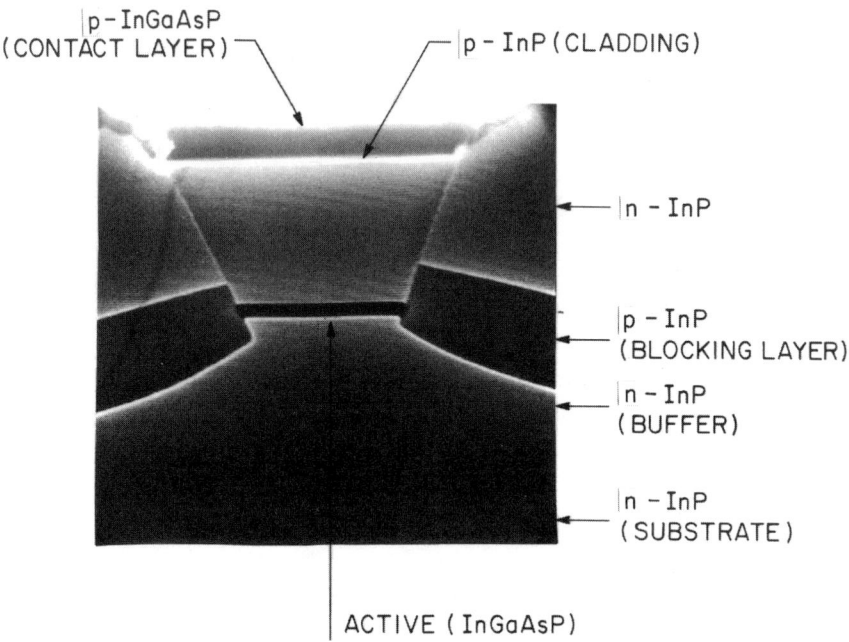

Figure 4.7: SEM photograph showing cross section of a buried waveguide. The active InGaAsP layer is about 2.5 μm wide and is immersed on all sides by InP cladding layers.

where $0 \leq y \leq 1$. By a suitable choice of the mixing fractions x and y, $In_{1-x}Ga_xAs_yP_{1-y}$ waveguides can be designed to work in a wide wavelength range of 1.0–1.65 μm that includes the region from 1.3–1.6 μm that is important for optical communication systems.

The fabrication of semiconductor waveguides requires epitaxial growth of multiple layers on a base substrate (GaAs or InP). The thickness and composition of each layer need to be controlled precisely. Several epitaxial growth techniques can be used for this purpose. The three primary techniques are known as liquid-phase epitaxy (LPE), vapor-phase epitaxy (VPE), and molecular-beam epitaxy (MBE), depending on whether the constituents of various layers are in the liquid form, vapor form, or in the form of a molecular beam. The VPE technique is also called chemical-vapor deposition. A variant of this technique is metal-organic chemical-vapor deposition (MOCVD), in which metal alkalis are used as the mixing compounds. Figure 4.7 shows a scanning-electron-microscope (SEM) photograph of an InGaAsP buried waveguide. This laser structure is made by etching a mesa that removes the active core layer on both sides of the mesa. Cladding layers are then regrown to bury the waveguide.

Both the MOCVD and MBE techniques provide an ability to control layer thickness to within 1 nm. In some cases, the thickness of the core layer is small enough that electrons and holes act as if they are confined to a quantum well. Such confinement leads to quantization of the energy bands into subbands. The main consequence is that the joint density of states acquires a staircase-like structure [8]. Such a modification

4.3 Materials and Waveguide Fabrication

of the density of states affects a semiconductor waveguide considerably and is often used to make modern *quantum-well* lasers [9]. Often, multiple core layers with a thickness of 5–10 nm, separated by transparent barrier layers of about 10 nm thickness, are used to improve the device performance. Such devices are called *multiquantum-well* (MQW) devices. Another feature that improves the device performance sometimes is the introduction of intentional, but controlled, strain within the core layers. The use of thin active layers permits a slight mismatch between lattice constants without introducing defects. The resulting strain changes the band structure and improves the device performance [8]. Such structures are called *strained* quantum wells. The concept of quantum-well layers has also been extended to make waveguides using quantum-wire- and quantum-dot materials in which electrons are confined in more than one dimension [9].

To make an active device such as a laser using semiconductor waveguides, one needs to inject current into the core layer. This can be accomplished through a *p–n* junction, formed by making the two cladding layers of the *p*- and *n*-types. A semiconductor is made *n*- or *p*-type by doping it with impurities whose atoms have an excess valence electron or one less electron compared with the semiconductor atoms. In the case of an *n*-type semiconductor, the excess electrons occupy the conduction-band states, normally empty in undoped (intrinsic) semiconductors. The Fermi level, lying in the middle of the bandgap for intrinsic semiconductors, moves toward the conduction band as the dopant concentration increases. In a heavily doped *n*-type semiconductor, the Fermi level E_{fc} lies inside the conduction band; such semiconductors are said to be degenerate. Similarly, the Fermi level E_{fv} moves toward the valence band for *p*-type semiconductors and lies inside it under heavy doping. In thermal equilibrium, the Fermi level must be continuous across the *p–n* junction. This is achieved through the diffusion of electrons and holes across the junction. The charged impurities left behind set up an electric field strong enough to prevent further diffusion of electrons and holds under equilibrium conditions. This field is referred to as the built-in electric field. Figure 4.8 shows the energy-band diagram of a *p–n* junction in thermal equilibrium (a) and under forward bias (b).

When a *p–n* junction is forward biased by applying an external voltage, the built-in electric field is reduced. This reduction results in diffusion of electrons and holes across the interface separating the core and cladding layers. An electric current begins to flow as a result of carrier diffusion. The current I increases exponentially with the applied voltage V according to the well-known relation [8]

$$I = I_s[\exp(qV/k_BT) - 1], \quad (4.3.3)$$

where I_s is the saturation current and depends on the diffusion coefficients associated with electrons and holes. As seen in Figure 4.8, within the waveguide region, electrons and holes are present simultaneously when the *p–n* junction is forward biased. These electrons and holes can recombine through spontaneous or stimulated emission and generate light in a semiconductor optical source.

The use of semiconductor materials for forming a planar waveguide is doubly beneficial when such a structure is used for making an optical source. The bandgap difference between the two semiconductors helps to confine electrons and holes to the

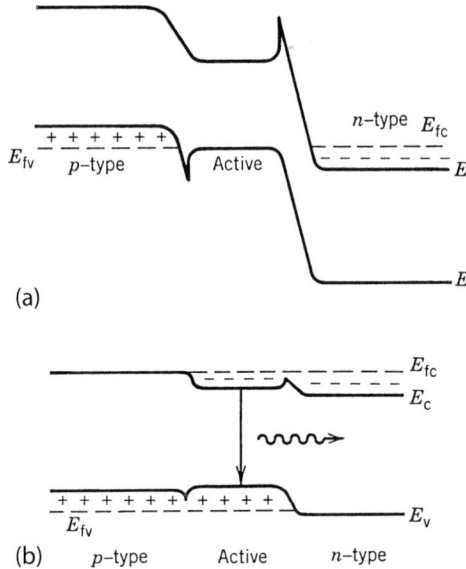

Figure 4.8: Energy-band diagram of a *p–n* junction in thermal equilibrium (a) and under forward bias (b).

middle core layer where they can recombine and generate light. At the same time, the active layer has a slightly larger refractive index than the surrounding *p*-type and *n*-type cladding layers. As a result of this refractive-index difference, any light generated must propagate in the form of a TE or TM mode of the waveguide. It is this feature that has made semiconductor waveguides practical for a wide variety of applications.

4.3.2 Electro-Optic Waveguides

Another group of solid-state materials often used to make planar waveguides belongs to the category of electro-optic crystals. Although such materials do not conduct electricity, they can still be used for making active devices because of the existence of an electro-optic effect, known as the *Pockels effect*; it manifests as a change in the refractive index of the material when an external voltage is applied across it. Crystals belonging to this category include $LiNbO_3$, lithium tantalate ($LiTaO_3$), and barium titanate ($BaTiO_3$), although $LiNbO_3$ is used almost exclusively for making optical modulators for lightwave applications (see Chapter 6).

All electro-optic materials are anisotropic. As a result, the constitutive relation in Eq. (1.3.5) between the field vectors **D** and **E** becomes tensorial and takes the form

$$D_i = \varepsilon_0 \sum_{j=1}^{3} \varepsilon_{ij} E_j. \tag{4.3.4}$$

The dielectric tensor ε_{ij} is governed by a real symmetric matrix ($\varepsilon_{ij} = \varepsilon_{ji}$) for any lossless material that is not optically active. This matrix can be diagonalized by rotating

4.3 Materials and Waveguide Fabrication

the coordinate system appropriately. The basis in which ε_{ij} is diagonal is known as the principal-axis system.

To describe the electro-optic properties, it is more convenient to introduce the impermeability tensor as $\eta_{ij} = 1/\varepsilon_{ij}$ and describe changes induced by an externally applied field $\mathbf{E}^a$ as [10]

$$\eta_{ij}(\mathbf{E}^a) = \eta_{ij}(0) + \sum_k r_{ijk}\mathbf{E}_k^a, \qquad (4.3.5)$$

where the tensor r_{ijk} describes the linear electro-optic effect. Since η is a symmetric matrix, r_{ijk} must be symmetric in its first two indices. For this reason, it is common to introduce a two-dimensional 6×3 matrix r_{hk} for which $h = 1$ to 3 for the three diagonal elements η_{11}, η_{22}, and η_{33} but $h = 4$ to 6 for the three off-diagonal elements η_{23}, η_{31}, and η_{12}, respectively. The 18 elements of the matrix r_{hk} represent the electro-optic coefficients of the crystal. Only a few of them are nonzero depending on the crystal's symmetry group [10]. For crystals with the symmetry $3m$ (e.g., LiNbO$_3$), the matrix has 8 nonzero elements but only 4 among them, namely r_{13}, r_{22}, r_{33}, and r_{42}, are independent. One can select a specific electro-optic coefficient by orienting the crystal along one of the principal axes and choosing the direction of the applied electric field appropriately.

In contrast to the case of semiconductor waveguides, LiNbO$_3$ waveguides do not require an epitaxial growth on the top of a substrate. Rather, the relatively high Curie temperature (about 1100°C) of LiNbO$_3$ allows for fabrication of low-loss waveguides through diffusion of metals directly into a LiNbO$_3$ substrate [11]. The most commonly used element for the in-diffusion process is titanium or Ti [12]. The presence of Ti atoms within the LiNbO$_3$ crystal increases the refractive index and thus forms a waveguide, referred to as a Ti:LiNbO$_3$ waveguide. Care must be taken to suppress the out-diffusion of Li atoms from the surface of the substrate. Surface flatness is also critical to ensure the uniformity of the waveguide along its entire length.

Another method for fabricating LiNbO$_3$ waveguides makes use of a process known as proton exchange [13]. This is a low-temperature (about 200°C) process in which Li ions within the substrate are replaced with protons when the substrate is placed in an acid bath. This technique does not work for y-cut wafers because the acid etches such a wafer chemically. The proton exchange increases the extraordinary part of the refractive index but leaves the ordinary part virtually unchanged. Such a waveguide supports only the TM modes and is useful for some applications because of its polarization selectivity. It is necessary in practice to follow the proton exchange with a high-temperature annealing process, which stabilizes the index difference. Accelerated aging tests predict a lifetime of over 25 years at a temperature as high as 95°C [14].

Before the electro-optic effect can be used to advantage, good-quality electrodes must be fabricated either directly on the surface of the LiNbO$_3$ wafer or on an optically transparent buffer layer that is often inserted to reduce losses. An adhesion layer (typically a Ti layer) is first deposited so that the following metal layer (e.g., a gold layer) sticks to the substrate (or the buffer layer). A photolithographic technique is then used to define the electrode pattern. Figure 4.9 shows schematically the design of an electro-optic modulator made using two LiNbO$_3$ channel waveguides in a Mach–Zehnder configuration. The two electrodes across each waveguide are used to change

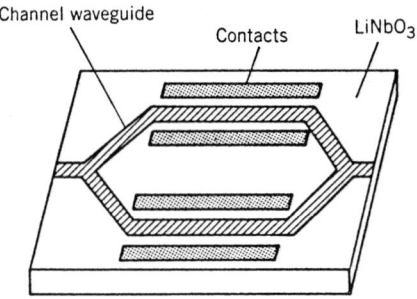

Figure 4.9: Schematic of an electro-optic modulator fabricated using two LiNbO$_3$ waveguides in a Mach–Zehnder configuration.

the refractive index by applying a voltage across them. Such modulators are discussed in Chapter 6.

4.3.3 Silica Waveguides

It is not necessary to use a crystalline material to make planar waveguides. Many passive optical devices make use of silica waveguides fabricated on top of a silicon substrate [15]–[20]. This technique is referred to as the silica-on-silicon technology and has been used in recent years to make planar lightwave circuits that are composed of many passive optical elements and can be used for numerous applications such as dispersion compensation and channel demultiplexing.

Silica-on-Silicon Technology

The basic advantage of forming silica waveguides on a silicon substrate is that one can employ the well-developed silicon technology used for electronic integrated circuits. In practice, silica waveguides are fabricated through a combination of flame hydrolysis deposition and reactive ion etching [17]. Figure 4.10 shows the main steps involved in the fabrication process. Two silica layers are first deposited on top of a silicon substrate using flame hydrolysis, a method similar to that used for making fiber preforms. The first layer acts as a cladding to the second layer, whose refractive index is increased by doping it with germania so that it can act as a waveguide. Both porous glass layers are solidified by heating them in an oven to a temperature close to 1300°C, a process referred to as *consolidation*. The core layer is converted into one or more ridges by removing parts of it using photolithography and reactive ion etching. The ridges are then covered with another silica layer, acting as a cladding and formed using flame hydrolysis followed by consolidation. A thermo-optic phase shifter is sometimes formed on top of the cladding layer.

The main design considerations are waveguide uniformity and low propagation losses. The silica-on-silicon technology is advanced enough that it can provide quite uniform waveguides. Losses depend on the core-cladding index difference quantified through the parameter Δ defined as in Eq. (1.1.4). In general, losses are low for smaller

4.3 Materials and Waveguide Fabrication

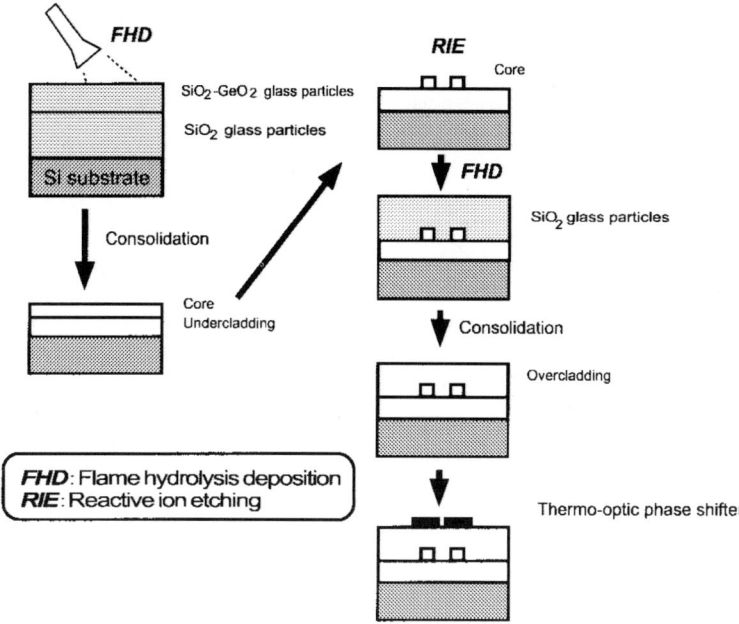

Figure 4.10: Steps involved in the fabrication of silica waveguides on a silicon substrate. (After Ref. [17]; ©1998 IEEE.)

values of Δ, and values as small as 0.017 dB/cm have been realized for $\Delta = 0.45\%$. Higher values of Δ are often used in practice for reducing the device length but losses then increase. Typically, propagation losses are 0.1 dB/cm for $\Delta = 2\%$.

The chief advantage of the silica-on-silicon technology is that it allows the fabrication of planar lightwave circuits (PLCs) in which multiple planar waveguides and many other optical components are integrated over a single silicon substrate. The main disadvantage is that when a PLC is connected to one or more optical fibers at its two ends, some coupling losses invariably occur. It is thus important to package the device in such a way that the insertion loss of a PLC is as small as possible.

Figure 4.11 shows schematically how a PLC device can be connected to a fiber array on both ends while minimizing insertion losses. All fibers in a fiber array are inserted into V-shaped grooves formed on a glass substrate. The end facets of the V grooves and the PLC chip are polished at an angle (typically 8°) to minimize any back reflection. The glass substrate containing fibers is connected to the PLC chip using an adhesive that is curable in ultraviolet light. A glass plate placed on top of V grooves and covering the fiber ends is also bonded to the PLC chip with the same adhesive. This design is found to be quite reliable in practice [17].

Silicon Oxynitride Waveguides

An alternative technology employs silicon substrates but makes use of silicon oxynitride (SiON) for forming the core layer of the planar waveguide [21]. This material is

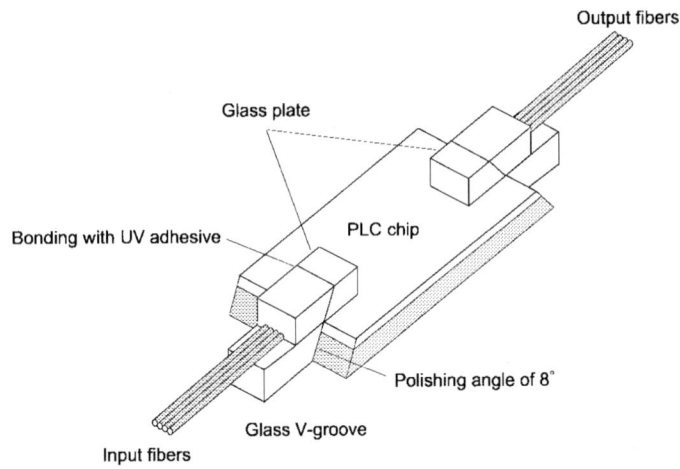

Figure 4.11: A packaged planar lightwave circuit. (After Ref. [17]; ©1998 IEEE.)

an alloy made by combining silica (SiO_2) with silicon nitride (Si_3N_4), two dielectrics with the refractive indices of 1.45 and 2.01, respectively. The refractive index of the SiON layer can be anywhere in the range 1.45–2.01 depending on the relative concentrations of oxygen and nitrogen atoms within the glass. This flexibility allows for fabrication of waveguides with large values of Δ without adding any extra loss.

SiON thin films can be deposited using several different techniques. The two most commonly employed are plasma-enhanced chemical vapor deposition and low-pressure chemical vapor deposition. The former uses SiH_4 as the base material together with several gases such as N_2O and NH_3. The latter makes use of SiH_2Cl_2 in combination with O_2 and NH_3. In practice, plasma-enhanced chemical vapor deposition is most suitable for forming a thin core layer with the core index $n_1 < 1.7$, while low-pressure chemical vapor deposition provides better results for $n_1 > 1.7$. Both processes provide excellent waveguide uniformity when used in their respective refractive-index ranges. Propagation losses are typically below 0.2 dB/cm but increase considerably in the spectral region near 1.5 μm because of the absorption caused by vibrational overtones of the N–H and Si–H bonds. The losses can be reduced in this spectral range by annealing the structure at high temperatures (up to 1150°C) and reducing the hydrogen content. Losses of less than 0.2 dB/cm at a wavelength of 1.55 μm have been realized by this approach.

Figure 4.12(a) shows the scanning electron microscope (SEM) photographs of a waveguide channel formed through reactive ion etching of a SiON layer [21]. To make the ridge, a 200-nm-thick chromium layer was first sputtered on top of the SiON layer, and a photolithographic technique was used to define the pattern on the chromium layer. The chromium mask was then removed using wet etching, and a upper cladding layer was deposited using plasma-enhanced chemical vapor deposition. The final structure is shown in Figure 4.12(b).

4.3 Materials and Waveguide Fabrication

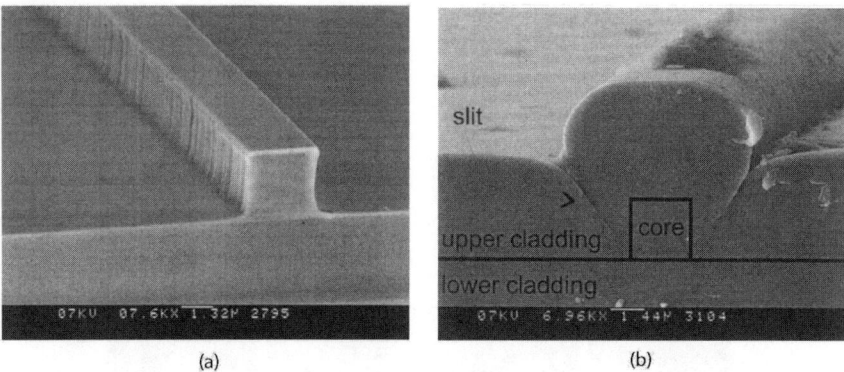

Figure 4.12: SEM photographs showing the ridge formed by reactive ion etching of a SiON layer (a) and the complete structure after the top cladding layer has been deposited (b). The position of the waveguide channel is marked by a black frame. (After Ref. [21]; ©1998 IEEE.)

Laser-Written Waveguides

In a relatively recent technique, CW or pulsed radiation from a laser is used for "writing" waveguides in silica and other glasses. Two distinct approaches have been followed for this purpose. In one approach, the photosensitivity of germanium-doped silica is exploited to enhance the refractive index in the region exposed to an ultraviolet (UV) laser beam [22]–[25]. The index-change mechanism is identical to that responsible for the formation of fiber gratings (see Section 2.2). The absorption of 244-nm light obtained from a KrF laser breaks the defect bonds and changes the refractive index by $\sim 10^{-4}$ only in the region that is exposed to laser light. Larger index changes of magnitude $>10^{-3}$ can be realized by using a 193-nm ArF laser or a 157-nm F_2 laser [25].

The formation of waveguides begins with a silica planar waveguide formed on a silicon substrate by depositing the core and two cladding layers through chemical vapor deposition. The core silica layer is doped with germania, which makes it photosensitive and also increases its refractive index. Sometimes, this layer is also doped with boron, a material that reduces the index of silica, so that the core and cladding layers have the same refractive index [24]. The three-layer structure is often exposed to H_2 or D_2 for a few hours to enhance its photosensitivity. Waveguides are then fabricated by exposing the sample to an ultraviolet CW beam, focused to a spot size of ~ 1 μm. The laser beam is slowly scanned to write one or more waveguides whose widths can be controlled by changing the beam diameter. The UV-written sample is then annealed at a temperature of around 80°C to remove the residual H_2 or D_2. A directional coupler was made with this technique as early as 1995 [22], followed by the fabrication of other components such as power splitters [23].

Another technique for writing waveguides directly into silica makes use of femtosecond laser pulses from a Ti:sapphire laser operating in the wavelength region near 800 nm [26]–[30]. The pulses used are so intense that they can modify the structure of silica glass through nonlinear effects such as multiphoton absorption. In fact, they

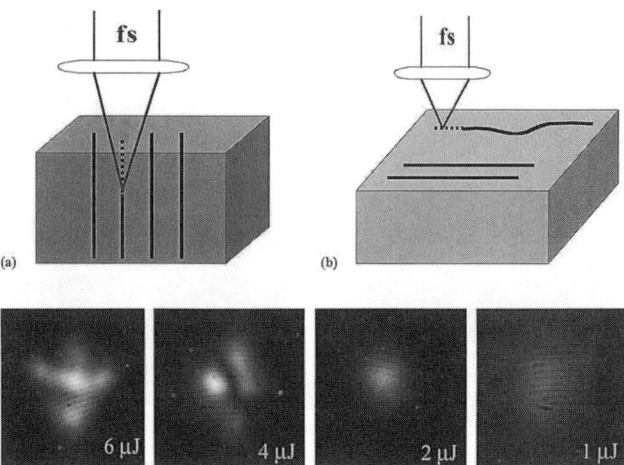

Figure 4.13: (a) Longitudinal and (b) transverse geometries for writing waveguides in bulk silica. The bottom four panels show the mode profiles at pulse energies in the range of 1–6 μJ. (After Ref. [30]; ©2003 IEEE.)

can produce refractive-index changes of $\sim 10^{-2}$ in the region where the laser beam is focused on the sample [28]. This technique does not require a prior growth of the core-cladding structure on a silicon substrate and can be used to write waveguides in bulk glasses by focusing the light anywhere into the glass. Figure 4.13 shows the (a) longitudinal and (b) transverse geometries used for writing waveguides in bulk silica.

In general, the quality of a laser-written waveguide depends on a large number of parameters such as the pulse energy, the repetition rate of pulses, and the speed at which the laser beam is scanned [30]. Pulse energies typically used vary in the range of 1 nJ to >1 μJ. The repetition rate of pulses can vary in the range of 100 Hz to >1 MHz. The scanning speed can vary from 5 μm/s to 50 mm/s. Figure 4.13 shows the sensitivity of a laser-written waveguide to pulse energy. This waveguide was written in a Nd-doped sodium-alumino-borosilicate glass by using laser pulses at a repetition rate of 250 Hz. The scan speed was 0.2 mm/s, and the beam was scanned 10 times to enhance the refractive index within the waveguide region. As seen in Figure 4.13, the single mode supported by the waveguide is very weakly confined at pulse energies of 1 μJ or less, while at energies above 3 μJ the waveguide appears to support multiple modes.

4.3.4 Silicon-on-Insulator Technology

Another approach to forming photonic integrated circuits on silicon substrates is known as the silicon-on-insulator (SOI) technology [31], the same technology used for fabrication of electronic integrated circuits. In this technique, the core layer acting as a waveguide is made of silicon itself and has a relatively large refractive index of 3.45. A silica layer under the core layer (refractive index 1.46) is used for lower cladding while air often acts as the top cladding layer, resulting in a huge core–cladding index difference and a tightly confined waveguide mode.

4.3 Materials and Waveguide Fabrication

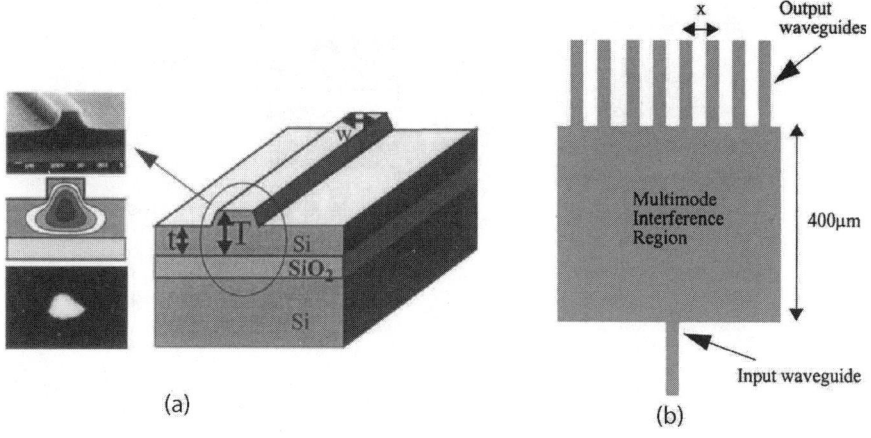

Figure 4.14: (a) Design of a silicon waveguide made with SOI technology. The three insets on the left show the SEM photograph of such a structure, the predicted mode profile, and the near-field image of the observed mode. (b) An MMI coupler made using such waveguides. (After Ref. [31]; ©1998 IEEE.)

Figure 4.14(a) shows the design of a SOI waveguide schematically. The most critical part is the SOI substrate in which a silica layer is buried within a silicon substrate. Several techniques can be used to create such an insulating layer. In one approach, oxygen is first implanted into the silicon substrate, followed by high-temperature annealing, which forms the buried silica layer. In another, a silicon wafer is first oxidized to form the silica layer on top, which is then bonded to another silicon wafer. The first wafer is thinned mechanically or chemically to the desired thickness. In all cases, the top silicon layer is made thicker in a central region to form the channel in which the optical mode propagates. Such a rib waveguide generally supports multiple modes in the lateral direction because of a relatively large channel width and a large refractive index difference between the core and cladding layers. However, higher-order modes have larger losses because they leak into the surrounding slab region as they propagate, resulting in a single-mode behavior.

The SOI technology has been used to make passive photonic devices whose operation requires interference among several modes. Figure 4.14(b) shows a multimode interference (MMI) coupler consisting of a single input and multiple output single-mode waveguides separated by a slab region in which many modes propagate simultaneously. The physics behind such devices is discussed later in Section 4.4.4. The SOI technology can also be used to make active optical devices that are integrated with the driving electronics. This is the chief advantage of this approach. In a variation of the SOI technique, porous silicon is used as a cladding material for a silicon waveguide [32].

In another related approach [33], the core layer is made of a material in which a certain fraction x of silicon atoms is replaced with germanium atoms, resulting in a binary alloy $Si_{1-x}Ge_x$. The refractive index of this material exceeds that of silicon by an amount that depends on x. This approach is similar to the GaAs and InP waveguides

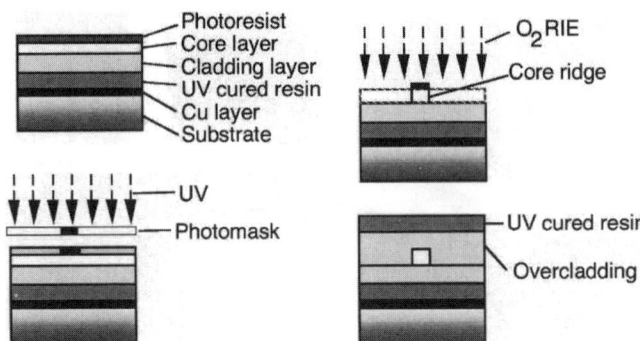

Figure 4.15: Steps involved in fabricating a polymer waveguide. (After Ref. [36]; ©1999 IEEE.)

discussed earlier. In fact, the waveguide is formed by growing the core and cladding layers epitaxially on top of a silicon substrate.

4.3.5 Polymer Waveguides

In a relatively recent technology, planar waveguides are made using using organic polymers [34]–[37]. The use of polymers offers a number of advantages, the low-cost potential being only one of them. Indeed, polymer waveguides have been used to make both the active and passive devices.

A large number of polymers, known by the names such as halogenated acrylate, fluorinated polyimide, and deuterated polymethylmethacrylate (PMMA), have been used to make planar waveguides. By copolymerizing specific combinations of monomers through exposure to UV or e-beam radiation, the refractive index of waveguide material can be varied in the range 1.3–1.6 while keeping losses relatively low in the wavelength range of 400–1,600 nm. For example, although PMMA has absorption losses of 0.5 dB/cm near 1,550 nm, the loss drops to only 0.07 dB/cm when PMMA is 80% fluorinated. These properties make polymers suitable for forming both the step-index and graded-index waveguides.

Polymer films can be fabricated on top of many substrates such as silicon, glass, quartz, or even plastic by using the technique of spin coating. Figure 4.15 shows schematically the main steps involved in making a polymer waveguide [36]. In this example, a resin layer is first formed on top of a copper-sputtered silicon substrate. The layer is exposed to UV light for curing the resin. The cladding layer is made of deuterated and fluorinated polymethacrylate while deuterated PMMA is used for the core layer. The top photoresist layer is used for reactive ion etching of the core layer to form a mesa by exposing UV light through a photomask. The overcladding layer is then deposited through spin coating. Finally, a UV-cured resin layer is formed on top of this layer.

Polymer waveguides can be used for a number of applications [37]. In the case of telecommunications, they can be employed for making power splitters, directional

4.4 Simple Passive Components

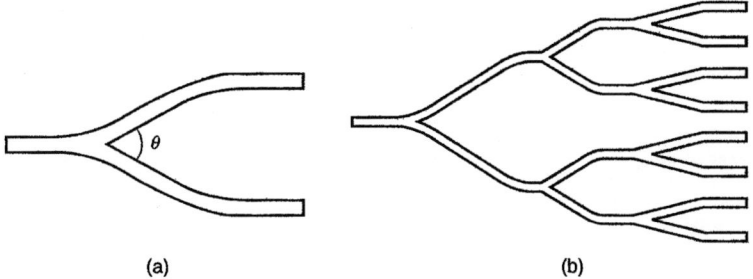

Figure 4.16: (a) A Y junction and (b) an eight-port branching circuit made by combining several Y junctions.

couplers, optical filters, and modulators. Some of these devices take advantage of the fact that a Bragg grating can be formed inside polymer waveguides by using techniques such as casting, embossing, molding, and e-beam writing. Active devices such as modulators require incorporation of chromophores within a polymer host, which can be aligned using a poling field and exhibit electro-optic properties. We discuss such modulators in Chapter 6.

4.4 Simple Passive Components

As mentioned earlier, planar waveguides can be used for making quite complicated passive and active components in the form of planar lightwave circuits. We discuss several such devices in Section 4.5 and later chapters. In this section we focus on a few basic components that act as building blocks for more complicated photonic integrated circuits.

4.4.1 Y Junctions

Perhaps the simplest waveguide component is the Y junction, a three-port device that acts as a power divider. Figure 4.16(a) shows a Y junction made by splitting a planar waveguide into two branches bifurcating at some angle θ. When light is injected on the input end, its power is divided equally between its two branches when the Y junction is perfectly symmetric around the input waveguide. In some sense, this device is similar to a fiber coupler, which can also act as a power splitter, except that it has only three ports. Conceptually, it differs considerably from a fiber coupler since there is no coupling region in which modes of two different waveguides overlap.

Functioning of Y junction can be understood as follows. In the junction region, the waveguide is thicker and supports higher-order modes. However, the geometrical symmetry forbids the excitation of asymmetric modes. If the thickness is changed gradually in an adiabatic manner, even higher-order symmetric modes are not excited, and the power is divided into two branches without much losses. In practice, a sudden opening of the gap violates the adiabatic condition, resulting in insertion losses associated with any Y junction. As one may expect, losses depend on the branching angle θ

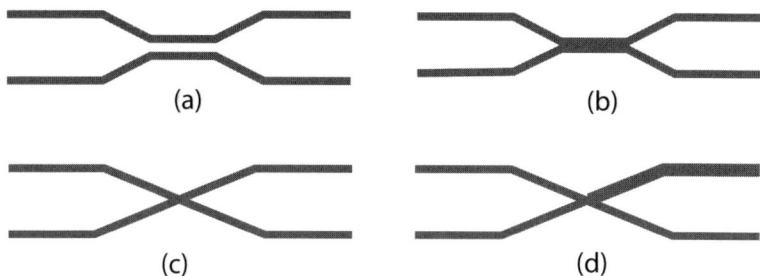

Figure 4.17: Four types of four-port devices that can be made by using two planar waveguides: (a) directional coupler; (b) coupled Y junctions; (c) symmetric X coupler, and (d) asymmetric X coupler.

and increase as θ does. In practice, θ should be maintained below 1° to keep insertion losses below 1 dB. Losses as small as 0.13 dB has been realized for waveguides with $\Delta = 0.3\%$ [38].

As shown in Figure 4.16(b), multiple Y junctions can be combined to form a device that splits the input power equally into eight output ports. In general, the number of ports doubles after each bifurcation and is given by 2^N after N bifurcation stages. Thus, 10 bifurcation stages will divide the power equally into 1,024 output ports. However, the insertion loss of a power divider also increases rapidly and often becomes intolerable after three or four bifurcation stages.

4.4.2 Four-Port Couplers

Directional couplers can also be made using planar waveguides, resulting in a four-port device in which a $\pi/2$ phase shift is added to the light emerging from the cross port (see Section 2.1). Figure 4.17(a) shows the design schematically. Similar to the case of fiber couplers, two waveguides are brought in close proximity in a central region where the tails of the modes associated with the two waveguides overlap and lead to power transfer between the two cores. The analysis of Section 2.1 can be applied to directional couplers made using planar waveguides. A complete power transfer can occur for symmetric couplers for which the two waveguides are identical in all respects. It is also easy to make 3-dB couplers that divide the input power equally among its two output ports.

A new kind of four-port device can be made by reducing the spacing between the waveguides to zero. In this device, two waveguides that need not be identical merge in a central region, and then separate again at the output side in an adiabatic fashion. Such a device, as shown in Figure 4.17(b), is similar to two Y junctions whose input ports have been joined together. Input from one port excites the symmetric and asymmetric modes in the central region. The launch conditions determine how the power is divided between the two output ports. Mathematically, the transfer matrix is still given by Eq. (2.1.21), but the coupling coefficient $\kappa \equiv k_0(\bar{n}_0 - \bar{n}_1)/2$ depends on the mode indices associated with the two modes excited in the central region.

4.4 Simple Passive Components

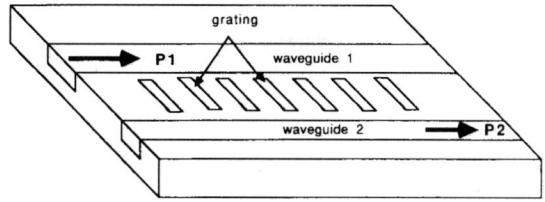

Figure 4.18: Schematic of a grating-assisted directional coupler.

The four-port structure shown in Figure 4.17(c) is referred to as the X coupler because of its shape. It is made by crossing two waveguides in the shape of an X in the central region. As the two guides overlap, it is possible, under appropriate conditions, to switch the input beam entering from one of the input ports so that it is transferred to the other waveguide. Another interesting four-port device shown in Figure 4.17(d) is the asymmetric X coupler in which two input waveguides are identical but the two output waveguides are made different in their sizes. The power splitting in such a device depends on the relative phase of the two inputs. If the two inputs are equal and in phase, most of the power is transferred to the port with a wider waveguide. In contrast, when the two inputs are out of phase, most of the power is transferred to the port with a narrow waveguide. This device is sometimes referred to as the "optical magic T" as it functions similar to the magic T well known in the field of microwaves [3].

4.4.3 Grating-Assisted Directional Couplers

As discussed in Section 2.1, complete power transfer between the two waveguides of a directional coupler ceases to occur when they are not exactly identical. In fact, as seen in Figure 2.2, little power is transferred when the asymmetry parameter δ_a exceeds the coupling coefficient κ by more than a factor of 2. The reason for this behavior is related to different propagation constants, β_{01} and β_{02}, for the two waveguides. In the beginning of each period in Figure 2.2, the two modes are in phase, and power transfer can occur. However, after a distance such that $(\beta_{01} - \beta_{02})z = \pi$, the modes go out of phase, and the direction of power transfer is reversed.

In practice, it is hard to make the two waveguides identical, and it is useful to find a scheme that permits significant power transfer even when the two waveguides are not identical. It turns out that this problem can be solved by fabricating an index grating in the coupling region of a directional coupler. The grating provides a mechanism to selectively turn the coupling on and off between the two waveguides. If the grating period is properly adjusted, it is possible to spoil the coupling whenever the modes are out of phase during each cycle. With this arrangement, power is transferred from one waveguide to another in small increments during each cycle. Such couplers are known as *grating-assisted* directional couplers and have been studied extensively [39]–[41].

Figure 4.18 shows schematically the design of a grating-assisted directional coupler. The coupled-mode theory of Section 2.1 developed for fiber couplers can be easily extended to the case of grating-assisted directional couplers [1]. The grating period $\Lambda = 2\pi/(\beta_{01} - \beta_{02})$ is related inversely to the difference between the propaga-

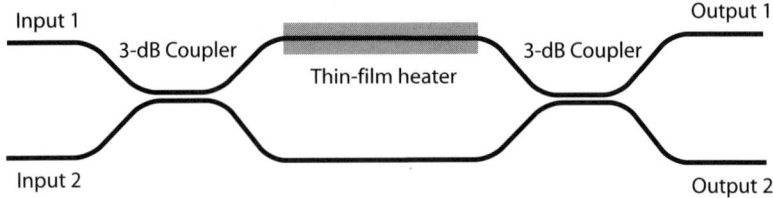

Figure 4.19: Schematic of a Mach–Zehnder interferometer with a thermo-optic phase shifter fabricated using silica-on-silicon technology.

tion constants of the fundamental modes supported by the two waveguides. Typically, $\Lambda \sim 10$ μm. Such long-period gratings can be easily fabricated using a variety of techniques. In the case of silica waveguides, a phase mask can be employed similar to the case of fiber gratings.

4.5 Advanced Passive Components

As an example of the more complicated passive devices that can be fabricated using the planar waveguide technology, we discuss several advanced devices in this section. More specifically, we focus on Mach–Zehnder interferometers, MMI couplers, star couplers, and arrayed waveguide gratings.

4.5.1 Mach–Zehnder Switches and Filters

Similar to the case of optical fibers, one can combine two directional couplers made using planar waveguides in series to form a Mach–Zehnder (MZ) interferometer. Figure 4.19 shows the basic design schematically. As was done for a fiber-based MZ interferometer, the two output ports of a 3-dB coupler on the left are connected to the two input ports of the 3-dB coupler on the right. The two arm lengths are made equal for a symmetric MZ interferometer. As discussed in Section 2.3.4, such a device transfers all of its input power coming from port 2 to the output port 1 (the so-called cross state). However, the device can be switched to the bar port so that the output appears from port 2 by changing the optical path length of one of the arms to induce an additional phase shift of π. When the MZ interferometer is formed using silica or polymer waveguides, the phase shift can be induced electrically through a thin-film heater integrated within one arm of the MZ interferometer as a thin layer of chromium. Heating of that arm changes the refractive index of silica (or polymer) through the thermo-optic effect, which in turn changes the phase of light passing through it.

It is evident that the MZ interferometer shown in Figure 4.19 acts as a thermo-optic switch because the input beam can be directed to different output ports by raising and lowering the temperature electrically. Because of its thermal nature, switching speed is not very fast. Such devices typically switch on a time scale ~ 1 ms that is fast enough for many applications.

4.5 Advanced Passive Components

Much faster switching can be accomplished using the electro-optic effect in which the refractive index of a suitable material is changed by applying an electric field across a waveguide fabricated using that material. The electro-optic material LiNbO$_3$ is commonly used for this purpose. In fact, a LiNbO$_3$ modulator uses the MZ interferometer configuration to modulate the input at a frequency as high as 40 GHz. Such modulators are discussed in Chapter 6.

An MZ interferometer also acts as an optical filter as transmission to the bar port (or the cross port) depends on the frequency ω of incident light. The transfer function of such a filter can be obtained from Eq. (2.3.22) and takes the simple form of $H(\omega) = \sin(\omega\tau)$, where τ is the additional delay in one arm of the MZ interferometer occurring because of its longer optical length. Clearly, such a filter is not sharp enough for practical applications. This problem can be solved by forming a cascaded chain of MZ interferometers, each designed with different delays and different power-splitting ratios for various directional couplers. For a chain of N cascaded MZ interferometers, one can choose $2N+1$ parameters, N delays, and $N+1$ splitting ratios. This freedom can be used to synthesize optical filters with virtually arbitrary amplitude and phase responses [42].

It is easy to calculate the transmission through a chain of N MZ interferometers using the transfer-matrix approach developed in Chapter 2. If F_{in} and F_{out} are the field column vectors at the input and output ends, they are related as

$$F_{\text{out}}(\omega) = T_{N+1} D_N T_N \cdots D_2 T_2 D_1 T_1 F_{\text{in}}, \quad (4.5.1)$$

where T_m is the transfer matrix of the mth coupler and D_m is the diagonal matrix representing phase shifts in the two arms of the mth MZ interferometer. These two matrices have the form

$$T_m = \begin{pmatrix} c_m & is_m \\ is_m & c_m \end{pmatrix}; \quad D_m = \begin{pmatrix} e^{i\phi_m} & 0 \\ 0 & e^{-i\phi_m} \end{pmatrix}, \quad (4.5.2)$$

where $c_m = \cos(\kappa_m l_m)$ and $s_m = \sin(\kappa_m l_m)$ for a coupler of length l_m with the coupling coefficient κ_m, and $2\phi_m = \omega \tau_m$ is the relative phase shift for the mth MZ interferometer with an additional delay of τ_m.

It is difficult to write a general expression for the transfer function of such a filter. However, it turns out that one can use the principle of the "sum over all possible optical paths" to simplify this task [42]. For example, the amplitude transmission coefficient through the cross port of a chain of two cascaded MZ interferometers is composed of four possible paths and can be written as [16]

$$t_b(\omega) = ic_1 c_2 s_3 e^{i(\phi_1+\phi_2)} + ic_1 s_2 s_3 e^{i(\phi_1-\phi_2)} + i^3 s_1 c_2 s_3 e^{i(-\phi_1+\phi_2)} + is_1 s_2 s_3 e^{-i(\phi_1+\phi_2)}, \quad (4.5.3)$$

where the factor of i accounts for the $\pi/2$ phase shift occurring every time light transfers to the cross port of a directional coupler. The number of terms increases rapidly as the number of MZ interferometers increases in the chain. Cascaded MZ interferometers have been to make useful devices such as gain-equalization filters and tunable dispersion compensators.

4.5.2 Multimode Interference Couplers

Directional couplers are four-port devices, denoted as 2×2 to indicate two input and two output ports. In many practical situations, one wants to split a single input into several outputs. Such devices are referred to as $1 \times N$ power splitters, where N denotes the number of output ports. Several techniques can be used to make such devices; this subsection focuses on couplers based on multimode interference and referred to as the MMI couplers.

MMI couplers are based on the well-known *Talbot effect*, which leads to self-imaging of objects when coherent light propagates through a bulk medium exhibiting some kind of periodicity [43]. In the case of planar waveguides, the same phenomenon occurs when one or more input waveguides are connected to a relatively thick central region capable of supporting many optical modes [44]–[47]. The length of the central coupling region is chosen such that the optical field entering from an input waveguide is self-imaged and forms an array of identical images at the location where the output waveguides are placed by design. Such a device can split the input power into several branches, resulting in an $1 \times N$ power splitter. An example of such a 1×8 coupler is shown in Figure 4.14(b).

The theory of MMI couplers is well known [44]. The basic idea is as follows. The input signal excites many optical modes in the coupling region, each of which propagates with a slightly different propagation constant whose value depends on both the mode number and the width of the central coupling region. If $\phi_m(x)$ denotes the mth mode with the propagation constant β_m, the field at any distance z can be written as

$$A(x,z) = \sum_{m=0}^{M-1} C_m \phi_m(x) \exp(i\beta_m z), \quad (4.5.4)$$

where M is the total number of excited modes and the expansion coefficients C_m depend on the input field $A(x,0)$. The propagation constant β_m for a slab of effective width W_e is obtained using $\beta_m^2 = n_s^2 k_0^2 - p_m^2$, where $p_m = (m+1)\pi/W_e$ and n_s is the refractive index of the slab material. Since the transverse part $p_m \ll k_0$, we can approximate the propagation constant as

$$\beta_m \approx n_s k_0 - \frac{(m+1)^2 \pi^2}{2n_s k_0 W_e^2} = \beta_0 - \frac{m(m+2)\pi}{3L_b}, \quad (4.5.5)$$

where the beat length between the two lowest-order modes is defined as

$$L_b = \frac{\pi}{\beta_0 - \beta_1} \approx \frac{4n_s W_e^2}{3\lambda}. \quad (4.5.6)$$

It is easy to see from Eqs. (4.5.4) and (4.5.5) that the input field will be reproduced at $3L_b$ because at that distance the relative phase $\beta_0 - \beta_m$ differs by an even or odd multiple of π, depending on whether m is even or odd. Thus, a 2×2 coupler designed to be symmetric around $x = 0$ (center of the central slab) will couple the light into the cross port for $z = 3L_b$ as it forms a mirror image at the output end. In contrast, it will transmit to the bar port when $z = 6L_b$. Depending on the slab length, multiple images of the input can also form on the output end for $L < 3L_b$. Figure 4.20 shows the intensity

4.5 Advanced Passive Components

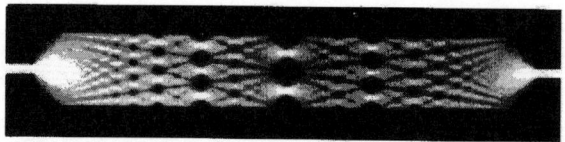

Figure 4.20: Numerically simulated intensity evolution for a 20-µm-wide slab over a length of $3L_b$ showing formation of multiple images at certain locations. (After Ref. [44]; ©1995 IEEE.)

pattern calculated numerically for a 20-µm-wide slab over a length of $3L_b/4$. Since the input field is symmetric around $x = 0$, only even modes of the slab waveguide are excited, restricting the sum in Eq. (4.5.4) to even values of m. As a result, the input field is self-imaged at a distance of only $3L_b/4$, reducing the required slab length by a factor of 4, an important feature from a design standpoint. Figure 4.20 shows that N images of the input field form at a distance of only $3L_b/4N$. Thus, choosing the coupler length appropriately, one can design $1 \times N$ power splitters based on the MMI concept.

4.5.3 Star Couplers

Some important WDM applications make use of $M \times N$ couplers designed with M input and N output ports, where M and N are arbitrary integers. Such couplers are known as star couplers. A simple way to make a star coupler with the same number of input and output ports ($M = N$) is to combine several 3-dB couplers. Figure 4.21 shows schematically how a 8×8 star coupler can be formed using twelve 2×2 directional couplers, each splitting its input into two equal parts. However, such a scheme becomes quickly impractical as the number of input and output ports increases. Moreover, directional couplers are wavelength-dependent, a feature that renders such device operable only over a limited wavelength range.

Compact and wavelength-independent star couplers have been made using several planar-waveguide technologies [48]–[50]. Figure 4.22 shows the basic design schematically. In this approach, many input and output waveguides are connected to a central

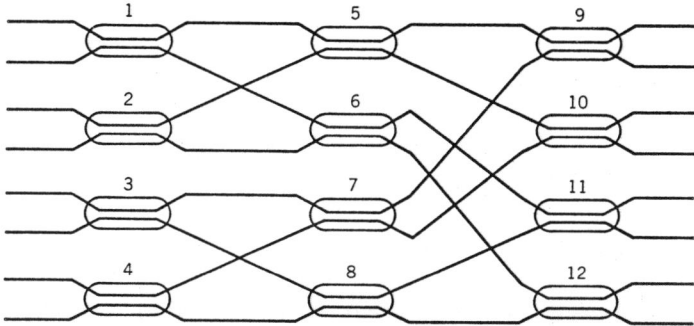

Figure 4.21: Schematic of an 8×8 star coupler formed by using twelve 2×2 directional couplers.

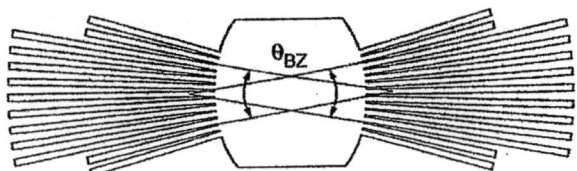

Figure 4.22: Schematic layout of a star coupler. Short waveguides at the edges are added to enlarge the periodic array. The angle θ_{BZ} is determined by the first Brillouin zone associated with the Bloch function of the periodic waveguide array. (After Ref. [42]; ©1999 Wiley.)

region that is so wide one can treat wave propagation inside it as if the optical field were propagating in a homogeneous dielectric medium. Moreover, waveguides on each side are arranged to have a constant angular separation and point radially to a focal point as shown in Figure 4.22. The two focal points are separated by the focal distance. The input and output boundaries of the central slab form arcs that are centered at the focal points and have a radius equal to the focal distance.

The power-splitting functionality of such a star coupler is a consequence of the periodic nature of the waveguide array at the input and output ends [16]. The number of waveguides in the array should be large. In fact, dummy waveguides are often added near edges to ensure a large periodic array on each end of the focusing slab (the short waveguides in Figure 4.22). Similar to the supermodes of two coupled modes found in Section 2.1, an infinite array of coupled waveguides supports supermodes, which take the form of Bloch functions because of the periodic nature of the array. Following the energy-band analogy in crystals, the optical field associated with a supermode can be written as

$$\psi(x, k_x) = \sum_m F(x - ma) \exp(imk_x a), \qquad (4.5.7)$$

where a is the period of the waveguide array, $F(x)$ is the optical mode associated with the waveguides, and the transverse part k_x of the propagation constant is restricted to the first Brillouin zone such that $-\pi/a < k_x < \pi/a$. Because of the coupling among the waveguides, light launched into one waveguide excites all supermodes within the first Brillouin zone with equal amplitudes. This can be seen by taking the inverse Fourier transform of Eq. (4.5.7).

As the waveguides approach the central slab, $\psi(x, k_x)$ evolves into a freely propagating wave with a curved wavefront moving toward its focus and making an angle $\theta \approx k_x/\beta_s$, where β_s is the propagation constant in the slab region. The maximum value of this angle is determined by the Brillouin zone and is given by

$$\theta_{BZ} \approx k_x^{\max}/\beta_s = \pi/(\beta_s a). \qquad (4.5.8)$$

A star coupler is designed such that all N output waveguides are within the illuminated region, that is, $Na/R = 2\theta_{BZ}$, where R is the radius of the arc equal to the focal distance. With this arrangement, the optical power entering from any input waveguide is equally divided among its N output waveguides.

Star couplers based on the periodic waveguide arrays were first made using the silica-on-silicon technology because it allows the fabrication of quite complicated in-

4.5 Advanced Passive Components

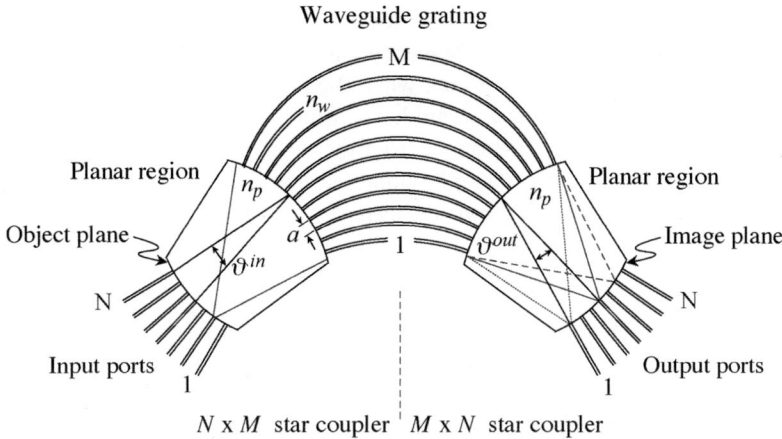

Figure 4.23: Schematic design of an $N \times N$ arrayed-waveguide grating. Two star couplers are connected by an array of M waveguides of different lengths. (After Ref. [62]; ©2002 IEEE.)

tegrated circuits. As early as 1989, a 19×19 star coupler was demonstrated using this technology [48]. By 1992, the technique had advanced enough that a 144×144 star coupler could be made with only 2-dB excess losses. However, additional insertion losses occur when such a device is connected to optical fibers at the input and output end. Star couplers have also been made using semiconductor waveguides because they are useful for making active devices.

4.5.4 Arrayed-Waveguide Gratings

A remarkable device that has been made using several planar-waveguide technologies and has found a variety of applications in WDM lightwave systems is the *arrayed-waveguide grating*, or AWG [51]–[62]. Figure 4.23 shows the design of an AWG schematically. It combines two $N \times M$ star couplers through an array of M waveguides whose lengths are chosen in such a way that the length difference δl between any two neighboring waveguides is constant. As a result, the phase difference between two neighboring waveguides is also constant as an input signal propagates through it. It is this constant phase difference that creates the grating-like behavior.

The wavelength dependence of an AWG can be understood in simple physical terms as follows. Consider a WDM signal consisting of multiple channels at different wavelengths with a constant channel spacing Δv. When this signal is launched into one of the input waveguides, the first star coupler splits its power into many parts and directs them into the waveguides forming the grating. At the output end of the grating array, the wavefront is tilted because of linearly varying phase shifts in waveguides of different lengths. The tilt is wavelength-dependent and it forces each channel to focus on a different output waveguide of the second coupler. This behavior is similar to a bulk grating that also directs different wavelengths to different locations.

One can write the "Bragg condition" for an AWG by considering the phase shift experience by a specific channel of frequency v as light propagates through the grating array. Constructive interference occurs when the phase difference between the optical paths through two adjacent waveguides is a multiple of 2π. If $\theta_{\text{in}} = pa_i/R$ and $\theta_{\text{out}} = qa_o/R$ are the angles, measured from the center of the waveguide array (see Figure 4.23), for the pth input and qth output waveguides with the pitches a_i and a_o, respectively, this condition can be written as [42]

$$k_0 n_w(\delta l) + k_0 n_p a_g(\theta_{\text{in}} + \theta_{\text{out}}) = 2\pi m, \qquad (4.5.9)$$

where $k_0 = 2\pi v/c$, m is an integer, n_w is the mode index of each waveguide in the array, n_p is the refractive index of the planar slab material, a_g is the pitch of the grating array, and δl is the length difference between two adjacent waveguides. This condition should hold for all channels.

It is easy to see from Eq. (4.5.9) that a multichannel signal with a constant channel spacing, entering from any input port, will be split into individual channels, which will appear at different output ports if the AWG is designed appropriately. Moreover, when multiple WDM signals with the same channel wavelengths enter through different input ports, the transmission spectrum from any output port will be periodic. The free spectral range (FSR) can be determined by finding the frequency difference $v' - v$, where v' is the frequency for which Eq. (4.5.9) is satisfied with m replaced by $m + 1$. At the center port of the AWG, $\theta_{\text{in}} = \theta_{\text{out}} = 0$, and the FSR takes the simple form

$$\Delta v_{\text{FSR}} = v' - v = \frac{c}{n_w \delta l}, \qquad (4.5.10)$$

where $n_w = n_w + v dn_g/dv$ is the group index. The FSR is slightly different for off-center ports because of the presence of angles θ_{in} and θ_{out} in Eq. (4.5.9) but this angle dependence is generally negligible.

AWGs based on the periodic waveguide arrays were proposed as early as 1988 [51]. By 1992, such devices have been fabricated using both the silica-on-silicon and semiconductor-waveguide technologies [52]–[57]. Figure 4.24 shows the layout of an AWG made with the silica-on-silicon technology [17]. This technology was used to produce in 1996 an AWG with 128 input and output ports capable of resolving WDM channels with only 25-GHz channel spacing [56]. Further advances led by 2001 to the realization of 400-channel AWGs with 25-GHz channel spacing [60].

The basic technique of fabricating silica waveguides on top of a silicon substrate has been discussed in Section 4.3.3. An AWG is a complex device requiring fabrication of hundreds, or even thousands, of such waveguides on the same substrate. Typically, the core of each rectangular waveguide is 6×6 μm^2 in size to facilitate coupling of light to optical fibers. The relative core-cladding index difference, the parameter Δ, plays an important role since it determines not only the size of the finished chip but also the magnitude of insertion losses. The reason can be understood by noting that losses of a waveguide depend on both Δ and the bending radius. Waveguides with larger Δ can be bent more tightly as Δ increases without incurring large losses. In practice, chip size can be reduced considerably by enhancing Δ from 0.75% to 1.5% [60]. Even then, the device size for a 256-channel AWG typically exceeds 7 cm × 5 cm and approaches

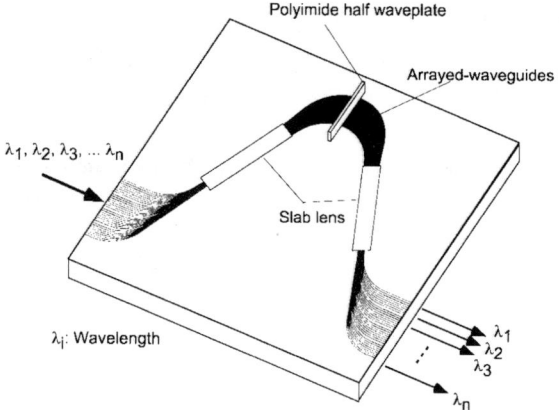

Figure 4.24: Layout of an arrayed-waveguide grating fabricated with silica-on-silicon technology. The curved black region shows the dense waveguide array with a half-wave plate in the middle to correct for the birefringence effects. (After Ref. [17]; ©1998 IEEE.)

12.5 cm × 6.5 cm for a 400-channel AWG. It is not practical to increase Δ beyond 1.5% because propagation losses for silica waveguides also increase with Δ, although they can be controlled to some extent by improving the uniformity of the core thickness and refractive index. In state-of-the-art devices, propagation losses are roughly 0.05 dB/cm for $\Delta = 1.5\%$ but increase to 0.09 dB/cm for $\Delta = 2.5\%$.

As was seen in Section 2.1, the effective mode index of any planar waveguide is different for the TE and TM modes. This index difference, or birefringence, for silica waveguides is $\sim 1 \times 10^{-4}$ and affects the performance of an AWG because it affects the Bragg condition in Eq. (4.5.9), making it different for TE and TM modes. A simple way to solve this problem is to insert a polyimide half-wave plate in the middle of the waveguide array, as shown in Figure 4.24. Such a wave plate converts TE modes into TM modes, and vice versa. This approach adds additional processing steps because it requires the formation of a groove that holds the wave plate in place. It also increases the insertion loss of device because light has to leave the silica waveguide at the groove position and reenter back into the waveguide after crossing the wave plate. The additional loss is tolerable for low-Δ devices (<0.4 dB for $\Delta = 1.5\%$) but becomes excessive for large values of Δ. Other methods have been developed to solve this problem [60].

AWGs have been produced using the InP-waveguide technology because they can be integrated within an active device to provide additional functionality [61]. Both the silica- and InP-based AWGs have found a number of applications discussed in Chapter 8 in the context of WDM components.

Problems

4.1 Starting from the Maxwell equations, derive the eigenvalue equations for the even and odd TE modes of a planar waveguide whose core of thickness d and index n_1 is surrounded by identical cladding layers of index n_2.

4.2 Repeat Problem 4.1 and derive the eigenvalue equations for the even and odd TM modes of the same symmetric planar waveguide.

4.3 Use the results of Problems 4.1 and 4.2 to find the single-mode condition for TE and TM modes. Use this condition to find the maximum allowed thickness of a single-mode waveguide at a wavelength of 1.55 μm? Assume $n_1 = 1.47$ and $n_2 = 1.45$.

4.4 Derive the eigenvalue equations for the TE modes of a planar waveguide whose core of thickness d and index n_1 is surrounded by different cladding layers of indices n_s and n_c assuming that $n_s > n_c$.

4.5 Assuming that $n_1 = 2.2$, $n_s = 2.1$, and $n_c = 2.0$, how thin can the asymmetric waveguide be before it will stop supporting any TE mode at a wavelength of 1.5 μm?

4.6 Consider a square-shape silica waveguide of 6-μm size supporting a single mode at the wavelength of 1.55 μm? Find the mode index, assuming that the core index is $n_1 = 1.47$ and that the core is buried inside a cladding material of index $n_2 = 1.45$.

4.7 Consider a rectangular InGaAsP waveguide whose core is 2 μm wide and 0.1 μm thick. Use the effective-index method to find the number of modes supported by this waveguide at the wavelength of 1.55 μm, assuming that the core index is $n_1 = 3.4$ and that the core is buried inside a cladding material of index $n_2 = 3.2$.

4.8 Consider again the rectangular InGaAsP waveguide of Problem 4.7. Use the effective-index method to find the maximum allowable width of the waveguide for single-mode operation at the wavelength of 1.55 μm. Calculate the effective index of this mode when the core width is 90% of this maximum value.

4.9 Find the composition of the quaternary alloy InGaAsP for making lasers operating at 1.3- and 1.55-μm wavelengths.

4.10 Consult any book and write down the electro-optic tensor in the form of a 6 × 3 matrix for $LiNbO_3$, KDP, and InP crystals. Also find the values of these coefficients in the wavelength region near 1.5 μm by searching the literature. You are allowed to use the Internet.

4.11 Describe how a grating-assisted coupler works. Use Ref. [1] or any other relevant paper to write the coupled-mode equations for a grating-assisted coupler and discuss their solution.

4.12 Consider a chain of M Mach–Zehnder interferometers made using using $M+1$ 3-dB couplers. Prove that the transmissivity to the cross port of such a device for an input signal of frequency v is given by $T(v) = \prod_{m=1}^{M} \cos^2(\pi v \tau_m)$, where τ_m is the relative time delay while crossing the mth member of the chain.

4.13 Describe how a multimode-interference coupler works. Use Eq. (4.5.4) to show that any input field is recovered at a distance $6L_b$. How would you design a 1×4 power splitter using this approach? Find the coupler length in terms of its width for such a device.

References

[1] D. Marcuse, *Theory of Dielectric Optical Waveguides*, 2nd ed., Academic Press, San Diego, CA, 1991.
[2] K. Okamoto, *Fundamentals of Optical Waveguides*, Academic Press, San Diego, CA, 2000.
[3] K. IIzuka, *Elements of Photonics*, Vol. II, Wiley, New York, 2002.
[4] E. A. J. Marcatili, *Bell Syst. Tech. J.* **48**, 2071 (1969).
[5] A. Kumar, K. Thyagarajan, and A. K. Ghatak, *Opt. Lett.* **8**, 63 (1983).
[6] K. S. Chiang, *Proc. SPIE* **2399**, 2 (1995).
[7] G. P. Agrawal and N. K. Dutta, *Semiconductor Lasers*, 2nd ed., Van Nostrand Reinhold, New York, 1993.
[8] S. L. Chuang, *Physics of Optoelectronic Devices*, Wiley, New York, 1995.
[9] E. Kapon, Ed., *Semiconductor Lasers*, Part I and II, Academic Press, San Diego, CA, 1999.
[10] R. W. Boyd, *Nonlinear Optics*, 2nd ed., Academic Press, San Diego, CA, 2003.
[11] E. L. Wooten, K. M. Kissa, A. Yi-Yan, E. J. Murphy, D. A. Lafaw, P. F. Hallemeier, D. Maack, D. V. Attanasio, D. J. Fritz, G. J. McBrien, and D. E. Bossi, *IEEE J. Sel. Topics Quantum Electron.* **6**, 69 (2000).
[12] R. V. Schmidt and I. P. Kamimow, *Appl. Phys. Lett.* **25**, 458 (1974).
[13] J. L. Jackel, C. E. Rice, and J. Veselka, *Appl. Phys. Lett.* **41**, 607 (1982).
[14] K. M. Kisa, P. G. Suchoski, and D. K. Lewis, *J. Lightwave Technol.* **13**, 1521 (1995).
[15] M. Kawachi, *Opt. Quantum Electron.* **22**, 391 (1990).
[16] Y. P. Li and C. H. Henry, *Optical Fiber Telecommunications*, Vol. 3B, I. P. Kaminow and T. L. Koch, Eds., Academic Press, San Diego, CA, 1997.
[17] A. Himemo, K. Kato, and T. Miya, *IEEE J. Sel. Topics Quantum Electron.* **4**, 913 (1998).
[18] A. Kaneko, T. Goh, H. Yamada, T. Tanaka, and I. Ogawa, *IEEE J. Sel. Topics Quantum Electron.* **5**, 1227 (1999).
[19] K. Kato and Y. Tohmori, *IEEE J. Sel. Topics Quantum Electron.* **6**, 4 (2000).
[20] T. Miya, *IEEE J. Sel. Topics Quantum Electron.* **6**, 38 (2000).
[21] R. M. de Riddler, K. Wörhoff, A. Driessen, P. V. Lambeck, and H. Albers, *IEEE J. Sel. Topics Quantum Electron.* **4**, 930 (1998).
[22] G. D. Maxwell and B. J. Ainslie, *Electron. Lett.* **31**, 95 (1995).
[23] M. Svalgaard, *Electron. Lett.* **33**, 1694 (1997).
[24] K. Faerch and M. Svalgaard, *IEEE Photon. Technol. Lett.* **14**, 173 (2002).
[25] K. P. Chen, P. R. Herman, and R. Taylor, *J. Lightwave Technol.* **21**, 140 (2003).
[26] K. Miura, J. Qiu, H. Inouye, T. Mitsuyu, and K. Hirao, *Appl. Phys. Lett.* **71**, 3329 (1997).
[27] D. Homoelle, S. Wielandy, A. L. Gaeta, N. F. Borrelli, and C. Smith, *Opt. Lett.* **24**, 1311 (1990).
[28] K. Minoshima, A. M. Kowalevicz, I. Hartl, E. P. Ippen, and J. G. Fujimoto, *Opt. Lett.* **26**, 1516 (2001).
[29] J. W. Chan, T. Huser, S. Risbud, and D. M. Krol, *Opt. Lett.* **26**, 1726 (2001); *Proc. SPIE* **4640**, 129 (2002).
[30] C. Florea and K. A. Winick, *J. Lightwave Technol.* **21**, 246 (2003).
[31] B. Jalali, S. Yegnanarayanan, T. Yoon, T. Yoshimoto, I. REndina, and F. Coppinger, *IEEE J. Sel. Topics Quantum Electron.* **4**, 938 (1998).
[32] H. F. Arrand, T. M. Benson, P. Sewell, A. Loni, R. J. Bozeat, R. Arens-Fischer, M. Krüger, M. Thönissen, and H. Lüth, *IEEE J. Sel. Topics Quantum Electron.* **4**, 975 (1998).

[33] S. Janz, J. M. Baribeau, A. Delage, H. Lafontaine, S. Mailhot, R. L. Williams, D. X. Xu, D. M. Bruce, P. E. Jessop, and M. Robillard, *IEEE J. Sel. Topics Quantum Electron.* **4**, 990 (1998).

[34] R. Yoshimura, M. Hikita, S. Tomaru, and S. Imamura, *J. Lightwave Technol.* **16**, 1030 (1998).

[35] J. Kobayashi, T. Matsuura, S. Sasaki, and T. Martino, *Appl. Opt.* **37**, 1032 (1998).

[36] M. Hikita, S. Tomaru, K. Enbutsu, N. Ooba, R. Yoshimura, M. Usui, T. Yoshida, and S. Imamura, *IEEE J. Sel. Topics Quantum Electron.* **5**, 1237 (1999).

[37] L. Eldada and L. W. Shacklette, *IEEE J. Sel. Topics Quantum Electron.* **6**, 54 (2000).

[38] Y. Hibino, F. Hanawa, H. Nakagome, M. Ishii, and N. Takato, *J. Lightwave Technol.* **13**, 1728 (1995).

[39] W. P. Huang and J. Wong, *J. Lightwave Technol.* **10**, 1367 (1992).

[40] N. H. Sun, J. K. Butler, G. A. Evans, L. Pang, and P. Congdon, *J. Lightwave Technol.* **15**, 2301 (1997).

[41] V. M. N. Passaro, *J. Lightwave Technol.* **18**, 973 (2000).

[42] C. M. Madsen and J. H. Zhao, *Optical Filter Design and Analysis*, Wiley, New York, 1999.

[43] H. F. Talbot, *Edinburg Philos. Mag.* **9**, 401 (1836).

[44] L. B. Soldano and E. C. M. Pennings, *J. Lightwave Technol.* **13**, 615 (1995).

[45] P. A. Besse, E. Gini, M. Bachmann, and H. Melchoir, *J. Lightwave Technol.* **14**, 2286 (1996).

[46] J. A. Besley, J. D. Love, and W. Langer, *J. Lightwave Technol.* **16**, 678 (1998).

[47] J. Leuthold and C. H. Joyner, *J. Lightwave Technol.* **19**, 700 (2001).

[48] C. Dragone, C. H. Henry, I. P. Kaminow, and R. C. Kistler, *IEEE Photon. Technol. Lett.* **1**, 241 (1989).

[49] K. Okamoto, H. Takahashi, M. Yasu, and Y. Hibino, *IEEE Photon. Technol. Lett.* **4**, 61 (1992).

[50] K. Okamoto, H. Okazaki, Y. Ohmori, and K. Kato, *IEEE Photon. Technol. Lett.* **4**, 1031 (1992).

[51] M. K. Smit, *Electron. Lett.* **24**, 385 (1988); C. Dragone, *Electron. Lett.* **24**, 385 (1988).

[52] C. Dragone, *IEEE Photon. Technol. Lett.* **3**, 812 (1991).

[53] M. Zirngibal, C. Dragone, and C. H. Joyner, *IEEE Photon. Technol. Lett.* **4**, 1250 (1992).

[54] R. Adar, C. H. Henry, C. Dragone, R. C. Kistler, and G. R. Weber, *J. Lightwave Technol.* **11**, 212 (1993).

[55] H. Takahashi, K. Oda. H. Toba, and Y. Inoue, *J. Lightwave Technol.* **13**, 447 (1995).

[56] K. Okamoto, K. Syuto, H. Takahashi, and Y. Ohmori, *Electron. Lett.* **32**, 1474 (1996).

[57] M. K. Smit and C. vam Dam, *IEEE J. Sel. Topics Quantum Electron.* **2**, 236 (1996).

[58] H. Tanobe, Y. Kondo, Y. Kadota, K. Okamoto, and Y. Yoshikuni, *IEEE Photon. Technol. Lett.* **10**, 235 (1998).

[59] C. R. Doerr, *Optical Fiber Telecommunications*, Vol. 4A, I. P. Kaminow and T. P. Lee, Eds., Academic Press, San Diego, CA, 2002, Chap. 9.

[60] Y. Hibino, *IEEE J. Sel. Topics Quantum Electron.* **8**, 1090 (2002).

[61] Y. Yoshikuni, *IEEE J. Sel. Topics Quantum Electron.* **8**, 1090 (2002).

[62] P. Bernasconi, L. Stulz, J. Bailey, M. Cappuzzo, E. Chen, L. Gomez, E. Laskowski, R. Long, and A. Wong-Foy, *IEEE J. Sel. Topics Quantum Electron.* **8**, 1115 (2002).

Chapter 5

Semiconductor Lasers and Amplifiers

Semiconductor lasers are used almost exclusively as an optical source for fiber-optic communication systems. They provide not only the optical carrier on which information is encoded but are also useful for pumping fiber amplifiers as was seen in Chapter 3. In fact, semiconductor lasers constitute a major active component for lightwave technology because they offer unique advantages such as compact size, high efficiency, good reliability, small emissive area compatible with fiber-core dimensions, and possibility of direct modulation at relatively high frequencies. Moreover, they can be converted through suitable design modifications into an amplifier that itself has many applications in the domain of lightwave technology. Although semiconductor lasers were first made in 1962, their use became practical only after 1970 when GaAs lasers operating continuously at room temperature became available [1]. Since then, semiconductor lasers have been developed for several spectral regions important for lightwave technology [2]–[8]. This chapter is organized as follows. After introducing the basic concepts in Section 5.1, we describe in Section 5.2 the techniques used for forcing the laser to operate in a single mode. The operating characteristics of semiconductor lasers are discussed in Section 5.3. Light-emitting diodes are covered in Section 5.4, whereas Section 5.5 is devoted to the properties of semiconductor optical amplifiers.

5.1 Basic Concepts

Similar to the case of fiber lasers, semiconductor lasers also emit light through stimulated emission and require population inversion. However, in contrast with fiber lasers, atoms participating in the process of stimulated emission are arranged in a crystal lattice. As a result, individual energy levels associated with a single atom merge and form energy bands. As shown schematically in Figure 5.1, light is produced whenever an electron in the conduction bands recombines with a hole in the valence band. The photon emitted during the recombination process carries an energy $h\nu \approx E_g$, where E_g is the bandgap energy of the semiconductor. Using $\nu = c/\lambda$, one can conclude that each

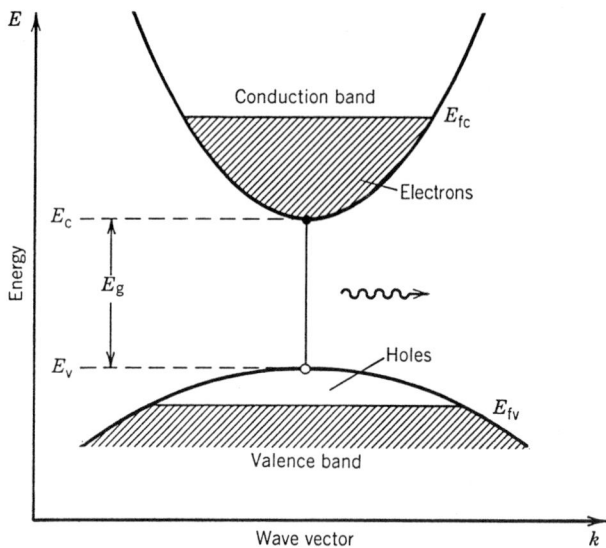

Figure 5.1: Conduction and valence bands in the active layer of a semiconductor laser. Electrons in the conduction band and holes in the valence band recombine to emit photons.

semiconductor can operate only in a certain wavelength region near $\lambda = hc/E_g$. For a semiconductor laser to emit light near 1.55 μm, its bandgap must be about 0.8 eV wide.

5.1.1 Optical Gain

Semiconductor lasers are pumped electrically using a *p–n* junction. As discussed in Section 4.3.1, such a pumping can be realized by using a three-layer structure in which a central core layer is sandwiched between the *p*-type and *n*-type cladding layers, both of which are doped so heavily that the Fermi-level separation $E_{fc} - E_{fv}$ exceeds the bandgap energy E_g (see Figure 5.1) under forward biasing of the *p–n* junction. Figure 5.2 shows the three-layer structure of a typical semiconductor laser together with its physical dimensions. The whole laser chip is typically under 1 mm in all three dimensions, resulting in an ultracompact design.

The central core layer in Figure 5.2 is made of the semiconductor that emits light and is called the "active" layer. The cladding layers are made using a semiconductor whose bandgap is larger than that of the active layer. The bandgap difference between the two semiconductors helps to confine electrons and holes to the active layer. At the same time, the active layer has a slightly larger refractive index than the surrounding cladding layers and acts as a planar waveguide whose number of modes can be controlled by changing the active-layer thickness. The main point is that such a heterostructure design helps to confine both the injected carriers (electrons and holes) and the light generated within the active layer through electron–hole recombination. A third feature is that the two cladding layers are transparent to the emitted light by virtue

5.1 Basic Concepts

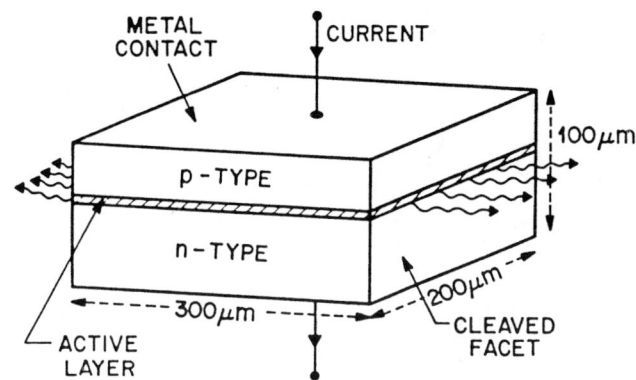

Figure 5.2: Schematic of a broad-area semiconductor laser. The active layer (hatched region) is sandwiched between *p*-type and *n*-type cladding layers with a larger bandgap.

of their higher bandgap, thereby resulting in a low-loss structure. These features have made semiconductor lasers practical for a wide variety of applications.

As discussed in Section 4.3.1, semiconductor lasers, operating in the wavelength range 1.1–1.6 μm of interest for lightwave technology, make use of the quaternary compound $In_{1-x}Ga_xAs_yP_{1-y}$ that is grown in a layer form on InP substrates using a suitable epitaxial growth technique. The lattice constant of each layer should remain matched to that of InP to maintain a well-defined lattice structure so that defects are not formed at the interfaces between any two layers with different bandgaps. The fractions x and y cannot be chosen arbitrarily but are related as $x/y = 0.45$ to ensure matching of the lattice constant. The bandgap of the quaternary compound can be expressed in terms of y only and is well approximated by Eq. (4.3.2).

When the injected carrier density in the active layer exceeds a certain value, population inversion is realized and the active region exhibits optical gain. An input signal propagating inside the active layer is then amplified by a a factor of $\exp(gL)$, where g is the *gain coefficient* and L is the active-layer length. The calculation of g requires the rates at which photons are absorbed and emitted through stimulated emission and depends on details of the band structure associated with the active material. In general, g is calculated numerically. Figure 5.3(a) shows the gain calculated for a 1.3-μm InGaAsP active layer at different values of the injected carrier density N. For $N = 1 \times 10^{18}$ cm^{-3}, $g < 0$, as population inversion has not yet occurred. As N increases, g becomes positive over a spectral range that increases with N. The peak value of the gain, g_p, also increases with N, together with a shift of the peak toward higher photon energies. The variation of g_p with N is shown in Figure 5.3(b). For $N > 1.5 \times 10^{18}$ cm^{-3}, g_p varies almost linearly with N. Figure 5.3 shows that the optical gain in semiconductors increases rapidly once population inversion is realized. It is because of such a high gain that semiconductor lasers can be made with physical dimensions of less than 1 mm.

The nearly linear dependence of g_p on N suggests an empirical approach in which

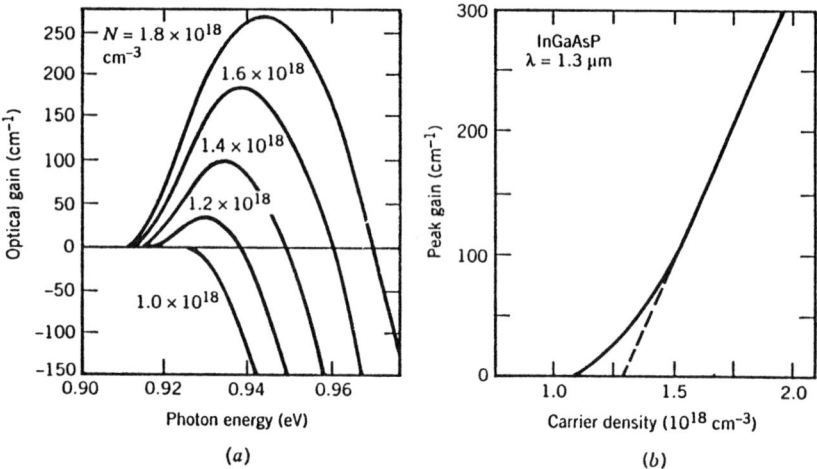

Figure 5.3: (a) Gain spectra at several carrier densities for a 1.3-μm laser. (b) Variation of peak gain g_p with N. Dashed line shows the linear fit in the high-gain region.

the peak gain is approximated by

$$g_p(N) = \sigma_g(N - N_T), \qquad (5.1.1)$$

where N_T is the transparency value of the carrier density and σ_g is the gain cross section; σ_g is also called the *differential gain*. Typical values of N_T and σ_g for InGaAsP lasers are in the range 1.0–1.5×10^{18} cm^{-3} and 2–3×10^{-16} cm^2, respectively [2]. As seen in Figure 5.3(b), the approximation (5.1.1) is reasonable in the high-gain region where g_p exceeds 100 cm^{-1}; most semiconductor lasers operate in this region. The use of Eq. (5.1.1) simplifies the analysis considerably, as band-structure details do not appear directly. The parameters σ_g and N_T can be estimated from numerical calculations such as those shown in Figure 5.3(b) or can be measured experimentally.

Semiconductor lasers with a larger value of σ_g generally perform better, since the same amount of gain can be realized at a lower carrier density or, equivalently, at a lower injected current. In quantum-well semiconductor lasers, σ_g is typically larger by about a factor of 2. The linear approximation in Eq. (5.1.1) for the peak gain can still be used in a limited range. A better approximation replaces Eq. (5.1.1) with $g_p(N) = g_0[1 + \ln(N/N_0)]$, where $g_p = g_0$ at $N = N_0$ and $N_0 = eN_T \approx 2.718 N_T$ by using the definition $g_p = 0$ at $N = N_T$ [3].

5.1.2 Feedback and Laser Threshold

The optical gain alone is not enough for laser operation. The other necessary ingredient is *optical feedback*, which converts any amplifier into an oscillator. In most lasers the feedback is provided by placing the gain medium inside a Fabry–Perot (FP) cavity formed by using two mirrors. Semiconductor lasers do not require external mirrors because the two cleaved facets can act as mirrors (see Figure 5.2) because of a relatively

5.1 Basic Concepts

large index difference across the air–semiconductor interface. The facet reflectivity normal to this interface is given by

$$R_m = \left(\frac{n-1}{n+1}\right)^2, \qquad (5.1.2)$$

where n is the refractive index of the gain medium. Typically, $n = 3.5$, resulting in 30% facet reflectivity. Even though the FP cavity formed by two cleaved facets is relatively lossy, the gain in a semiconductor laser is large enough that high losses can be tolerated.

A simple way to obtain the threshold condition is to study how the amplitude of an optical mode changes during one round trip inside the FP cavity. Assume that the mode has initially an amplitude A_0, frequency v, and propagation constant $\beta = \bar{n}(2\pi v)/c$, where $\bar{n}$ is the mode index. After one round trip, its amplitude increases by $\exp[2(g/2)L]$ because of gain (g is the power gain) and its phase changes by $2\beta L$, where L is the length of the laser cavity. At the same time, its amplitude decreases by $\sqrt{R_1 R_2} \exp(-\alpha_{\text{int}} L)$ because of reflection at the laser facets and because of internal losses α_{int} resulting from free-carrier absorption and interface scattering. The facet reflectivities R_1 and R_2 can be different if facets are coated to change their natural reflectivity. In the steady state, the mode should remain unchanged after one round trip, that is,

$$A_0 \exp(gL)\sqrt{R_1 R_2} \exp(-\alpha_{\text{int}} L) \exp(2i\beta L) = A_0. \qquad (5.1.3)$$

By equating the amplitude and the phase on two sides, we obtain

$$g = \alpha_{\text{int}} + \frac{1}{2L}\ln\left(\frac{1}{R_1 R_2}\right) = \alpha_{\text{int}} + \alpha_{\text{mir}} = \alpha_{\text{cav}}, \qquad (5.1.4)$$

$$2\beta L = 2m\pi \quad \text{or} \quad v = v_m = mc/2\bar{n}L, \qquad (5.1.5)$$

where m is an integer. Equation (5.1.4) shows that the gain g equals total cavity loss α_{cav} at the threshold and beyond. It is important to note that g is not the same as the material gain g_m shown in Figure 5.3. The optical mode extends beyond the active layer while the gain exists only inside it. As a result, $g = \Gamma g_m$, where Γ is the confinement factor of the active region with typical values <0.4.

5.1.3 Longitudinal Modes

The phase condition in Eq. (5.1.5) shows that the laser frequency v must match one of the frequencies in the set v_m, where m is an integer. These frequencies correspond to the *longitudinal modes* and are determined by the optical length $\bar{n}L$. The spacing Δv_L between the longitudinal modes is constant. In fact, as discussed in Section 2.3.2, it is the same as the free spectral range associated with any FP resonator and is given by $\Delta v_L = c/2n_g L$ when material dispersion is included [2], where n_g is the group index. Typically, $\Delta v_L = 150$ GHz for $L = 250$ μm.

A semiconductor laser generally emits light in several longitudinal modes of the cavity simultaneously. As seen in Figure 5.4, the gain spectrum $g(\omega)$ of semiconductor lasers is wide enough (bandwidth ~ 10 THz) that many longitudinal modes of the FP cavity experience gain simultaneously. The mode closest to the gain peak becomes

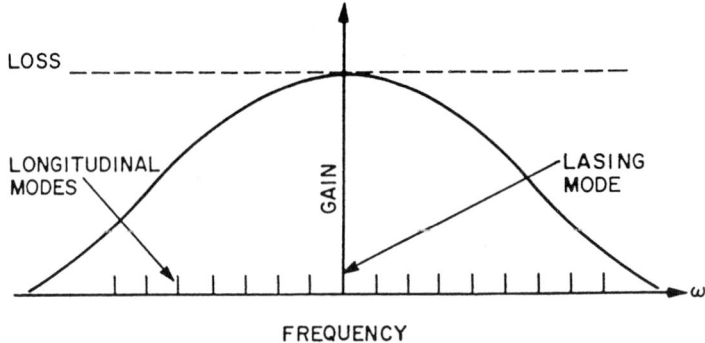

Figure 5.4: Schematic of gain and loss profiles in semiconductor lasers. Vertical bars show the location of longitudinal modes.

the dominant mode. Under ideal conditions, the other modes should not reach threshold since their gain always remains less than that of the main mode. In practice, the difference is extremely small (~ 0.1 cm^{-1}) and one or two neighboring modes on each side of the main mode carry a significant portion of the laser power together with the main mode. Since each mode propagates inside the fiber at a slightly different speed because of group-velocity dispersion, the multimode nature of a semiconductor laser often limits the bit rate of lightwave systems operating near 1.55 μm. Performance can be improved by designing lasers such that they oscillate in a single longitudinal mode. Such lasers are discussed in Section 5.2.

5.1.4 Laser Structures

The simplest structure of a semiconductor laser consists of a thin active layer (with a thickness of 0.1 μm or less) sandwiched between the p-type and n-type cladding layers of another semiconductor with a higher bandgap. Such lasers are called *broad-area* lasers since the current is injected over a relatively broad area covering the entire width of the laser chip (see Figure 5.2). The laser light is emitted from the two cleaved facets in the form of an elliptic spot of dimensions $\sim 1 \times 100$ μm^2. In the transverse direction perpendicular to the junction plane, the spot size is ~ 1 μm because the active layer supports only the fundamental TE$_0$ and TM$_0$ modes. In practice, the gain is slightly larger for the TE$_0$ mode, and the laser light is polarized in the junction plane. Since no confinement mechanism exists in the lateral direction (parallel to the junction plane), emitted light spreads over the entire width of a broad-area laser, resulting in a highly elliptical beam. Such lasers suffer from a number of deficiencies and are rarely used in practice. The major drawbacks are a relatively high threshold current and a spatial pattern that changes in an uncontrollable manner with the current. These problems can be solved by introducing a mechanism for light confinement in the lateral direction.

The light-confinement problem is solved in the so-called *index-guided* semiconductor lasers by introducing an index step Δn_L in the lateral direction so that a rectangular waveguide is formed. Any design shown in Figure 4.5 can be used for this purpose.

5.1 Basic Concepts

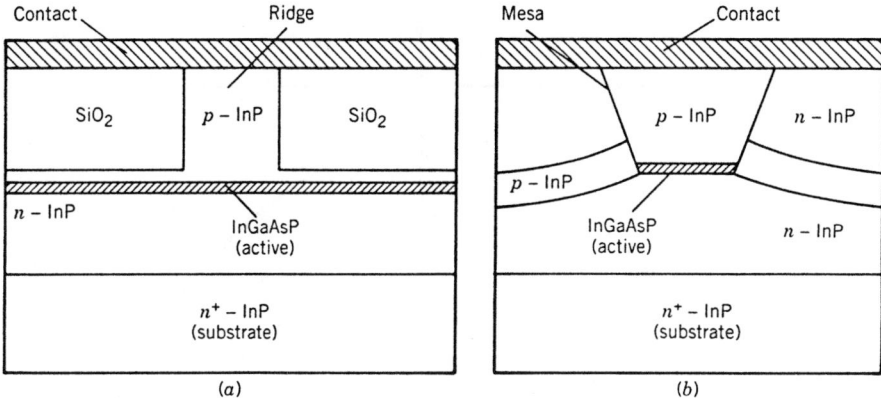

Figure 5.5: Schematic cross section of (a) a ridge-waveguide laser and (b) a buried-heterostructure laser.

Figure 5.5 shows two examples. In a *ridge-waveguide laser* (Figure 5.5a), a ridge is formed by etching most of the top cladding layer [2]. A silica layer is then deposited to block the current flow so that the current enters only through the ridge. Since the cladding material used for the ridge has a much larger refractive index than silica, the mode index is also higher under the ridge, resulting in an index step $\Delta n_L \sim 0.01$. This index difference guides the optical mode in the lateral direction. The magnitude of the index step is sensitive to many fabrication details, such as the ridge width and the proximity of the silica layer to the active layer. Although this scheme offers only weak lateral confinement, the relative simplicity of the ridge-waveguide design and the resulting low cost make this design attractive for some applications.

In strongly index-guided semiconductor lasers, the active region is buried on all sides by several layers of lower refractive index (typical dimensions $\sim 0.1 \times 1 \ \mu m^2$). Such lasers are called *buried heterostructure* (BH) lasers (see Figure 5.5b). Several different kinds of BH lasers have been developed. They are known under names such as etched-mesa BH, planar BH, double-channel planar BH, and V-grooved or channelled substrate BH lasers, depending on the fabrication method used to realize the laser structure [2]. They all allow a relatively large index step ($\Delta n_L > 0.1$) in the lateral direction and, as a result, permit strong mode confinement. Because of a large built-in index step, the spatial distribution of the emitted light is inherently stable, provided that the laser is designed to support a single spatial mode.

As the active region of a BH laser is in the form of a rectangular waveguide, its spatial modes can be obtained by following a method similar to that discussed in Section 4.2. The single-mode condition can be obtained from Eq. (4.2.6). All semiconductor lasers support a single mode in the transverse direction because of the use of a thin active layer but they may support multiple lateral modes depending on the waveguide width. In practice, a BH laser operates in a single mode if the active-region width is reduced to below 2 μm. The spot size still remains elliptical with typical dimensions of $2 \times 1 \ \mu m^2$. Because of small spot-size dimensions, the output beam diffracts considerably in both the lateral and transverse directions as it leaves the laser. An elliptic

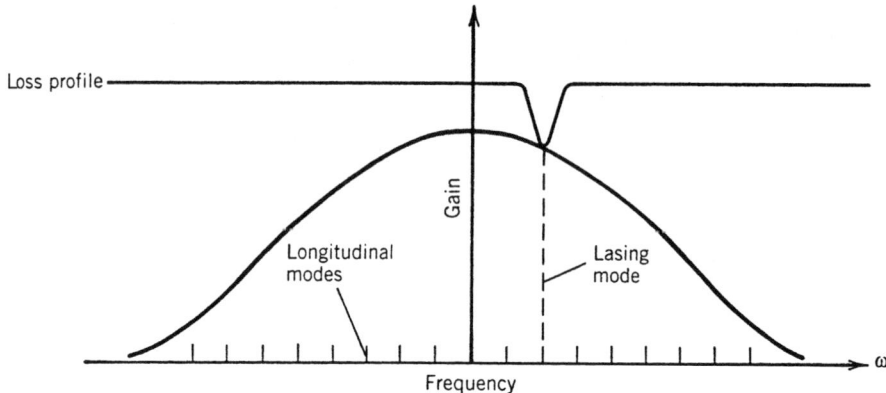

Figure 5.6: Gain and loss profiles for semiconductor lasers oscillating predominantly in a single longitudinal mode.

spot size and a large divergence angle make it difficult to couple light into the fiber efficiently. A spot-size converter is sometimes used to improve the coupling efficiency.

5.2 Control of Longitudinal Modes

As discussed earlier, FP semiconductor lasers oscillate in several longitudinal modes simultaneously because of a relatively small gain difference (~ 0.1 cm^{-1}) between two neighboring modes of the FP cavity. The resulting spectral width (2–4 nm) is acceptable for some applications but becomes a concern for many others. This section is devoted to techniques that can be used to design semiconductor lasers such that they emit light predominantly in a single longitudinal mode [9]–[16].

The basic idea is to design the laser such that losses are different for different longitudinal modes of the cavity, in contrast with FP lasers whose losses are mode-independent. Figure 5.6 shows the gain and loss profiles schematically for such a laser. The longitudinal mode with the smallest cavity loss reaches threshold first and becomes the dominant mode. Other neighboring modes are discriminated by their higher losses. The power carried by these side modes is usually a small fraction (<1%) of the total emitted power. The performance of a single-mode laser is often characterized by the *mode-suppression ratio* (MSR), defined as MSR = P_{mm}/P_{sm}, where P_{mm} is the main-mode power and P_{sm} is the power of the most dominant side mode. The MSR should exceed 1,000 (or 30 dB) for a good single-mode laser.

5.2.1 Distributed Feedback Lasers

Distributed feedback (DFB) semiconductor lasers were developed during the 1980s and are used routinely for WDM lightwave systems [11]–[16]. The feedback in DFB lasers, as the name implies, is not localized at the facets but is distributed throughout the cavity length. This is achieved through an internal built-in grating that leads to a

5.2 Control of Longitudinal Modes

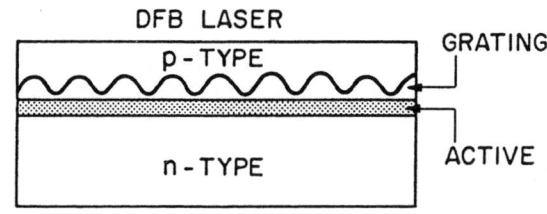

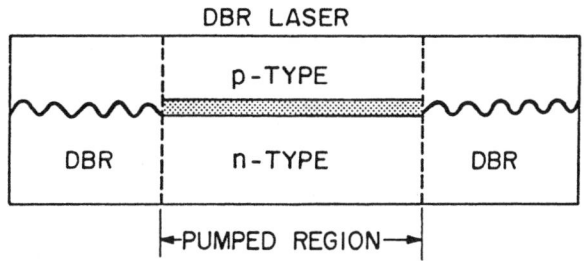

Figure 5.7: DFB and DBR laser structures. The shaded area shows the active region and the wavy line indicates the presence of a Bragg gratin.

periodic variation of the mode index. Feedback occurs by means of *Bragg diffraction*, a phenomenon that couples the waves propagating in the forward and backward directions. Mode selectivity of the DFB mechanism results from the *Bragg condition*: The coupling occurs only for wavelengths λ_B satisfying

$$\Lambda = m(\lambda_B/2\bar{n}), \tag{5.2.1}$$

where Λ is the grating period, $\bar{n}$ is the average mode index, and the integer m represents the order of Bragg diffraction. The coupling between the forward and backward waves is strongest for the first-order Bragg diffraction ($m = 1$). For a DFB laser operating at $\lambda_B = 1.55$ μm, Λ is about 235 nm if we use $m = 1$ and $\bar{n} = 3.3$ in Eq. (5.2.1). Such gratings can be made by using a holographic technique [2].

From the standpoint of device operation, semiconductor lasers employing the DFB mechanism can be classified into two broad categories: DFB lasers and *distributed Bragg reflector* (DBR) lasers. Figure 5.7 shows two kinds of laser structures. Though the feedback occurs throughout the cavity length in DFB lasers, it does not take place inside the active region of a DBR laser. In effect, the end regions of a DBR laser act as mirrors whose reflectivity is maximum for a wavelength λ_B satisfying Eq. (5.2.1). The cavity losses are therefore minimum for the longitudinal mode closest to λ_B and increase substantially for other longitudinal modes (see Figure 5.6). The MSR is determined by the gain margin defined as the excess gain required by the most dominant side mode to reach threshold. A gain margin of 3–5 cm^{-1} is generally enough to realize an MSR >30 dB for DFB lasers operating continuously [12]. However, a larger gain margin is needed (>10 cm^{-1}) when DFB lasers are modulated directly. *Phase-shifted DFB lasers* [11], in which the grating is shifted by $\lambda_B/4$ in the middle of the laser to produce a $\pi/2$ phase shift, are often used, since they are capable of providing a much larger gain margin than that of conventional DFB lasers. Another design that has led to improvements in the device performance is known as the *gain-coupled DFB laser* [17]. In these lasers, both the optical gain and the mode index vary periodically along the cavity length.

Fabrication of DFB semiconductor lasers requires advanced technology with multiple epitaxial growths [14]. The principal difference from FP lasers is that a grating is etched onto one of the cladding layers surrounding the active layer. A thin *n*-type waveguide layer with a refractive index intermediate to that of the active layer and the substrate acts as a grating. The periodic variation of the thickness of the waveguide layer translates into a periodic variation of the mode index $\bar{n}$ along the cavity length and leads to a coupling between the forward and backward propagating waves through Bragg diffraction.

A holographic technique is often used to form a grating with a pitch of ~ 0.2 μm. It works by forming a fringe pattern on a photoresist deposited on the wafer surface by interfering two optical beams and then etching the pattern chemically. In the alternative electron-beam lithographic technique, an electron beam writes the desired pattern on the electron-beam resist. Both methods use chemical etching to form grating corrugations, with the patterned resist acting as a mask. Once the grating has been etched onto the substrate, multiple layers are grown by using an epitaxial growth technique. A second epitaxial regrowth is needed to make a BH device such as that shown in Figure 5.5(b). Despite the technological complexities, DFB lasers are routinely produced commercially. They are used in nearly all 1.55-μm optical communication systems operating at bit rates of 2.5 Gb/s or more. DFB lasers are reliable enough that they have been used since 1992 in all transoceanic lightwave systems.

5.2.2 Coupled-Cavity Semiconductor Lasers

In a *coupled-cavity* semiconductor laser [2], single-mode operation is realized by coupling the laser cavity to an external cavity, which feeds a portion of the exiting light back into the laser cavity. The feedback from the external cavity is not necessarily in phase with the field inside the laser cavity because of the phase shift occurring in the external cavity. The in-phase feedback occurs only for those laser modes whose wavelength nearly coincides with one of the longitudinal modes of the external cavity. In effect, the effective reflectivity of the laser facet facing the external cavity becomes wavelength-dependent and leads to low losses for certain wavelengths. The longitudinal mode that is closest to the gain peak and has the lowest cavity loss becomes the dominant mode.

Several kinds of coupled-cavity schemes have been developed for making single-mode laser; Figure 5.8 shows three of them. A simple scheme couples the light from a semiconductor laser to an external grating [Figure 5.8(a)]. It is necessary to reduce the natural reflectivity of the cleaved facet facing the grating through an antireflection coating to provide a strong coupling. Such lasers are called *external-cavity* semiconductor lasers and have attracted considerable attention because of their tunability [9]. The wavelength of the single mode selected by the coupled-cavity mechanism can be tuned over a wide range (typically 50 nm) simply by rotating the grating. Wavelength

5.2 Control of Longitudinal Modes

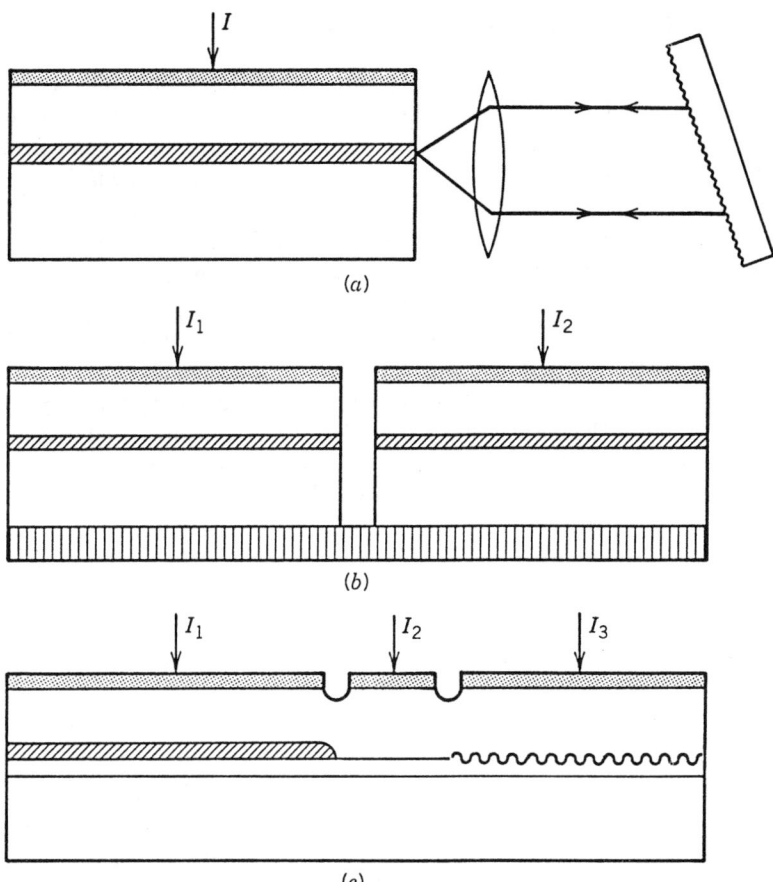

Figure 5.8: Coupled-cavity laser structures: (a) external-cavity laser; (b) cleaved-coupled-cavity laser; (c) multisection DBR laser.

tunability is a desirable feature for lasers used in WDM lightwave systems. A drawback of the laser shown in Figure 5.8(a) from the system standpoint is its nonmonolithic nature, which makes it difficult to realize the mechanical stability required of optical transmitters.

A monolithic design for coupled-cavity lasers is offered by the cleaved-coupled-cavity laser [10] shown in Figure 5.8(b). Such lasers are made by cleaving a conventional multimode semiconductor laser in the middle so that the laser is divided into two sections of about the same length but separated by a narrow air gap (with a width of ~ 1 μm). The reflectivity of cleaved facets ($\sim 30\%$) allows enough coupling between the two sections as long as the gap is not too wide. It is even possible to tune the wavelength of such a laser over a tuning range ~ 20 nm by varying the current injected into one of the cavity sections acting as a mode controller. However, tuning is not continuous, since it corresponds to successive mode hops of about 2 nm.

5.2.3 Tunable Semiconductor Lasers

Modern WDM lightwave systems require single-mode, narrow-linewidth lasers whose wavelength remains fixed over time. DFB lasers satisfy this requirement but their wavelength stability comes at the expense of tunability. The large number of DFB lasers used inside a WDM transmitter make the design and maintenance of such a lightwave system expensive and impractical. The availability of semiconductor lasers whose wavelength can be tuned over a wide range would solve this problem [6].

Multisection DFB and DBR lasers were developed during the 1990s to meet the somewhat conflicting requirements of stability and tunability [18]–[25] and were reaching the commercial stage in 2001. Figure 5.8(c) shows a typical laser structure. It consists of three sections, referred to as the active section, the phase-control section, and the Bragg section. Each section can be biased independently by injecting different amounts of currents. The current injected into the Bragg section is used to change the Bragg wavelength ($\lambda_B = 2n\Lambda$) through carrier-induced changes in the refractive index n. The current injected into the phase-control section is used to change the phase of the feedback from the DBR through carrier-induced index changes in that section. The laser wavelength can be tuned almost continuously over the range 10–15 nm by controlling the currents in the phase and Bragg sections. By 1997, such lasers exhibited a tuning range of 17 nm and output powers of up to 100 mW with high reliability [24].

Several other designs of tunable DFB lasers have been developed in recent years. In one scheme, the built-in grating inside a DBR laser is chirped by varying the grating period Λ or the mode index $\bar{n}$ along the cavity length. As seen from Eq. (5.2.1), the Bragg wavelength itself then changes along the cavity length. Since the laser wavelength is determined by the Bragg condition, such a laser can be tuned over a wavelength range determined by the grating chirp. In a simple implementation of the basic idea, the grating period remains uniform, but the waveguide is bent to change the effective mode index $\bar{n}$. Such multisection DFB lasers can be tuned over 5–6 nm while maintaining a single longitudinal mode with high side-mode suppression [20].

In another scheme, a *superstructure* or sampled grating is used for the DBR section of a multisection laser [21]–[23]. In such gratings, the amplitude or the phase of the coupling coefficient is modulated in a periodic fashion along the grating length. As a result, the reflectivity peaks at several wavelengths whose interval is determined by the modulation period. Such multisection DBR lasers can be tuned discretely over a wavelength range exceeding 100 nm. By controlling the current in the phase-control section, a quasi-continuous tuning range of 40 nm was realized in 1995 with a superstructure grating [21]. The tuning range can be extended considerably by using a four-section device in which another DBR section is added to the left side of the device shown in Figure 5.8(c). Each DBR section supports its own comb of wavelengths but the spacing in each comb is not the same. The coinciding wavelength in the two combs becomes the output wavelength that can be tuned over a wide range (analogous to the Vernier effect).

In a related approach, a fourth section is added between the gain and phase sections in Figure 5.8(c): It consist of a grating-assisted directional coupler with a sampled grating. Figure 5.9 shows the design of the device schematically. The coupler section has two vertically separated waveguides having different thickness and material composi-

5.2 Control of Longitudinal Modes

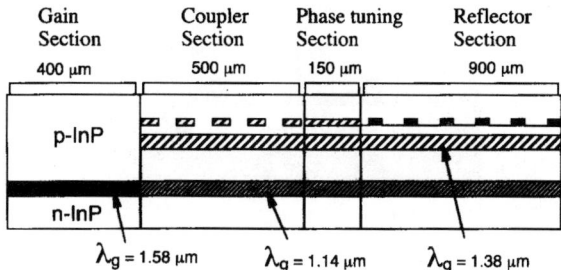

Figure 5.9: Schematic of a tunable DBR laser with an integrated grating-assisted directional coupler. (After Ref. [22]; ©1995 IEEE.)

tion so that they form an asymmetric directional coupler (see Section 2.1). The grating can selectively transfer a single wavelength from the wavelength comb supported by the DBR section with a sampled grating. The largest tuning range of 114 nm was produced in 1995 by this kind of device [22]. Such widely tunable DBR lasers are likely to find applications in many WDM lightwave systems.

5.2.4 Vertical-Cavity Surface-Emitting Lasers

A new class of semiconductor lasers, known as *vertical-cavity surface-emitting lasers* (VCSELs), emerged during the 1990s with many potential applications [26]–[33]. VCSELs operate in a single longitudinal mode by virtue of an extremely small cavity length (~ 1 μm), for which the mode spacing exceeds the gain bandwidth (see Figure 5.4). They emit light in a direction normal to the active-layer plane in a manner analogous to that of a surface-emitting LED (see Section 5.4). Moreover, the emitted light is in the form of a circular beam that can be coupled into a single-node fiber with high efficiency. These properties result in a number of advantages that are leading to rapid adoption of VCSELs for lightwave communications.

As seen in Figure 5.10, fabrication of VCSELs requires growth of multiple thin layers on a substrate. The active region, in the form of one or several quantum wells, is surrounded by two high-reflectivity (>99.5%) DBR mirrors that are grown epitaxially on both sides of the active region to form a high-Q microcavity [28]. Each DBR mirror is made by growing many pairs of alternating GaAs and AlAs layers, each $\lambda/4$ thick, where λ is the wavelength emitted by the VCSEL. A wafer-bonding technique is sometimes used for VCSELs operating in the 1.55-μm wavelength region to accommodate the InGaAsP active region [31]. Chemical etching or a related technique is used to form individual circular disks (each corresponding to one VCSEL) whose diameter can be varied over a wide range (typically 5–20 μm). The entire two-dimensional array of VCSELs can be tested without requiring separation of lasers because of the vertical nature of light emission. As a result, the cost of a VCSEL can be much lower than that of an edge-emitting laser. VCSELs also exhibit a relatively low threshold (~ 1 mA or less). Their only disadvantage is that they cannot emit more than a few milliwatts of power because of a small active volume. For this reason, they are mostly used for local-area and data-communication applications and have virtually replaced

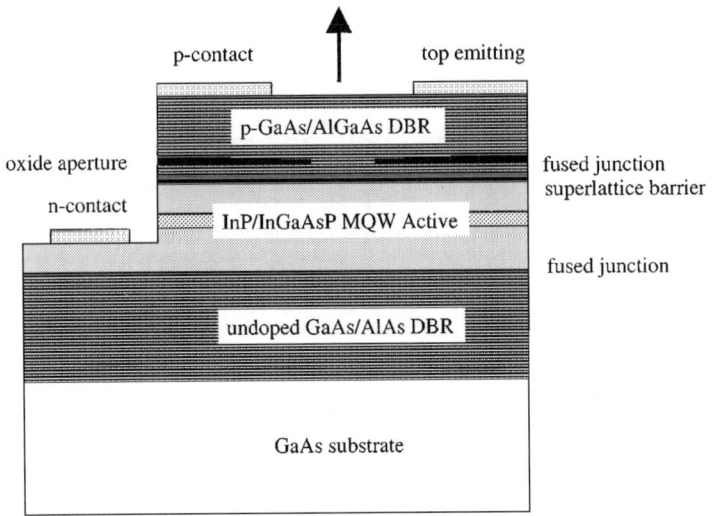

Figure 5.10: Schematic of a 1.55-μm VCSEL made by using the wafer-bonding technique. (After Ref. [31]; ©2000 IEEE.)

LEDs. Early VCSELs were designed to emit near 0.8 μm and operated in multiple transverse modes because of their relatively large diameters ($\sim$10 μm).

In recent years, VCSEL technology has advanced enough that VCSELs can be designed to operate in a wide wavelength range extending from 650 to 1,600 nm [28]. Their applications in the 1.3- and 1.55-μm wavelength windows require that VCSELs operate in a single transverse mode. By 2001, several techniques had emerged for controlling the transverse modes of a VCSEL, the most common being the oxide-confinement technique in which an insulating aluminum-oxide layer, acting as a dielectric aperture, confines both the current and the optical mode to a <3-μm-diameter region. Such VCSELs operate in a single mode with narrow linewidth and can replace a DFB laser in many lightwave applications as long as their low output power is acceptable. They are especially useful for data transfer and local-loop applications because of their low-cost packaging. VCSELs are also well suited for WDM applications for two reasons. First, their wavelengths can be tuned over a wide range (>50 nm) using the micro-electro-mechanical system (MEMS) technology [29]. Second, one can make two-dimensional VCSEL arrays such that each laser operates at a different wavelength [33]. WDM sources, containing multiple monolithically integrated lasers, are required for modern lightwave systems.

5.3 Laser Characteristics

The operating characteristics of semiconductor lasers are well described by a set of rate equations that govern the interaction of photons and electrons inside the active region. In this section we use the rate equations to discuss first the continuous-wave (CW)

5.3 Laser Characteristics

properties and then consider small- and large-signal modulation characteristics. The last two subsections focus on the intensity noise and spectral bandwidth of semiconductor lasers.

5.3.1 CW Characteristics

A rigorous derivation of the rate equations generally starts from Maxwell's equations. The rate equations can also be written heuristically by considering various physical phenomena through which the number of photons, P, and the number of electrons, N, change with time inside the active region. For a single-mode laser, these equations take the form [2]

$$\frac{dP}{dt} = GP + R_{sp} - \frac{P}{\tau_p}, \qquad (5.3.1)$$

$$\frac{dN}{dt} = \frac{I}{q} - \frac{N}{\tau_c} - GP, \qquad (5.3.2)$$

where the net rate of stimulated emission G is defined as

$$G = \Gamma v_g g_m = G_N(N - N_0), \qquad (5.3.3)$$

and R_{sp} is the rate of spontaneous emission into the lasing mode. Note that R_{sp} is much smaller than the total spontaneous-emission rate. The reason is that spontaneous emission occurs in all directions over a wide spectral range (40–50 nm) but only a small fraction of it, propagating along the cavity axis and emitted at the laser frequency, actually contributes to Eq. (5.3.1). In fact, R_{sp} and G are related as $R_{sp} = n_{sp}G$, where n_{sp} is the spontaneous-emission factor introduced in Section 3.1 and is about 2 for semiconductor lasers [2]. The variable N in the rate equations represents the number of electrons rather than the carrier density; the two are related by the active volume V. In Eq. (5.3.3), v_g is the group velocity, Γ is the confinement factor, and g_m is the material gain at the mode frequency. From Eq. (5.1.1), G varies linearly with N with $G_N = \Gamma v_g \sigma_g / V$ and $N_0 = N_T V$.

The last term in Eq. (5.3.1) takes into account the loss of photons inside the cavity. The parameter τ_p is referred to as the *photon lifetime*. It is related to the *cavity loss* α_{cav} introduced in Eq. (5.1.4) as

$$\tau_p^{-1} = v_g \alpha_{cav} = v_g(\alpha_{mir} + \alpha_{int}). \qquad (5.3.4)$$

The three terms in Eq. (5.3.2) indicate the rates at which electrons are created or destroyed inside the active region. The carrier lifetime τ_c includes the loss of electrons owing to both spontaneous emission and nonradiative recombination.

The P–I curve characterizes the emission properties of a semiconductor laser, as it indicates not only the threshold level but also the current that needs to be applied to obtain a certain amount of power. Figure 5.11 shows the P–I curves of a 1.3-μm InGaAsP laser at temperatures in the range 10–130°C. At room temperature, the threshold is reached near 20 mA, and the laser can emit 10 mW of output power from each

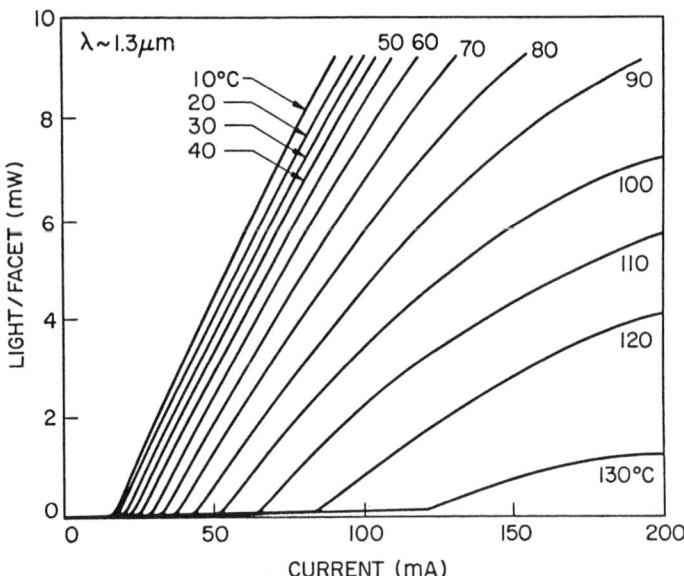

Figure 5.11: P–I curves at several temperatures for a 1.3-μm buried heterostructure laser.

facet at 100 mA of applied current. The laser performance degrades at high temperatures. The threshold current is found to increase exponentially with temperature, that is,

$$I_{\text{th}}(T) = I_0 \exp(T/T_0), \tag{5.3.5}$$

where I_0 is a constant and T_0 is a characteristic temperature often used to express the temperature sensitivity of threshold current. For InGaAsP lasers T_0 is typically in the range 50–70 K. By contrast, T_0 exceeds 120 K for GaAs lasers. Because of the temperature sensitivity of InGaAsP lasers, it is often necessary to control their temperature through a built-in thermoelectric cooler.

The rate equations can be used to understand most of the features seen in Figure 5.11. In the case of CW operation at a constant current I, the time derivatives in Eqs. (5.3.1) and (5.3.2) can be set to zero. The solution takes a particularly simple form if spontaneous emission is neglected by setting $R_{\text{sp}} = 0$. For currents such that $G\tau_p < 1$, $P = 0$ and $N = \tau_c I/q$. The threshold is reached at a current for which $G\tau_p = 1$. The carrier population is then clamped to the threshold value $N_{\text{th}} = N_0 + (G_N \tau_p)^{-1}$. The threshold current is given by

$$I_{\text{th}} = \frac{q N_{\text{th}}}{\tau_c} = \frac{q}{\tau_c}\left(N_0 + \frac{1}{G_N \tau_p}\right). \tag{5.3.6}$$

For $I > I_{\text{th}}$, the photon number P increases linearly with I as

$$P = (\tau_p/q)(I - I_{\text{th}}). \tag{5.3.7}$$

The emitted power P_e is related to P by the relation

$$P_e = \tfrac{1}{2}(v_g \alpha_{\text{mir}})\hbar \omega P. \tag{5.3.8}$$

5.3 Laser Characteristics

The derivation of Eq. (5.3.8) is intuitively obvious if we note that $v_g \alpha_{mir}$ is the rate at which photons of energy $\hbar\omega$ escape from the two facets. The factor of $\frac{1}{2}$ makes P_e the power emitted from each facet for a FP laser with equal facet reflectivities. For FP lasers with coated facets or for DFB lasers, Eq. (5.3.8) needs to be suitably modified [2]. By using Eqs. (5.3.4) and (5.3.7) in Eq. (5.3.8), the emitted power is given by

$$P_e = \frac{\hbar\omega}{2q} \frac{\eta_{int}\alpha_{mir}}{\alpha_{mir} + \alpha_{int}} (I - I_{th}), \tag{5.3.9}$$

where the internal quantum efficiency η_{int} is introduced phenomenologically to indicate the fraction of injected electrons that is converted into photons through stimulated emission. In the above-threshold regime, η_{int} is almost 100% for most semiconductor lasers.

A quantity of practical interest is the slope of the P–I curve for $I > I_{th}$; it is called the *slope efficiency* and is defined as

$$\frac{dP_e}{dI} = \frac{\hbar\omega}{2q} \eta_d \quad \text{with} \quad \eta_d = \frac{\eta_{int}\alpha_{mir}}{\alpha_{mir} + \alpha_{int}}. \tag{5.3.10}$$

The quantity η_d is called the *differential quantum efficiency*, as it is a measure of the efficiency with which light output increases with an increase in the injected current. One can define the external quantum efficiency η_{ext} as

$$\eta_{ext} = \frac{\text{photon-emission rate}}{\text{electron-injection rate}} = \frac{2P_e/\hbar\omega}{I/q} = \frac{2q}{\hbar\omega} \frac{P_e}{I}. \tag{5.3.11}$$

By using Eqs. (5.3.9) through (5.3.11), η_{ext} and η_d are found to be related by

$$\eta_{ext} = \eta_d (1 - I_{th}/I). \tag{5.3.12}$$

Generally, $\eta_{ext} < \eta_d$ but becomes nearly the same for $I \gg I_{th}$. Similar to the case of LEDs, one can define the total quantum efficiency (or wall-plug efficiency) as $\eta_{tot} = 2P_e/(V_0 I)$, where V_0 is the applied voltage. It is related to η_{ext} as

$$\eta_{tot} = \frac{\hbar\omega}{qV_0} \eta_{ext} \approx \frac{E_g}{qV_0} \eta_{ext}, \tag{5.3.13}$$

where E_g is the bandgap energy. Generally, $\eta_{tot} < \eta_{ext}$ as the applied voltage exceeds E_g/q. For GaAs lasers, η_d can exceed 80% and η_{tot} can approach 50%. The InGaAsP lasers are less efficient with $\eta_d \sim 50\%$ and $\eta_{tot} \sim 20\%$.

The exponential increase in the threshold current with temperature can be understood from Eq. (5.3.6). The carrier lifetime τ_c is generally N-dependent because of Auger recombination and decreases with N as N^2. The rate of Auger recombination increases exponentially with temperature and is responsible for the temperature sensitivity of InGaAsP lasers. Figure 5.11 also shows that the slope efficiency decreases with an increase in the output power (bending of the P–I curves). This decrease can be attributed to junction heating occurring under CW operation. It can also result from an increase in internal losses or current leakage at high operating powers. Despite these problems, the performance of DFB lasers improved substantially during the 1990s [14].

DFB lasers emitting >100 mW of power at room temperature in the 1.55-μm spectral region were fabricated by 1996 using a strained MQW design [34]. Such lasers exhibited <10 mA threshold current at 20°C and emitted ∼20 mW of power at 100°C while maintaining a MSR of >40 dB. By 2001, DFB lasers capable of delivering more than 200 mW of power were available commercially.

5.3.2 Small-Signal Modulation

The modulation response of semiconductor lasers is studied by solving the rate equations (5.3.1) and (5.3.2) with a time-dependent current of the form

$$I(t) = I_b + I_m f_p(t), \tag{5.3.14}$$

where I_b is the bias current, I_m is the current, and $f_p(t)$ represents the shape of the current pulse. Two changes are necessary for a realistic description. First, Eq. (5.3.3) for the gain G must be modified to become [2]

$$G = G_N(N - N_0)(1 - \varepsilon_{NL} P), \tag{5.3.15}$$

where ε_{NL} is a nonlinear-gain parameter that leads to a slight reduction in G as P increases. The physical mechanism behind this reduction can be attributed to several phenomena, such as spatial hole burning, spectral hole burning, carrier heating, and two-photon absorption [35]–[38]. Typical values of ε_{NL} are $\sim 10^{-7}$. Equation (5.3.15) is valid for $\varepsilon_{NL} P \ll 1$. The factor $1 - \varepsilon_{NL} P$ should be replaced by $(1 + P/P_s)^{-b}$, where P_s is a material parameter, when the laser power exceeds far above 10 mW. The exponent b equals $\frac{1}{2}$ for spectral hole burning [36] but can vary over the range 0.2–1 because of the contribution of carrier heating [38].

The second change is related to an important property of semiconductor lasers. It turns out that whenever the optical gain changes as a result of changes in the carrier population N, the refractive index also changes. From a physical standpoint, amplitude modulation in semiconductor lasers is always accompanied by phase modulation because of carrier-induced changes in the mode index $\bar{n}$. Phase modulation can be included through the equation [2]

$$\frac{d\phi}{dt} = \frac{1}{2}\beta_c \left[G_N(N - N_0) - \frac{1}{\tau_p} \right], \tag{5.3.16}$$

where β_c is the amplitude-phase coupling parameter, commonly called the *linewidth enhancement factor*, as it leads to an enhancement of the spectral width associated with a single longitudinal mode. Typical values of β_c for InGaAsP lasers are in the range 4–8, depending on the operating wavelength [39]. Lower values of β_c occur in MQW lasers, especially for strained quantum wells [3].

In general, the nonlinear nature of the rate equations makes it necessary to solve them numerically. A useful analytic solution can be obtained for the case of small-signal modulation in which the laser is biased above threshold ($I_b > I_{th}$) and modulated such that $I_m \ll I_b - I_{th}$. The rate equations can be linearized in that case and solved analytically, using the Fourier transform technique, for an arbitrary form of $f_p(t)$. The

5.3 Laser Characteristics

small-signal modulation bandwidth can be obtained by considering the response of semiconductor lasers to sinusoidal modulation at the frequency ω_m so that $f_p(t) = \sin(\omega_m t)$. The laser output is also modulated sinusoidally. The general solution of Eqs. (5.3.1) and (5.3.2) is given by

$$P(t) = P_b + |p_m|\sin(\omega_m t + \theta_m), \tag{5.3.17}$$
$$N(t) = N_b + |n_m|\sin(\omega_m t + \psi_m), \tag{5.3.18}$$

where P_b and N_b are the steady-state values at the bias current I_b, $|p_m|$ and $|n_m|$ are small changes occurring because of current modulation, and θ_m and ψ_m govern the phase lag associated with the small-signal modulation. In particular, $p_m \equiv |p_m|\exp(i\theta_m)$ is given by [2]

$$p_m(\omega_m) = \frac{P_b G_N I_m/q}{(\Omega_R + \omega_m - i\Gamma_R)(\Omega_R - \omega_m + i\Gamma_R)}, \tag{5.3.19}$$

where

$$\Omega_R = [GG_N P_b - (\Gamma_P - \Gamma_N)^2/4]^{1/2}, \quad \Gamma_R = (\Gamma_P + \Gamma_N)/2, \tag{5.3.20}$$
$$\Gamma_P = R_{\text{sp}}/P_b + \varepsilon_{\text{NL}}GP_b, \quad \Gamma_N = \tau_c^{-1} + G_N P_b. \tag{5.3.21}$$

Ω_R and Γ_R are the frequency and the damping rate of relaxation oscillations. These two parameters play an important role in governing the dynamic response of semiconductor lasers. In particular, the efficiency is reduced when the modulation frequency exceeds Ω_R by a large amount.

Similar to the case of LEDs, one can introduce the transfer function as

$$H(\omega_m) = \frac{p_m(\omega_m)}{p_m(0)} = \frac{\Omega_R^2 + \Gamma_R^2}{(\Omega_R + \omega_m - i\Gamma_R)(\Omega_R - \omega_m + i\Gamma_R)}. \tag{5.3.22}$$

The modulation response is flat [$H(\omega_m) \approx 1$] for frequencies such that $\omega_m \ll \Omega_R$, peaks at $\omega_m = \Omega_R$, and then drops sharply for $\omega_m \gg \Omega_R$. These features are observed experimentally for all semiconductor lasers [40]–[43]. Figure 5.12 shows the modulation response of a 1.55-μm DFB laser at several bias levels [43]. The 3-dB modulation bandwidth, f_{3dB}, is defined as the frequency at which $|H(\omega_m)|$ is reduced by 3 dB (by a factor of 2) compared with its direct-current (dc) value. Equation (5.3.22) provides the following analytic expression for f_{3dB}:

$$f_{\text{3dB}} = \frac{1}{2\pi}\left[\Omega_R^2 - \Gamma_R^2 + 2(\Omega_R^4 + \Omega_R^2\Gamma_R^2 + \Gamma_R^4)^{1/2}\right]^{1/2}. \tag{5.3.23}$$

For most lasers, $\Gamma_R \ll \Omega_R$, and f_{3dB} can be approximated by

$$f_{\text{3dB}} \approx \frac{\sqrt{3}\,\Omega_R}{2\pi} \approx \left(\frac{3G_N P_b}{4\pi^2 \tau_p}\right)^{1/2} = \left[\frac{3G_N}{4\pi^2 q}(I_b - I_{\text{th}})\right]^{1/2}, \tag{5.3.24}$$

where Ω_R was approximated by $(GG_N P_b)^{1/2}$ in Eq. (5.3.21) and G was replaced by $1/\tau_p$ since gain equals loss in the above-threshold regime. The last expression was obtained by using Eq. (5.3.7) at the bias level.

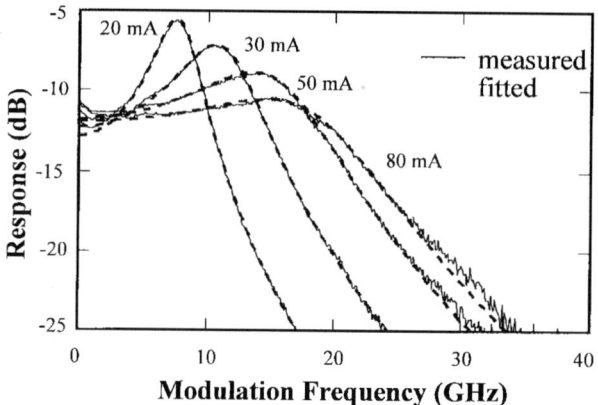

Figure 5.12: Measured (solid curves) and fitted (dashed curves) modulation response of a 1.55-μm DFB laser as a function of modulation frequency at several bias levels. (After Ref. [43]; ©1997 IEEE.)

Equation (5.3.24) provides a remarkably simple expression for the modulation bandwidth. It shows that $f_{3\text{dB}}$ increases with an increase in the bias level as $\sqrt{P_b}$ or as $(I_b - I_{\text{th}})^{1/2}$. This square-root dependence has been verified for many DFB lasers exhibiting a modulation bandwidth of up to 30 GHz [40]–[43]. Figure 5.12 shows how $f_{3\text{dB}}$ can be increased to 24 GHz for a DFB laser by biasing it at 80 mA [43]. A modulation bandwidth of 25 GHz was realized in 1994 for a packaged 1.55-μm InGaAsP laser specifically designed for high-speed response [41].

5.3.3 Large-Signal Modulation

The small-signal analysis, although useful for a qualitative understanding of the modulation response, is not generally applicable to optical communication systems where the laser is typically biased close to threshold and modulated considerably above threshold to obtain optical pulses representing digital bits. In this case of large-signal modulation, the rate equations should be solved numerically. Figure 5.13 shows, as an example, the shape of the emitted optical pulse for a laser biased at $I_b = 1.1 I_{\text{th}}$ and modulated at 2 Gb/s using rectangular current pulses of duration 500 ps and amplitude $I_m = I_{\text{th}}$. The optical pulse does not have sharp leading and trailing edges because of a limited modulation bandwidth and exhibits a rise time of ∼100 ps and a fall time of ∼300 ps. The initial overshoot near the leading edge is a manifestation of relaxation oscillations. Even though the optical pulse is not an exact replica of the applied electrical pulse, deviations are small enough that semiconductor lasers can be used in practice.

As mentioned before, amplitude modulation in semiconductor lasers is accompanied by phase modulation governed by Eq. (5.3.16). A time-varying phase is equivalent to transient changes in the mode frequency from its steady-state value v_0. Such a pulse is called chirped. The frequency chirp $\delta v(t)$ is obtained by using Eq. (5.3.16) and is

5.3 Laser Characteristics

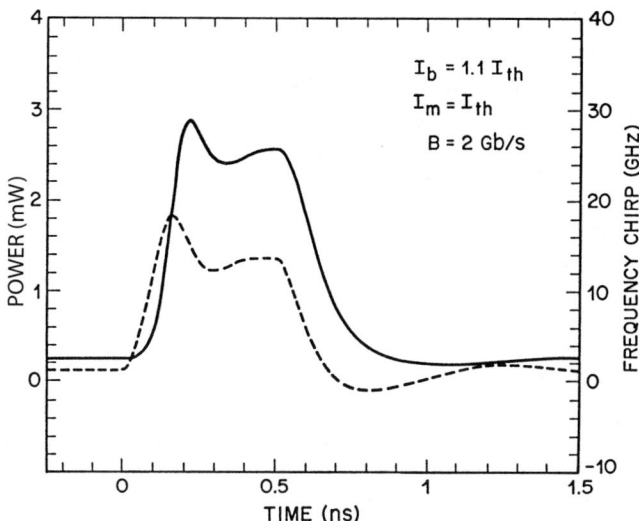

Figure 5.13: Simulated modulation response of a semiconductor laser to 500-ps rectangular current pulses. Solid curve shows the pulse shape and the dashed curve shows the frequency chirp imposed on the pulse ($\beta_c = 5$).

given by

$$\delta v(t) = \frac{1}{2\pi} \frac{d\phi}{dt} = \frac{\beta_c}{4\pi} \left[G_N(N - N_0) - \frac{1}{\tau_p} \right]. \quad (5.3.25)$$

The dashed curve in Figure 5.12 shows the frequency chirp across the optical pulse. The mode frequency shifts toward the blue side near the leading edge and toward the red side near the trailing edge of the optical pulse [44]. Such a frequency shift implies that the pulse spectrum is considerably broader than that expected in the absence of frequency chirp.

Since frequency chirp is often the limiting factor for lightwave systems operating near 1.55 μm, several methods have been used to reduce its magnitude [46]–[50]. These include pulse-shape tailoring, injection locking, and coupled-cavity schemes. A direct way to reduce the frequency chirp is to design semiconductor lasers with small values of the linewidth enhancement factor β_c. The use of quantum-well design reduces β_c by about a factor of about 2. A further reduction occurs for strained quantum wells [49]. Indeed, $\beta_c \approx 1$ has been measured in modulation-doped strained MQW lasers [50]. Such lasers exhibit low chirp under direct modulation. The frequency chirp resulting from current modulation can be avoided altogether if the laser is continuously operated, and an external modulator is used to modulate the laser output [51]. In practice, lightwave systems operating at 10 Gb/s or more use either a monolithically integrated electroabsorption modulator or an external LiNbO$_3$ modulator (see Chapter 6).

Lightwave systems designed using optical time-division multiplexing or solitons require mode-locked lasers capable of generating short optical pulses (with a width of

<10 ps) at a high repetition rate equal to the bit rate. External-cavity semiconductor lasers can be used for this purpose, and are especially practical if a fiber grating is used for an external mirror. The technique of gain switching has also been used to generate short pulses from a semiconductor laser. Mode-locked fiber lasers covered in Section 3.4 can also be used for the same purpose [52].

5.3.4 Relative Intensity Noise

The output of a semiconductor laser exhibits fluctuations in its intensity, phase, and frequency even when the laser is biased at a constant current with negligible current fluctuations. The two fundamental noise mechanisms are spontaneous emission and electron–hole recombination (shot noise). Noise in semiconductor lasers is dominated by spontaneous emission. Each spontaneously emitted photon adds to the coherent field (established by stimulated emission) a small field component whose phase is random, and thus perturbs both the amplitude and phase in a random manner. Moreover, such spontaneous-emission events occur randomly at a high rate ($\sim 10^{12}$ s^{-1}) because of a relatively large value of R_{sp} in semiconductor lasers. The net result is that the intensity and the phase of the emitted light exhibit fluctuations over a time scale as short as 100 ps. Intensity fluctuations lead to a limited *signal-to-noise ratio* (SNR), whereas phase fluctuations lead to a finite spectral linewidth when semiconductor lasers are operated at a constant current. Since such fluctuations can affect the performance of lightwave systems, it is important to estimate their magnitude [53].

The rate equations can be used to study laser noise by adding a noise term, known as the *Langevin force*, to each of them [54]. Equations (5.3.1), (5.3.2), and (5.3.16) then become

$$\frac{dP}{dt} = \left(G - \frac{1}{\tau_p}\right)P + R_{sp} + F_P(t), \qquad (5.3.26)$$

$$\frac{dN}{dt} = \frac{I}{q} - \frac{N}{\tau_c} - GP + F_N(t), \qquad (5.3.27)$$

$$\frac{d\phi}{dt} = \frac{1}{2}\beta_c\left[G_N(N-N_0) - \frac{1}{\tau_p}\right] + F_\phi(t), \qquad (5.3.28)$$

where $F_P(t)$, $F_N(t)$, and $F_\phi(t)$ are the Langevin forces. They are assumed to be Gaussian random processes with zero mean and to have a correlation function of the form (the Markoffian approximation)

$$\langle F_i(t)F_j(t')\rangle = 2D_{ij}\delta(t-t'), \qquad (5.3.29)$$

where $i, j = P$, N, or ϕ, angle brackets denote the ensemble average, and D_{ij} is called the *diffusion coefficient*. The dominant contribution to laser noise comes from only two diffusion coefficients $D_{PP} = R_{sp}P$ and $D_{\phi\phi} = R_{sp}/4P$; others can be assumed to be nearly zero [55].

The intensity-autocorrelation function is defined as

$$C_{pp}(\tau) = \langle \delta P(t)\delta P(t+\tau)\rangle/\bar{P}^2, \qquad (5.3.30)$$

5.3 Laser Characteristics

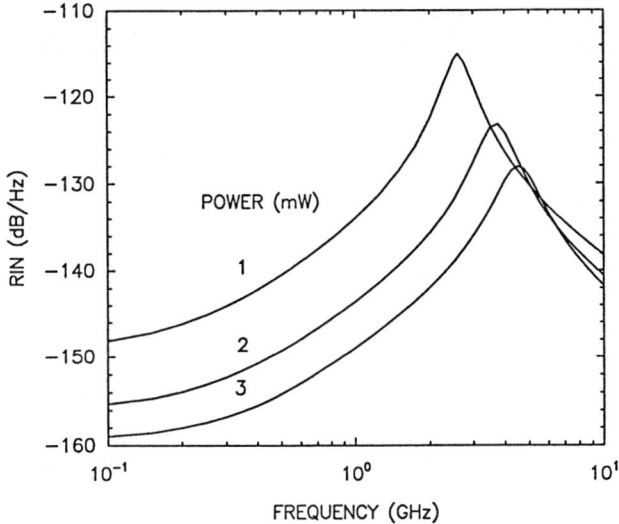

Figure 5.14: RIN spectra at several power levels for a typical 1.55-μm semiconductor laser.

where $\bar{P} \equiv \langle P \rangle$ is the average value and $\delta P = P - \bar{P}$ represents a small fluctuation. The Fourier transform of $C_{pp}(\tau)$ is known as the *relative-intensity-noise* (RIN) spectrum and is given by

$$\text{RIN}(\omega) = \int_{-\infty}^{\infty} C_{pp}(\tau) \exp(-i\omega t)\, dt. \tag{5.3.31}$$

The RIN can be calculated by linearizing Eqs. (5.3.26) and (5.3.27) in δP and δN, solving the linearized equations in the frequency domain, and performing the average with the help of Eq. (5.3.29). It is given approximately by [2]

$$\text{RIN}(\omega) = \frac{2R_{\text{sp}}\{(\Gamma_N^2 + \omega^2) + G_N \bar{P}[G_N \bar{P}(1 + N/\tau_c R_{\text{sp}} \bar{P}) - 2\Gamma_N]\}}{\bar{P}[(\Omega_R - \omega)^2 + \Gamma_R^2][(\Omega_R + \omega)^2 + \Gamma_R^2]}, \tag{5.3.32}$$

where Ω_R and Γ_R are the frequency and the damping rate of relaxation oscillations. They are given by Eq. (5.3.21), with P_b replaced by $\bar{P}$.

Figure 5.14 shows the calculated RIN spectra at several power levels for a typical 1.55-μm InGaAsP laser. The RIN is considerably enhanced near the relaxation-oscillation frequency Ω_R but decreases rapidly for $\omega \gg \Omega_R$, since the laser is not able to respond to fluctuations at such high frequencies. In essence, the semiconductor laser acts as a bandpass filter of bandwidth Ω_R to spontaneous-emission fluctuations. At a given frequency, RIN decreases with an increase in the laser power as P^{-3} at low powers, but this behavior changes to P^{-1} dependence at high powers.

The autocorrelation function $C_{pp}(\tau)$ is calculated using Eqs. (5.3.31) and (5.3.32). The calculation shows that $C_{pp}(\tau)$ follows relaxation oscillations and approaches zero for $\tau > \Gamma_R^{-1}$ [56]. This behavior indicates that intensity fluctuations do not remain correlated for times longer than the damping time of relaxation oscillations. The quantity of practical interest is the SNR defined as $\bar{P}/\sigma_p$, where σ_p is the root-mean-square

(RMS) noise. From Eq. (5.3.30), SNR $= [C_{pp}(0)]^{-1/2}$. At power levels above a few milliwatts, the SNR exceeds 20 dB and improves linearly with the power as

$$\text{SNR} = \left(\frac{\varepsilon_{\text{NL}}}{R_{\text{sp}}\tau_p}\right)^{1/2} \bar{P}. \tag{5.3.33}$$

The presence of ε_{NL} indicates that the nonlinear form of the gain in Eq. (5.3.15) plays a crucial role. This form needs to be modified at high powers. Indeed, a more accurate treatment shows that the SNR eventually saturates at a value of about 30 dB and becomes power-independent [56].

So far, the laser has been assumed to oscillate in a single longitudinal mode. In practice, even DFB lasers are accompanied by one or more side modes. Even though side modes remain suppressed by more than 20 dB on the basis of the average power, their presence can affect the RIN significantly. In particular, the main and side modes can fluctuate in such a way that individual modes exhibit large intensity fluctuations, but the total intensity remains relatively constant. This phenomenon is called *mode-partition noise* (MPN) and occurs due to an anticorrelation between the main and side modes [2]. It manifests through the enhancement of RIN for the main mode by 20 dB or more in the low-frequency range 0–1 GHz; the exact value of the enhancement factor depends on the MSR [57]. In the case of a VCSEL, the MPN involves two transverse modes [58]. In the absence of fiber dispersion, MPN would be harmless for optical communication systems, as all modes would remain synchronized during transmission and detection. However, in practice all modes do not arrive simultaneously at the receiver because they travel at slightly different speeds. Such a desynchronization not only degrades the SNR of the received signal but also leads to intersymbol interference.

5.3.5 Spectral Linewidth

The spectrum of emitted light is related to the field-autocorrelation function $\Gamma_{EE}(\tau)$ through a Fourier-transform relation similar to Eq. (5.3.31), that is,

$$S(\omega) = \int_{-\infty}^{\infty} \Gamma_{EE}(t) \exp[-i(\omega - \omega_0)\tau] d\tau, \tag{5.3.34}$$

where $\Gamma_{EE}(t) = \langle E^*(t)E(t+\tau)\rangle$ and $E(t) = \sqrt{P}\exp(i\phi)$ is the optical field. If intensity fluctuations are neglected, $\Gamma_{EE}(t)$ is given by

$$\Gamma_{EE}(t) = \langle \exp[i\Delta\phi(t)]\rangle = \exp[-\langle\Delta\phi^2(\tau)\rangle/2], \tag{5.3.35}$$

where the phase fluctuation $\Delta\phi(\tau) = \phi(t+\tau) - \phi(t)$ is taken to be a Gaussian random process. The phase variance $\langle\Delta\phi^2(\tau)\rangle$ can be calculated by linearizing Eqs. (5.3.26) through (5.3.28) and solving the resulting set of linear equations. The result is [55]

$$\langle\Delta\phi^2(\tau)\rangle = \frac{R_{\text{sp}}}{2\bar{P}}\left[(1+\beta_c^2 b)\tau + \frac{\beta_c^2 b}{2\Gamma_R \cos\delta}[\cos(3\delta) - e^{-\Gamma_R\tau}\cos(\Omega_R\tau - 3\delta)]\right], \tag{5.3.36}$$

5.3 Laser Characteristics

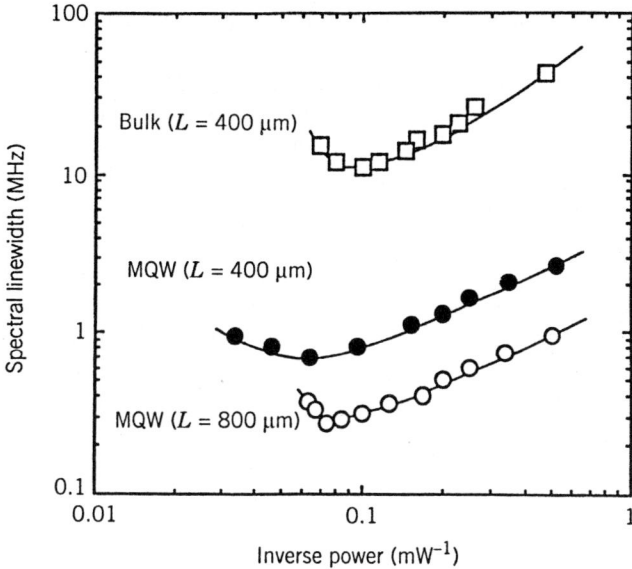

Figure 5.15: Measured linewidth as a function of emitted power for several 1.55-μm DFB lasers. Active layer is 100 nm thick for the bulk laser and 10 nm thick for MQW lasers. (After Ref. [59]; ©1991 IEEE.)

where

$$b = \Omega_R/(\Omega_R^2 + \Gamma_R^2)^{1/2} \quad \text{and} \quad \delta = \tan^{-1}(\Gamma_R/\Omega_R). \quad (5.3.37)$$

The spectrum is obtained by using Eqs. (5.3.34) through (5.3.36). It is found to consist of a dominant central peak located at ω_0 and multiple satellite peaks located at $\omega = \omega_0 \pm m\Omega_R$, where m is an integer. The amplitude of satellite peaks is typically less than 1% of that of the central peak. The physical origin of the satellite peaks is related to relaxation oscillations, which are responsible for the term proportional to b in Eq. (5.3.36). If this term is neglected, the autocorrelation function $\Gamma_{EE}(\tau)$ decays exponentially with τ. The integral in Eq. (5.3.34) can then be performed analytically, and the spectrum is found to be Lorentzian. The spectral linewidth Δv is defined as the full-width at half-maximum (FWHM) of this Lorentzian line and is given by [55]

$$\Delta v = R_{\text{sp}}(1 + \beta_c^2)/(4\pi \bar{P}), \quad (5.3.38)$$

where $b = 1$ was assumed as $\Gamma_R \ll \Omega_R$ under typical operating conditions. The linewidth is enhanced by a factor of $1 + \beta_c^2$ as a result of the amplitude-phase coupling governed by β_c in Eq. (5.3.28); β_c is called the linewidth enhancement factor for this reason.

Equation (5.3.38) shows that Δv should decrease as $\bar{P}^{-1}$ with an increase in the laser power. Such an inverse dependence is observed experimentally at low power levels (<10 mW) for most semiconductor lasers. However, often the linewidth is found to saturate to a value in the range of 1–10 MHz at a power level above 10 mW. Figure 5.15 shows such linewidth-saturation behavior for several 1.55-μm DFB lasers [59]. It also shows that the linewidth can be reduced considerably by using a MQW design for the DFB laser. The reduction is due to a smaller value of the parameter β_c realized by such a design.

The linewidth can also be reduced by increasing the cavity length L, since R_{sp} decreases and P increases at a given output power as L is increased. Although not obvious from Eq. (5.3.38), Δv can be shown to vary as L^{-2} when the length dependence of R_{sp} and P is incorporated. As seen in Figure 5.15, Δv is reduced by about a factor of 4 when the cavity length is doubled. The 800-μm-long MQW-DFB laser is found to exhibit a linewidth as small as 270 kHz at a power output of 13.5 mW [59]. It is further reduced in strained MQW lasers because of relatively low values of β_c, and a value of about 100 kHz has been measured in lasers with $\beta_c \approx 1$ [50]. It should be stressed, however, that the linewidth of most DFB lasers is typically 5–10 MHz when operating at a power level of 10 mW. Figure 5.15 shows that as the laser power increases, the linewidth not only saturates but also begins to rebroaden. Several mechanisms such as current fluctuations, $1/f$ noise, nonlinear gain and index changes, and interaction with weak side modes have been invoked to explain this saturation. The linewidth of most DFB lasers is small enough that it is not a limiting factor for lightwave systems.

5.4 Light-Emitting Diodes

For some applications, a coherent source is not required, and one can employ a light-emitting diode (LED), a less expensive and longer-lasting optical source with a relatively wide optical spectrum [60]. The basic structure of an LED is similar to that of semiconductor lasers in the sense that both employ an active layer sandwiched between two cladding layers and pumped using a forward-biased p–n junction. The main difference is that stimulations emission does not occur because the cladding layers are doped such that population inversion is not realized. Radiative recombination of electron–hole pairs in the active layer generates light through spontaneous emission, some of which escapes from the device and can be coupled into an optical fiber. The emitted light is incoherent with a relatively wide spectral width (30–60 nm) and a relatively large angular spread.

5.4.1 CW Characteristics

It is easy to estimate the internal power generated by spontaneous emission. At a given current I the carrier-injection rate is I/q. In the steady state, the rate of electron–hole pairs recombining through radiative and nonradiative processes is equal to the carrier-injection rate I/q. Since the internal quantum efficiency η_{int} determines the fraction of electron–hole pairs that recombine through spontaneous emission, the rate of photon generation is simply $\eta_{int} I/q$. The internal optical power is thus given by

$$P_{int} = \eta_{int}(\hbar\omega/q)I, \quad (5.4.1)$$

where $\hbar\omega$ is the photon energy, assumed to be nearly the same for all photons. If η_{ext} is the fraction of photons escaping from the device, the emitted power is given by

$$P_e = \eta_{ext} P_{int} = \eta_{ext}\eta_{int}(\hbar\omega/q)I. \quad (5.4.2)$$

5.4 Light-Emitting Diodes

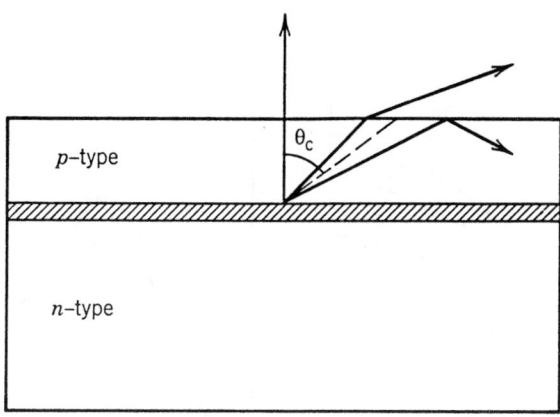

Figure 5.16: Total internal reflection at the output facet of an LED. Only light emitted within a cone of angle θ_c is transmitted, where θ_c is the critical angle for the semiconductor–air interface.

The quantity η_{ext} is called the *external quantum efficiency*. It can be calculated by taking into account internal absorption and the total internal reflection at the semiconductor–air interface. As seen in Figure 5.16, only light emitted within a cone of angle θ_c, where $\theta_c = \sin^{-1}(1/n)$ is the critical angle and n is the refractive index of the semiconductor material, escapes from the LED surface. Internal absorption can be avoided by using heterostructure LEDs in which the cladding layers surrounding the active layer are transparent to the radiation generated. The external quantum efficiency can then be written as

$$\eta_{\text{ext}} = \frac{1}{4\pi} \int_0^{\theta_c} T_f(\theta)(2\pi \sin\theta)\,d\theta, \tag{5.4.3}$$

where we have assumed that the radiation is emitted uniformly in all directions over a solid angle of 4π. The Fresnel transmissivity T_f depends on the incidence angle θ. In the case of normal incidence ($\theta = 0$), $T_f(0) = 4n/(n+1)^2$. If we replace for simplicity $T_f(\theta)$ by $T_f(0)$ in Eq. (5.4.3), η_{ext} is given approximately by

$$\eta_{\text{ext}} = n^{-1}(n+1)^{-2}. \tag{5.4.4}$$

By using Eq. (5.4.4) in Eq. (5.4.2), we obtain the power emitted from one facet (see Figure 5.16). If we use $n = 3.5$ as a typical value, $\eta_{\text{ext}} = 1.4\%$, indicating that only a small fraction of the internal power becomes the useful output power. A further loss in useful power occurs when the emitted light is coupled into an optical fiber. Because of the incoherent nature of the emitted light, an LED acts as a *Lambertian source* with an angular distribution $S(\theta) = S_0 \cos\theta$, where S_0 is the intensity in the direction $\theta = 0$. The coupling efficiency for such a source [60] scales with the numerical aperture (NA) as $(\text{NA})^2$. Since NA for optical fibers falls typically in the range 0.1–0.3, only a few percent of the emitted power is coupled into the fiber. Normally, the launched power for LEDs is 100 μW or less, even though the internal power can easily exceed 10 mW.

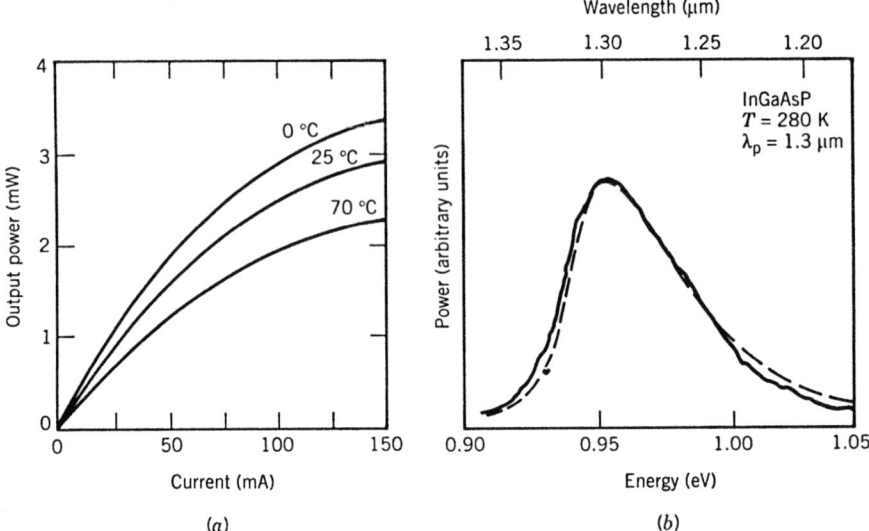

Figure 5.17: (a) Power–current curves at several temperatures; (b) spectrum of the emitted light for a typical 1.3-μm LED. The dashed curve shows the theoretically calculated spectrum. (After Ref. [61]; ©1981 American Institute of Physics.)

A measure of the LED performance is the total quantum efficiency η_{tot}, defined as the ratio of the emitted optical power P_e to the applied electrical power, $P_{elec} = V_0 I$, where V_0 is the voltage drop across the device. By using Eq. (5.4.2), η_{tot} is given by

$$\eta_{tot} = \eta_{ext}\eta_{int}(\hbar\omega/qV_0). \tag{5.4.5}$$

Typically, $\hbar\omega \approx qV_0$, and $\eta_{tot} \approx \eta_{ext}\eta_{int}$. The total quantum efficiency η_{tot}, also called the *power-conversion efficiency* or the *wall-plug efficiency*, is a measure of the overall performance of the device.

Another quantity sometimes used to characterize the LED performance is the *responsivity* defined as the ratio $R_{LED} = P_e/I$. From Eq. (5.4.2),

$$R_{LED} = \eta_{ext}\eta_{int}(\hbar\omega/q). \tag{5.4.6}$$

A comparison of Eqs. (5.4.5) and (5.4.6) shows that $R_{LED} = \eta_{tot}V_0$. Typical values of R_{LED} are ∼0.01 W/A. The responsivity remains constant as long as the linear relation between P_e and I holds. In practice, this linear relationship holds only over a limited current range [61]. Figure 5.17(a) shows the power–current (P–I) curves at several temperatures for a typical 1.3-μm LED. The responsivity of the device decreases at high currents above 80 mA because of the bending of the P–I curve. One reason for this decrease is related to the increase in the active-region temperature. The internal quantum efficiency η_{int} is generally temperature-dependent because of an increase in the nonradiative recombination rates at high temperatures.

5.4 Light-Emitting Diodes

The LED spectrum is related to the rate of spontaneous emission given approximately by

$$R_{\text{spon}}(\omega) = A_0(\hbar\omega - E_g)^{1/2}\exp[-(\hbar\omega - E_g)/k_B T], \quad (5.4.7)$$

where A_0 is a constant, k_B is the Boltzmann constant, and E_g is the bandgap. It is easy to deduce that $R_{\text{spon}}(\omega)$ peaks when $\hbar\omega = E_g + k_B T/2$ and has a full-width at half-maximum (FWHM) $\Delta\nu \approx 1.8 k_B T/h$. At room temperature ($T = 300$ K) the FWHM is about 11 THz. In practice, the spectral width is expressed in nanometers by using $\Delta\nu = (c/\lambda^2)\Delta\lambda$ and increases as λ^2 with an increase in the emission wavelength λ. As a result, $\Delta\lambda$ is larger for InGaAsP LEDs emitting at 1.3 µm by about a factor of 1.7 compared with GaAs LEDs. Figure 5.17(b) shows the output spectrum of a typical 1.3-µm LED and compares it with the theoretical curve obtained by using Eq. (5.4.7). Because of a large spectral width ($\Delta\lambda = 50$–60 nm), LEDs are suitable primarily for local-area networks and are often used in combination with plastic fibers to reduce the overall system cost.

5.4.2 Modulation Response

The modulation response of LEDs depends on carrier dynamics and is limited by the carrier lifetime τ_c. It can be determined by using the carrier rate equation (5.3.2) after dropping the last term resulting from stimulated emission. The resulting equation is

$$\frac{dN}{dt} = \frac{I}{q} - \frac{N}{\tau_c}. \quad (5.4.8)$$

It can be easily solved in the Fourier domain owing to its linear nature. Consider sinusoidal modulation of the injected current in the form

$$I(t) = I_b + I_m \exp(i\omega_m t), \quad (5.4.9)$$

where I_b is the bias current, I_m is the modulation current, and ω_m is the modulation frequency. Since Eq. (5.4.8) is linear, its general solution can be written as

$$N(t) = N_b + N_m \exp(i\omega_m t), \quad (5.4.10)$$

where $N_b = \tau_c I_b/q$ and N_m is given by

$$N_m(\omega_m) = \frac{\tau_c I_m/q}{1 + i\omega_m \tau_c}. \quad (5.4.11)$$

The modulated power P_m is related to $|N_m|$ linearly. One can define the LED transfer function $H(\omega_m)$ as

$$H(\omega_m) = \frac{N_m(\omega_m)}{N_m(0)} = \frac{1}{1 + i\omega_m \tau_c}. \quad (5.4.12)$$

The 3-dB modulation bandwidth $f_{3\text{dB}}$ is defined as the frequency at which $|H(\omega_m)|$ is reduced by 3 dB, or by a factor of 2. The result is

$$f_{3\text{dB}} = \sqrt{3}\,(2\pi\tau_c)^{-1}. \quad (5.4.13)$$

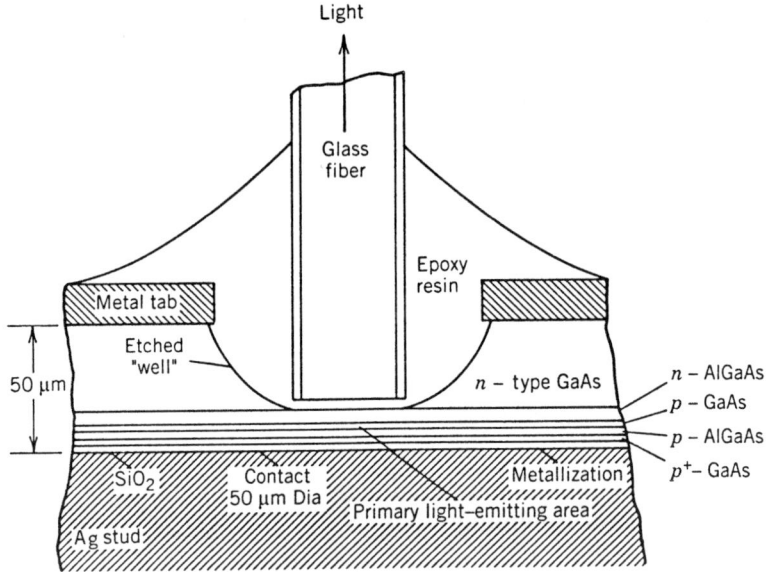

Figure 5.18: Schematic of a surface-emitting LED with a double-heterostructure geometry.

Typically, τ_c is in the range of 2–5 ns for InGaAsP LEDs. The corresponding LED modulation bandwidth is in the range of 50–140 MHz. Note that Eq. (5.4.13) provides the optical bandwidth because f_{3dB} is defined as the frequency at which optical power is reduced by 3 dB. The corresponding electrical bandwidth is the frequency at which $|H(\omega_m)|^2$ is reduced by 3 dB and is given by $(2\pi\tau_c)^{-1}$.

5.4.3 LED Structures

The LED structures can be classified as surface-emitting or edge-emitting, depending on whether the LED emits light from a surface that is parallel to the junction plane or from the edge of the junction region. Both types can be made using either a p–n homojunction or a heterostructure design in which the active region is surrounded by p- and n-type cladding layers. The heterostructure design leads to superior performance, as it provides a control over the emissive area and eliminates internal absorption because of the transparent cladding layers.

Figure 5.18 shows schematically a surface-emitting LED design referred to as the *Burrus-type* LED [62]. The emissive area of the device is limited to a small region whose lateral dimension is comparable to the fiber-core diameter. The use of a gold stud avoids power loss from the back surface. The coupling efficiency is improved by etching a well and bringing the fiber close to the emissive area. The power coupled into the fiber depends on many parameters, such as the numerical aperture of the fiber and the distance between fiber and LED. The addition of epoxy in the etched well tends to increase the external quantum efficiency as it reduces the refractive-index mismatch. Several variations of the basic design exist in the literature. In one variation, a truncated

5.5 Semiconductor Optical Amplifiers

spherical microlens fabricated inside the etched well is used to couple light into the fiber. In another variation, the fiber end is itself made in the form of a spherical lens. With a proper design, surface-emitting LEDs can couple up to 1% of the internally generated power into an optical fiber.

The edge-emitting LEDs employ a design identical to that used for semiconductor lasers. In fact, a semiconductor laser is converted into an LED by depositing an antireflection coating on its output facet to suppress lasing action. Beam divergence of edge-emitting LEDs differs from surface-emitting LEDs because of waveguiding in the plane perpendicular to the junction. Surface-emitting LEDs operate as a Lambertian source with angular distribution $S_e(\theta) = S_0 \cos\theta$ in both directions. The resulting beam divergence has a FWHM of 120° in each direction. In contrast, edge-emitting LEDs have a divergence of only about 30° in the direction perpendicular to the junction plane. Considerable light can be coupled into a fiber of even low numerical aperture (<0.3) because of reduced divergence and high radiance at the emitting facet. The modulation bandwidth of edge-emitting LEDs is generally larger (∼200 MHz) than that of surface-emitting LEDs because of a reduced carrier lifetime at the same applied current. The choice between the two designs is dictated, in practice, by a compromise between cost and performance.

In spite of a relatively low output power and a low bandwidth of LEDs compared with those of lasers, LEDs are useful for low-cost applications requiring data transmission at a bit rate of 100 Mb/s or less over a few kilometers. For this reason, several new LED structures were developed during the 1990s [63]–[68]. In one design, known as resonant-cavity LED [63], two metal mirrors are fabricated around the epitaxially grown layers, and the device is bonded to a silicon substrate. In a variant of this idea, the bottom mirror is fabricated epitaxially by using a stack of alternating layers of two different semiconductors, while the top mirror consists of a deformable membrane suspended by an air gap [64]. The operating wavelength of such an LED can be tuned over 40 nm by changing the air-gap thickness. In another scheme, several quantum wells with different compositions and bandgaps are grown to form a MQW structure [65]. Since each quantum well emits light at a different wavelength, such LEDs can have an extremely broad spectrum (extending over a 500-nm wavelength range) and are useful for local-area WDM networks.

5.5 Semiconductor Optical Amplifiers

All lasers act as amplifiers close to but before reaching threshold, and semiconductor lasers are no exception. Indeed, research on semiconductor optical amplifiers (SOAs) started soon after the invention of semiconductor lasers in 1962. However, it was only during the 1980s that SOAs were developed for practical applications, motivated largely by their potential applications in lightwave systems [69]–[73]. Although, with the advent of the doped-fiber and Raman amplifiers, SOAs are rarely used for fiber-loss compensation in lightwave systems, they are often used for signal processing. This section focuses on the amplification characteristics of SOAs and their applications.

5.5.1 Amplifier Design

Several kinds of fiber amplifiers discussed in Chapter 3 are examples of *traveling-wave* amplifiers as the amplified signal travels in the forward direction only. Semiconductor lasers can be used as amplifiers when biased below threshold, but multiple reflections at the facets must be included. Such amplifiers are called *FP amplifiers*. The amplification factor is obtained by using the standard theory of FP interferometers (see Section 2.3.3) and is given by [2]

$$G_{FP}(\nu) = \frac{(1-R_1)(1-R_2)G(\nu)}{(1-G\sqrt{R_1R_2})^2 + 4G\sqrt{R_1R_2}\sin^2[\pi(\nu-\nu_m)/\Delta\nu_L]}, \qquad (5.5.1)$$

where R_1 and R_2 are the facet reflectivities, ν_m represents the cavity-resonance frequencies given in Eq. (5.1.5), and $\Delta\nu_L$ is the longitudinal-mode spacing, or the free spectral range, of the FP cavity. The single-pass amplification factor $G = \exp(gL)$ corresponds to that of a TW amplifier when gain saturation is negligible; G_{FP} reduces to G when $R_1 = R_2 = 0$.

As evident from Eq. (5.5.1), $G_{FP}(\nu)$ peaks whenever ν coincides with one of the cavity-resonance frequencies and drops sharply in between them. The amplifier bandwidth is thus determined by the sharpness of the FP resonance. One can calculate the amplifier bandwidth from the detuning $\nu - \nu_m$ for which G_{FP} drops by 3 dB from its peak value. The result is given by

$$\Delta\nu_A = \frac{2\Delta\nu_L}{\pi}\sin^{-1}\left(\frac{1-G\sqrt{R_1R_2}}{(4G\sqrt{R_1R_2})^{1/2}}\right). \qquad (5.5.2)$$

To achieve a large amplification factor, $G\sqrt{R_1R_2}$ should be quite close to 1. As seen from Eq. (5.5.2), the amplifier bandwidth is then a small fraction of the free spectral range of the FP cavity (typically, $\Delta\nu_L \sim 100$ GHz and $\Delta\nu_A < 10$ GHz). Such a small bandwidth makes FP amplifiers unsuitable for most lightwave system applications.

Traveling-wave operation of SOAs can be realized if the reflection feedback from the end facets is suppressed. A simple way to reduce the reflectivity is to coat both facets with an antireflection coating. However, the residual reflectivity of the coated facet must be extremely small ($<0.1\%$) for an SOA to act as a traveling-wave amplifier. Furthermore, the minimum reflectivity depends on the amplifier gain itself. One can estimate the tolerable value of the facet reflectivity by considering the maximum and minimum values of G_{FP} from Eq. (5.5.1) near a cavity resonance. It is easy to verify that their ratio is given by

$$\Delta G = \frac{G_{FP}^{\max}}{G_{FP}^{\min}} = \left(\frac{1+G\sqrt{R_1R_2}}{1-G\sqrt{R_1R_2}}\right)^2. \qquad (5.5.3)$$

If ΔG exceeds 3 dB, the amplifier bandwidth is set by the cavity resonances rather than by the gain spectrum. To keep $\Delta G < 2$, the facet reflectivities should satisfy the condition

$$G\sqrt{R_1R_2} < 0.17. \qquad (5.5.4)$$

5.5 Semiconductor Optical Amplifiers

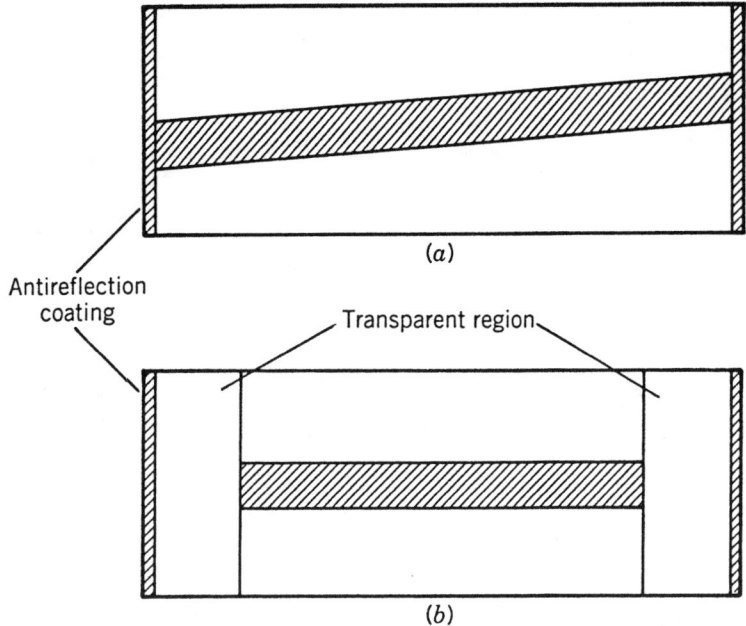

Figure 5.19: (a) Tilted-stripe and (b) buried-facet structures for nearly TW semiconductor optical amplifiers.

It is customary to characterize the SOA as a TW amplifier when Eq. (5.5.4) is satisfied. An SOA designed to provide a 30-dB amplification factor ($G = 1,000$) should have facet reflectivities such that $\sqrt{R_1 R_2} < 1.7 \times 10^{-4}$.

Considerable effort is required to produce antireflection coatings with reflectivities less than 0.1%. Even then, it is difficult to obtain low facet reflectivities in a predictable and regular manner. For this reason, alternative techniques have been developed to reduce the reflection feedback in SOAs. In one method, the active-region stripe is tilted from the facet normal, as shown in Figure 5.19(a). Such a structure is referred to as the *angled-facet* or *tilted-stripe* structure [74]. The reflected beam at the facet is physically separated from the forward beam because of the angled facet. Some feedback can still occur, as the optical mode spreads beyond the active region in all semiconductor laser devices. In practice, the combination of an antireflection coating and the tilted stripe can produce reflectivities below 10^{-3} (as small as 10^{-4} with design optimization). In an alternative scheme shown in Figure 5.19(b), a transparent region is inserted between the active-layer ends and the facets [76]. The optical beam spreads in this window region before arriving at the semiconductor–air interface. The reflected beam spreads even further on the return trip and does not couple much light into the thin active layer. Such a structure is called buried-facet or window-facet structure and has provided reflectivities as small as 10^{-4} when used in combination with antireflection coatings.

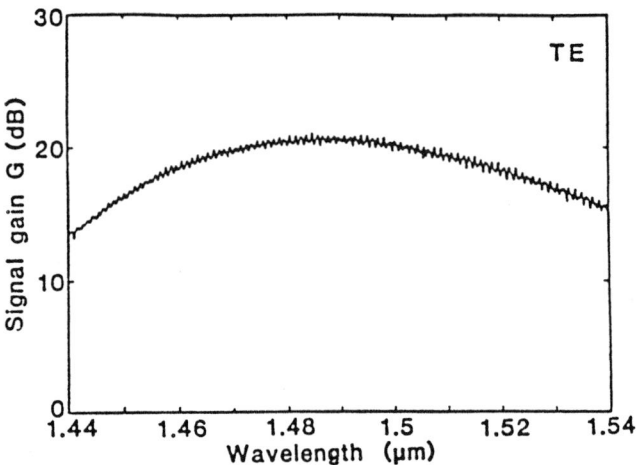

Figure 5.20: Amplifier gain versus signal wavelength for a semiconductor optical amplifier whose facets are coated to reduce reflectivity to about 0.04%. (After Ref. [69]; ©1987 IEEE.)

5.5.2 Amplifier Characteristics

The amplification factor of SOAs is given by Eq. (5.5.1). Its frequency dependence results mainly from the frequency dependence of $G(v)$ when condition (5.5.4) is satisfied. The measured amplifier gain exhibits ripples reflecting the effects of residual facet reflectivities. Figure 5.20 shows the wavelength dependence of the amplifier gain measured for a SOA with the facet reflectivities of about 4×10^{-4}. Gain ripples are negligibly small as the condition (5.5.4) is well satisfied ($G\sqrt{R_1 R_2} \approx 0.04$). The 3-dB amplifier bandwidth is about 70 nm for this amplifier.

To discuss gain saturation, we use the carrier rate equation given in Eq. (5.3.2). Expressing the photon number in terms of the optical power, this equation can be written as

$$\frac{dN}{dt} = \frac{I}{q} - \frac{N}{\tau_c} - \frac{\sigma_g(N-N_0)}{\sigma_m h v} P, \quad (5.5.5)$$

where τ_c is the carrier lifetime and σ_m is the cross-sectional area of the waveguide mode. In the case of a CW beam or pulses much longer than τ_c, the steady-state value of N is obtained by setting $dN/dt = 0$ in Eq. (5.5.5). When the solution is substituted in Eq. (5.3.3), the optical gain is found to saturate as

$$g = \frac{g_0}{1 + P/P_s}, \quad (5.5.6)$$

where the small-signal gain g_0 is given by

$$g_0 = (\Gamma \sigma_g/V)(I\tau_c/q - N_0), \quad (5.5.7)$$

and the saturation power P_s is defined as

$$P_s = h v \sigma_m / (\sigma_g \tau_c). \quad (5.5.8)$$

5.5 Semiconductor Optical Amplifiers

Typical values of P_s are in the range of 5–10 mW.

The noise figure F_n of SOAs is larger than the minimum value of 3 dB for several reasons. The dominant contribution comes from the spontaneous-emission factor n_{sp}. For SOAs, n_{sp} is obtained from Eq. (3.1.18) after replacing N_2 and N_1 by N and N_0, respectively. An additional contribution results from internal losses (such as free-carrier absorption or scattering loss), which reduce the available gain from g to $g - \alpha_{int}$. Using Eq. (3.1.23) and including this additional contribution, we can write the noise figure as [71]

$$F_n = 2\left(\frac{N}{N - N_0}\right)\left(\frac{g}{g - \alpha_{int}}\right). \tag{5.5.9}$$

Typical values of F_n for SOAs fall in the range of 5–7 dB.

5.5.3 Practical Issues

Although SOAs were used before 1990 for compensating fiber losses in some system experiments [75], they suffer from several drawbacks that make their use as in-line amplifiers impractical. A few among them are polarization sensitivity, interchannel crosstalk, a relatively high noise figure, and large coupling losses occurring as the signal is coupled into and out of an SOA. For this reason, they have been rarely used commercially for this purpose. This may however change as SOAs also being pursued for metropolitan-area networks as a low-cost alternative to fiber amplifiers. We focus on the undesirable features of SOAs in this subsection.

An undesirable characteristic of SOAs is their *polarization sensitivity*. The amplifier gain G differs for the TE and TM modes by as much as 5–8 dB simply because both G and σ_g are different for the two orthogonally polarized modes. This feature makes the amplifier gain sensitive to the polarization state of the input beam, a property undesirable for lightwave systems in which the state of polarization changes with propagation along the fiber (unless fibers are used). Several schemes have been devised to reduce the polarization sensitivity [76]–[81]. In one scheme, the amplifier is designed such that the width and the thickness of the active region are comparable. A gain difference of less than 1.3 dB between TE and TM polarizations has been realized by making the active layer 0.26 μm thick and 0.4 μm wide [76]. Another scheme makes use of a large-optical-cavity structure; a gain difference of less than 1 dB has been obtained with such a structure [77].

Several other schemes reduce the polarization sensitivity by using two amplifiers or two passes through the same amplifier. Figure 5.21 shows three such configurations. In Figure 5.21(a), the TE-polarized signal in one amplifier becomes TM-polarized in the second amplifier, and vice versa. If both amplifiers have identical gain characteristics, the twin-amplifier configuration provides signal gain that is independent of the signal polarization. A drawback of the *series configuration* is that residual facet reflectivities lead to mutual coupling between the two amplifiers. In the *parallel configuration* shown in Figure 5.21(b), the incident signal is split into a TE- and a TM-polarized signal, each of which is amplified by separate amplifiers. The amplified TE and TM signals are then combined to produce the amplified signal with the same polarization as that of the input beam [78]. The *double-pass configuration* of Figure 5.21(c) passes the

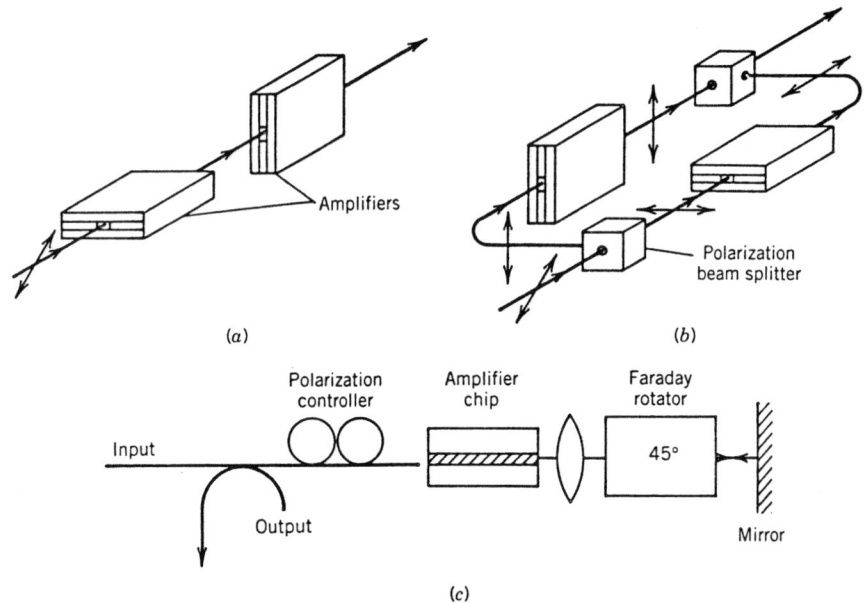

Figure 5.21: Three configurations used to reduce the polarization sensitivity of semiconductor optical amplifiers: (a) twin amplifiers in series; (b) twin amplifiers in parallel; and (c) double pass through a single amplifier.

signal through the same amplifier twice, but the polarization is rotated by 90° between the two passes [79]. Since the amplified signal propagates in the backward direction, a 3-dB fiber coupler is needed to separate it from the incident signal. Despite a 6-dB loss occurring at the fiber coupler (3 dB for the input signal and 3 dB for the amplified signal), this configuration provides high gain from a single amplifier, as the same amplifier supplies gain on the two passes.

Although SOAs can be used to amplify several channels simultaneously, they suffer from a fundamental problem related to their relatively fast response. Ideally, the signal in each channel should be amplified by the same amount. In practice, several nonlinear phenomena in SOAs induce *interchannel crosstalk*, an undesirable feature that should be minimized for practical lightwave systems. Two such nonlinear phenomena are *cross-gain saturation* and *four-wave mixing* (FWM). Both of them originate from the stimulated-emission term in Eq. (5.5.5). In the case of multichannel amplification, the power P in this equation is replaced with

$$P = \frac{1}{2}\left|\sum_{j=1}^{M} A_j \exp(-i\omega_j t) + \text{c.c.}\right|^2, \qquad (5.5.10)$$

where c.c. stands for the complex conjugate, M is the number of channels, A_j is the amplitude, and ω_j is the carrier frequency of the jth channel. Because of the coherent addition of individual channel fields, Eq. (5.5.10) contains time-dependent terms

5.5 Semiconductor Optical Amplifiers

resulting from beating of the signal in different channels, that is,

$$P = \sum_{j=1}^{M} P_j + \sum_{j=1}^{M}\sum_{k \neq j}^{M} 2\sqrt{P_j P_k}\cos(\Omega_{jk}t + \phi_j - \phi_k), \qquad (5.5.11)$$

where $A_j = \sqrt{P_j}\exp(i\phi_j)$ was assumed together with $\Omega_{jk} = \omega_j - \omega_k$. When Eq. (5.5.11) is substituted in Eq. (5.5.5), the carrier population is also found to oscillate at the beat frequency Ω_{jk}. Since the gain and the refractive index both depend on N, they are also modulated at the frequency Ω_{jk}; such a modulation creates gain and index gratings, which induce interchannel crosstalk by scattering a part of the signal from one channel to another. This phenomenon can also be viewed as FWM [82]. FWM can occur even for widely spaced channels through intraband nonlinearities [83] occurring at fast time scales (<1 ps).

The origin of cross-gain saturation is also evident from Eq. (5.5.11). The first term on the right side shows that the power P in Eq. (5.5.6) should be replaced by the total power in all channels. Thus, the gain of a specific channel is saturated not only by its own power but also by the power of neighboring channels, a phenomenon known as cross-gain saturation. It is undesirable in WDM systems since the amplifier gain changes with time, depending on the bit pattern of neighboring channels. As a result, the amplified signal appears to fluctuate more or less randomly. Such fluctuations degrade the effective SNR at the receiver. The interchannel crosstalk occurs for any channel spacing or the modulation format. It can be avoided only by reducing the channel powers to low enough values that the SOA operates in the unsaturated regime [84]–[87].

5.5.4 SOA as a Nonlinear Device

The same nonlinear effects that limit the usefulness of SOAs in lightwave systems as an optical amplifier also render them quite useful for nonlinear signal processing (while amplifying the signal simultaneously). SOAs can be used for all-optical switching, FWM, wavelength conversion, and logic operations, and thus constitute an important active component of lightwave technology [88]–[91]. SOAs are not only extremely compact (active volume <1 mm^3), they can also be integrated monolithically with other devices on the same chip. Since they additionally provide amplification, SOAs allow features such as fan-out and cascadability, both of which are general requirements for large-scale photonic circuits.

The most important feature of SOAs is that they exhibit a strong carrier-induced third-order nonlinearity with effective values of $n_2 \sim 10^{-9}$ cm^2/W [92] that are seven orders of magnitude larger than that of silica fibers. Although this nonlinearity does not respond on a femtosecond time scale, it is fast enough that it can be used to make devices operating at bit rates as fast as 40 Gb/s. The origin of this nonlinearity lies in gain saturation and the fact that any change in the carrier density affects not only the optical gain but also the refractive index within the active region of an SOA.

A simple way to understand the nonlinear response of an SOA is to consider what happens when a short optical pulse is launched into it. The amplitude $A(z,t)$ of the pulse envelope inside the SOA evolves as [92]

$$\frac{\partial A}{\partial z} + \frac{1}{v_g}\frac{\partial A}{\partial t} = \frac{1}{2}(1 - i\beta_c)gA, \qquad (5.5.12)$$

where $v - G$ is the group velocity, g is the gain, and carrier-induced index changes are included through the linewidth enhancement factor β_c. The time dependence of g is governed by Eq. (5.5.5), which can be written in simple form as

$$\frac{\partial g}{\partial t} = \frac{g_0 - g}{\tau_c} - \frac{g|A|^2}{E_{\text{sat}}}, \qquad (5.5.13)$$

where the saturation energy E_{sat} is defined as

$$E_{\text{sat}} = h\nu(\sigma_m/\sigma_g), \qquad (5.5.14)$$

and g_0 is given by Eq. (5.5.7). Typically, $E_{\text{sat}} \sim 1$ pJ.

Equations (5.5.12) and (5.5.13) govern amplification of optical pulses in SOAs. They can be solved analytically for pulses whose duration is short compared with the carrier lifetime ($\tau_p \ll \tau_c$). The first term on the right side of Eq. (5.5.13) can then be neglected during pulse amplification. By introducing the reduced time $\tau = t - z/v_g$ together with $A = \sqrt{P}\exp(i\phi)$, Eqs. (5.5.12) and (5.5.13) can be written as [92]

$$\frac{\partial P}{\partial z} = g(z,\tau)P(z,\tau), \qquad (5.5.15)$$

$$\frac{\partial \phi}{\partial z} = -\tfrac{1}{2}\beta_c g(z,\tau), \qquad (5.5.16)$$

$$\frac{\partial g}{\partial \tau} = -g(z,\tau)P(z,\tau)/E_{\text{sat}}. \qquad (5.5.17)$$

Equation (5.5.15) can easily be integrated over the amplifier length L to yield

$$P_{\text{out}}(\tau) = P_{\text{in}}(\tau)\exp[h(\tau)], \qquad (5.5.18)$$

where $P_{\text{in}}(\tau)$ is the input power and $h(\tau)$ is the total integrated gain defined as

$$h(\tau) = \int_0^L g(z,\tau)\,dz. \qquad (5.5.19)$$

If Eq. (5.5.17) is integrated over the amplifier length after replacing gP by $\partial P/\partial z$, $h(\tau)$ satisfies [92]

$$\frac{dh}{d\tau} = -\frac{1}{E_{\text{sat}}}[P_{\text{out}}(\tau) - P_{\text{in}}(\tau)] = -\frac{P_{\text{in}}(\tau)}{E_{\text{sat}}}(e^h - 1). \qquad (5.5.20)$$

Equation (5.5.20) can easily be solved to obtain $h(\tau)$. The amplification factor $G(\tau)$ is related to $h(\tau)$ as $G = \exp(h)$ and is given by

$$G(\tau) = \frac{G_0}{G_0 - (G_0 - 1)\exp[-E_0(\tau)/E_{\text{sat}}]}, \qquad (5.5.21)$$

5.5 Semiconductor Optical Amplifiers

where G_0 is the unsaturated amplifier gain and $E_0(\tau) = \int_{-\infty}^{\tau} P_{\text{in}}(\tau)\, d\tau$ is the partial energy of the input pulse defined such that $E_0(\infty)$ equals the input pulse energy E_{in}.

The solution (5.5.21) shows that the amplifier gain is different for different parts of the pulse. The leading edge experiences the full gain G_0 as the amplifier is not yet saturated. The trailing edge experiences the least gain since the whole pulse has saturated the amplifier gain. As seen from Eq. (5.5.16), gain saturation leads to a time-dependent phase shift across the pulse. This phase shift is found by integrating Eq. (5.5.16) over the amplifier length and is given by

$$\phi(\tau) = -\tfrac{1}{2}\beta_c \int_0^L g(z,\tau)\, dz = -\tfrac{1}{2}\beta_c h(\tau) = -\tfrac{1}{2}\beta_c \ln[G(\tau)]. \tag{5.5.22}$$

Since the pulse modulates its own phase through gain saturation, this phenomenon is referred to as *saturation-induced* self-phase modulation [92]. The frequency chirp is related to the phase derivative as

$$\Delta \nu_c = -\frac{1}{2\pi}\frac{d\phi}{d\tau} = \frac{\beta_c}{4\pi}\frac{dh}{d\tau} = -\frac{\beta_c P_{\text{in}}(\tau)}{4\pi E_{\text{sat}}}[G(\tau) - 1], \tag{5.5.23}$$

where Eq. (5.5.20) was used.

Self-phase modulation and the associated frequency chirp are similar to the phenomena that occur when an optical pulse propagates through a fiber (see Section 1.6.1). Just as in optical fibers, the spectrum of the amplified pulse broadens and contains several peaks of different amplitudes [92]. Figure 5.22 shows the expected shape (a) and spectrum (b) of amplified pulses when a Gaussian pulse of energy such that $E_{\text{in}}/E_{\text{sat}} = 0.1$ is amplified by a SOA. The dominant spectral peak is shifted toward the red side and is broader than the input spectrum. It is also accompanied by one or more satellite peaks. The temporal and spectral changes depend on the level of amplifier gain. The experiments performed by using picosecond pulses from mode-locked semiconductor lasers have confirmed the behavior seen in Figure 5.22.

It turns out that the frequency chirp imposed by the SOA is opposite in nature compared with that imposed by directly modulated semiconductor lasers. Moreover, the chirp is nearly linear over a considerable portion of the amplified pulse. For this reason, the amplified pulse can be compressed when it is propagated through an optical fiber of appropriate length in the anomalous-dispersion region. Such a compression was observed in an experiment [93] in which 40-ps optical pulses were first amplified in a 1.52-μm SOA and then propagated through 18 km of single-mode fiber with $\beta_2 = -18$ ps^2/km. This compression mechanism can be used to design fiber-optic communication systems in which SOAs are used to compensate simultaneously for both fiber loss and dispersion [94].

The preceding analysis considered a single pulse. The signal in a lightwave system consists of a random sequence of 1 and 0 bits. If the energy of each 1 bit is large enough to saturate the gain partially, the following bit will experience less gain. The gain will recover partially if the bit 1 is preceded by one or more 0 bits. In effect, the gain of each bit in an SOA depends on the bit pattern. This phenomenon becomes quite problematic for WDM systems in which several pulse trains pass through the amplifier simultaneously. It is possible to implement a gain-control mechanism that keeps the

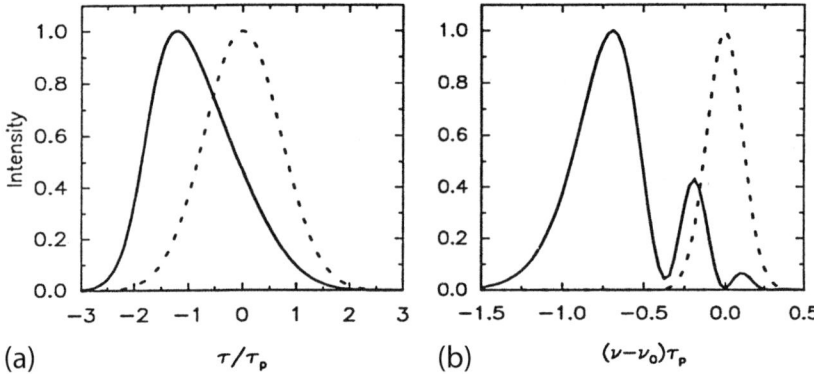

Figure 5.22: (a) Shape and (b) spectrum at the output of a semiconductor optical amplifier with $G_0 = 30$ dB and $\beta_c = 5$ for a Gaussian input pulse of energy $E_{in}/E_{sat} = 0.1$. The dashed curves show for comparison the shape and spectrum of the input pulse.

amplifier gain pinned at a constant value. The basic idea is to make the SOA oscillate at a controlled wavelength outside the range of interest (typically below 1.52 μm). Since the gain remains clamped at the threshold value for a laser, the signal is amplified by the same factor for all pulses.

Problems

5.1 Explain how a semiconductor laser can be forced to operate in a single longitudinal mode. Sketch two techniques that can be used to design such lasers.

5.2 Solve the rate equations in the steady state and obtain the analytic expressions for P and N as a function of the injection current I. Neglect spontaneous emission for simplicity.

5.3 A 250-μm-long InGaAsP laser has an internal loss of 40 cm^{-1}. It operates at 1.55 μm in a single mode, with the modal index 3.3 and the group index 3.4. Calculate the photon lifetime. What is the threshold value of the electron population? Assume that the gain varies as $G = G_N(N - N_0)$ with $G_N = 6 \times 10^3$ s^{-1} and $N_0 = 1 \times 10^8$.

5.4 Determine the threshold current for the semiconductor laser of Problem 5.3 by taking 2 ns as the carrier lifetime. How much power is emitted from one facet when the laser is operated twice above threshold?

5.5 Consider the laser of Problem 5.3 operating twice above threshold. Calculate the differential quantum efficiency and the external quantum efficiency for the laser. What is the device (wall-plug) efficiency if the external voltage is 1.5 V? Assume that the internal quantum efficiency is 90%.

5.6 Calculate the frequency (in GHz) and the damping time of the relaxation oscillations for the laser of Problem 5.3 operating twice above threshold. Assume that $G_P = -4 \times 10^4$ s^{-1}, where G_P is the derivative of G with respect to P. Also assume that $R_{sp} = 2/\tau_p$.

5.7 A semiconductor laser is operating continuously at a certain current. Its output power changes slightly because of a transient current fluctuation. Show that the laser power will attain its original value through an oscillatory approach. Obtain the frequency and the damping time of such relaxation oscillations.

5.8 The threshold current of a semiconductor laser doubles when the operating temperature is increased by 50°C. What is the characteristic temperature of the laser?

5.9 Derive an expression for the 3-dB modulation bandwidth by assuming that the gain G in the rate equations varies with N and P as

$$G(N,P) = G_N(N - N_0)(1 + P/P_s)^{-1/2}.$$

Show that the bandwidth saturates at high operating powers.

5.10 Solve the rate equations (5.3.1) and (5.3.2) numerically by using $I(t) = I_b + I_m f_p(t)$, where $f_p(t)$ represents a rectangular pulse of 200-ps duration. Assume that $I_b/I_{th} = 0.8$, $I_m/I_{th} = 3$, $\tau_p = 3$ ps, $\tau_c = 2$ ns, and $R_{sp} = 2/\tau_p$. Use Eq. (5.3.15) for the gain G with $G_N = 10^4$ s^{-1}, $N_0 = 10^8$, and $\varepsilon_{NL} = 10^{-7}$. Plot the optical pulse shape and the frequency chirp. Why is the optical pulse much shorter than the applied current pulse?

5.11 Calculate the autocorrelation $C_{pp}(\tau)$ by using Eqs. (5.3.31) and (5.3.32). Use it to derive an expression for the SNR of the laser output.

5.12 Show that the external quantum efficiency of a planar LED is given approximately by $\eta_{ext} = n^{-1}(n+1)^{-2}$, where n is the refractive index of the semiconductor–air interface. Consider Fresnel reflection and total internal reflection at the output facet. Assume that the internal radiation is uniform in all directions.

5.13 Prove that the 3-dB optical bandwidth of a LED is related to the 3-dB electrical bandwidth by the relation f_{3dB}(optical) $= \sqrt{3} f_{3dB}$(electrical).

5.14 A 250-μm-long semiconductor laser is used as an FP amplifier by biasing it below threshold. Calculate the amplifier bandwidth by assuming 32% reflectivity for both facets and 30-dB peak gain. The group index $n_g = 4$. How much does the bandwidth change when both facets are coated to reduce the facet reflectivities to 1%?

5.15 Starting from Eq. (5.5.1), prove that $G\sqrt{R_1 R_2} < 0.17$ is required to keep the amplifier gain uniform to within 3 dB for a traveling-wave amplifier. What should be the facet reflectivities to ensure traveling-wave operation of an SOA designed to provide 20-dB gain. Assume $R_1 = 2R_2$.

References

[1] Z. Alferov, *IEEE J. Sel. Topics Quantum Electron.* **6**, 832 (2000).
[2] G. P. Agrawal and N. K. Dutta, *Semiconductor Lasers*, 2nd ed., Van Nostrand Reinhold, New York, 1993.

[3] S. L. Chuang, *Physics of Optoelectronic Devices*, Wiley, New York, 1995.
[4] L. A. Coldren and S. W. Corzine, *Diode Lasers and Photonic Integrated Circuits*, Wiley, New York, 1995.
[5] G. P. Agrawal, Ed., *Semiconductor Lasers: Past, Present, and Future*, AIP Press, Woodbury, NY, 1995.
[6] M. C. Aman and J. Buus, *Tunable Semiconductor Lasers*, Artech House, Norwood, MA, 1998.
[7] E. Kapon, Ed., *Semiconductor Lasers*, Part I and II, Academic Press, San Diego, CA, 1999.
[8] D. A. Ackerman, J. E. Johnson, L. J. P. Ketelsen, L. E. Eng, P. A. Kiely, and T. G. B. Mason, *Optical Fiber Telecommunications*, Vol. 4A, I. P. Kaminow and T. P. Lee, Eds., Academic Press, San Diego, CA, 2002, Chap. 12.
[9] R. Wyatt and W. J. Devlin, *Electron. Lett.* **19**, 110 (1983).
[10] W. T. Tsang, in *Semiconductors and Semimetals*, Vol. 22B, W. T. Tsang, Ed., Academic Press, San Diego, CA, 1985, Chap. 4.
[11] S. Akiba, M. Usami, and K. Utaka, *J. Lightwave Technol.* **5**, 1564 (1987).
[12] G. P. Agrawal, in *Progress in Optics*, Vol. 26, E. Wolf, Ed., North-Holland, Amsterdam, 1988, Chap. 3.
[13] J. Buus, *Single Frequency Semiconductor Lasers*, SPIE Press, Bellingham, WA, 1991.
[14] N. Chinone and M. Okai, in *Semiconductor Lasers: Past, Present, and Future*, G. P. Agrawal, Ed., AIP Press, Woodbury, NY, 1995, Chap. 2.
[15] G. Morthier and P. Vankwikelberge, *Handbook of Distributed Feedback Laser Diodes*, Artech House, Norwood, MA, 1995.
[16] J. E. Carroll, J. E. Whiteaway, and R. G. Plumb, *Distributed Feedback Semiconductor Lasers*, INSPEC, London, 1998.
[17] J. Hong, C. Blaauw, R. Moore, S. Jatar, and S. Doziba, *IEEE J. Sel. Topics Quantum Electron.* **5**, 442 (1999).
[18] K. Kobayashi and I. Mito, *J. Lightwave Technol.* **6**, 1623 (1988).
[19] T. L. Koch and U. Koren, *J. Lightwave Technol.* **8**, 274 (1990).
[20] H. Hillmer, A. Grabmaier, S. Hansmann, H.-L. Zhu, H. Burkhard, and K. Magari, *IEEE J. Sel. Topics Quantum Electron.* **1**, 356 (1995).
[21] H. Ishii, F. Kano, Y. Tohmori, Y. Kondo, T. Tamamura, and Y. Yoshikuni, *IEEE J. Sel. Topics Quantum Electron.* **1**, 401 (1995).
[22] P.-J. Rigole, S. Nilsson, I. Bäckbom, T. Klinga, J. Wallin, B. Stålnacke, E. Berglind, and B. Stoltz, *IEEE Photon. Technol. Lett.* **7**, 697 (1995); **7**, 1249 (1995).
[23] G. Albert, F. Delorme, S. Grossmaire, S. Slempkes, A. Ougazzaden, and H. Nakajima, *IEEE J. Sel. Topics Quantum Electron.* **3**, 598 (1997).
[24] F. Delorme, G. Albert, P. Boulet, S. Grossmaire, S. Slempkes, and A. Ougazzaden, *IEEE J. Sel. Topics Quantum Electron.* **3**, 607 (1997).
[25] L. Coldren, *IEEE J. Sel. Topics Quantum Electron.* **6**, 988 (2000).
[26] C. J. Chang-Hasnain, in *Semiconductor Lasers: Past, Present, and Future*, G. P. Agrawal, Ed., AIP Press, Woodbury, NY, 1995, Chap. 5.
[27] A. E. Bond, P. D. Dapkus, and J. D. O'Brien, *IEEE J. Sel. Topics Quantum Electron.* **5**, 574 (1999).
[28] C. Wilmsen, H. Temkin, and L.A. Coldren, Eds., *Vertical-Cavity Surface-Emitting Lasers*, Cambridge University Press, New York, 1999.
[29] C. J. Chang-Hasnain, *IEEE J. Sel. Topics Quantum Electron.* **6**, 978 (2000).
[30] K. Iga, *IEEE J. Sel. Topics Quantum Electron.* **6**, 1201 (2000).

[31] A. Karim, S. Björlin, J. Piprek, and J. E. Bowers, *IEEE J. Sel. Topics Quantum Electron.* **6**, 1244 (2000).

[32] H. Li and K. Iga, *Vertical-Cavity Surface-Emitting Laser Devices*, Springer, New York, 2001.

[33] A. Karim, P. Abraham, D. Lofgreen, Y. J. Chiu, J. Piprek, and J. E. Bowers, *Electron. Lett.* **37**, 431 (2001).

[34] T. R. Chen, J. Ungar, J. Iannelli, S. Oh, H. Luong, and N. Bar-Chaim, *Electron. Lett.* **32**, 898 (1996).

[35] G. P. Agrawal, *IEEE J. Quantum Electron.* **23**, 860 (1987).

[36] G. P. Agrawal, *IEEE J. Quantum Electron.* **26**, 1901 (1990).

[37] G. P. Agrawal and G. R. Gray, *Proc. SPIE* **1497**, 444 (1991).

[38] C. Z. Ning and J. V. Moloney, *Appl. Phys. Lett.* **66**, 559 (1995).

[39] M. Osinski and J. Buus, *IEEE J. Quantum Electron.* **23**, 9 (1987).

[40] H. Ishikawa, H. Soda, K. Wakao, K. Kihara, K. Kamite, Y. Kotaki, M. Matsuda, H. Sudo, S. Yamakoshi, S. Isozumi, and H. Imai, *J. Lightwave Technol.* **5**, 848 (1987).

[41] P. A. Morton, T. Tanbun-Ek, R. A. Logan, N. Chand, K. W. Wecht, A. M. Sergent, and P. F. Sciortino, *Electron. Lett.* **30**, 2044 (1994).

[42] E. Goutain, J. C. Renaud, M. Krakowski, D. Rondi, R. Blondeau, and D. Decoster, *Electron. Lett.* **32**, 896 (1996).

[43] S. Lindgren, H. Ahlfeldt, L Backlin, L. Forssen, C. Vieider, H. Elderstig, M. Svensson, L. Granlund, L. Andersson, B. Kerzar, B. Broberg, O. Kjebon, R. Schatz, E. Forzelius, and S. Nilsson, *IEEE Photon. Technol. Lett.* **9**, 306 (1997).

[44] R. A. Linke, *Electron. Lett.* **20**, 472 (1984); *IEEE J. Quantum Electron.* **21**, 593 (1985).

[45] G. P. Agrawal and M. J. Potasek, *Opt. Lett.* **11**, 318 (1986).

[46] R. Olshansky and D. Fye, *Electron. Lett.* **20**, 928 (1984).

[47] G. P. Agrawal, *Opt. Lett.* **10**, 10 (1985).

[48] N. A. Olsson, C. H. Henry, R. F. Kazarinov, H. J. Lee, and K. J. Orlowsky, *IEEE J. Quantum Electron.* **24**, 143 (1988).

[49] H. D. Summers and I. H. White, *Electron. Lett.* **30**, 1140 (1994).

[50] F. Kano, T. Yamanaka, N. Yamamoto, H. Mawatan, Y. Tohmori, and Y. Yoshikuni, *IEEE J. Quantum Electron.* **30**, 533 (1994).

[51] D. M. Adams, C. Rolland, N. Puetz, R. S. Moore, F. R. Shepard, H. B. Kim, and S. Bradshaw, *Electron. Lett.* **32**, 485 (1996).

[52] G. P. Agrawal, *Applications of Nonlinear Fiber Optics*, Academic Press, San Diego, CA, 2001.

[53] G. P. Agrawal, *Proc. SPIE* **1376**, 224 (1991).

[54] M. Lax, *Rev. Mod. Phys.* **38**, 541 (1966); *IEEE J. Quantum Electron.* **3**, 37 (1967).

[55] C. H. Henry, *IEEE J. Quantum Electron.* **18**, 259 (1982); **19**, 1391 (1983); *J. Lightwave Technol.* **4**, 298 (1986).

[56] G. P. Agrawal, *Electron. Lett.* **27**, 232 (1991).

[57] G. P. Agrawal, *Phys. Rev. A* **37**, 2488 (1988).

[58] J. Y. Law and G. P. Agrawal, *IEEE Photon. Technol. Lett.* **9**, 437 (1997).

[59] M. Aoki, K. Uomi, T. Tsuchiya, S. Sasaki, M. Okai, and N. Chinone, *IEEE J. Quantum Electron.* **27**, 1782 (1991).

[60] J. Gower, *Optical Communication Systems*, 2nd ed., Prentice-Hall, Upper Saddle River, NJ, 1993.

[61] H. Temkin, G. V. Keramidas, M. A. Pollack, and W. R. Wagner, *J. Appl. Phys.* **52**, 1574 (1981).
[62] C. A. Burrus and R. W. Dawson, *Appl. Phys. Lett.* **17**, 97 (1970).
[63] S. T. Wilkinson, N. M. Jokerst, and R. P. Leavitt, *Appl. Opt.* **34**, 8298 (1995).
[64] M. C. Larson and J. S. Harris, Jr., *IEEE Photon. Technol. Lett.* **7**, 1267 (1995).
[65] I. J. Fritz, J. F. Klem, M. J. Hafich, A. J. Howard, and H. P. Hjalmarson, *IEEE Photon. Technol. Lett.* **7**, 1270 (1995).
[66] T. Whitaker, *Compound Semicond.* **5**, 32 (1999).
[67] P. Bienstman and R. Baets, *IEEE J. Quantum Electron.* **36**, 669 (2000).
[68] P. Sipila, M. Saarinen, M. Guina, V. Vilokkinen, M. Toivonen, and M. Pessa, *Semicond. Sci. Technol.* **15**, 418 (2000).
[69] T. Saitoh and T. Mukai, *IEEE J. Quantum Electron.* **23**, 1010 (1987).
[70] N. A. Olsson, *J. Lightwave Technol.* **7**, 1071 (1989).
[71] T. Saitoh and T. Mukai, in *Coherence, Amplification, and Quantum Effects in Semiconductor Lasers*, Y. Yamamoto, Ed., Wiley, New York, 1991, Chap. 7.
[72] G.-H. Duan, in *Semiconductor Lasers: Past, Present, and Future*, G. P. Agrawal, Ed., AIP Press, Woodbury, NY, 1995, Chap. 10.
[73] L. H. Spiekman, *Optical Fiber Telecommunications*, Vol. 4A, I. P. Kaminow and T. P. Lee, Eds., Academic Press, San Diego, CA, 2002, Chap. 14.
[74] C. E. Zah, J. S. Osinski, C. Caneau, S. G. Menocal, L. A. Reith, J. Salzman, F. K. Shokoohi, and T. P. Lee, *Electron. Lett.* **23**, 990 (1987).
[75] N. A. Olsson, M. G. Öberg, L. A. Koszi, and G. J. Przybylek, *Electron. Lett.* **24**, 36 (1988).
[76] I. Cha, M. Kitamura, H. Honmou, and I. Mito, *Electron. Lett.* **25**, 1241 (1989).
[77] S. Cole, D. M. Cooper, W. J. Devlin, A. D. Ellis, D. J. Elton, J. J. Isaak, G. Sherlock, P. C. Spurdens, and W. A. Stallard, *Electron. Lett.* **25**, 314 (1989).
[78] G. Großkopf, R. Ludwig, R. G. Waarts, and H. G. Weber, *Electron. Lett.* **23**, 1387 (1987).
[79] N. A. Olsson, *Electron. Lett.* **24**, 1075 (1988).
[80] M. Sumida, *Electron. Lett.* **25**, 1913 (1989).
[81] M. Koga and T. Mutsumoto, *J. Lightwave Technol.* **9**, 284 (1991).
[82] G. P. Agrawal, *Opt. Lett.* **12**, 260 (1987).
[83] G. P. Agrawal, *Appl. Phys. Lett.* **51**, 302 (1987); *J. Opt. Soc. Am. B* **5**, 147 (1988).
[84] G. P. Agrawal, *Electron. Lett.* **23**, 1175 (1987).
[85] I. M. I. Habbab and G. P. Agrawal, *J. Lightwave Technol.* **7**, 1351 (1989).
[86] S. Ryu, K. Mochizuki, and H. Wakabayashi, *J. Lightwave Technol.* **7**, 1525 (1989).
[87] G. P. Agrawal and I. M. I. Habbab, *IEEE J. Quantum Electron.* **26**, 501 (1990).
[88] H. Kawaguchi, *IEE Proc.* **140**, Pt. J, 3 (1993).
[89] R. J. Manning, A. D. Ellis, A. J. Poustie, and K. J. Blow, *J. Opt. Soc. Am. B* **14**, 3204 (1997).
[90] K. E. Stubkjaer, *IEEE J. Sel. Topics Quantum Electron.* **6**, 1428 (2000).
[91] G. P. Agrawal and D. N. Maywar, in *Nonlinear Photonic Crystals*, Springer Series in Photonics, Vol. 10, Eds. R. E. Slusher and B. H. Eggleton, Springer, New York, 2003, Chap. 13.
[92] G. P. Agrawal and N. A. Olsson, *IEEE J. Quantum Electron.* **25**, 2297 (1989).
[93] G. P. Agrawal and N. A. Olsson, *Opt. Lett.* **14**, 500 (1989).
[94] N. A. Olsson, G. P. Agrawal, and K. W. Wecht, *Electron. Lett.* **25**, 603 (1989).

Chapter 6

Optical Modulators

Although, as discussed in Chapter 5, direct modulation of semiconductor lasers can be employed while designing a lightwave system, amplitude modulation of the optical carrier is always accompanied by phase modulation when such a scheme is used [1]. At bit rates of 10 Gb/s or higher, the frequency chirp imposed by current modulation becomes so large that direct modulation of semiconductor lasers is rarely used in practice. For such high-speed applications, a DFB laser is biased at a constant current to provide a continuous-wave (CW) output, and an active component placed next to the laser converts the CW light into a data-coded pulse train with the right modulation format. This chapter focuses on this new active component known as the optical modulator. In Section 6.1 we discuss the basic concepts and the physical mechanisms commonly used for making modulators and then focus in Sections 6.2 to 6.4 on three kinds of modulators fabricated using lithium-niobate, polymer, and semiconductor materials.

6.1 Physics Behind Modulators

The first step in the design of an optical communication system is to decide how the electrical data would be converted into an optical signal that carries the same information with it. The original electrical data can be in an analog form, but it is invariably converted into a digital bit stream consisting of a pseudorandom sequence of 0 and 1 bits. In the digital case, two choices exist for formatting the electrical bit stream. These are shown in Figures 6.1(a) and (b), and are known as the *return-to-zero* (RZ) and *nonreturn-to-zero* (NRZ) formats. In the RZ format, each electrical pulse representing bit 1 is shorter than the bit slot, and its amplitude returns to zero before the bit duration is over. In the NRZ format, the pulse remains on throughout the bit slot and its amplitude does not drop to zero between two or more successive 1 bits. As a result, pulse width varies depending on the bit pattern, whereas it remains the same in the case of RZ format. An advantage of the NRZ format is that the bandwidth associated with the bit stream is smaller than that of the RZ format by about a factor of 2 simply because on–off transitions occur fewer times. For this reason, the NRZ format is often used in practice although the RZ format is gaining acceptance at high bit rates.

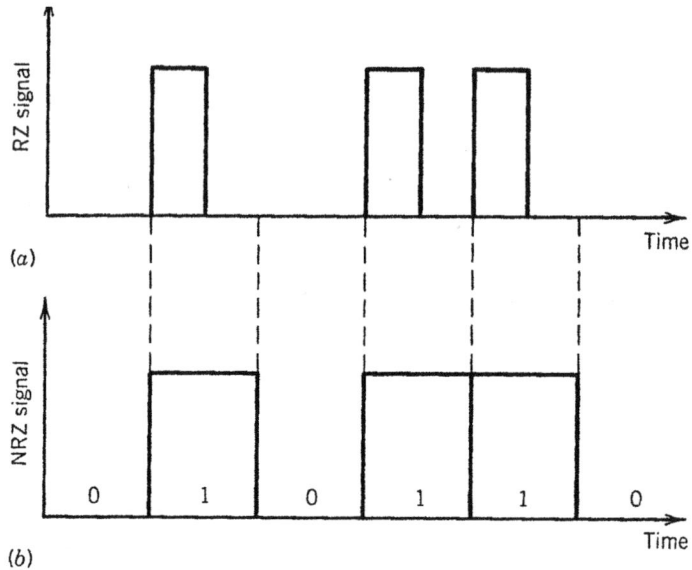

Figure 6.1: Digital bit stream 010110... coded by using (a) return-to-zero (RZ) and (b) nonreturn-to-zero (NRZ) formats.

6.1.1 Modulation Schemes

An important issue is related to the choice of the physical quantity that is modulated to encode the data on an optical carrier. The electromagnetic carrier before modulation has its electric field in the form

$$\mathbf{E}(t) = \hat{\mathbf{e}} A \cos(\omega_0 t + \phi), \qquad (6.1.1)$$

where $\mathbf{E}$ is the electric field vector, $\hat{\mathbf{e}}$ is the polarization unit vector, A is the amplitude, ω_0 is the carrier frequency, and ϕ is the phase. The modulation of polarization is not used in practice since telecommunication fibers do not preserve the state of polarization during propagation of an optical signal. However, one may choose to modulate the amplitude A, the frequency ω_0, or the phase ϕ. The three modulation choices are known as amplitude modulation (AM), frequency modulation (FM), and phase modulation (PM). In the digital case, they are also referred to as amplitude-shift keying (ASK), frequency-shift keying (FSK), and phase-shift keying (PSK), depending on whether the amplitude, frequency, or phase of the carrier wave is shifted between the two fixed levels to form a binary digital signal. The simplest technique consists of simply changing the signal power between two levels, one of which is set to zero. This scheme is often called *on–off keying* to reflect the on–off nature of the resulting optical signal. Most digital lightwave systems employ on–off keying through AM or ASK.

The simplest solution for generating an optical bit stream consists of applying the electrical signal directly to the semiconductor laser used as an optical source. As discussed in Section 5.3.3, direct modulation of semiconductor lasers suffers from the

6.1 Physics Behind Modulators

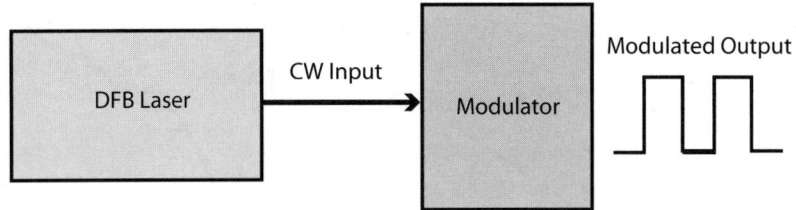

Figure 6.2: Schematic illustration of the external-modulation scheme. Laser is biased at a constant current while the time-dependent electrical signal is applied to the modulator.

problem of frequency chirping, whose origin lies in the physical fact that any change in the carrier density not only affects the optical gain but also the refractive index within the active region of a semiconductor laser. While changes in the gain modulate the amplitude, changes in the refractive index modulate the phase. In other words, AM is always accompanied by FM when direct modulation is used. It turns out that the chirping problem becomes so severe at bit rates above 5 Gb/s that it is better to employ an external modulator for modulating the optical carrier.

Figure 6.2 shows the external-modulation scheme schematically. Constant-intensity light from a DFB semiconductor laser is coupled into a modulator. The role of the optical modulator is to convert this CW input beam of constant intensity into an optical signal that mimics the time dependence of the electrical signal applied to the modulator. It should do so with high fidelity and with low insertion losses. In fact, a good modulator must meet a number of requirements including high extinction ratio, large bandwidth, low chirp, linear response, and a low bias voltage. As discussed next, several physical mechanisms can be used for making modulators.

6.1.2 Electroabsorption

Electroabsorption, as the name applies, is the phenomenon whereby the absorption coefficient of an optical material is changed by applying an electrical voltage across it. This phenomenon occurs in semiconductors and is known as the *Franz–Keldysh effect* [2]–[5]. Physically speaking, the bandgap of a semiconductor decreases when an electric field is applied across it. As a result, although photons of energy slightly smaller than the bandgap can pass through the semiconductor in the absence of the electric field, they are absorbed when the electric field is applied.

Mathematically, when the Schrödinger equation for electrons (or holes) is solved in the presence of an electric field, one finds the solution in terms of the Airy functions, rather than plane waves [5]. As a result, the wave function associated with electrons and holes develops an exponentially decaying tail into the bandgap region, in effect reducing the bandgap. Figure 6.3(a) shows how photons of energy less than the bandgap energy can be absorbed because of this modification. Changes in the absorption spectrum with the applied field are shown in Figure 6.3(b) for a typical semiconductor. The sharp absorption edge is replaced with a gradual increase when the field is applied, and the semiconductor has a finite value of the absorption coefficient in the region close

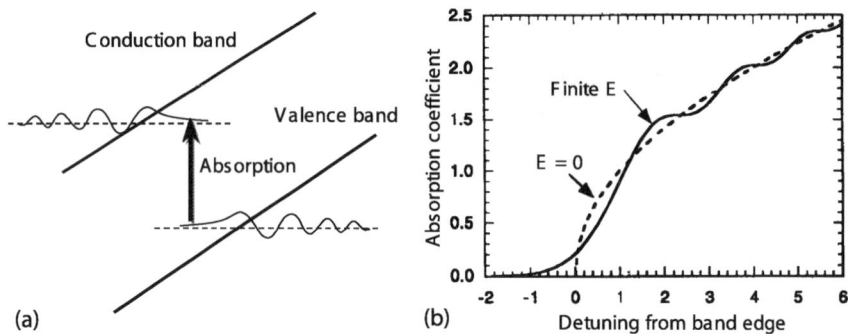

Figure 6.3: (a) Schematic illustration of the Franz–Keldysh effect. Tilting of the conduction and valence bands in the presence of an applied electric field allows absorption in the bandgap region. (b) Absorption spectra of a quantum well with (solid line) and without (dashed line) applied field.

to but below the bandgap. The tilt of the conduction and valence bands in Figure 6.3 depends on the magnitude of the applied field. As the electric field strength increases, the tilt increases, and the probability of finding an electron within the bandgap also increases, resulting in more absorption.

In practice, it is desirable to use a semiconductor waveguide for making modulators based on the electroabosrption phenomenon. When a thin layer of the semiconductor material is used, one should consider both the excitonic and quantum-well effects. The effects of excitons dominate in quantum wells in the spectral region just below the bandgap. In the presence of an electric field, excitons can be ionized, resulting in a Stark shift and a broadening of the main excitonic absorption peak. This change in the absorption spectrum is referred to as the *quantum-confined Stark effect* and has been studied extensively [6]–[8]. In practice, multiple quantum-well layers are used to enhance the magnitude of the effect. Figure 6.4 shows, as an example, changes in the absorption spectra of a multiquantum-well (MQW) GaAs sample as the applied voltage across it is varied from 0 to 8 V. The absorption tail shifts to longer wavelengths (smaller photon energies) as the electric field is enhanced by increasing the applied voltage. As a result, the value of the absorption coefficient can be changed from nearly zero to >1000 cm^{-1} for an optical signal containing photons of energy slightly smaller than the bandgap energy.

To make use of the quantum-confined Stark effect for modulation of an optical carrier, one needs a semiconductor whose bandgap is slightly larger than the photon energy associated with the optical carrier so that the semiconductor layer is transparent to the incident optical beam in the absence of an applied voltage. The same layer becomes absorptive when its bandgap is reduced electronically by applying an external voltage. Such a device can convert a CW signal into an optical bit stream mimicking the applied electrical signal as long as changes in absorption occur on a time scale faster than the bit slot.

6.1 Physics Behind Modulators

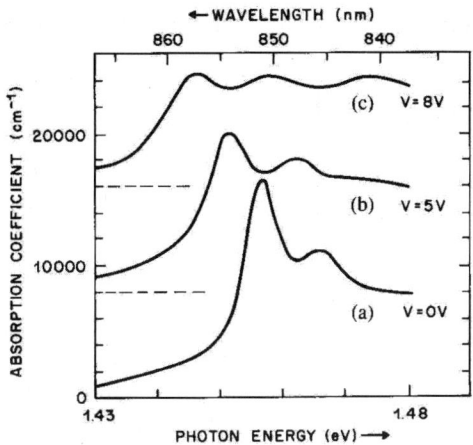

Figure 6.4: Absorption spectra of a MQW GaAs sample at three voltages in the range 0–8 V. Dashed lines indicate a shift in the vertical scale for clarity. (After Ref. [7]; ©1985 APS.)

6.1.3 Electrorefraction

Electrorefraction, as the name applies, is the phenomenon in which the refractive index of an optical material is changed by applying an electrical voltage across it. This phenomenon occurs in some low-symmetry crystals and is known as the Pockels effect or the linear electro-optic effect. As discussed in Section 4.3.2, the electro-optic effect is described in terms of the impermeability tensor whose elements η_{ij} are related to the dielectric-constant tensor elements as $\eta_{ij} = 1/\varepsilon_{ij}$. The impermeability tensor changes in response to an external electric field $\mathbf{E}$ as [9]

$$\eta_{ij}(\mathbf{E}) = \eta_{ij}^{(0)} + \sum_k r_{ijk} E_k + \sum_{kl} s_{ijkl} E_k E_l, \qquad (6.1.2)$$

where the tensor r_{ijk} and s_{ijkl} describe the linear and quadratic electro-optic effects, respectively. The Pockels effect is governed by the tensor r_{ijk}.

As discussed in Section 4.3.2, the tensor r_{ijk} is symmetric in its first two indices, and it is common to introduce a two-dimensional 6×3 matrix r_{hk}, where the index h runs from 1 to 3 for the three diagonal elements of η_{ij} and from 4 to 6 for the three off-diagonal elements in the order η_{23}, η_{31}, and η_{12}, respectively. With this notation, changes in the dielectric constant ε, or the refractive index n ($\varepsilon = n^2$), in response to an external electric field are governed by

$$\Delta\left(\frac{1}{\varepsilon_h}\right) \equiv \Delta\left(\frac{1}{n_h^2}\right) = \sum_{k=1}^{3} r_{hk} E_k. \qquad (6.1.3)$$

The 18 elements of the matrix r_{hk} describe the electro-optic properties of a crystal. Fortunately, only a few of them are nonzero depending on the crystal's symmetry group. Materials commonly used for making waveguide modulators are crystals with the point symmetry $3m$ (such as LiNbO$_3$) and with the point symmetry $\bar{4}3m$ (such as

GaAs and InP). For these two types of crystals, the matrix r_{hk} takes the form [9]

$$r_{hk}(3m) = \begin{pmatrix} 0 & -r_{22} & r_{13} \\ 0 & r_{22} & r_{13} \\ 0 & 0 & r_{33} \\ 0 & r_{42} & 0 \\ r_{42} & 0 & 0 \\ r_{22} & 0 & 0 \end{pmatrix}; \quad r_{hk}(\bar{4}3m) = \begin{pmatrix} 0 & 0 & 0 \\ 0 & 0 & 0 \\ 0 & 0 & 0 \\ r_{41} & 0 & 0 \\ 0 & r_{41} & 0 \\ 0 & 0 & r_{41} \end{pmatrix}. \quad (6.1.4)$$

Whereas LiNbO$_3$ has four independent electro-optic coefficients, the response of GaAs and InP semiconductors is completely specified by a single parameter r_{41}. Numerical values of the nonzero elements depend on the spectral region in which a modulator operates. For example, $r_{41} = 1.1$ pm/V for GaAs at a wavelength close to 0.9 μm, but this value increases to 1.43 pm/V in the wavelength region near 1.3 μm [10]. For InP, $r_{41} = 1.45$ pm/V at a wavelength close to 1.06 μm, but this value decreases to 1.3 pm/V in the wavelength region near 1.3 μm [5].

Consider the case of a GaAs waveguide in which light propagates along the crystallographic z axis denoted as [001]. One can apply the electric field along any direction. Let us assume that the electric field is also oriented along the z axis pointing normal to the waveguide plane. The index ellipsoid then takes the form [9]

$$\frac{x^2}{n^2} + \frac{y^2}{n^2} + \frac{z^2}{n^2} + 2r_{41}Exy = 0. \quad (6.1.5)$$

Because of the last term, GaAs acquires birefringence even though it is an isotropic material when no field is applied. We can remove the last term by rotating the x and y axes by 45°, which then become the principal axes of the crystal (together with the z axis along which the field is applied). Denoting the rotated axes as x' and y', we obtain

$$\frac{x'^2}{n_x^2} + \frac{y'^2}{n_y^2} + \frac{z^2}{n^2} = 0, \quad (6.1.6)$$

where n_x and n_y are the modified refractive indices along the principal axes and are given by

$$\frac{1}{n_x^2} = \frac{1}{n^2} - r_{41}E, \quad \frac{1}{n_y^2} = \frac{1}{n^2} + r_{41}E. \quad (6.1.7)$$

In practice, field-induced index changes are relatively small, and we can simplify the preceding equation to obtain

$$n_x \approx n + \tfrac{1}{2}n^3 r_{41}E, \quad n_y \approx n - \tfrac{1}{2}n^3 r_{41}E. \quad (6.1.8)$$

The index changes induced by an electric field produce an additional phase shift whose magnitudes are of opposite signs for the TE and TM modes. As a result, the state of polarization at the output changes, and the change can be controlled by the applied field. The polarization changes can be translated into amplitude modulation by placing the waveguide between two polarizers [9]. Bulk electro-optic modulators often use this technique with a KDP crystal. Figure 6.5 shows the operating principle

6.1 Physics Behind Modulators

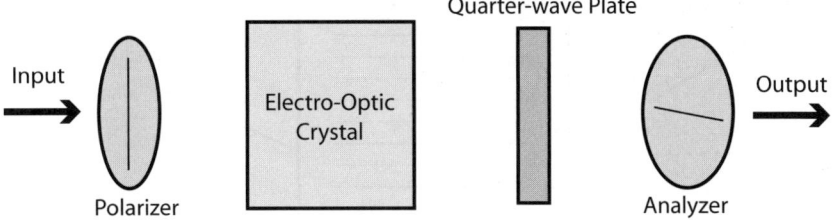

Figure 6.5: Schematic illustration of a bulk electro-optic modulator. The quarter-wave plate is used to bias the modulator.

schematically. The crystal rotates linearly polarized input light such that it becomes elliptically polarized at the output. The combination of a quarter-wave plate and an analyzer (a polarizer oriented orthogonal to the input polarizer) is used to produce intensity modulation. In the absence of any applied field, 50% of input power exits from the analyzer. This amount can be increased or decreased in a controlled fashion by applying a varying voltage across the modulator crystal.

In the domain of lightwave technology, $LiNbO_3$ waveguides are most commonly used for making optical modulators. As seen from Eq. (6.1.4), the electro-optic response of $LiNbO_3$ is governed by four parameters. The numerical values of these four electro-optic coefficients in the spectral region near 1.5 μm are $r_{13} = 8.6$, $r_{22} = 3.4$, $r_{33} = 30.9$, and $r_{42} = 28.0$ (all in units of pm/V). In practice, it is common to apply the electric field along the crystallographic z axis to select the largest electro-optic coefficient r_{33}. An additional advantage is that only the refractive index along the z axis is modified by the electric field. In fact, it follows from Eq. (6.1.3) that

$$\Delta n = -\tfrac{1}{2} n^3 r_{33} E. \tag{6.1.9}$$

Using $n \approx 2.2$ and $r_{33} = 30.9$ pm/V in the spectral region near 1.5 μm, we find that $\Delta n/E \sim 10^{-4}$ μm/V, that is, refractive-index changes of $\sim 10^{-4}$ can be realized by applying a few volts across a distance ~ 10 μm. Since the phase of an input beam of wavelength λ changes in response to index changes as $\Delta\phi = (2\pi/\lambda)\Delta n L$ for a modulator of length L, such index changes are large enough to produce a π phase shift for $\lambda = 1.55$ μm and $L \sim 1$ cm. Such phase shifts can be converted into amplitude modulation using two $LiNbO_3$ waveguides so they form either a directional coupler or a Mach–Zehnder (MZ) interferometer. Both configurations are discussed in Section 6.2 devoted to $LiNbO_3$ modulators.

6.1.4 Photoelastic Effect

The third mechanism exploited for making modulators is known as the *photoelastic effect*. It refers to the possibility of changing the refractive index of a material under elastic strain. The magnitude of the index change is governed by the photoelastic tensor as [9]

$$\Delta\left(\frac{1}{n^2}\right)_{ij} = \sum_{k=1}^{3}\sum_{l=1}^{3} p_{ijkl} S_{kl}, \tag{6.1.10}$$

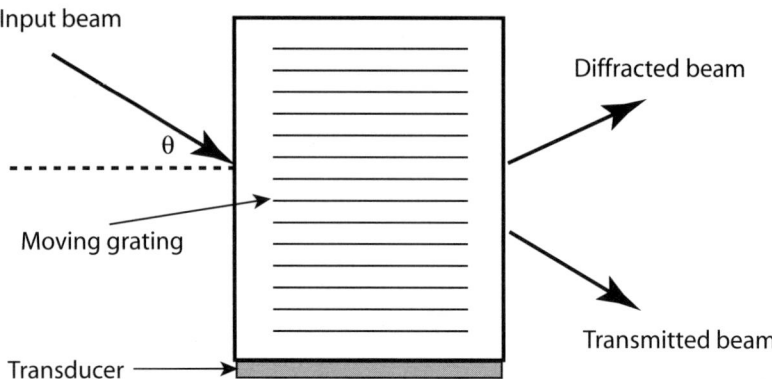

Figure 6.6: Schematic illustration of a bulk acousto-optic modulator. An electrically induced acoustic wave creates a moving index grating that deflects the input beam whenever a voltage is applied to the crystal.

where S_{kl} is the strain tensor.

A simple way to introduce strain into a crystal is to propagate a high-frequency acoustic wave (ultrasound) through it. Although a general analysis is quite complicated because of the tensorial nature of Eq. (6.1.10), the index change induced by an acoustic wave of power P_a can be written after some simplifications as [10]

$$\Delta n = \sqrt{n^6 p^2 P_a/(2\rho v_a^3 S_a)} \equiv \sqrt{MP_a/(2S_a)}, \qquad (6.1.11)$$

where p is the appropriate element of the photoelastic tensor, ρ is the density of the medium, v_a is the acoustic velocity, and S_a is the cross-sectional area through which the acoustic wave travels. The quantity $M = n^6 p^2/(\rho v_a^3)$ is the *figure of merit* that depends only on material parameters. Its numerical values can vary in a wide range from 1 to 500×10^{-15} m²/W; for a LiNbO$_3$ crystal, $M \approx 7 \times 10^{-15}$ m²/W but this value increases by a factor of 15 for a GaAs crystal.

For an acoustic intensity $\sim$1 kW/cm², index changes $\Delta n \sim 10^{-4}$ occur for LiNbO$_3$ over each oscillation of the acoustic wave. Even though this is a relatively small index change, small reflections occurring during each cycle can add up to a large reflection over the length of the crystal under proper phase matching. This situation is similar to a fiber grating (see Section 2.2). In fact, one can understand the physics in terms of an index grating created by the acoustic wave. The only difference is that the grating is not stationary but moves with the acoustic velocity v_a. As a result, the reflected light experiences a small frequency shift of $\sim$100 MHz.

Figure 6.6 shows the design of a bulk acousto-optic modulator based on the photoelastic effect. A transducer is used to excite an acoustic wave at a frequency $v_a \sim$ 100 MHz, which creates an index grating with the period $\Lambda = v_a/v_a$. The Bragg condition (2.2.2) can be used to find that the input light of wavelength λ, incident at an angle θ, will be reflected by the grating when the condition $m\lambda = 2\Lambda \sin\theta$ is satisfied, where m is an integer representing the order of Bragg diffraction. For first-order diffraction ($m = 1$), the frequency of acoustic wave should be chosen such that $v_a = 2v_a \sin\theta/\lambda$.

6.2 Lithium Niobate Modulators

Using $\lambda = 1.55$ μm and $v_a = 6.57$ km/s for LiNbO$_3$, we find that the required frequency is about 300 MHz for $\theta = 2°$.

Another configuration for acousto-optic modulators makes use of Raman–Nath diffraction [9]. In this case, the light propagates normal to the index grating and the interaction between the two occurs over a relatively short length l. In the Raman–Nath regime, $l < \Lambda^2/\lambda$. The incident beam is diffracted by the index grating into several directions. The angle for the mth-order diffraction is given by $\theta_m \approx m\lambda/\Lambda$ and the diffracted powers at various angles are obtained from $P_m = P_{in}J_m^2(\delta)$, where P_{in} is the input power and $\delta = (2\pi l/\lambda)\Delta n$. By choosing device parameters such that $\delta \approx 2.405$, one can ensure that no light comes out in the direction of the incident light ($m = 0$). A digital acousto-optic modulator works by turning on and off the transducer electrically for 0 and 1 bits, respectively. The input light is deflected for all 0 bits but passes nearly unchanged for all 1 bits.

LiNbO$_3$ waveguides can also be used for making acousto-optic modulators through the use of surface acoustic waves. The modulation bandwidth of all acousto-optic devices is limited by the phonon lifetime (typically ~10 ns). As a result, such modulators cannot operate at bit rates much above 100 Mb/s and are rarely used in modern lightwave systems for this purpose. Nevertheless, as discussed in Chapter 8, the acousto-optic phenomenon is quite useful for making tunable optical filters suitable for WDM applications.

6.2 Lithium Niobate Modulators

As discussed in Section 4.3.2, LiNbO$_3$ waveguides can be made by diffusing titanium into selected regions of the substrate. This ease of fabrication of Ti:LiNbO$_3$ waveguides has made this electro-optic crystal the material of choice for making many active and passive components. LiNbO$_3$ modulators were developed during the 1980s, and the technology had matured enough by 2002 that they were available commercially at bit rates as high as 40 Gb/s [11]–[18].

6.2.1 Phase and Amplitude Modulation

The simplest kind of LiNbO$_3$ modulator is a phase modulator in which the effective mode index of a single waveguide is changed by applying an external voltage, as shown schematically in Figure 6.7(a). An important design parameter is the voltage V_π required to produce a π phase shift. Since one of the design objectives is to produce a low-voltage device, it is common to apply the external electric field along the crystallographic z axis to make use of the largest electro-optic coefficient r_{33} of the crystal with the value 30.9 pm/V. If we assume that a voltage V is applied across the two electrodes separated by a distance d_e so that an electric field $E_a = V/d_e$ exists along the z axis, the index change from Eq. (6.1.9) is given by

$$\Delta n = -n^3 r_{33} \Gamma V/(2d_e), \qquad (6.2.1)$$

where Γ is introduced to account for the partial overlap that occurs in practice between the optical and electric fields; typically, $\Gamma = 0.5$.

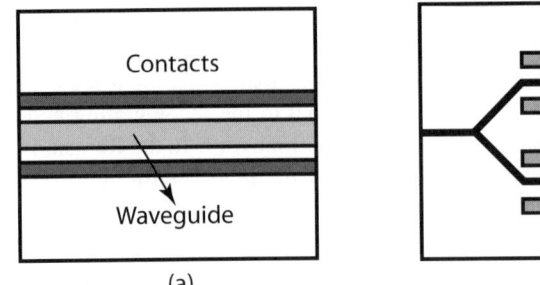 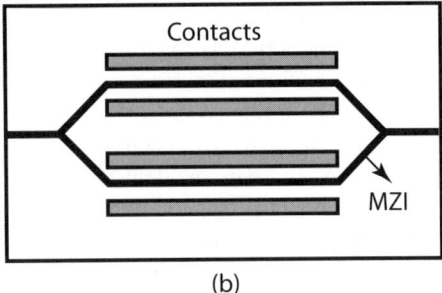

Figure 6.7: Schematic of (a) a phase modulator and (b) an amplitude modulator fabricated using LiNbO$_3$ waveguides. Electrodes surrounding each waveguide are also shown.

Since the optical phase acquired by an optical signal while propagating through a waveguide is proportional to the mode index, the additional voltage-dependent phase shift for an optical signal of wavelength λ can be written as

$$\Delta\phi = (2\pi/\lambda)|\Delta n|L, \qquad (6.2.2)$$

where L is the length of the region over which the electric field is applied. The voltage required for π phase shift is obtained by setting $\Delta\phi = \pi$. From Eq. (6.2.1) and (6.2.2) it is found to be

$$V_\pi L = \frac{\lambda d_e}{\Gamma n^3 r_{33}}. \qquad (6.2.3)$$

Using $r_{33} = 30.9$ pm/V, $n \approx 2.2$, $\Gamma = 0.5$, $d_e = 15$ μm, and $\lambda = 1.55$ μm, we find that $V_\pi L = 14$ V-cm. Thus, only a few volts are required for a device 3–4 cm long. Voltage can be further reduced by bringing the electrodes closer and improving the overlap factor Γ.

Phase modulation is not sufficient for most lightwave applications. Rather, one needs an amplitude modulator in which the optical power changes in response to an external voltage. Two basic designs are used for this purpose. In one approach, two LiNbO$_3$ waveguides are brought together in a central region to make a directional coupler, and the input signal is launched into one of the input ports. The waveguides are designed to be identical and long enough that all power is transferred to the cross port in the absence of an external voltage. When the mode index is changed in one of the waveguides by applying the electric field, phase matching is destroyed, and most of the power exits from the bar port. Several variants of this basic ideas have been used to make different types of LiNbO$_3$ modulators [11].

The most common approach for converting phase modulation to amplitude modulation makes use of the MZ configuration shown schematically in Figure 6.7(b). The MZ interferometer can be made using two directional couplers or two Y junctions. In the absence of external voltage, the optical fields in the two arms of a symmetric MZ interferometer experience identical phase shifts and interfere constructively. The output comes through the cross port when two directional couplers are employed (see Section 2.3.4). The additional phase shift introduced through voltage-induced index changes

6.2 Lithium Niobate Modulators

destroys the constructive nature of the interference. In particular, light is transmitted through the bar port when the phase difference between the two arms equals π. We focus on MZ-based LiNbO$_3$ modulators in this section.

6.2.2 Temporal Response

The temporal response of a LiNbO$_3$ modulator designed using an MZ interferometer can be obtained from Eq. (2.3.22). Typically, the input power is split equally between the two branches using 3-dB couplers with $\rho_1 = \rho_2 = \frac{1}{2}$, and the power transmitted through the bar port of the modulator is given by

$$P(t) = P_{\max} \sin^2 \tfrac{1}{2}[\phi_b + \kappa V(t)], \tag{6.2.4}$$

where $P_{\max}$ is the maximum power, ϕ_b is the bias phase shift introduced either by mismatching the two arms or by applying a constant bias voltage, $V(t)$ is the voltage used for modulation, and the constant κ is related to the modulation efficiency through Eq. (6.2.2). The temporal response of the modulator is governed by this equation as it shows how the output power changes with time as $V(t)$ changes. A LiNbO$_3$ modulator can respond to voltage changes instantaneously because of the electronic nature of the nonlinearity associated with the electro-optic effect.

For digital applications, the applied voltage $V(t)$ is changed between two values, 0 and V_π, for 0 and 1 bits, respectively. The modulator is biased in this case such that $\phi_b = 0$ so $P(t)$ changes between 0 and $P_{\max}$ for 0 and 1 bits, respectively. In the case of an analog signal, $V(t)$ changes with time in a continuous fashion, and the objective is that $P(t)$ follow the same variations as closely as possible, that is, $P(t) \propto V(t)$. As seen from Eq. (6.2.4), output power does not vary linearly with voltage changes. However, if the modulator is biased such that $\phi_b = \pi/2$, a nearly linear response can be obtained in some voltage range. This can be seen from Eq. (6.2.4) by writing it as

$$P(t) = \tfrac{1}{2} P_{\max} \{1 - \sin[\kappa V(t)]\} \approx \tfrac{1}{2} P_{\max}[1 - \kappa V(t)], \tag{6.2.5}$$

where it is assumed that $\kappa V(t) \ll 1$ remains valid at all times.

The temporal response in Eq. (6.2.4) is obtained under ideal conditions. In practice, several factors limit the modulator performance and need to be understood. For example, the maximum power $P_{\max}$ may not necessarily be equal to the input power P_{in} because of the insertion losses associated with the modulator. The insertion loss can thus be defined as (in decibels)

$$\alpha_{\text{ins}} = -10 \log_{10}(P_{\max}/P_{\text{in}}). \tag{6.2.6}$$

Similarly, the minimum power $P_{\min}$ in the off state may not be zero for a digital modulator. This is quantified through the *extinction ratio* defined in decibels as

$$r_{\text{ex}} = -10 \log_{10}(P_{\min}/P_{\max}). \tag{6.2.7}$$

The extinction ratio of an external modulator is also called the on–off ratio. Modern LiNbO$_3$ modulators provide an on–off ratio in excess of 20 dB, while insertion losses are typically below 5 dB.

The third factor that one must consider is the frequency chirp. To see the origin of frequency chirping, we extend the analysis of Section 2.3.4 to the case in which a time-dependent phase shift occurs in both arms of the MZ interferometer. If we assume a balanced MZ interferometer such that $\beta_1 L_1 = \beta_2 L_2$, the optical field transmitted through the bar port is obtained from Eq. (2.3.21) as

$$A_3(t) = A_0\{\sqrt{\rho_1 \rho_2}\exp[i\phi_1(t)] - \sqrt{(1-\rho_1)(1-\rho_2)}\exp[i\phi_2(t)]\}, \qquad (6.2.8)$$

where $\phi_1(t)$ and $\phi_2(t)$ are the voltage-induced phase shifts in the two arms. In the ideal case of 3-dB couplers, $\rho_1 = \rho_2 = \frac{1}{2}$, and the output of the modulator has the form

$$A_3(t) = A_0\{\exp[i(\phi_1 + \phi_2 + \pi)/2]\sin[(\phi_1 - \phi_2)/2]. \qquad (6.2.9)$$

This equation shows that if a time-dependent phase shift is introduced in only one of the arms so that either $\phi_1 = 0$ or $\phi_2 = 0$, the modulator output is always chirped. In contrast, when $\phi_2 = -\phi_1$, the time-dependent phase disappears, resulting in chirp-free modulation. At the same time, the relative phase shift is doubled, and a π phase shift can be obtained at half the voltage. Because of these two advantages, it is common to apply the electric field in opposite directions in the two MZ-interferometer arms. Such modulators are referred to as *push–pull modulators*.

For some applications it is useful to have a modulator whose chirp can be controlled electronically. In the context of semiconductor lasers, a chirp parameter β_c was introduced in Section 5.3.2 to represent the amplitude-phase coupling. We can use the same parameter for modulators. If we neglect the spontaneous emission term in Eq. (5.3.1) and use Eq. (5.3.16), β_c can be written in the form

$$\beta_c = \frac{d\phi/dt}{(2P)^{-1}(dP/dt)}, \qquad (6.2.10)$$

where $P(t)$ and $\phi(t)$ are the power and phase of the output field $A_3(t)$ in Eq. (6.2.9). In general, β_c has a complicated form and varies with time. However, when the modulator is biased at the midpoint of the switching curve and is driven by two identical synchronized signals to produce phase shifts in the two arms of the MZ interferometer, it can be written in the simple form [12]

$$\beta_c = \frac{\Delta V_1 + \Delta V_2}{\Delta V_1 - \Delta V_2}, \qquad (6.2.11)$$

where ΔV_1 and ΔV_2 are the peak values of the voltages applied to the two arms. As before, chirp-free operation is realized for $\Delta V_2 = -\Delta V_1$. For other choices, β_c can be made to take any value in the range $-\infty < \beta_c < \infty$, resulting in a modulator whose chirp is completely controllable.

How fast can one modulate the output of a LiNbO$_3$ modulator? Equation (6.2.9) shows that there is no intrinsic limitation imposed by the LiNbO$_3$ material as the electro-optic response of this material is nearly instantaneous. In practice, the speed of a modulator is limited by electronics. We discuss in the next subsection a traveling-wave design often used for high-speed modulators.

6.2 Lithium Niobate Modulators

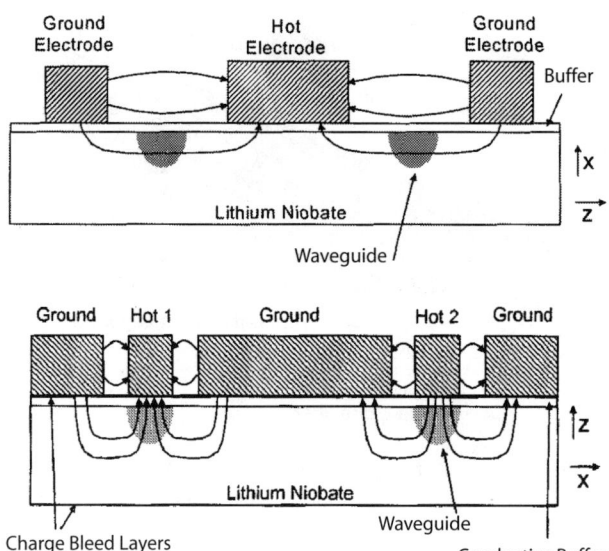

Figure 6.8: Schematic of x-cut and z-cut LiNbO$_3$ modulators showing electrode configuration. Shaded regions show two Ti-diffused waveguides. (After Ref. [15]; ©2000 IEEE.)

6.2.3 Modulator Design

The design of LiNbO$_3$ modulators requires attention to many details. Among the design objectives are high modulation efficiency, low applied voltage, low insertion losses, high on–off ratio, large modulation bandwidth, negligible or controllable frequency chirp, long lifetime, and low cost. As mentioned earlier, modulation efficiency is maximized by using the largest electro-optic coefficient r_{33} occurring along the crystallographic z axis, that is, the applied electric field should point along this axis. This can be realized in both the x-cut and z-cut LiNbO$_3$ substrates. Figure 6.8 shows how the electrodes and waveguides are oriented for the two types of modulators.

In the case of x-cut LiNbO$_3$, two waveguides (shaded regions) run parallel to the y direction (out of the plane in Figure 6.8) and are separated by a distance ~ 10 μm along the z axis. Three electrodes on the top are used to apply an electric field across the two waveguides and induce changes in the refractive index through the electro-optic effect. The central electrode acts as a positive terminal, while the two side electrodes are grounded. This configuration applies electric fields in both waveguides in the opposite directions so that if the mode index increases in one arm, it decreases by the same amount in the other arm. As discussed earlier, the dual-drive approach improves efficiency by lowering the required voltage by a factor of 2 compared with the single-drive case. At the same time, such push–pull modulators produce chirp-free output. A buffer layer is sometimes inserted between the electrodes and the waveguides. It is designed to be transparent to the optical signal propagating inside the modulator. The role of the buffer layer would become clear later. It is especially needed for modulators designed to work at high frequencies.

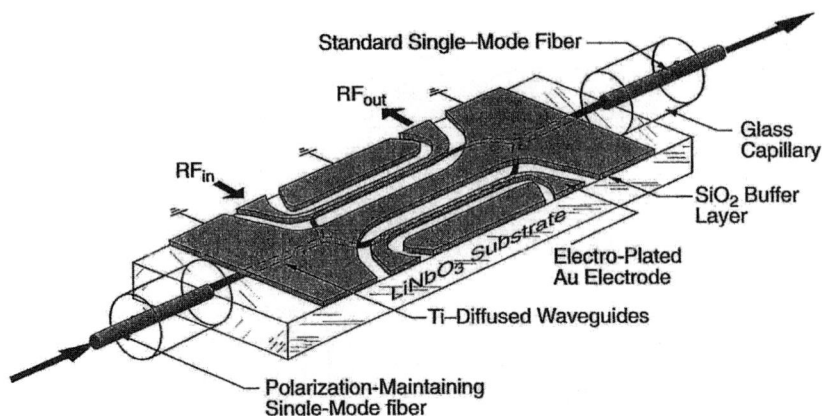

Figure 6.9: Schematic of a dual-drive traveling-wave LiNbO$_3$ modulator. Dashed lines show the MZ interferometer formed using Y junctions on both the input and output ends connected to optical fibers. Shaded regions show the hot and grounded electrodes together with the input and output ports for the RF signal. (After Ref. [12]; ©1997 Elsevier.)

In the case of z-cut LiNbO$_3$, two waveguides still run parallel to the y direction but are separated by a distance ~ 10 μm along the x axis. The electric field is oriented along the z axis by placing the hot electrodes directly over the two waveguides as shown in Figure 6.8. With this configuration, the electric field points perpendicular to the substrate surface. It turns out that the overlap between the optical and electric fields is enhanced in this geometry compared with an x-cut substrate, reducing the required V_π voltage by 20–40%. The buffer layer between the electrodes and the waveguides is essential to ensure that the optical mode does not penetrate the lossy metallic layer (typically gold) used for the electrodes. A disadvantage of z-cut devices is that piezoelectric and pyroelectric effects lead to accumulation of electric charges at the substrate surfaces. It is common to employ a a charge-bleed layer at the bottom of the substrate to solve this problem. The buffer layer is also made conductive for the same reason.

There are two possibilities for biasing an optical modulator. In the conventional design, the electric field is applied along the entire length of the MZ interferometer arms and the applied voltage is changed with time to generate the modulated optical signal. In this lumped-RC design, the modulator response is limited by the *RC* bandwidth of the driving circuit. LiNbO$_3$ modulators designed to operate at high speeds require a different design. In a commonly used traveling-wave approach, the radio-frequency (RF) electrical signal propagates on a miniature transmission line in the form of a coplanar waveguide along the length of the LiNbO$_3$ waveguide so that it overlaps with the optical signal throughout the device length. Such modulators are referred to as traveling-wave modulators. Figure 6.9 shows the design of a dual-drive traveling-wave modulator. The MZ interferometer is formed using two Y junctions while the RF signal propagates with the optical field over the entire length in each MZ-interferometer arm. The input and output ends of the MZ interferometer are connected to two fiber pigtails so that the modulator acts as an all-fiber device from a practical standpoint.

6.3 Polymer-Based Modulators

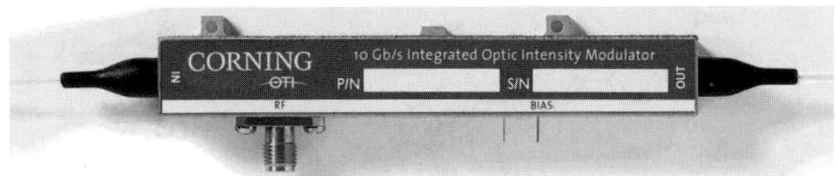

Figure 6.10: Photograph of a commercially available LiNbO$_3$ modulator capable of operating at 10 Gb/s. (Courtesy Corning Inc.)

The design of a traveling-wave modulator requires optimization of several parameters such as the width and height of the electrodes, their spacing, and the thickness of the buffer layer used for velocity matching. The main problem is that the refractive index of LiNbO$_3$ is about 2.2 at optical frequencies but increases to more than 6 at microwave frequencies. The buffer layer is made of silica and thus has a lower refractive index of about 1.9 at microwave frequencies. Its main function is to lower the effective index of the transmission line in which the RF signal propagates so that its velocity matches that of the optical signal. Typically, the thickness of the buffer layer is ∼1 μm, while the width and the spacing of electrodes exceed 10 μm. Even then, losses of the RF signal limit the speed of all traveling-wave LiNbO$_3$ modulators. Microwave losses α_m scale with the modulation frequency f_m as $\alpha_m = C_m\sqrt{f_m}$, where $C_m \approx 0.5$ dB/(cm-GHz$^{1/2}$) under typical conditions [12]. As a result, microwave power is reduced by 3 dB even for a 2-cm-long transmission line at frequencies close to 9 GHz.

The bandwidth of a modulator can be defined as the highest microwave frequency at which the modulated optical power is reduced by 3 dB compared with the value obtained at low modulation frequencies. It is easy to realize modulation bandwidths of 8–10 GHz with a proper design, and digital modulators capable of operating at 10 Gb/s have been available commercially since 1995 or so. Figure 6.10 depicts a commercially available 10-Gb/s packaged modulator. Such modulators are quite compact (width about 1.5 cm and length around 12 cm). The drive voltage is typically 5 V but can be reduced to below 3 V with a suitable design [17]. In laboratory experiments, especially designed LiNbO$_3$ modulators have been operated at frequencies exceeding 100 GHz using z-cut substrates [14]. Such high frequencies are difficult to realize in commercial devices. Nonetheless, digital modulators operating at bit rates of 40 Gb/s became available commercially around 2000, although drive voltage typically falls in the range of 10–20 V at such high frequencies. The voltage can be reduced to near 5 V by realizing nearly perfect velocity and impedance matching [16]. It is important to note that such modulators do not exhibit a 3-dB bandwidth of 40 GHz, and typically optical response drops by 6 dB or so at frequencies close to 40 GHz.

6.3 Polymer-Based Modulators

Several other electro-optic materials have been used to make external modulators. Among them, modulators making use of polymer waveguides have attracted the most attention [19]–[25]. Such modulators offer a number of potential advantages over

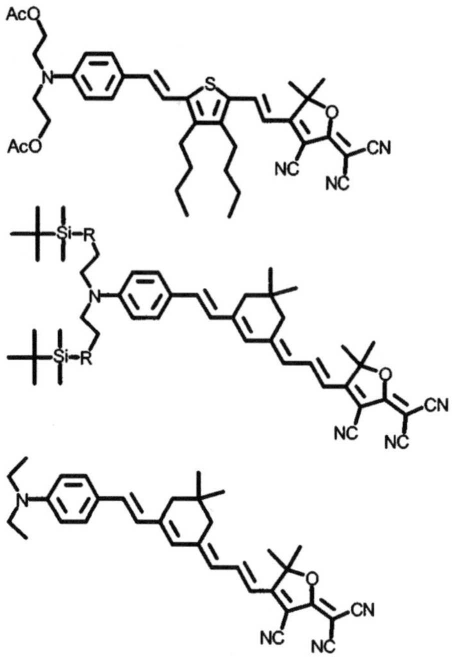

Figure 6.11: Chemical structure of three commonly used chromophores known as FTC, CLD-1, and CLD-72. (After Ref. [22]; ©2001 IEEE.)

LiNbO$_3$ modulators including a low dielectric constant at microwave frequencies, high electro-optic coefficient, low operating voltage, high modulation bandwidth, and low cost. This section describes the properties of polymer-based modulators.

Most polymers do not exhibit the Pockels effect in their natural state and thus require incorporation of optically nonlinear *chromophores* into a polymer material before it can be used for making a modulator. The two form a guest–host system somewhat similar to erbium-doped fibers in the sense that the chromophore molecules are not attached to the polymer backbone. Figure 6.11 shows the chemical structure of the three most commonly used chromophores known as FTC, CLD-1, and CLD-72. The electro-optic coefficient r_{33} for these chromophores has values of about 57, 90, and 126 pm/V at 1,060 nm [22], respectively, all of them exceeding the 31-pm/V value occurring for LiNbO$_3$ crystals. However, before these chromophores exhibit the electro-optic effect, they must be aligned by applying a poling field. The poling process aligns the dipole moments of individual chromophore molecules and produces a permanent dipole moment responsible for the electro-optic effect.

6.3.1 Device Fabrication

As discussed in Section 4.3.5, polymer waveguides are fabricated using a three-layer structure in which the central core layer made with one polymer is sandwiched between two cladding layers made using another polymer with a lower refractive index. The

6.3 Polymer-Based Modulators

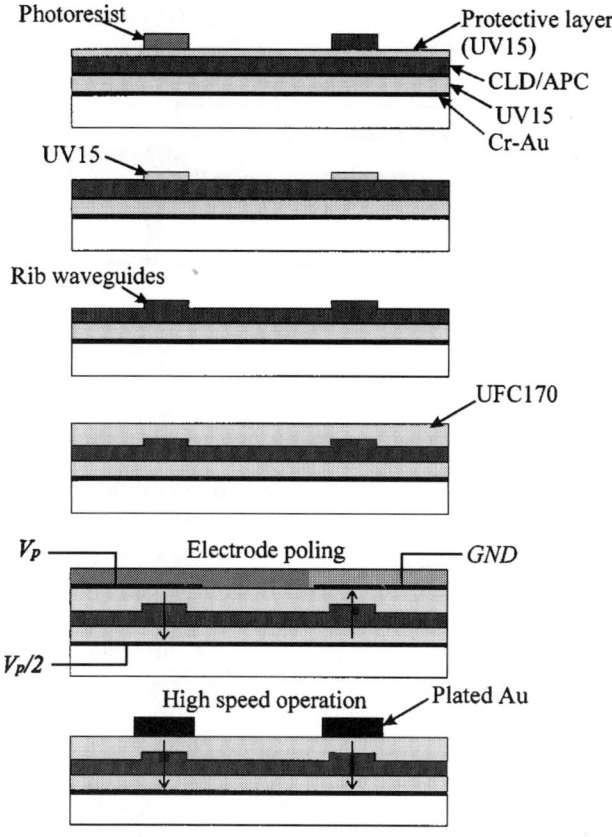

Figure 6.12: Schematic illustration of the main steps involved in fabricating a polymeric modulator. (After Ref. [22]; ©2001 IEEE.)

core layer in the case of a modulator is made by doping one of the chromophores into a polymer host, typically PMMA (the same polymer used for plastic fibers) or an amorphous polycarbonate. The bottom cladding layer is made with a UV-curable epoxy (UV15). The upper cladding layer requires a different polymer because it should not contain any solvent or small molecules, both of which tend to dissolve the core layer. A polymer known as UFC170 has been found suitable for this purpose [22]. It can be cured with a relatively small dosage of UV light to form a thin but hard polymer film on top of the core layer. The refractive indices for such a three-layer waveguide are 1.504, 1.612, and 1.488 starting from the bottom cladding layer.

Figure 6.12 shows the main steps involved in fabricating a polymeric modulator based on such a waveguide [22]. First, two metal layers (made of chromium and gold) are deposited on top of the substrate; these layers constitute the bottom electrode. The polymer UV15 is then spin-coated on top of the electrode and cured using a UV lamp to form the bottom cladding layer. The core layer (with a thickness of 2–3 μm) is made by spin-coating the solution containing the mixture of the chromophore and the host

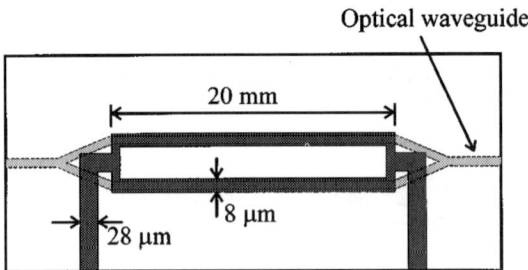

Figure 6.13: Schematic illustration of the RF strip line (black region) together with the polymeric waveguides (shaded gray) used to form the MZ interferometer. (After Ref. [22]; ©2001 IEEE.)

polymer, drying it in a vacuum, and baking it at 120°C. A thin layer of UV15 polymer (with a thickness of about 0.5 μm) is deposited on top of the core layer to protect it during further processing. A photoresist layer is then placed on top of this layer.

A photolithographic technique in combination with reactive ion etching is used to etch the core layer partially, except in a central region whose width controls the modes associated with the resulting rib waveguide. The rib width and height are typically 5 μm and 0.5 μm, respectively. The rib waveguides are patterned to form the MZ configuration needed for amplitude modulation. After etching, a top cladding layer (with a thickness of 2–3 μm) is formed on top of the entire structure using the UFC170 polymer. In a variation of the rib design, a narrow channel is first formed in the bottom cladding layer before depositing the core layer containing chromophore molecules. The resulting structure is referred to as the inverted-rib waveguide [25]. Its use simplifies the fabrication procedure and also helps to reduce insertion losses.

The next step consists of poling the core layer to align the chromophore molecules. For this purpose, two gold electrodes are placed on top of the cladding layer and covered with a polymer layer. For the push–pull operation, the two arms of the MZ interferometer are poled in opposite directions by reversing the direction of the poling field. This can be realized through the voltage scheme shown in Figure 6.12. One of the top electrodes is kept at the poling voltage V_p, while the other is grounded. At the same time, the bottom electrode has a voltage of $V_p/2$, resulting in equal but opposite electric fields in the two waveguides. Poling is usually performed in a nitrogen atmosphere to eliminate the losses incurred when poling is done in the air and carried out for a duration of 30 minutes or so at a temperature of about 150°C [22]. Poling also needs a significant electric field strength ($\sim$50 V/μm) in each waveguide, thus requiring an external voltage as high as 800 V. After poling is finished, the poling electrodes are removed through chemical etching.

The final step requires the construction of an RF strip line for realizing traveling-wave operation of the modulator. Again, photolithography is used to etch the pattern into a photoresist and then the top electrode is formed using a gold layer (with a thickness of 3 μm or so). Figure 6.13 shows the structure of the RF strip line together with the polymeric waveguides (shaded gray) in the MZ interferometer configuration. Similar to the case of LiNbO$_3$ modulators, microwave losses along the strip line determine

6.3 Polymer-Based Modulators

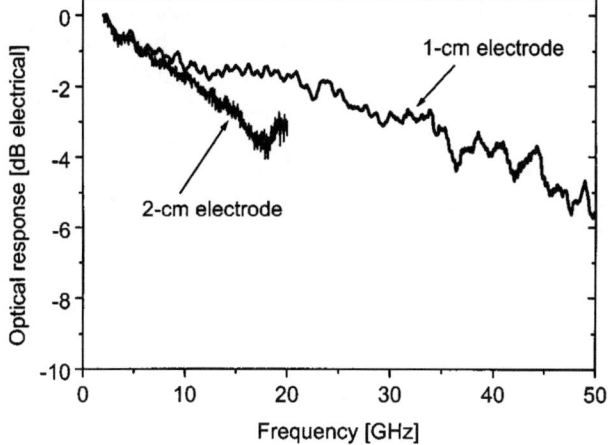

Figure 6.14: Frequency response of a polymeric modulator for two devices of different lengths. (After Ref. [22]; ©2001 IEEE.)

the modulation bandwidth of a traveling-wave polymer modulator.

6.3.2 Device Performance

An important parameter for any optical device is the insertion loss. In general, polymer-based modulators exhibit higher insertion losses than those made with LiNbO$_3$. Insertion losses depend not only on the length but also on the width of the waveguide because they represent a combination of coupling losses (which depend on the width through mode size) and propagation losses (which scale with the length). For a 3-μm-wide and 3-cm-long core, total losses typically exceed 9 dB. However, insertion losses of less than 5 dB have been realized by reducing the length to 2 cm and increasing the core width to 5 μm. Such a wide waveguide helps in reducing coupling losses because the mode size becomes comparable to that of a fiber, but supports multiple optical modes, an undesirable feature from the device standpoint. However, the higher-order modes are discriminated by the rib nature of the rectangular waveguide. As a result, the propagation of input light occurs predominantly through the fundamental mode.

Another important device parameter for any modulator is the voltage V_π required to induce a relative π phase shift in two arms of the MZ interferometer. A larger value of the electro-optic coefficient for polymer modulators translates into a lower value of V_π for them compared with LiNbO$_3$ modulators. The product $V_\pi L$ acts as a figure of merit for both types of modulators. Its values were measured to be as small as 2.8 V-cm at a wavelength of 1.3 μm and 4.2 V-cm at a wavelength of 1.55 μm [22]. In a 2001 experiment, a polymeric modulator required only 1.8 V for shifting the phase of a 1.55-μm signal by π in one of the MZ interferometer arms [21]. The device was 3 cm long and exhibited about 5-dB chip losses. The extinction ratio for digital applications is typically better than 20 dB for such modulators.

The modulator bandwidth is the most important parameter for high-speed modulators. Similar to the case of LiNbO$_3$ modulators, the bandwidth is limited only by electronics. An advantage of polymeric materials is that their refractive indices at optical and microwave frequencies are not as far apart as for LiNbO$_3$. This feature makes it much easier to match the velocities of microwave and optical signals. Figure 6.14 shows the measured frequency response over a range of 2–50 GHz for two devices with the electrode length of 1 and 2 cm [22]. The 1-cm-long device has a bandwidth of 34 GHz but this bandwidth is reduced to below 15 GHz for 2-cm-long devices. These results show that the bandwidth is limited by microwave losses of $\sim$0.7 dB/(cm-GHz$^{1/2}$). Several designs have been proposed for reducing microwave losses [24]. For a well-designed strip line, the losses should be below 0.5 dB/(cm-GHz$^{1/2}$), resulting in much higher bandwidths [22]. Such modulators can be operated at frequencies approaching 100 GHz if one is willing to tolerate higher losses. As early as 1997, such a device was modulated at frequencies as high as 110 GHz [20].

6.4 Electroabsorption Modulators

Electroabsorption modulators (EAMs) make use of the quantum-confined Stark effect discussed in Section 6.1 and developed since the 1908s [26]–[36]. The basic design of such a modulator, shown schematically in Figure 6.15, is similar to that of a semiconductor laser. Both devices make use of a semiconductor core layer, often consisting of multiple quantum wells, and sandwiched between the p-type and n-type cladding layers with higher bandgaps. The main difference is that the bandgap of quantum wells in the case of an EAM is chosen to be slightly larger than the photon energy at the incident wavelength. As a result, light is almost completely transmitted when the p–n junction is not biased. However, when the bandgap of the core layer is reduced electronically by applying an external voltage in the reverse-bias configuration, the input signal is absorbed in the core layer, and virtually no light comes out of the modulator. By changing the voltage rapidly at the bit rate, the output optical signal takes the form of a bit stream that mimics the electrical bit stream. With proper design, an extinction ratio of 15 dB or more can be realized in modern EAMs for an applied reverse bias of a few volts at bit rates of up to 40 Gb/s. Although some chirp is still imposed on coded pulses, it can be made small enough not to be detrimental to the system performance.

6.4.1 Temporal Response

An EAM acts as an optical device whose absorbtion α can be modified by applying an external voltage V across it. Thus, if a CW optical signal with power P_0 is incident at the input end of an EAM of length L_m, the output power is given by

$$P(V) = P_0 \exp[-\alpha(V)L_m]. \qquad (6.4.1)$$

The temporal response of the modulator is governed by this equation as it shows how the output power changes with time as $V(t)$ changes. An EAM can respond to voltage changes almost instantaneously, because of the fast response associated with the underlying quantum-confined Stark effect, as long as the charge carriers generated during

6.4 Electroabsorption Modulators

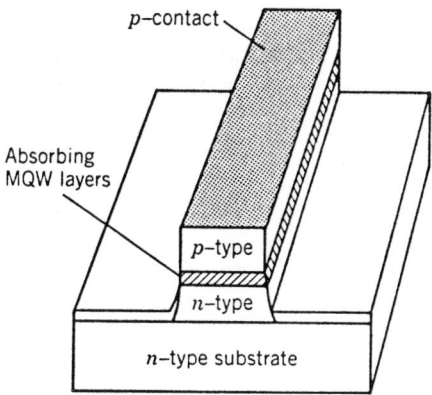

Figure 6.15: Schematic of an electroabsorption modulator designed using InGaAsP quantum wells.

absorption can be removed from the active region fast enough. Under typical biasing conditions, this can occur for electrons at time scales below 1 ps. In MQW devices, charge pileup can increase this time to 10 ps for holes, but this is still fast enough that EAMs can be operated at bit rates of 40 Gb/s, provided they are not limited by the electronics of the driving circuit.

Figure 6.16 shows how the transmissivity $P(V)/P_0$ changes with V for an EAM designed to operate in the 1.55-μm spectral region using InGaAsP quantum wells. The wavelength dependence of α limits the useful wavelength range of an EAM to 20 nm or so. However, within this wavelength range, the transmissivity of an EAM can be changed by more than 20 dB by applying 3 V across the device.

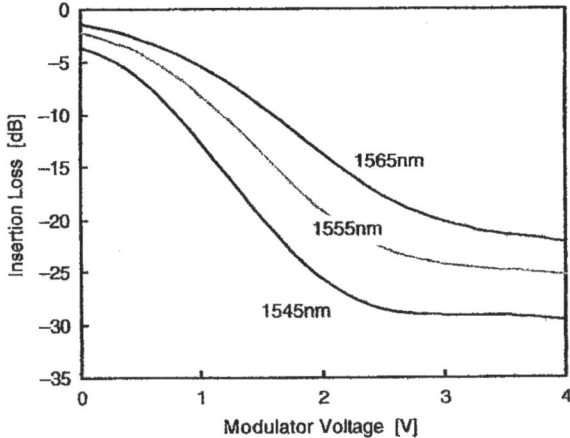

Figure 6.16: Transmissivity of an electroabsorption modulator as a function of the applied reverse bias V at three different wavelengths. (After Ref. [35]; ©2002 Elsevier.)

Equation (6.4.1) can be used to define several important performance parameters for EAMs. In the digital case, $V(t)$ is changed between two values, say, V_0 and V_1, to produce outputs P_0 and P_1 for 0 and 1 bits, respectively. The on-state power P_1 is not necessarily equal to the input power P_{in} because of insertion losses associated with the modulator. The amount of insertion loss can be written in decibels as

$$\alpha_{ins} = -10\log_{10}(P_1/P_{in}). \tag{6.4.2}$$

Similarly, the minimum power P_0 in the off state may not be zero for a digital modulator. The extinction ratio or the on–off ratio can be written in decibels as

$$r_{ex} = -10\log_{10}(P_0/P_1) \approx 4.343 L_m[\alpha(V_0) - \alpha(V_1)]. \tag{6.4.3}$$

Modern EAM can provide an on–off ratio in excess of 100 (20 dB) while maintaining insertion losses below 5 dB. Both the insertion loss and the extinction ratio of an EAM are strongly wavelength-dependent, as seen clearly from Figure 6.16.

Equation (6.4.1) only accounts for the voltage-induced power variations. As was the case for semiconductor lasers in Section 5.3.3, amplitude modulation in all EAMs is accompanied by phase modulation even though the physical origin is somewhat different. In the case of lasers, time-dependent changes in the carrier density lead to changes in the refractive index. In the case of an EAM, changes in the absorption spectrum produce changes in the refractive index through the Kramers–Kronig relation

$$\Delta n(\omega) = \frac{c}{\pi}\int_0^\infty \frac{\Delta\alpha(\omega')}{\omega'^2 - \omega^2}d\omega', \tag{6.4.4}$$

where Δ denotes voltage-induced changes. Even though the origins are different, we can still introduce the chirp parameter β_c using the general definition

$$\beta_c = \Delta\beta_r/\Delta\beta_i, \tag{6.4.5}$$

where β_r and β_i are the real and imaginary parts of the propagation constant at the optical frequency ω, defined as $\beta = \bar{n}\omega/c + i\alpha/2$, and $\Delta\beta$ denotes voltage-induced changes in β. Fortunately, $|\beta_c|$ is not very large for EAMs, with typical values being close to 1. Moreover, it can be controlled by using strained quantum wells [36].

6.4.2 Design and Performance

Similar to the case of LiNbO$_3$ modulators, the modulation speed of EAMs is generally limited by electronics. For lightwave systems operating at bit rates of 10 Gb/s or less, one can apply the electric field along the entire length of the MQW waveguide and modulate the applied voltage to generate the optical bit stream. In this conventional design, the modulation bandwidth is limited by the RC bandwidth of the driving circuit. Much higher bandwidths can be realized by using the traveling-wave design in which the microwave signal propagates along the length of the waveguide over a strip line.

Figure 6.17 shows the design of an EAM fabricated with the lumped RC-circuit approach [28]. The central modulation region is designed to be relatively short ($\sim$100 μm)

6.4 Electroabsorption Modulators

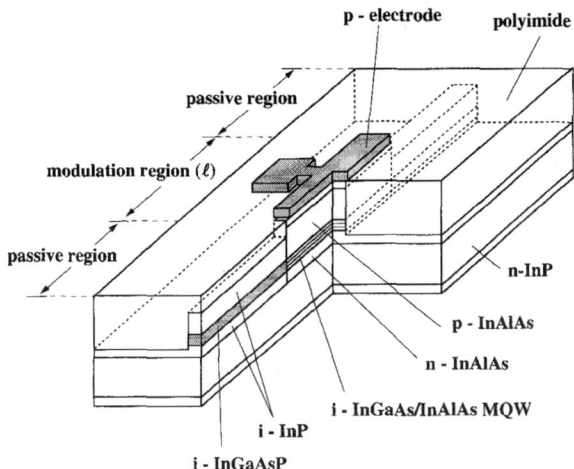

Figure 6.17: Schematic of a MQW electroabsorption modulator integrated with two passive waveguides at the input and output ends. (After Ref. [28]; ©1996 IEEE.)

to reduce the capacitance of the device that limits the modulator bandwidth. It is integrated on each side with the input and output passive waveguides so that the total length of the modulator is close to 1 mm.

The MQW region in Figure 6.17 consists of ten 8-nm-thick $In_{0.48}Ga_{0.52}As$ quantum-well layers separated by 5-nm barrier layers of $In_{0.60}Ga_{0.40}As$. The quantum wells exist under tensile strain (because of 0.35% lattice mismatch), which helps to reduce the applied voltage [28]. The barrier layers, in contrast, are under compressive strain (0.5% lattice mismatch), which helps to increase the saturation power. The passive-waveguide layers are regrown after removing portions of the epitaxial layers through reactive ion etching. Polymide is used to reduced the residual capacitance of the bonding pad. Both facets of the modulator are coated with antireflection layers to avoid forming a Fabry–Perot cavity. Insertion losses of such a modulator depend on both the width w_m and length l_m of the MQW region and were measured to be in the range of 6–9 dB for $w_m = 2.5$ to 3.5 μm and $l_m = 100$ μm. They scale linearly with l_m and inversely with w_m.

Both the extinction ratio and the modulator bandwidth for such an EAM also depend on the length l_m of the modulation section. Figure 6.18(a) shows how the transmissivity changes with the applied voltage, while Figure 6.18(b) shows the frequency response for $l_m = 50$, 100, and 150 μm for EAMs designed with $w_m = 2.5$ μm. As one may expect, the extinction ratio is reduced from more than 30 dB to close to 10 dB as l_m is reduced from 150 to 50 μm. At the same time, however, the modulation bandwidth increases from 20 GHz to 33 GHz. The chirp parameter was also found to be relatively small ($\beta_c = 0.7$) for short-length EAMs. Clearly, such modulators can be operated at speeds as high as 40 Gb/s at the expense of a reduced extinction ratio.

The traveling-wave EAMs can provide large bandwidth without sacrificing the extinction ratio [30]–[33]. In this approach, the microwave signal is copropagated

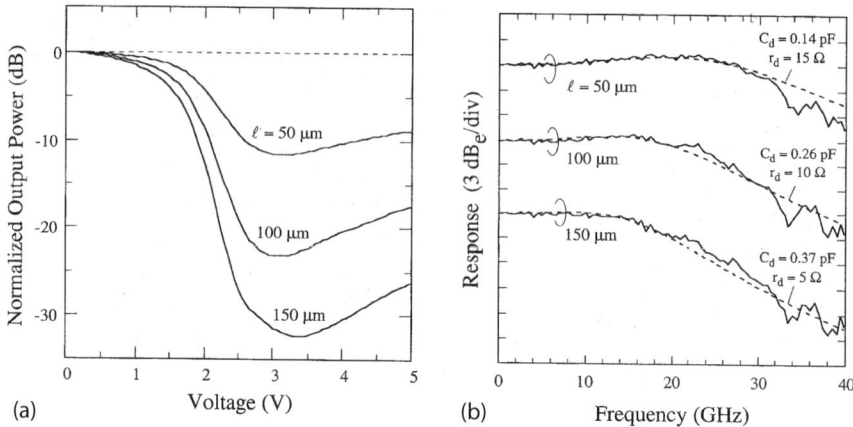

Figure 6.18: Changes in (a) extinction ratio and (b) frequency response with the length of the modulation region for EAMs with integrated waveguides. Frequency response was measured at a reverse bias of 1.5 V. Dashed curves show the calculated frequency response for the given values of device resistance and capacitance. (After Ref. [28]; ©1996 IEEE.)

with the optical signal using a coplanar-waveguide (CPW) electrode design. Figure 6.19 shows two scanning-electron-microscope (SEM) photographs of such a modulator [31]. Figure 6.19(a) shows the CPW line used to guide the microwave signal over the optical waveguide that carries the optical field. Figure 6.19(b) shows the 2-μm-wide ridge that contains the optical waveguide. The active region consists of 10 MQWs of 10.4-nm thickness (under 0.37% tensile strain) separated by 7.6-nm-thick barrier layers under 0.5% compressive strain, all made using the InGaAsP quaternary alloy. The bandgap of MWQs correspond to a wavelength close to 1.5 μm in the absence of the microwave signal so that little light is absorbed in the 1.55-μm region. Such a 3-μm-long device required only 1.2 V for realizing a 20-dB extinction ratio and had a bandwidth of about 25 GHz with a load of 32 Ω. Bandwidths larger than 50 GHz have been realized for traveling-wave EAMs while keeping the driving voltage below 2 V [30].

6.4.3 Integration with Lasers

The chief advantage of EAMs is that they are made using the same semiconductor material that is used for making lasers, and thus the two can easily be integrated on the same chip. This approach reduces the insertion losses associated with the modulator since it is much easier to couple light between two semiconductor waveguides than from an optical fiber to a semiconductor waveguide or vice versa. Moreover, a semiconductor optical amplifier can also be integrated on the same chip to compensate for residual losses. Motivated by these reasons, the development of modulator-integrated DFB or DBR lasers was pursued actively during the 1990s [37]–[48]. By 1999, 10-Gb/s optical transmitters with an integrated EAM were available commercially and were used routinely for WDM lightwave systems [43]. By 2001, such integrated mod-

6.4 Electroabsorption Modulators

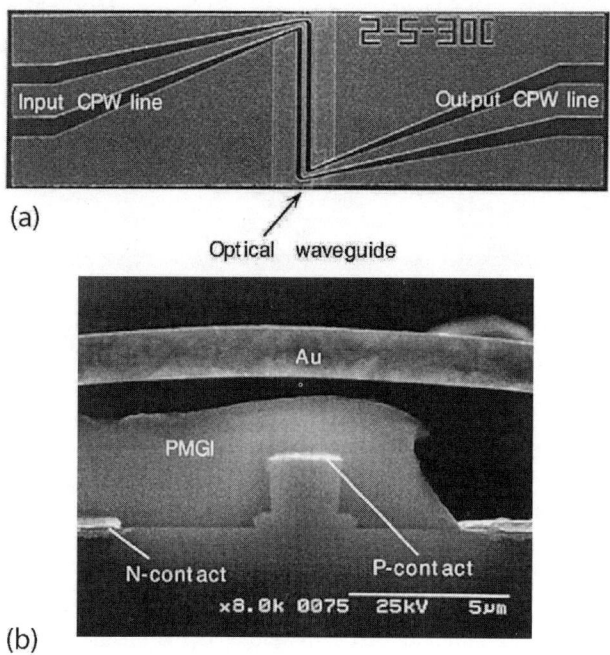

Figure 6.19: SEM photographs showing (a) top view of the CPW electrode structure that guides the microwave signal and (b) facet view of the EAM showing the 2-μm-wide ridge that contains optical waveguide. (After Ref. [31]; ©1999 IEEE.)

ulators exhibited a bandwidth of more than 50 GHz and had the potential to operate at bit rates of up to 100 Gb/s [44].

EAMs can also be used to generate ultrashort pulses suitable for optical time-division multiplexing. A DFB laser, integrated monolithically with a MQW modulator, was used as early as 1993 to generate a 20-GHz pulse train [37]. The 7-ps output pulses were nearly transform-limited because of an extremely low chirp associated with the modulator. By 1999, a 40-GHz train of 1.6 ps pulses was produced using an EAM. Such short pulses can be used for time-division-multiplexed systems operating at bit rates of up to 160 Gb/s [40].

Figure 6.20 shows the basic idea behind modulator-integrated DFB lasers schematically. The DFB laser on the left provides the CW signal at a fixed wavelength (determined by the grating) that is modulated by the EAM on the right. The middle section is designed to isolate the two devices electrically while inducing minimum losses. The facets of the whole device are coated such that the left facet has a high reflectivity (>90%), while the right facet has as low reflectivity as possible (<1%).

The fabrication of modulator-integrated lasers requires attention to many details. In general, the active core layers in the laser and modulator sections should be made using different compositions with different bandgaps so that they can be optimized for each device separately. Two different approaches are used for this purpose. In one scheme,

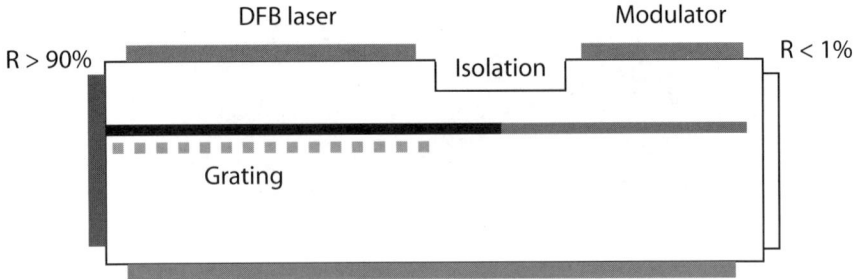

Figure 6.20: Schematic of a modulator-integrated DFB laser. The DFB laser on the left provides the CW light that is modulated by the EAM on the right. The middle section is designed to isolate the two devices electrically with minimum losses.

the waveguides for the laser and modulator are butt-joined using separate epitaxial growth steps for each of them. First, the layers are grown for one device, say the laser. Then a mask is used to remove the epitaxial layers from the modulator region, and new layers are regrown. Although this approach offers the maximum flexibility for optimizing each device separately, the vertical alignment of the layers in the two sections is relatively difficult and affects the yield. In the much simpler selective-area-growth technique, both devices (laser and modulator) are formed in a single epitaxial growth, but the oxide pads placed on the wafer prior to growth allow one to shift the laser wavelength toward the red side by more than 100 nm. The wavelength shift results from a change in the Bragg wavelength of the laser grating occurring because of changes in the effective mode index induced by the oxide pads. This technique is commonly used in practice for making modulator-integrated lasers.

The performance of modulator-integrated DFB lasers is limited by the optical and electrical crosstalk between the laser and modulator sections. Typically, the separation between the electrical contacts used for the two devices is less than 0.2 mm. Any leakage from the modulator contact to the laser contact can change the dc bias of the laser in a periodic manner. Such unwanted laser-current changes shift the laser wavelength and produce chirp because the laser frequency changes with time. Since laser frequency can shift by more than 200 MHz/mA, the middle section should provide an isolation impedance of 800 Ω or more [35]. Although such values are easily realized at low modulation frequencies, this level of isolation is difficult to achieve at microwave frequencies approaching 40 GHz. In one approach, the FM efficiency of the laser is controlled by reducing the chirp parameter β_c for the laser.

The optical crosstalk between the laser and modulator results from the residual reflectivity of the output facet (see Figure 6.20). This residual reflectivity is seen by the laser only when the modulator is on because in the off state, the laser light is totally absorbed by the modulator before it reaches the output facet. As a result, the laser gain, and hence the emission wavelength, are slightly different during each on–off cycle of the modulator. This is an additional source of frequency chirping. It can be nearly eliminated if the front facet has a residual reflectivity of less than 0.01%. As discussed in Section 5.5 within the context of semiconductor optical amplifiers, it is

6.4 Electroabsorption Modulators

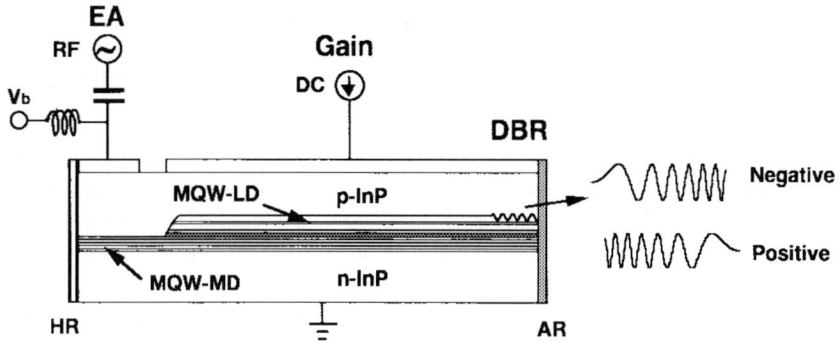

Figure 6.21: Schematic of a modulator-integrated mode-locked laser consisting of an MQW modulator section (left), an MQW amplifier section (middle), and a chirped-grating section on the right. (After Ref. [41]; ©1999 IEEE.)

hard to realize antireflection coatings of this quality in practice.

In general, frequency chirp associated with the laser and modulator sections is a limiting factor for modulator-integrated DFB lasers. Typically, the chirp parameter β_c exceeds 2 in the on state and changes to below -2 in the off state when the reverse voltage of about 3 V is applied. In a novel approach, chirp was reduced by designing the quantum wells of the modulator to be relatively shallow [47]. More specifically, the bandgap difference between the barrier and quantum-well layers was reduced from 0.2 to close to 0.1 eV. The measured values of β_c were below 0.7 for such devices in the entire 0–3 V range of the reverse bias, resulting in improved performance when the device was used in a lightwave system operating at 10 Gb/s. Physically speaking, frequency chirp is due to changes in the refractive index occurring because of the pileup of electrons and holes inside the quantum wells. Since the carrier escape time is reduced considerably in shallow quantum wells, the carrier density does not build up to high values, resulting in a lower chirp.

Integration of DBR lasers with an EAM provides certain advantages and is being pursued for realizing tunable optical sources. In one experiment, a four-section DBR laser fabricated with a sampled grating was integrated with a modulator and an amplifier, resulting in a six-section device that was tunable over 40 nm while maintaining an extinction ratio better than 10 dB [48]. The integration of a modulator with the DBR laser can be used for other other purposes as well. Figure 6.21 shows the design of a DBR laser in which the integrated modulator is used to actively mode-lock the laser for generating short optical pulses at a 40-GHz repetition rate [41]. The device has three sections working from left to right as a modulator, an amplifier, and a chirped grating. The grating is chirped such that it nearly cancels the chirp provided by the amplifier and modulator. The total device length is about 1.1 mm. The grating section in such a mode-locked laser provides both the feedback and dispersion while the amplifier section provides both the gain and self-phase modulation. The device can generate chirp-free pulses as short as 4 ps, while its repetition rate is tunable over the range of 38–41 GHz.

Problems

6.1 What is the difference between direct and external modulation? Sketch an example of each modulation scheme.

6.2 Consider an optical carrier with the electric field $E(t) = A\cos(\omega_0 t + \phi)$, where A, ω_0, and ϕ are constants. The amplitude A is modulated at a frequency ω_m such that $A = A_0 + a_m \cos(\omega_m t)$. Prove that the CW carrier develops modulation sidebands on each side of the carrier frequency. What are the amplitudes and frequencies of these sidebands?

6.3 In Problem 6.2, the phase ϕ (rather than amplitude) is modulated at a frequency ω_m such that $\phi = \phi_0 + a_m \cos(\omega_m t)$. Show that the CW carrier develops multiple modulation sidebands on each side of the carrier frequency. Find expressions for the amplitudes and frequencies of these sidebands.

6.4 Solve the Schrödinger equation for a free electron moving in the presence of an electric field applied along the z axis and write the solution in terms of the Airy functions.

6.5 In a LiNbO$_3$ crystal, the electric field E is applied along the crystallographic x axis. Use the concept of index ellipsoid to determine changes in the refractive index along the three crystallographic axes. What changes, if any, occur when the field is applied along the y axis?

6.6 Consider an InP waveguide in which 1.55-μm light propagates and an electric field E is applied along the crystallographic z axis along which light propagates. How much does the mode index change along the three axes? How would you use these index changes for making a modulator?

6.7 In a LiNbO$_3$ modulator designed to operate at 1.55 μm, an electric field E is applied along the crystallographic z axis. Find an expression for the index change and use it to calculate the voltage V_π for a 4-cm-long waveguide, assuming that the electrodes are separated by 10 μm. Use $r_{33} = 30.9$ pm/V and 2.2 for the mode index. Also assume perfect overlap between the optical end electric fields.

6.8 Consider a LiNbO$_3$ modulator in the symmetric Mach–Zehnder configuration realized using two 3-dB directional couplers. A voltage V is applied along one arm of length L. Derive an expression for the power $P(V)$ exiting from the bar port using the transfer-matrix approach.

6.9 Repeat Problem 6.8 for the push–pull design of a LiNbO$_3$ modulator. How would you use such a modulator in the chirp-free mode?

6.10 What is meant by a traveling-wave modulator? Sketch a design for such a modulator and explain how it works.

6.11 How can a polymer exhibit the electro-optic effect? Discuss the fabrication steps involved in making polymer modulators.

6.12 Explain how an electroabsorption modulator functions. What physical mechanisms can change the absorption of a semiconductor under the action of an applied electric field? Consider both the bulk and quantum-well designs.

References

[1] G. P. Agrawal and N. K. Dutta, *Semiconductor Lasers*, 2nd ed., Van Nostrand Reinhold, New York, 1993.
[2] W. Franz, *Z. Naturforsch.* **13A**, 484 (1958).
[3] L. V. Keldysh, *Sov. Phys. JETP* **7**, 788 (1958).
[4] H. Shen and F. H. Pollak, *Phys. Rev. B* **42**, 7097 (1990).
[5] S. L. Chuang, *Physics of Optoelectronic Devices*, Wiley, New York, 1995.
[6] J. D. Dow and D. Redfield, *Phys. Rev. B* **1**, 3358 (1970).
[7] D. A. B. Miller, D. S. Chemla, T. C. Damen, A. C. Gossard, W. Wiegmann, T. H. Wood, and C. A. Burrus, *Phys. Rev. B* **32**, 1043 (1985).
[8] S. Schmitt-Rink, D. S. Chemla, and D. A. B. Miller, *Adv. Phys.* **38**, 89 (1989).
[9] R. W. Boyd, *Nonlinear Optics*, 2nd ed., Academic Press, San Diego, CA, 2003.
[10] C. R. Pollock, *Fundamentals of Optoelectronics*, Irwin, Chicago, 1995.
[11] L. Thylen, *J. Lightwave Technol.* **6**, 847 (1988).
[12] F. Heismann, S. K. Korotky, and J. J. Veselka, in *Optical Fiber Telecommunications*, Vol. 3B, I. P. Kaminow and T. L. Loch, Eds., Academic Press, San Diego, CA, 1997, Chap. 8.
[13] S. Hopfer, Y. Shani, and D. Nir, *J. Lightwave Technol.* **16**, 73 (1998).
[14] K. Noguchi, O. Mitomi, and H. Miyazawa, *J. Lightwave Technol.* **16**, 615 (1998).
[15] E. L. Wooten, K. M. Kissa, A. Yi-Yan, E. J. Murphy, D. A. Lafaw, P. F. Hallemeier, D. Maack, D. V. Attanasio, D. J. Fritz, G. J. McBrien, and D. E. Bossi, *IEEE J. Sel. Topics Quantum Electron.* **6**, 69 (2000).
[16] M. M. Howerton, R. P. Moeller, A. S. Greenblatt, and R. Krahenbuhl, *IEEE Photon. Technol. Lett.* **12**, 792 (2000).
[17] A. Mahapatra and E. J. Murphy, *Optical Fiber Telecommunications*, Vol. 4A, I. P. Kaminow and T. P. Lee, Eds., Academic Press, San Diego, CA, 2002, Chap. 6.
[18] J. Kondo, A. Kondo, K. Aoki, M. Imaeda, T. Mori, Y. Mizuno, S. Takatsuji, Y. Kozuka, O. Mitomi, and M. Minakata, *J. Lightwave Technol.* **20**, 2110 (2002).
[19] Y. Shi, W. Wang, J. Bechtel, A. Chen, S. Garner, S. Kalluri, W. H. Steier, D. Chen, H. R. Fetterman, L. R. Dalton, and L. Yu, *IEEE J. Sel. Topics Quantum Electron.* **2**, 289 (1996).
[20] D. Chen, H. R. Fetterman, A. Chen, W. H. Steier, L. R. Dalton, W. Wang, and Y. Shi, *Appl. Phys. Lett.* **70**, 3335 (1997).
[21] H. Zhang, M. C. Oh, A. Szep, W. H. Steier, C. Zhang, L. R. Dalton, H. Erlig, Y. Chang, D. H. Chang, and H. R. Fetterman, *Appl. Phys. Lett.* **78**, 3136 (2001).
[22] M. C. Oh, H. Zhang, C. Zhang, H. Erlig, Y. Chang, B. Tsap, D. Chang, A. Szep, W. H. Steier, H. R. Fetterman, and L. R. Dalton, *IEEE J. Sel. Topics Quantum Electron.* **7**, 826 (2001).
[23] M. Lee, H. Katz, C. Erben, D. M. Gill, P. Gopalan, and J. D. Heber, *Science* **298**, 1401 (2002).
[24] D. M. Gill and A. Chowdhury, *J. Lightwave Technol.* **20**, 2145 (2002).
[25] S. K. Kim, H. Zhang, D. H. Chang, C. Zhang, C. Wang, W. H. Steier, and H. R. Fetterman, *IEEE Photon. Technol. Lett.* **15**, 218 (2003).
[26] T. H. Wood, *J. Lightwave Technol.* **6**, 783 (1988).
[27] F. Devaux, S. Chelles, A. Ougazzaden, A. Mircea, and J. C. Harmand, *Semiconduct. Sci. Technol.* **10**, 887 (1995).
[28] T. Ido, S. Tanaka, M. Suzuki, M. Koizumi, H. Sano, and H. Inoue, *J. Lightwave Technol.* **14**, 2026 (1996).

[29] S. Kaneko, M. Noda, Y. Miyazaki, H. Watanabe, K. Kasahara, and T. Tajime, *J. Lightwave Technol.* **17**, 669 (1999).

[30] K. Kawano, M. Kohtoku, M. Ueki, T. Ito, S. Kondoh, Y. Noguchi, and Y. Hasumi, *Electron. Lett.* **33**, 1580 (1997).

[31] S. Z. Zhang, Y. J. Chiu, P. Abraham, and J. E. Bowers, *IEEE Photon. Technol. Lett.* **11**, 191 (1999).

[32] G. L. Li, S. A. Pappert, P. Mages, C. K. Sun, W. S. C. Chang, and P. K. L. Yu, *IEEE Photon. Technol. Lett.* **13**, 1076 (2001).

[33] Y. J. Chiu, H. F. Chou, V. Kaman. P. Abraham, and J. E. Bowers, *IEEE Photon. Technol. Lett.* **14**, 792 (2002).

[34] S. Irmscher, R. Lewen, and U. Eriksson, *IEEE Photon. Technol. Lett.* **14**, 923 (2002).

[35] D. A. Ackerman, J. E. Johnson, L. J. P. Ketelsen, L. E. Eng, P. A. Kiely, and T. G. B. Mason, *Optical Fiber Telecommunications*, Vol. 4A, I. P. Kaminow and T. P. Lee, Eds., Academic Press, San Diego, CA, 2002, Chap. 12.

[36] Y. Miyazaki, H. Tada, S. Tokizaki, K. Takagi, T. Aoyagi, and Y. Mitsui, *IEEE J. Quantum Electron.* **39**, 813 (2003).

[37] M. Aoki, M. Suzuki, H. Sano, T. Kawano, T. Ido, T. Taniwatari, K. Uomi, and A. Takai, *IEEE J. Quantum Electron.* **29**, 2088 (1993).

[38] H. Takeuchi, K. Tsuzuki, K. Sato, H. Yamamoto, M. Itaya, A. Sano, M. Yoneyama, and T. Otsuji, *IEEE J. Sel. Topics Quantum Electron.* **3**, 336 (1997).

[39] T. Tanbun-Ek, L. E. Adams, G. Nykolak, C. Bethea, R. People, A. M. Sergent, P. W. Wisk, P. F. Sciortino, S. N. G. Chu, T. Fullowan, R. Pawelek, and W. T. Tsang, *IEEE J. Sel. Topics Quantum Electron.* **3**, 960 (1997).

[40] A. D. Ellis, R. J. Manning, I. D. Phillips, and D. Nesset, *Electron. Lett.* **35**, 645 (1999).

[41] K. Sato, A. Hirano, and H. Ishii, *IEEE J. Sel. Topics Quantum Electron.* **5**, 590 (1999).

[42] Y. Miyazaki, E. Ishimura, T. Aoyagi, H. Tada, K. Matsumoto, T. Takiguchi, K. Shimizu, M. Noda, T. Mizuochi, T. Nishimura, and E. Omura, *IEEE J. Quantum Electron.* **36**, 909 (2000).

[43] Y. Kim, S. K. Kim, J. Lee, Y. Kim, J. Kang, W. Choi, and J. Jeong, *Opt. Fiber Technol.* **7**, 84 (2001).

[44] Y. Akage, K. Kawano, S. Oku, R. Iga, H. Okamoto, Y. Miyamoto, and H. Takeuchi, *Electron. Lett.* **37**, 299 (2001).

[45] H. Kawanishi, Y. Yamauchi, N. Mineo, Y. Shibuya, H. Mural, K. Yamada, and H. Wada, *IEEE Photon. Technol. Lett.* **13**, 954 (2001).

[46] R. A. Salvatore, R. T. Sahara, M. A. Bock, and I. Libenzon, *IEEE J. Quantum Electron.* **38**, 464 (2002).

[47] Y. Miyazaki, H. Tada, T. Aoyagi, T. Nashimuraand, and Y. Mitsui, *IEEE J. Quantum Electron.* **38**, 1075 (2002).

[48] Y. A. Akulova, G. A. Fish, P. C. Koh, C. L. Schow, P. Kozodoy, A. P. Dahl, S. Nakagawa, M. C. Larson, M. P. Mack, T. A. Strand, C. W Coldren, E. Hegblom, S. K. Penniman, T. Wipiejewski, and L. C. Coldren, *IEEE J. Sel. Topics Quantum Electron.* **8**, 1349 (2002).

Chapter 7

Photodetectors

The role of an optical receiver is to convert the optical signal back into electrical form and recover the data transmitted through the lightwave system. Its main component is a photodetector that converts light into electricity through the photoelectric effect. The requirements for a photodetector are similar to those of an optical source. It should have high sensitivity, fast response, low noise, low cost, and high reliability. Its size should be compatible with the fiber-core size. These requirements are best met by photodetectors made using semiconductor waveguides. Such photodetectors have been developed since the 1970s and are used routinely for optical communication systems [1]–[8]. Section 7.1 introduces the basic concepts behind the photodetection process such as quantum efficiency, responsivity, and rise time. Section 7.2 focuses on photodetectors that make use of a reversed-biased p–n junction and are commonly used in practice. Avalanche photodiodes are considered in Section 7.3 where we discuss both the advantages and disadvantages of such devices. Section 7.4 is devoted to another design of photodetectors in which a semiconductor-absorbing layer is brought into contact directly with finger-shaped electrodes. The integration of a photodetector with other components required to make an optical receiver is considered in Section 7.5 with an emphasis on the role played by each component.

7.1 Basic Concepts

The fundamental mechanism behind the photodetection process is absorption of light by atoms under suitable conditions. This section introduces basic concepts such as responsivity, quantum efficiency, rise time, and bandwidth that are common to all photodetectors and are used to characterize them.

7.1.1 Responsivity and Quantum Efficiency

To introduce the basic concepts, it is useful to consider the simplest semiconductor-based photodetector shown schematically in Figure 7.1. It consists of a semiconductor slab with ohmic contacts deposited on its two facets so that an external voltage can

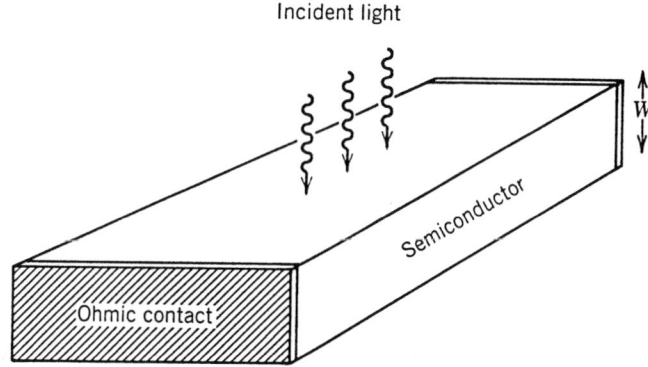

Figure 7.1: A semiconductor slab used as a photodetector.

be applied across the slab. Light to be detected is incident on the slab from the top as shown in Figure 7.1 and is partially absorbed along its width W. If the energy $h\nu$ of incident photons exceeds the bandgap energy, an electron–hole pair is generated each time a photon is absorbed by the semiconductor. Under the influence of the electric field created by the applied voltage, electrons and holes are swept across the semiconductor, resulting in a flow of electric current. The photocurrent I_p is directly proportional to the incident optical power P_{in}, that is,

$$I_p = R_d P_{\text{in}}. \tag{7.1.1}$$

The constant R_d is referred to as the *responsivity* of the photodetector because more current is produced at a given input power for larger values of R_d; it is expressed in units of ampere/watts (A/W).

The responsivity R can be written in terms of a more fundamental quantity η, called the *quantum efficiency* and defined as

$$\eta = \frac{\text{electron generation rate}}{\text{photon incidence rate}} = \frac{I_p/q}{P_{\text{in}}/h\nu} = \frac{h\nu}{q} R_d, \tag{7.1.2}$$

where Eq. (7.1.1) was used. The responsivity R_d is thus given by

$$R_d = \frac{\eta q}{h\nu} \approx \frac{\eta \lambda}{1.24}, \tag{7.1.3}$$

where the wavelength $\lambda \equiv c/\nu$ of incident light is expressed in micrometers. The responsivity of a photodetector increases with the wavelength λ simply because more photons are present for the same optical power. Such a linear dependence on λ is not expected to continue forever because eventually the photon energy becomes too small to generate electrons. In semiconductors, this happens for $h\nu < E_g$, where E_g is the bandgap. The quantum efficiency η then drops to zero.

The dependence of η on λ enters through the absorption coefficient α. If the facets of the semiconductor slab in Figure 7.1 are assumed to have an antireflection coating,

7.1 Basic Concepts

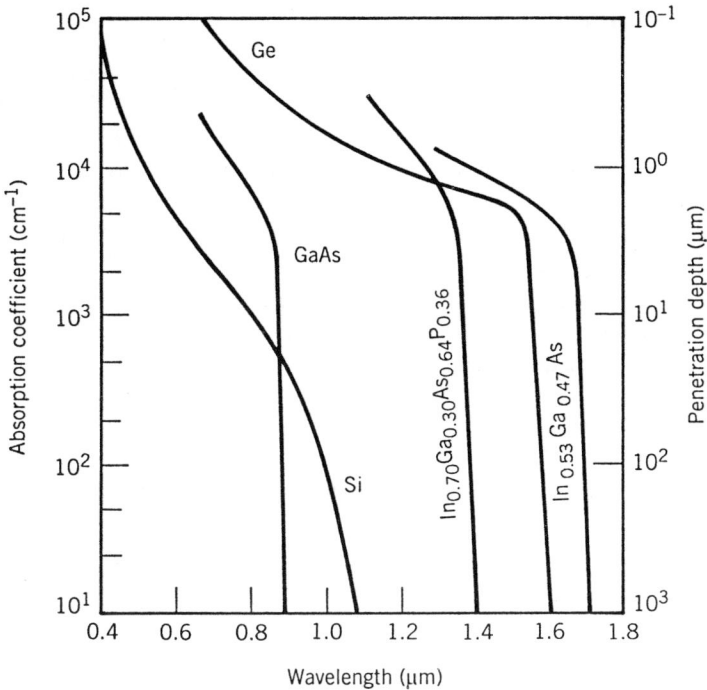

Figure 7.2: Wavelength dependence of the absorption coefficient for several semiconductor materials. (After Ref. [1]; ©1979 Academic Press.)

the power transmitted through the slab of width W is $P_{tr} = \exp(-\alpha W) P_{in}$. The absorbed power can be written as

$$P_{abs} = P_{in} - P_{tr} = [1 - \exp(-\alpha W)] P_{in}. \qquad (7.1.4)$$

Since each absorbed photon creates an electron–hole pair, the quantum efficiency η is given by

$$\eta = P_{abs}/P_{in} = 1 - \exp(-\alpha W). \qquad (7.1.5)$$

As expected, η becomes zero when $\alpha = 0$. On the other hand, η approaches 1 if $\alpha W \gg 1$.

Figure 7.2 shows the wavelength dependence of α for several semiconductor materials commonly used to make photodetectors for lightwave systems. The wavelength λ_c at which α becomes zero is called the cutoff wavelength, as that material can be used for a photodetector only for $\lambda < \lambda_c$. As seen in Figure 7.2, indirect-bandgap semiconductors such as Si and Ge can be used to make photodetectors even though the absorption edge is not as sharp as for direct-bandgap materials. Large values of α ($\sim 10^4$ cm^{-1}) can be realized for most semiconductors, and η can approach 100% for $W \sim 10$ μm. This feature illustrates the efficiency of semiconductors for the purpose of photodetection.

7.1.2 Rise Time and Bandwidth

The *bandwidth* of a photodetector is determined by the speed with which it responds to variations in the incident optical power. It is useful to introduce the concept of *rise time* T_r, defined as the time over which the current builds up from 10 to 90% of its final value when the incident optical power is changed abruptly. Clearly, T_r will depend on the time taken by electrons and holes to travel to the electrical contacts. It also depends on the response time of the electrical circuit used to process the photocurrent.

The rise time T_r of a linear electrical circuit is defined as the time during which the response increases from 10 to 90% of its final output value when the input is changed abruptly (a step function). When the input voltage across an RC circuit changes instantaneously from 0 to V_0, the output voltage changes as

$$V_{\text{out}}(t) = V_0[1 - \exp(-t/RC)], \tag{7.1.6}$$

where R is the resistance and C is the capacitance of the RC circuit. The rise time is found to be given by

$$T_r = (\ln 9)RC \approx 2.2\tau_{RC}, \tag{7.1.7}$$

where $\tau_{RC} = RC$ is the time constant of the RC circuit.

The rise time of a photodetector can be written by extending Eq. (7.1.7) as

$$T_r = (\ln 9)(\tau_{\text{tr}} + \tau_{RC}), \tag{7.1.8}$$

where τ_{tr} is the transit time and τ_{RC} is the time constant of the equivalent RC circuit. The transit time is added to τ_{RC} because it takes some time before the carriers are collected after their generation through absorption of photons. The maximum collection time is just equal to the time an electron takes to traverse the absorption region. Clearly, τ_{tr} can be reduced by decreasing W. However, as seen from Eq. (7.1.5), the quantum efficiency η begins to decrease significantly for $\alpha W < 3$. Thus, there is a trade-off between the bandwidth and the responsivity (speed versus sensitivity) of a photodetector. Often, the RC time constant τ_{RC} limits the bandwidth because of electrical parasitics. The numerical values of τ_{tr} and τ_{RC} depend on the detector design and can vary over a wide range.

The bandwidth of a photodetector is defined in a manner analogous to that of a RC circuit and is given by

$$\Delta f = [2\pi(\tau_{\text{tr}} + \tau_{RC})]^{-1}. \tag{7.1.9}$$

As an example, when $\tau_{\text{tr}} = \tau_{RC} = 100$ ps, the bandwidth of the photodetector is below 1 GHz. Clearly, both τ_{tr} and τ_{RC} should be reduced below 10 ps for photodetectors needed for lightwave systems operating at bit rates of 10 Gb/s or more.

Together with the bandwidth and the responsivity, the dark current I_d of a photodetector is the third important parameter. Here, I_d is the current generated in a photodetector in the absence of any optical signal and originates from stray light or from thermally generated electron–hole pairs. For a good photodetector, the dark current should be negligible ($I_d < 10$ nA).

7.2 Reverse-Biased p–n Junctions

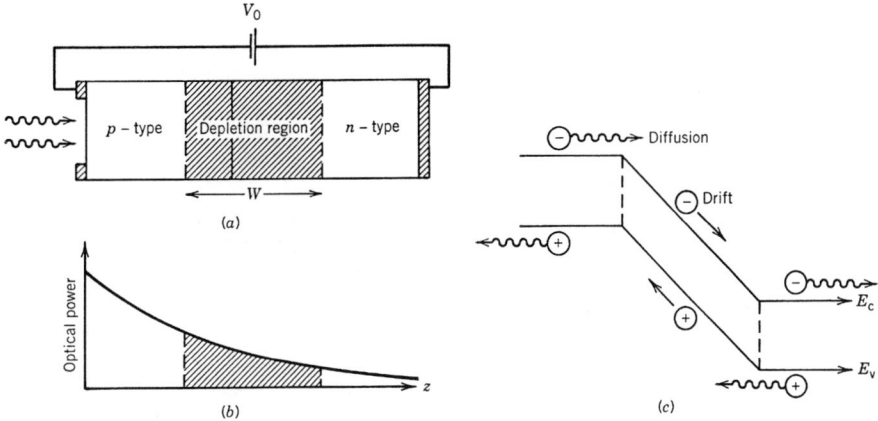

Figure 7.3: (a) A *p–n* photodiode under reverse bias; (b) variation of optical power inside the photodiode; (c) energy-band diagram showing carrier movement through drift and diffusion.

7.2 Reverse-Biased *p–n* Junctions

The semiconductor slab of Figure 7.1 is useful for illustrating the basic concepts but such a simple device is rarely used in practice. This section focuses on reverse-biased *p–n* and *p–i–n* junctions that are commonly used for making photodetectors known as *photodiodes*.

7.2.1 *p–n* Photodiodes

A reverse-biased *p–n* junction consists of a region, known as the *depletion region*, that is essentially devoid of free charge carriers and where a large built-in electric field opposes the flow of electrons from the *n*-side to the *p*-side (and of holes from *p* to *n*). When such a *p–n* junction is illuminated with light on one side, say, the *p*-side (see Figure 7.3), electron–hole pairs are created through absorption. Because of the large built-in electric field, electrons and holes generated inside the depletion region accelerate in opposite directions and drift to the *n*- and *p*-sides, respectively. The resulting flow of current is proportional to the incident optical power. Thus, a reverse-biased *p–n* junction acts as a photodetector and is referred to as the *p–n* photodiode.

Figure 7.3(a) shows the structure of a *p–n* photodiode. As shown in Figure 7.3(b), optical power decreases exponentially as the incident light is absorbed inside the depletion region. The electron–hole pairs generated inside the depletion region experience a large electric field and drift rapidly toward the *p*- or *n*-side, depending on the electric charge [Figure 7.3(c)]. The resulting current flow constitutes the photodiode response to the incident optical power in accordance with Eq. (7.1.1). The responsivity of a photodiode is quite high ($R_d \sim 1$ A/W) because of a high quantum efficiency.

The bandwidth of a *p–n* photodiode is often limited by the transit time τ_{tr} in Eq. (7.1.9). If W is the width of the depletion region and v_d is the drift velocity, the transit

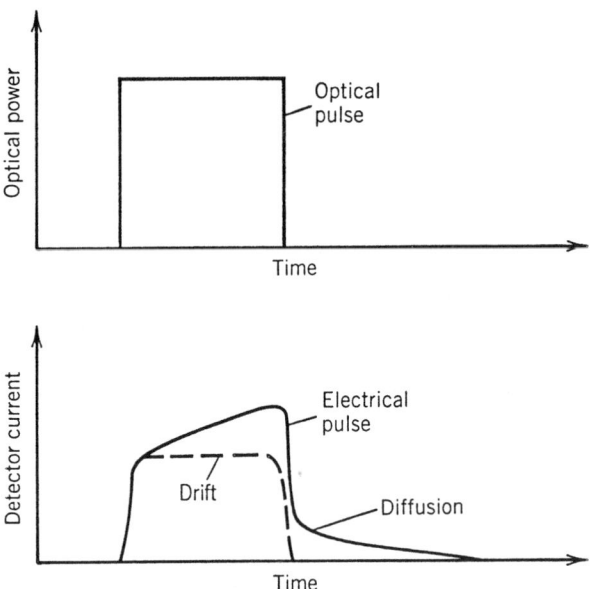

Figure 7.4: Response of a *p–n* photodiode to a rectangular optical pulse when both drift and diffusion contribute to the detector current.

time is given by

$$\tau_{tr} = W/v_d. \tag{7.2.1}$$

Typically, $W \sim 10$ μm, $v_d \sim 10^5$ m/s, and $\tau_{tr} \sim 100$ ps. Both W and v_d can be optimized to minimize τ_{tr}. The depletion-layer width depends on the acceptor and donor concentrations and can be controlled through them. The velocity v_d depends on the applied voltage but attains a maximum value (called the *saturation velocity*) of $\sim 10^5$ m/s that depends on the material used for the photodiode. The RC time constant τ_{RC} can be written as

$$\tau_{RC} = (R_L + R_s)C_p, \tag{7.2.2}$$

where R_L is the external load resistance, R_s is the internal series resistance, and C_p is the parasitic capacitance. Typically, $\tau_{RC} \sim 100$ ps, although lower values are possible with a proper design. Indeed, modern *p–n* photodiodes are capable of operating at bit rates of up to 40 Gb/s.

The limiting factor for the bandwidth of *p–n* photodiodes is the presence of a diffusive component in the photocurrent. The physical origin of the diffusive component is related to the absorption of incident light outside the depletion region. Electrons generated in the *p*-region have to diffuse to the depletion-region boundary before they can drift to the *n*-side; similarly, holes generated in the *n*-region must diffuse to the depletion-region boundary. Diffusion is an inherently slow process; carriers take a nanosecond or longer to diffuse over a distance of about 1 μm. Figure 7.4 shows how the presence of a diffusive component can distort the temporal response of a photodiode. The diffusion contribution can be reduced by decreasing the widths of the *p*- and

7.2 Reverse-Biased p–n Junctions

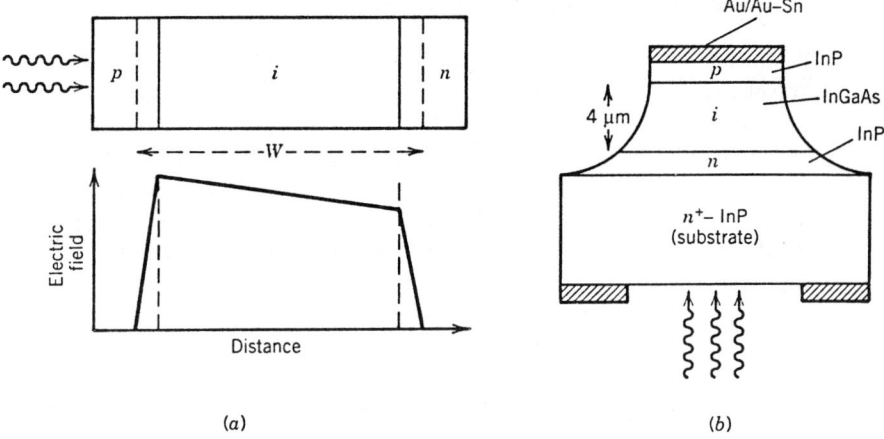

Figure 7.5: (a) A p–i–n photodiode together with the electric-field distribution under reverse bias; (b) design of an InGaAs p–i–n photodiode.

n-regions and increasing the depletion-region width so that most of the incident optical power is absorbed inside it. This is the approach adopted for p–i–n photodiodes, discussed next.

7.2.2 p–i–n Photodiodes

A simple way to increase the depletion-region width is to insert a layer of undoped (or lightly doped) semiconductor material between the p–n junction. Since the middle layer consists of nearly intrinsic material, such a structure is referred to as the p–i–n photodiode. Figure 7.5(a) shows the device structure together with the electric-field distribution inside it under reverse-bias operation. Because of its intrinsic nature, the middle i-layer offers high resistance, and most of the voltage drop occurs across it. As a result, a large electric field exists in the i-layer. In essence, the depletion region extends throughout the i-region, and its width W can be controlled by changing the middle-layer thickness. The main difference from the p–n photodiode is that the drift component of the photocurrent dominates over the diffusion component simply because most of the incident power is absorbed inside the i-region of a p–i–n photodiode.

Since the depletion width W can be tailored in p–i–n photodiodes, a natural question is how large W should be. As discussed in Section 4.1, the optimum value of W depends on a compromise between speed and sensitivity. The responsivity can be increased by increasing W so that the quantum efficiency η approaches 100% [see Eq. (7.1.5)]. However, the response time also increases, as it takes longer for carriers to drift across the depletion region. For indirect-bandgap semiconductors such as Si and Ge, typically W must be in the range of 20–50 μm to ensure reasonable quantum efficiency. The bandwidth of such photodiodes is then limited by a relatively long transit time ($\tau_{tr} > 200$ ps). By contrast, W can be as small as 3–5 μm for photodiodes that use direct-bandgap semiconductors, such as InGaAs. The transit time for such photodiodes

Table 4.1 Characteristics of common p–i–n photodiodes

Parameter	Symbol	Unit	Si	Ge	InGaAs
Wavelength	λ	μm	0.4–1.1	0.8–1.8	1.0–1.7
Responsivity	R	A/W	0.4–0.6	0.5–0.7	0.6–0.9
Quantum efficiency	η	%	75–90	50–55	60–70
Dark current	I_d	nA	1–10	50–500	1–20
Rise time	T_r	ns	0.5–1	0.1–0.5	0.02–0.5
Bandwidth	Δf	GHz	0.3–0.6	0.5–3	1–10
Bias voltage	V_b	V	50–100	6–10	5–6

is $\tau_{tr} \sim 10$ ps. Such values of τ_{tr} correspond to a detector bandwidth of $\Delta f \sim 10$ GHz if we use Eq. (7.1.9) with $\tau_{tr} \gg \tau_{RC}$.

The performance of p–i–n photodiodes can be improved considerably by using a double-heterostructure design. Similar to the case of semiconductor lasers, the middle i-type layer is sandwiched between the p-type and n-type layers of a different semiconductor whose bandgap is chosen such that light is absorbed only in the middle i-layer. Most p–i–n photodiodes used for lightwave applications employ InGaAs for the middle layer and InP for the surrounding p-type and n-type layers [9]. Figure 7.5(b) shows such an InGaAs p–i–n photodiode. Since the bandgap of InP is 1.35 eV, InP is transparent for light whose wavelength exceeds 0.92 μm. By contrast, the bandgap of lattice-matched $In_{1-x}Ga_xAs$ material with $x = 0.47$ is about 0.75 eV, a value that corresponds to a cutoff wavelength of 1.65 μm. The middle InGaAs layer thus absorbs strongly in the wavelength region 1.3–1.6 μm. The diffusive component of the detector current is eliminated completely in such a heterostructure photodiode simply because photons are absorbed only inside the depletion region. The front facet is often coated using suitable dielectric layers to minimize reflections. The quantum efficiency η can be made almost 100% by using an InGaAs layer 4–5 μm thick. InGaAs photodiodes are quite useful for lightwave systems and are often used in practice. Table 4.1 lists the operating characteristics of three common p–i–n photodiodes.

Considerable effort was directed during the 1990s toward developing high-speed p–i–n photodiodes capable of operating at bit rates exceeding 10 Gb/s [9]–[19]. Bandwidths of up to 70 GHz were realized as early as 1986 by using a thin absorption layer (<1 μm) and by reducing the parasitic capacitance C_p with a small size, but only at the expense of a lower quantum efficiency and responsivity [9]. By 1995, p–i–n photodiodes exhibited a bandwidth of 110 GHz for devices designed to reduce τ_{RC} to near 1 ps [14].

7.2.3 Advanced Designs

Several techniques have been developed to improve the efficiency of high-speed photodiodes. In one approach, a Fabry–Perot (FP) cavity is formed around the p–i–n structure to enhance the quantum efficiency [10]–[13], resulting in a laser-like structure. As discussed in Section 5.1, an FP cavity has a set of longitudinal modes at which the

7.2 Reverse-Biased p–n Junctions

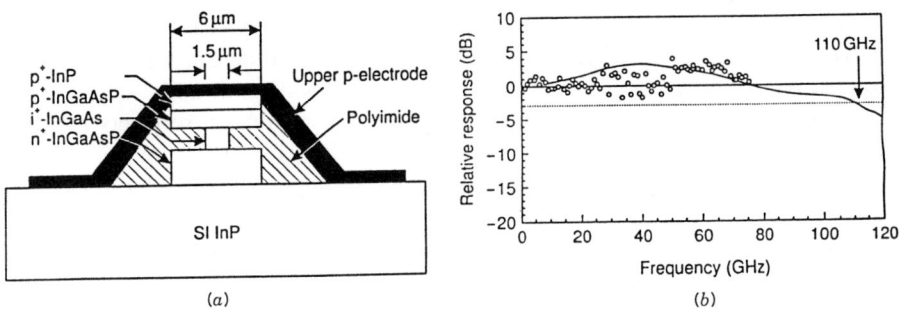

Figure 7.6: (a) Schematic cross section of a mushroom-mesa waveguide photodiode and (b) its measured frequency response. (After Ref. [16]; ©1994 IEEE.)

internal optical field is resonantly enhanced through constructive interference. As a result, when the incident wavelength is close to a longitudinal mode, such a photodiode exhibits high sensitivity. The wavelength selectivity can even be used to advantage in WDM applications. A nearly 100% quantum efficiency was realized in a photodiode in which one mirror of the FP cavity was a distributed Bragg reflector, formed with a stack of AlGaAs/AlAs layers [11] in a way similar to the Bragg mirrors used for VCSELs. This approach was extended to InGaAs photodiodes by inserting a 90-nm-thick InGaAs absorbing layer into a microcavity composed of a GaAs/AlAs Bragg mirror and a dielectric mirror. The device exhibited 94% quantum efficiency at the cavity resonance with a bandwidth of 14 nm [12]. By using an air-bridged metal waveguide together with an undercut mesa structure, a bandwidth of 120 GHz has been realized [13]. The use of such a structure within an FP cavity should provide a *p–i–n* photodiode with a high bandwidth and high efficiency.

Another approach to realizing efficient high-speed photodiodes makes use of an optical waveguide into which the optical signal is edge coupled [15]–[19]. The structure of such a *waveguide photodiode* resembles an unpumped semiconductor laser except that various epitaxial layers are optimized differently. In contrast with a semiconductor laser, the waveguide can be made relatively wide to support multiple transverse modes in order to improve the coupling efficiency [15]. Since absorption takes place along the length of the optical waveguide (~ 10 μm), the quantum efficiency can be nearly 100% even for an ultrathin absorption layer. At the same time, the transit time can be reduced considerably by collecting the photogenerated carriers vertically across a much shorter distance. The bandwidth of such *waveguide photodiodes* is limited by τ_{RC} in Eq. (7.1.9), which can be decreased by controlling the waveguide cross-section area. Indeed, a 50-GHz bandwidth was realized in 1992 for a waveguide photodiode [15].

The bandwidth of waveguide photodiodes can be increased to 110 GHz by adopting a mushroom-mesa waveguide structure [16]. Such a device is shown schematically in Figure 7.6(a). In this structure, the width of the *i*-type absorbing layer was reduced to 1.5 μm, while the *p*- and *n*-type cladding layers were made 6 μm wide. In this way, both the parasitic capacitance and the internal series resistance were minimized, reducing τ_{RC} to about 1 ps. The frequency response of such a device at the 1.55-

μm wavelength is also shown in Figure 7.6(b). It was measured using a spectrum analyzer (circles) as well as taking the Fourier transform of the short-pulse response (solid curve). Clearly, waveguide photodiodes can provide both a high responsivity and a large bandwidth. They have been used for 40-Gb/s optical receivers [18] and have the potential for operating at bit rates as high as 100 Gb/s [19].

The performance of waveguide photodiodes can be improved further by adopting an electrode structure designed to support traveling RF waves with matching impedance to avoid reflections. Such photodiodes are called *traveling-wave photodetectors*. In a GaAs-based implementation of this idea, a bandwidth of 172 GHz with 45% quantum efficiency was realized in a traveling-wave photodetector designed with a 1-μm-wide waveguide [20]. By 2000, such an InP/InGaAs photodetector exhibited a bandwidth of 310 GHz in the 1.55-μm spectral region [21].

7.3 Avalanche Photodiodes

All detectors require a certain minimum current to operate reliably. The current requirement translates into a minimum power requirement through $P_{in} = I_p/R_d$. Detectors with a large responsivity R_d are preferred since they require less optical power. The responsivity of p–i–n photodiodes is limited by Eq. (7.1.3) and takes its maximum value $R_d = q/h\nu$ for $\eta = 1$. Avalanche photodiode (APDs) can have much larger values of R_d, as they are designed to provide an internal current gain in a way similar to photomultiplier tubes. They are used when the amount of optical power that can be spared for the receiver is limited. In this section we first discuss the physical mechanism that leads to the enhancement of responsivity and then consider the advantages and disadvantages of APDs in practice.

7.3.1 Impact Ionization

The physical phenomenon behind the internal current gain is known as *impact ionization* [22]. Under certain conditions, an accelerating electron can acquire sufficient energy to generate a new electron–hole pair. In the energy-band picture, an energetic electron in the conduction band gives a part of its kinetic energy to another electron in the valence band that ends up in the conduction band, leaving behind a hole. The net result of impact ionization is that a single primary electron, generated through the absorption of a photon, creates many secondary electrons and holes, all of which contribute to the photodiode current. Of course, the primary hole can also generate secondary electron–hole pairs that contribute to the current. The generation rate is governed by two parameters, α_e and α_h, the *impact-ionization coefficients* of electrons and holes, respectively. Their numerical values depend on the semiconductor material and on the electric field that accelerates electrons and holes. Figure 7.7 shows α_e and α_h for several semiconductors [23]. Values of $\sim 1 \times 10^4$ cm^{-1} are obtained for electric fields in the range 2–4×10^5 V/cm. Such large fields can be realized by applying a high voltage ($\sim$100 V) to the APD.

APDs differ in their design from that of p–i–n photodiodes mainly in one respect: An additional layer is added in which secondary electron–hole pairs are generated

7.3 Avalanche Photodiodes

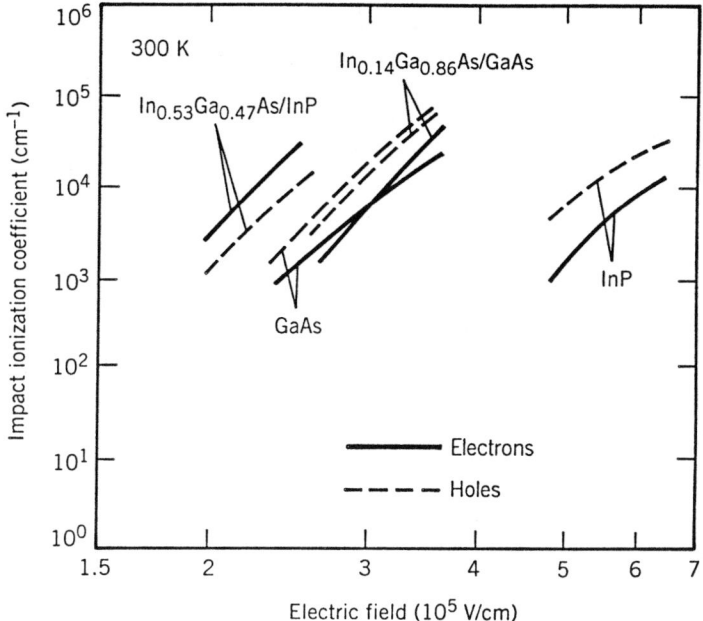

Figure 7.7: Impact-ionization coefficients of several semiconductors as a function of the electric field for electrons (solid line) and holes (dashed line). (After Ref. [23]; ©1977 Elsevier.)

through impact ionization. Figure 7.8(a) shows the APD structure together with the variation of electric field in various layers. Under reverse bias, a high electric field exists in the p-type layer sandwiched between i-type and n^+-type layers. This layer is referred to as the *multiplication layer*, since secondary electron–hole pairs are generated here through impact ionization. The i-layer still acts as the depletion region in which most of the incident photons are absorbed and primary electron–hole pairs are generated. Electrons generated in the i-region cross the gain region and generate secondary electron–hole pairs responsible for the current gain.

7.3.2 APD Gain

The APD gain can be calculated by using the following two simple rate equations governing current flow within the multiplication layer [22]:

$$\frac{di_e}{dx} = \alpha_e i_e + \alpha_h i_h, \qquad (7.3.1)$$

$$-\frac{di_h}{dx} = \alpha_e i_e + \alpha_h i_h, \qquad (7.3.2)$$

where i_e is the electron current and i_h is the hole current. The minus sign in Eq. (7.3.2) is due to the opposite direction of the hole current. The total current,

$$I = i_e(x) + i_h(x), \qquad (7.3.3)$$

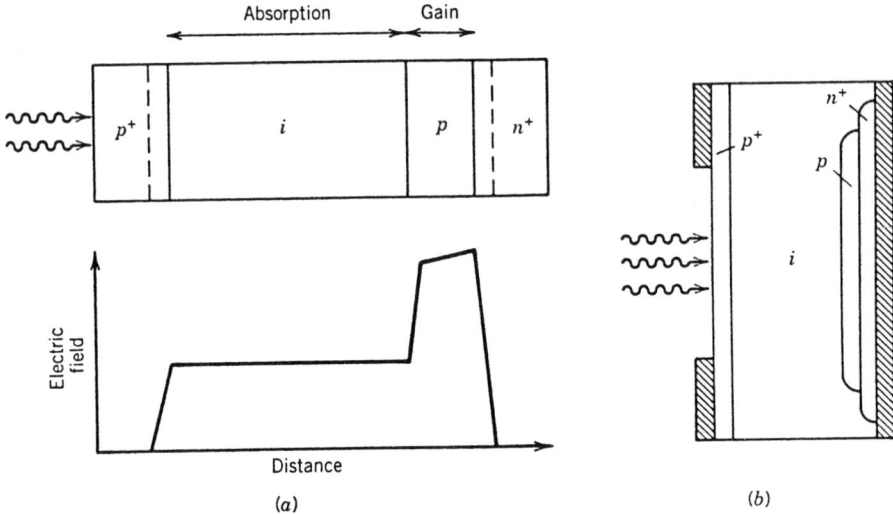

Figure 7.8: (a) An APD together with the electric-field distribution inside various layers under reverse bias; (b) design of a silicon reach-through APD.

remains constant at every point inside the multiplication region. If we replace i_h in Eq. (7.3.1) with $I - i_e$, we obtain

$$\frac{di_e}{dx} = (\alpha_e - \alpha_h)i_e + \alpha_h I. \quad (7.3.4)$$

In general, α_e and α_h are x-dependent if the electric field across the gain region is nonuniform. The analysis is considerably simplified if we assume a uniform electric field and treat α_e and α_h as constants. We also assume that $\alpha_e > \alpha_h$. The avalanche process is initiated by electrons that enter the gain region of thickness d at $x = 0$. By using the condition $i_h(d) = 0$ (only electrons cross the boundary to enter the n-region), the boundary condition for Eq. (7.3.4) is $i_e(d) = I$. By integrating this equation, the *multiplication factor* defined as $M = i_e(d)/i_e(0)$ is given by

$$M = \frac{1 - k_A}{\exp[-(1 - k_A)\alpha_e d] - k_A}, \quad (7.3.5)$$

where $k_A = \alpha_h/\alpha_e$. The APD gain is quite sensitive to the ratio of the impact-ionization coefficients. When $\alpha_h = 0$ so that only electrons participate in the avalanche process, $M = \exp(\alpha_e d)$, and the APD gain increases exponentially with d. On the other hand, when $\alpha_h = \alpha_e$, so that $k_A = 1$ in Eq. (7.3.5), $M = (1 - \alpha_e d)^{-1}$. The APD gain then becomes infinite for $\alpha_e d = 1$, a condition known as *avalanche breakdown*.

Because of the current gain, the responsivity of an APD is enhanced by the multiplication factor M and is given by

$$R_{APD} = MR = M(\eta q/h\nu), \quad (7.3.6)$$

7.3 Avalanche Photodiodes

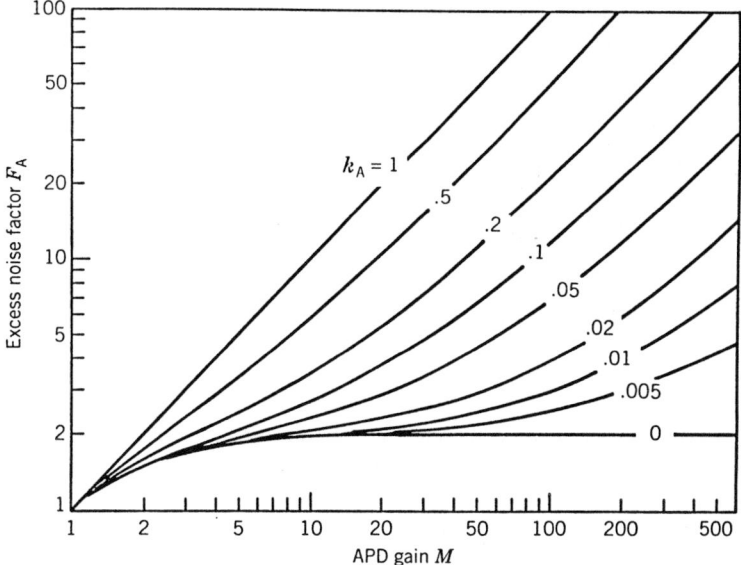

Figure 7.9: Excess noise factor F_A as a function of the average APD gain M for several values of the ionization-coefficient ratio k_A.

where Eq. (7.1.3) was used. It should be stressed that the avalanche process in APDs is intrinsically noisy and results in a gain factor that fluctuates around an average value. The quantity M in Eq. (7.3.6) refers to the average APD gain.

Although higher APD gain can be realized with a smaller gain region when α_e and α_h are comparable, the performance is better in practice for APDs in which either $\alpha_e \gg \alpha_h$ or $\alpha_h \gg \alpha_e$ so that the avalanche process is dominated by only one type of charge carrier. The reason behind this requirement is related to the excess shot noise generated in all APDs, whose magnitude is minimum for $k_A = 0$ and increases rapidly as $k_A \rightarrow 1$. The excess noise factor depends on the APD gain M as [24]

$$F_A(M) = k_A M + (1 - k_A)(2 - 1/M), \tag{7.3.7}$$

where $k_A = \alpha_h/\alpha_e$ or α_e/α_h defined such that $0 < k_A < 1$. Figure 7.9 shows the gain dependence of F_A for several values of k_A. In general, F_A increases with M. However, although F_A is at most 2 for $k_A = 0$, it keeps on increasing linearly ($F_A = M$) when $k_A = 1$. The ratio k_A should be as small as possible for achieving the best performance from an APD.

7.3.3 APD Bandwidth

The intrinsic bandwidth of an APD depends on the multiplication factor M. This is easily understood by noting that the transit time τ_{tr} for an APD is no longer given by Eq. (7.2.1) but increases considerably simply because the generation and collection of secondary electron–hole pairs take additional time. The APD gain decreases at high

Table 4.2 Characteristics of common APDs

Parameter	Symbol	Unit	Si	Ge	InGaAs
Wavelength	λ	μm	0.4–1.1	0.8–1.8	1.0–1.7
Responsivity	R_{APD}	A/W	80–130	3–30	5–20
APD gain	M	—	100–500	50–200	10–40
k-factor	k_A	—	0.02–0.05	0.7–1.0	0.5–0.7
Dark current	I_d	nA	0.1–1	50–500	1–5
Rise time	T_r	ns	0.1–2	0.5–0.8	0.1–0.5
Bandwidth	Δf	GHz	0.2–1	0.4–0.7	1–10
Bias voltage	V_b	V	200–250	20–40	20–30

frequencies because of such an increase in the transit time and limits the bandwidth. The decrease in $M(\omega)$ can be written as [23]

$$M(\omega) = M_0[1 + (\omega \tau_e M_0)^2]^{-1/2}, \qquad (7.3.8)$$

where $M_0 = M(0)$ is the low-frequency gain and τ_e is the effective transit time that depends on the ionization coefficient ratio $k_A = \alpha_h/\alpha_e$. For the case $\alpha_h < \alpha_e$, $\tau_e = c_A k_A \tau_{tr}$, where c_A is a constant ($c_A \sim 1$). If we assume that $\tau_{RC} \ll \tau_e$, the APD bandwidth is given approximately by $\Delta f = (2\pi \tau_e M_0)^{-1}$. This relation shows the *trade-off* between the APD gain M_0 and the bandwidth Δf (speed versus sensitivity). It also shows the advantage of using a semiconductor material for which $k_A \ll 1$.

Table 4.2 compares the operating characteristics of Si, Ge, and InGaAs APDs. As $k_A \ll 1$ for Si, silicon APDs can be designed to provide high performance and are useful for lightwave systems operating near 0.8 μm at bit rates of $\sim$100 Mb/s. A particularly useful design, shown in Figure 7.8(b), is known as reach-through APD because the depletion layer reaches to the contact layer through the absorption and multiplication regions. It can provide high gain ($M \approx 100$) with low noise and a relatively large bandwidth. For lightwave systems operating in the wavelength range of 1.3–1.6 μm, Ge or InGaAs APDs must be used. The improvement in sensitivity for such APDs is limited to a factor below 10 because of a relatively low APD gain ($M \sim 10$) that must be used to reduce the noise.

The performance of InGaAs APDs can be improved through suitable design modifications to the basic APD structure shown in Figure 7.8. The main reason for a relatively poor performance of InGaAs APDs is related to the comparable numerical values of the impact-ionization coefficients α_e and α_h (see Figure 7.7). As a result, the bandwidth is considerably reduced, and the noise is also relatively high (see Section 4.4). Furthermore, because of a relatively narrow bandgap, InGaAs undergoes tunneling breakdown at electric fields of about 1×10^5 V/cm, a value that is below the threshold for avalanche multiplication. This problem can be solved in heterostructure APDs by using an InP layer for the gain region because quite high electric fields ($>5 \times 10^5$ V/cm) can exist in InP without tunneling breakdown. Since the absorption region (*i*-type InGaAs layer) and the multiplication region (*n*-type InP layer) are separate in such a device, this structure is known as SAM, where SAM stands for *separate absorption and multiplication*

7.3 Avalanche Photodiodes

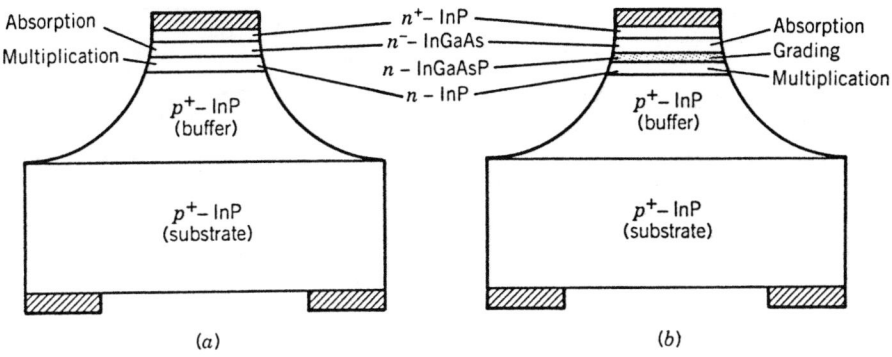

Figure 7.10: Design of (a) SAM and (b) SAGM APDs containing separate absorption, multiplication, and grading regions.

regions. As $\alpha_h > \alpha_e$ for InP (see Figure 7.7), the APD is designed such that holes initiate the avalanche process in an *n*-type InP layer, and k_A is defined as $k_A = \alpha_e/\alpha_h$. Figure 7.10(a) shows a mesa-type SAM APD structure.

One problem with the SAM APD is related to the large bandgap difference between InP ($E_g = 1.35$ eV) and InGaAs ($E_g = 0.75$ eV). Because of a valence-band step of about 0.4 eV, holes generated in the InGaAs layer are trapped at the heterojunction interface and are considerably slowed before they reach the multiplication region (InP layer). Such an APD has an extremely slow response and a relatively small bandwidth. The problem can be solved by using another layer between the absorption and multiplication regions whose bandgap is intermediate to those of InP and InGaAs layers. The quaternary material InGaAsP, the same material used for semiconductor lasers, can be tailored to have a bandgap anywhere in the range of 0.75–1.35 eV and is ideal for this purpose. It is even possible to grade the composition of InGaAsP over a region of 10–100 nm thickness. Such APDs are called SAGM APDs, where SAGM indicates *separate absorption, grading, and multiplication* regions [25]. Figure 7.10(b) shows the design of an InGaAs APD with the SAGM structure. The use of an InGaAsP grading layer improves the bandwidth considerably. As early as 1987, a SAGM APD exhibited a gain–bandwidth product of $M\Delta f = 70$ GHz for $M > 12$ [26]. This value was increased to 100 GHz in 1991 by using a charge region between the grading and multiplication regions [27]. In such SAGCM APDs, the InP multiplication layer is undoped, while the InP charge layer is heavily *n*-doped. Holes accelerate in the charge layer because of a strong electric field, but the generation of secondary electron–hole pairs takes place in the undoped InP layer. SAGCM APDs improved considerably during the 1990s [28]–[32]. A gain–bandwidth product of 140 GHz was realized in 2000 using a 0.1-μm-thick multiplication layer that required <20 V across it [32]. Such APDs are quite suitable for making a compact 10-Gb/s APD receiver.

A different approach to the design of high-performance APDs makes use of a superlattice structure [33]–[38]. The major limitation of InGaAs APDs results from comparable values of α_e and α_h. A superlattice design offers the possibility of reducing the ratio $k_A = \alpha_h/\alpha_e$ from its standard value of nearly unity. In one scheme, the absorption

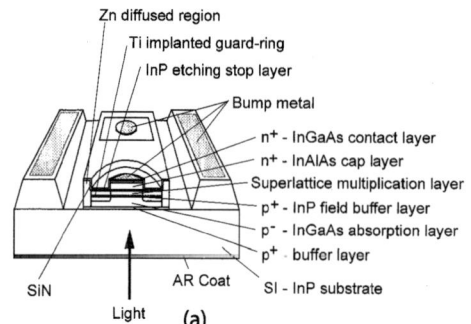 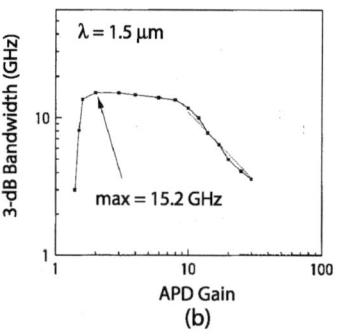

Figure 7.11: (a) Device structure and (b) measured 3-dB bandwidth as a function of M for a superlattice APD. (After Ref. [38]; ©2000 IEEE.)

and multiplication regions alternate and consist of thin layers ($\sim$10 nm) of semiconductor materials with different bandgaps. This approach was first demonstrated for GaAs/AlGaAs multiquantum-well (MQW) APDs and resulted in a considerable enhancement of the impact-ionization coefficient for electrons [33]. Its use is less successful for the InGaAs/InP material system. Nonetheless, considerable progress has been made through the so-called *staircase* APDs, in which the InGaAsP layer is compositionally graded to form a sawtooth kind of structure in the energy-band diagram that looks like a staircase under reverse bias. Another scheme for making high-speed APDs uses alternate layers of InP and InGaAs for the grading region [33]. However, the ratio of the widths of the InP to InGaAs layers varies from zero near the absorbing region to almost infinity near the multiplication region. Since the effective bandgap of a quantum well depends on the quantum-well width (InGaAs layer thickness), a graded "pseudo-quaternary" compound is formed as a result of variation in the layer thickness.

The most successful design for InGaAs APDs uses a superlattice structure for the multiplication region of a SAM APD. A superlattice consists of a periodic structure such that each period is made using two ultrathin ($\sim$10-nm) layers with different bandgaps. In the case of 1.55-μm APDs, alternate layers of InAlGaAs and InAlAs are used, the latter acting as a barrier layer. An InP field-buffer layer often separates the InGaAs absorption region from the superlattice multiplication region. The thickness of this buffer layer is quite critical for APD performance. For a 52-nm-thick field-buffer layer, the gain–bandwidth product was limited to $M\Delta f = 120$ GHz [34] but increased to 150 GHz when the thickness was reduced to 33.4 nm [37]. These early devices used a mesa structure. During the late 1990s, a planar structure was developed for improving device reliability [38]. Figure 7.11(a) sketches the device structure schematically; Figure 7.11(b) shows the 3-dB bandwidth measured as a function of the APD gain. The gain–bandwidth product of 110 GHz is large enough for making APDs operating at 10 Gb/s. Indeed, such an APD receiver was used for a 10-Gb/s lightwave system with excellent performance.

The gain–bandwidth limitation of InGaAs APDs results primarily from using the InP material system for the generation of secondary electron–hole pairs. A hybrid approach in which a Si multiplication layer is incorporated next to an InGaAs absorption

7.3 Avalanche Photodiodes

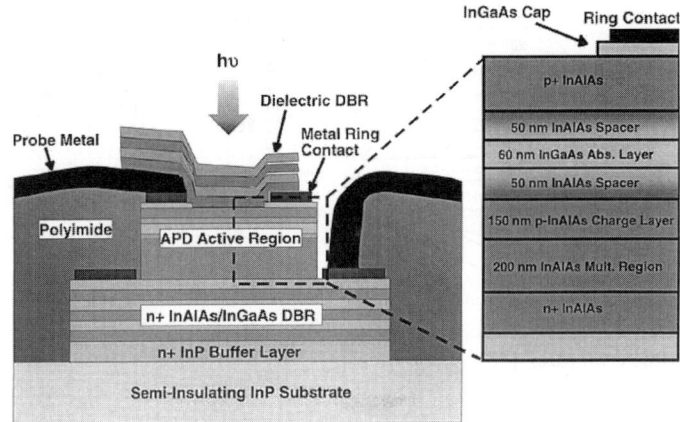

Figure 7.12: Device structure for a resonant-cavity APD. The gain layer is 200-nm thick and is accompanied by a 150-nm-thick charge layer. (After Ref. [42]; ©1999 IEEE.)

layer may be useful, provided the heterointerface problems can be overcome. In a 1997 experiment, a gain-bandwidth product of more than 300 GHz was realized by using such a hybrid approach [39]. The APD exhibited a 3-dB bandwidth of over 9 GHz for values of M as high as 35 while maintaining 60% quantum efficiency. In a somewhat different approach, the InGaAs absorbing layer grown on an InP substrate was brought into contact with a p-type gain layer made of Si and grown on a Si wafer [40]. The two wafers were fused together using a well-known technique [41]. Since $k_A = 0.02 \ll 1$ for Si, such a wafer-fused APD structure offers the potential of low noise. Indeed, an excess noise factor of only 2.3 was measured at $M = 20$.

7.3.4 Advanced APD Designs

Most APDs use an absorbing layer thick enough (about 1 μm) that the quantum efficiency exceeds 50%. The thickness of the absorbing layer affects the transit time τ_{tr} and the bias voltage V_b. In fact, both of them can be reduced significantly by using a thin absorbing layer ($\sim$0.1 μm), resulting in improved APDs provided that a high quantum efficiency can be maintained. Similar to the case of p-i-n photodiodes, two approaches have been used to meet these somewhat conflicting design requirements.

In the resonant-cavity design, an FP cavity is formed to enhance the absorption within a thin layer through multiple round trips. Figure 7.12 shows the device structure schematically [42]. The ultrathin (60 nm) InGaAs absorption layer is sandwiched between two 50-nm-thick InAlAs spacer layers. The gain layer is 200-nm thick and is accompanied by a 150-nm-thick charge layer. The entire APD structure is placed between two distributed Bragg reflectors (DBRs) formed with multilayer stacks. The DBRs have a reflectivity of about 95% and form a resonant microcavity in which the incident light is forced to bounce back and forth multiple times. Such a 1.55-μm APD exhibited an external quantum efficiency of $\sim$70% while maintaining a gain–bandwidth product of 290 GHz.

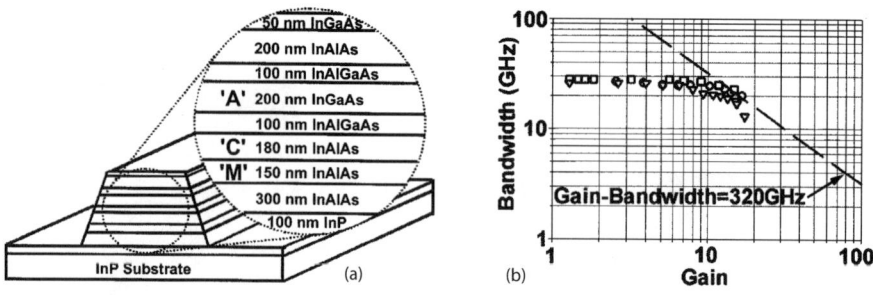

Figure 7.13: (a) Device structure and (b) measured 3-dB bandwidth as a function of M for a waveguide APD. (After Ref. [46]; ©2001 IEEE.)

In a different approach capable of enhancing the quantum efficiency of an APD, an optical waveguide is used into which the incident light is edge-coupled [43]–[46]. Propagation inside the waveguide provides long absorption lengths, thus improving quantum efficiency. At the same time, the external voltage is applied across the waveguide thickness, which can be quite small. This feature enhances the bandwidth by reducing the transit time to ~1 ps. At the same time, high electric fields can be generated at a lower bias voltage of ~10 V.

Figure 7.13(a) shows the structure of a waveguide APD schematically [46]. The entire structure was grown on a sulphur-doped InP substrate using the MOCVD technique. The layers marked A, C, and M denote the absorption, charge, and multiplication (gain) layers, respectively. The 200-nm-thick absorption layer is made of $In_{0.53}Ga_{0.47}As$, while the 150-nm-thick multiplication layer is made of $In_{0.52}Al_{0.48}As$. The 180-nm-thick charge layer has the same composition as the multiplication layer but was doped with zinc. The waveguide width was in the range 4–20 μm while its length was varied from 10 to 100 μm. The parasitic capacitance was low enough that the RC bandwidth was estimated to be about 57 GHz for such devices, but the transit time limited the bandwidth to near 30 GHz. Figure 7.13(b) shows the 3-dB bandwidth measured as a function of the APD gain for a waveguide whose width and lengths were equal to 10 μm. The APD had a bandwidth of 28 GHz for $M < 10$, and the gain–bandwidth product was as large as 320 GHz.

The excess noise factor of APDs with thin gain regions does not follow Eq. (7.3.7) that is found to hold for relatively thick multiplication layers [47]. The reason can be understood as follows. After an ionization event initiated by an electron or hole, the electron–hole pair created needs to travel a certain distance before it can gain sufficient energy from the electric field to have an appreciable impact-ionization probability. When the thickness of the multiplication region is below < 1 μm or so, one must include this effect to calculate the excess noise factor. It turns out that Eq. (7.3.7) overestimates the noise for thin gain layers. Several designs have been used for realizing APDs with a relatively low noise [48]–[50]. With a proper design, APDs can provide high quantum efficiency and low noise while exhibiting simultaneously a bandwidth of more than 30 GHz. Such devices are useful for lightwave systems operating at 40 Gb/s.

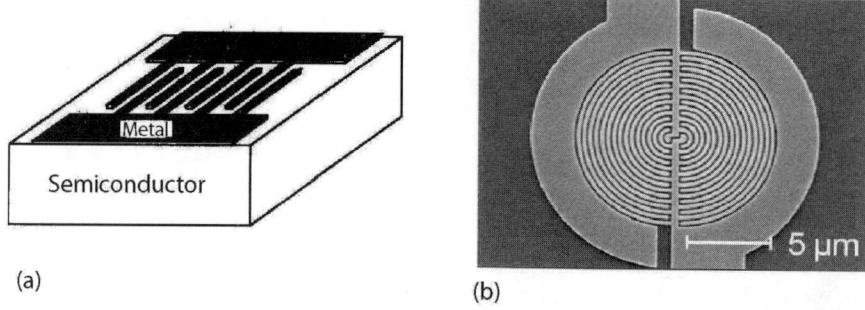

Figure 7.14: (a) Finger-shaped and (b) ring-shaped electrode structures used for MSM photodetectors. (After Ref. [61]; ©1999 IEEE.)

7.4 MSM Photodetectors

In a different kind of photodetector, known as a metal–semiconductor–metal (MSM) photodetector, a semiconductor absorbing layer is sandwiched between two metal electrodes. As a result, a Schottky barrier is formed at each metal–semiconductor interface that prevents the flow of electrons from the metal to the semiconductor. Similar to a *p–i–n* photodiode, electron–hole pairs generated through the absorption of light flow toward the metal contacts, resulting in a photocurrent that is a measure of the incident optical power, as indicated in Eq. (7.1.1). However, in contrast with a *p–i–n* photodiode or APD, no *p–n* junction is required. In this sense, an MSM photodetector employs the simplest design, as shown in Figure 7.1.

For practical reasons, it is difficult to sandwich a thin semiconductor layer between two metal electrodes. This problem can be solved by placing the two metal contacts on the same (top) side of an epitaxially grown absorbing layer using an *interdigited* electrode structure with a finger spacing of about 1 μm [51]. Figure 7.14(a) show the basic design. In modern devices, the concentric ring structure shown in Figure 7.14(b) is often used in place of finger-shaped electrodes. The resulting planar structure has an inherently low parasitic capacitance and thus allows high-speed operation (up to 300 GHz) of MSM photodetectors. If the light is incident from the electrode side, the responsivity of a MSM photodetector is reduced because some light is blocked by the opaque electrodes. This problem can be solved through back illumination if the substrate is transparent to the incident light.

GaAs-based MSM photodetectors were developed throughout the 1980s, and they exhibit excellent operating characteristics [51]. The development of InGaAs-based MSM photodetectors, suitable for lightwave systems operating in the range of 1.3–1.6 μm, started in the late 1980s, with most progress made during the 1990s [52]–[61]. The major problem with the InGaAs material is its relatively low *Schottky-barrier height* (about 0.2 eV) compared with GaAs-based devices. This problem was solved by introducing a thin layer of InP or InAlAs between the InGaAs layer and the metal contact. Such a layer, called the *barrier-enhancement layer*, improves the performance of InGaAs-based MSM photodetectors drastically. The use of a 20-nm-thick InAlAs

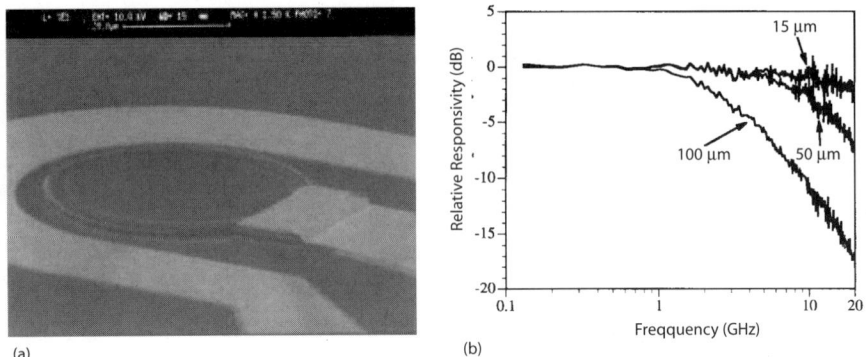

Figure 7.15: (a) SEM photograph and (b) frequency response of MSM photodetectors of three different diameters. (After Ref. [59]; ©1997 IEEE.)

barrier-enhancement layer resulted in 1992 in 1.3-μm MSM photodetectors exhibiting 92% quantum efficiency (through back illumination) with a low dark current [53]. A packaged device had a bandwidth of 4 GHz despite a large 150-μm diameter. If top illumination is desirable for processing or packaging reasons, the responsivity can be enhanced by using semitransparent metal contacts. In one experiment, the responsivity at 1.55 μm increased from 0.4 to 0.7 A/W when the thickness of gold contact was reduced from 100 to 10 nm [54]. In another approach, the structure is separated from the host substrate and bonded to a silicon substrate with the interdigited contact on bottom. Such an "inverted" MSM photodetector then exhibits high responsivity when illuminated from the top [55].

Figure 7.15(a) shows a scanning electron microscope (SEM) photograph of a 25-μm-diameter MSM photodetector [59]. Several epitaxial layers were grown on a semi-insulating InP substrate using the MOCVD technique. The InGaAs absorption layer was only 1 μm thick. The 50-nm-thick InAlAs layer acted as a barrier-enhancement layer. An indium–tin-oxide film was deposited on top of the barrier-enhancement layer because it has low resistivity, is nearly transparent in the wavelength range of 1.3–1.6 μm, and forms good Schottky-barrier contacts. Its thickness was chosen to be $\lambda/2$ at the operating wavelength λ so that the same film acted as an antireflection coating. The responsivity of the device varied from 0.55 to 0.6 at wavelengths near 1.3 and 1.55 μm. Figure 7.15(b) shows the frequency response of such devices at an applied bias of 10 V for three different diameters. As expected, the bandwidth was larger for smaller devices. The 3-dB bandwidth was only 2.5 GHz for 100-μm-diameter devices but increased to 25 GHz when the diameter was reduced to 15 μm.

The temporal response of MSM photodetectors is generally different under back and top illuminations [56]. In particular, the bandwidth Δf is larger by about a factor of 2 for top illumination, although the responsivity is reduced because of metal shadowing. The performance of a MSM photodetector can be further improved by using a graded superlattice structure. Such devices exhibit a low dark-current density, a responsivity of about 0.6 A/W at 1.3 μm, and a rise time of about 16 ps [58]. In 1998, a 1.55-μm MSM photodetector exhibited a bandwidth of 78 GHz [60].

7.4 MSM Photodetectors

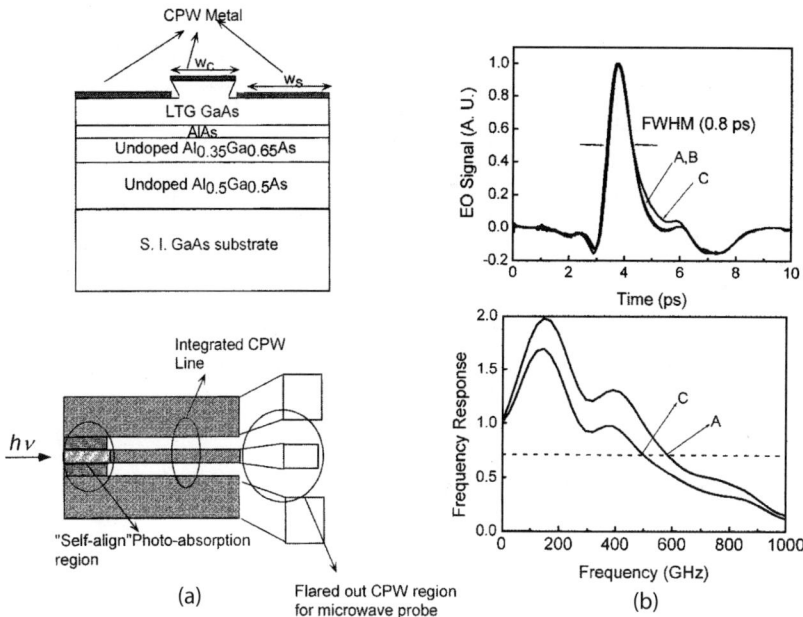

Figure 7.16: (a) Cross-sectional and top views of a traveling-wave MSM photodetector. Acronyms LTG and S.I. stand for low-temperature growth and semi-insulating, respectively. (b) Temporal and frequency response at optical power levels of 1, 1.8, and 2.2 mW for traces A, B, and C, respectively. (After Ref. [62]; ©2001 IEEE.)

Further improvement in the bandwidth of MSM photodetectors can be realized by employing a traveling-wave configuration [62], the same used for enhancing the bandwidth of modulators (see Section 6.2.3). Similar to the case of LiNbO$_3$ modulators, the input light is propagated inside a waveguide whose core is made of the absorbing layer. Metal electrodes are deposited on top of the waveguide such that they form a coplanar waveguide (CPW) line for the microwave signal generated through the absorption of a propagating optical field. Figure 7.16(a) shows the structure of a GaAs-based MSM photodetector together with the top view displaying the CPW line. The absorbing GaAs layer was grown at low temperature and was only 500-nm thick. The hot electrode is on top of the etched mesa formed to guide the optical mode laterally and is separated from the ground electrodes by <300 nm. Light absorption takes place over the initial 10-μm-long region.

The temporal response of such a device was measured by propagating 100-fs optical pulses obtained from a mode-locked Ti:sapphire laser and using an electro-optic sampling technique. Figure 7.16(b) shows the transient response at a bias of 5 V for three different average input power levels. It also shows the frequency response obtained by calculating the Fourier transform of the temporal data. The 0.8-ps width of the response function corresponds to a 3-dB bandwidth of more than 500 GHz. Clearly, MSM photodetectors are capable of an ultrafast response. The planar structure of MSM photodetectors is also suitable for monolithic integration.

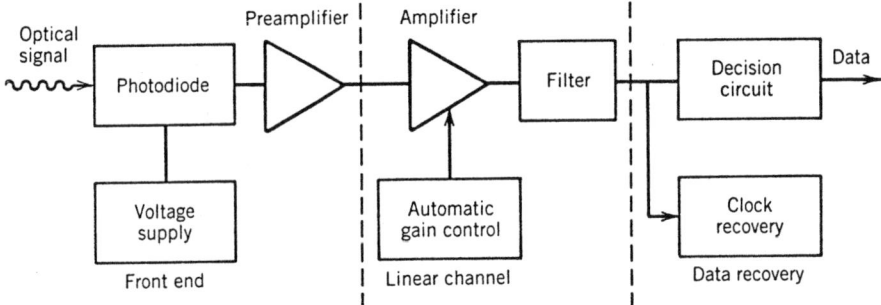

Figure 7.17: Diagram of a digital optical receiver showing various components. Vertical dashed lines divide receiver components into three sections.

7.5 Integrated Optical Receivers

An optical receiver consists of many other components in addition to a photodetector. Figure 7.17 shows the block diagram of a digital receiver used to recover the optical bit stream incident on it. Its components can be arranged into three groups—the front end, the linear channel, and the decision circuit. We first focus on these three groups and then discuss the techniques used to integrate as many components as possible on a single chip.

7.5.1 Front End

The front end of a receiver consists of a photodiode followed by a preamplifier. The optical signal is coupled onto the photodiode by using a coupling scheme similar to that used for optical transmitters. The photodiode converts the optical bit stream into an electrical time-varying signal. The role of the preamplifier is to amplify the electrical signal for further processing.

The design of the front end requires a trade-off between speed and sensitivity. Since the input voltage to the preamplifier can be increased by using a large load resistor R_L, a high-impedance front end is often used [see Figure 7.18(a)]. Furthermore, a large R_L reduces the thermal noise and improves the receiver sensitivity. The main drawback of high-impedance front end is its low bandwidth given by $\Delta f = (2\pi R_L C_T)^{-1}$, where $R_s \ll R_L$ is assumed in Eq. (7.2.2) and $C_T = C_p + C_A$ is the total capacitance, which includes the contributions from the photodiode (C_p) and the transistor used for amplification (C_A). The receiver bandwidth is limited by its slowest component. A high-impedance front end cannot be used if Δf is considerably less than the bit rate. An equalizer is sometimes used to increase the bandwidth. The equalizer acts as a filter that attenuates low-frequency components of the signal more than the high-frequency components, thereby effectively increasing the front-end bandwidth. If the receiver sensitivity is not of concern, one can simply decrease R_L to increase the bandwidth, resulting in a low-impedance front end.

7.5 Integrated Optical Receivers

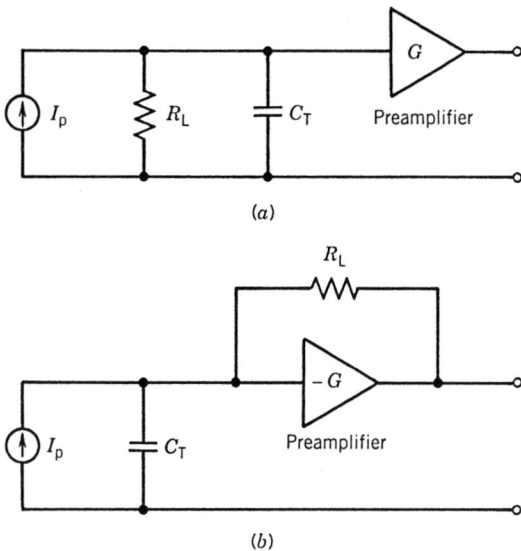

Figure 7.18: Equivalent circuit for (a) high-impedance and (b) transimpedance front ends in optical receivers. The photodiode is modeled as a current source in both cases.

Transimpedance front ends provide a configuration that has high sensitivity together with a large bandwidth. Its dynamic range is also improved compared with high-impedance front ends. As seen in Figure 7.18(b), the load resistor is connected as a feedback resistor around an inverting amplifier. Even though R_L is large, the *negative feedback* reduces the effective input impedance by a factor of G, where G is the amplifier gain. The bandwidth is thus enhanced by a factor of G compared with high-impedance front ends. Transimpedance front ends are often used in optical receivers because of their improved characteristics. A major design issue is related to the stability of the feedback loop. More details can be found in Refs. [4]–[8].

7.5.2 Linear Channel

The linear channel in optical receivers consists of a high-gain amplifier (the main amplifier) and a low-pass filter. An equalizer is sometimes included just before the amplifier to correct for the limited bandwidth of the front end. The amplifier gain is controlled automatically to limit the average output voltage to a fixed level irrespective of the incident average optical power at the receiver. The low-pass filter shapes the voltage pulse. Its purpose is to reduce the noise without introducing much *intersymbol interference* (ISI). As discussed in Section 4.4, the receiver noise is proportional to the receiver bandwidth and can be reduced by using a low-pass filter whose bandwidth Δf is smaller than the bit rate. Since other components of the receiver are designed to have a bandwidth larger than the filter bandwidth, the receiver bandwidth is determined by the low-pass filter used in the linear channel. For $\Delta f < B$, the electrical pulse spreads beyond the allocated bit slot. Such a spreading can interfere with the detection of neighboring bits, a phenomenon referred to as ISI.

It is possible to design a low-pass filter in such a way that ISI is minimized. Since the combination of preamplifier, main amplifier, and the filter acts as a linear system (hence the name *linear channel*), the output voltage can be written as

$$V_{\text{out}}(t) = \int_{-\infty}^{\infty} z_T(t-t') I_p(t') \, dt', \tag{7.5.1}$$

where $I_p(t)$ is the photocurrent generated in response to the incident optical power ($I_p = R P_{\text{in}}$). In the frequency domain,

$$\tilde{V}_{\text{out}}(\omega) = Z_T(\omega) \tilde{I}_p(\omega), \tag{7.5.2}$$

where Z_T is the total impedance at the frequency ω and a tilde represents the Fourier transform. Here, $Z_T(\omega)$ is determined by the transfer functions associated with various receiver components and can be written as [2]

$$Z_T(\omega) = G_p(\omega) G_A(\omega) H_F(\omega) / Y_{\text{in}}(\omega), \tag{7.5.3}$$

where $Y_{\text{in}}(\omega)$ is the input admittance and $G_p(\omega)$, $G_A(\omega)$, and $H_F(\omega)$ are transfer functions of the preamplifier, the main amplifier, and the filter. It is useful to isolate the frequency dependence of $\tilde{V}_{\text{out}}(\omega)$ and $\tilde{I}_p(\omega)$ through normalized spectral functions $H_{\text{out}}(\omega)$ and $H_p(\omega)$, which are related to the Fourier transform of the output and input pulse shapes, respectively, and write Eq. (7.5.2) as

$$H_{\text{out}}(\omega) = H_T(\omega) H_p(\omega), \tag{7.5.4}$$

where $H_T(\omega)$ is the total transfer function of the linear channel and is related to the total impedance as $H_T(\omega) = Z_T(\omega) / Z_T(0)$. If the amplifiers have a much larger bandwidth than the low-pass filter, $H_T(\omega)$ can be approximated by $H_F(\omega)$.

The ISI is minimized when $H_{\text{out}}(\omega)$ corresponds to the transfer function of a *raised-cosine filter* and is given by [2]

$$H_{\text{out}}(f) = \begin{cases} \frac{1}{2}[1 + \cos(\pi f / B)], & f < B, \\ 0, & f \geq B, \end{cases} \tag{7.5.5}$$

where $f = \omega / 2\pi$ and B is the bit rate. The impulse response, obtained by taking the Fourier transform of $H_{\text{out}}(f)$, is given by

$$h_{\text{out}}(t) = \frac{\sin(2\pi B t)}{2\pi B t} \frac{1}{1 - (2Bt)^2}. \tag{7.5.6}$$

The functional form of $h_{\text{out}}(t)$ corresponds to the shape of the voltage pulse $V_{\text{out}}(t)$ received by the decision circuit. At the decision instant $t = 0$, $h_{\text{out}}(t) = 1$, and the signal is maximum. At the same time, $h_{\text{out}}(t) = 0$ for $t = m/B$, where m is an integer. Since $t = m/B$ corresponds to the decision instant of the neighboring bits, the voltage pulse of Eq. (7.5.6) does not interfere with the neighboring bits.

7.5 Integrated Optical Receivers

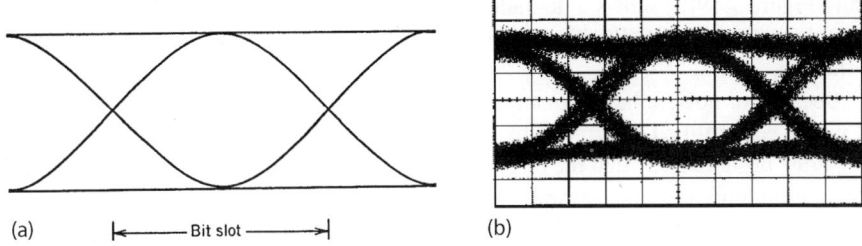

Figure 7.19: Ideal (a) and degraded (b) eye patterns for the NRZ format.

The linear-channel transfer function $H_T(\omega)$ that will result in output pulse shapes of the form (7.5.6) is obtained from Eq. (7.5.4) and is given by

$$H_T(f) = H_{\text{out}}(f)/H_p(f). \tag{7.5.7}$$

For an ideal bit stream in the nonreturn-to-zero (NRZ) format (rectangular input pulses of duration $T_B = 1/B$), $H_p(f) = B\sin(\pi f/B)/\pi f$, and $H_T(f)$ becomes

$$H_T(f) = (\pi f/2B)\cot(\pi f/2B). \tag{7.5.8}$$

Equation (7.5.8) determines the frequency response of the linear channel that would produce output pulse shape given by Eq. (7.5.6) under ideal conditions. In practice, the input pulse shape is far from being rectangular. The output pulse shape also deviates from Eq. (7.5.6), and some ISI invariably occurs.

7.5.3 Decision Circuit

The data-recovery section of optical receivers consists of a decision circuit and a clock-recovery circuit. The purpose of the latter is to isolate a spectral component at $f = B$ from the received signal. This component provides information about the bit slot ($T_B = 1/B$) to the decision circuit and helps to synchronize the decision process. In the case of RZ (return-to-zero) format, a spectral component at $f = B$ is present in the received signal; a narrow-bandpass filter such as a surface-acoustic-wave filter can isolate this component easily. Clock recovery is more difficult in the case of NRZ format because the signal received lacks a spectral component at $f = B$. A commonly used technique generates such a component by squaring and rectifying the spectral component at $f = B/2$ that can be obtained by passing the received signal through a high-pass filter.

The decision circuit compares the output from the linear channel to a threshold level, at sampling times determined by the clock-recovery circuit, and decides whether the signal corresponds to bit 1 or bit 0. The best sampling time corresponds to the situation in which the signal level difference between 1 and 0 bits is maximum. It can be determined from the *eye diagram* formed by superposing 2–3-bit-long electrical sequences in the bit stream on top of each other. The resulting pattern is called an eye diagram because of its appearance. Figure 7.19(a) illustrates an ideal eye diagram, while Figure 7.19(b) shows a degraded eye diagram in which the noise and the timing jitter close the eye partially. The best sampling time corresponds to maximum opening of the eye.

Because of noise inherent in any receiver, there is always a finite probability that a bit would be identified incorrectly by the decision circuit. Digital receivers are designed to operate in such a way that the error probability is quite small (typically $<10^{-9}$). The eye diagram provides a visual way of monitoring the receiver performance: Closing of the eye is an indication that the receiver is not performing properly.

7.5.4 Optoelectronic Integration

All receiver components shown in Figure 7.17, with the exception of the photodiode, are standard electrical components and can be easily integrated on the same chip by using the integrated-circuit (IC) technology developed for microelectronic devices. Integration is particularly necessary for receivers operating at high bit rates. By 1988, both Si and GaAs IC technologies have been used to make integrated receivers up to a bandwidth of more than 2 GHz [63]. Since then, the bandwidth has been extended to 10 GHz.

Considerable effort has been directed at developing monolithic optical receivers that integrate all components, including the photodetector, on the same chip by using the *optoelectronic integrated-circuit* (OEIC) technology [64]–[89]. Such a complete integration is relatively easy for GaAs receivers, and the technology behind GaAs-based OEICs is quite advanced. The use of MSM photodiodes has proved especially useful as they are structurally compatible with the well-developed *field-effect-transistor* (FET) technology. This technique was used as early as 1986 to demonstrate a four-channel OEIC receiver chip [66].

For lightwave systems operating in the wavelength range of 1.3–1.6 μm, InP-based OEIC receivers are needed. Since the IC technology for GaAs is much more mature than for InP, a hybrid approach is sometimes used for InGaAs receivers. In this approach, called *flip-chip OEIC technology* [67], the electronic components are integrated on a GaAs chip, whereas the photodiode is made on top of an InP chip. The two chips are then connected by flipping the InP chip on the GaAs chip, as shown in Figure 7.20. The advantage of the flip-chip technique is that the photodiode and the electrical components of the receiver can be independently optimized while keeping the parasitics (e.g., effective input capacitance) to a bare minimum.

The InP-based IC technology advanced considerably during the 1990s, making it possible to develop InGaAs OEIC receivers [68]–[89]. Several kinds of transistors have been used for this purpose. In one approach, a p–i–n photodiode is integrated with the FETs or high-electron-mobility transistors (HEMTs) side by side on the same InP substrate [69]–[73]. By 1993, HEMT-based receivers were capable of operating at 10 Gb/s while exhibiting high sensitivity [72]. The bandwidth of such receivers has been increased to >40 GHz, making it possible to use them at bit rates above 40 Gb/s [73]. A waveguide p–i–n photodiode has also been integrated with HEMTs to develop a two-channel OEIC receiver.

In another approach [74]–[79], the heterojunction-bipolar transistor (HBT) technology is used to fabricate the p–i–n photodiode within the HBT structure itself through a

7.5 Integrated Optical Receivers

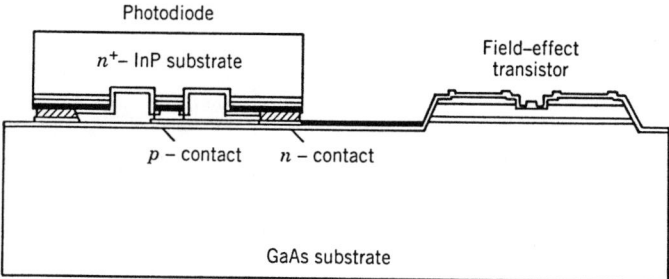

Figure 7.20: Flip-chip OEIC technology for integrated receivers. The InGaAs photodiode is fabricated on an InP substrate and then bonded to the GaAs chip through common electrical contacts. (After Ref. [67]; ©1988 IEE.)

common-collector configuration. Such transistors are often called *heterojunction phototransistors*. OEIC receivers operating at 5 Gb/s (with a bandwidth of 3 GHz) were made by 1993 [74]. By 1995, OEIC receivers making use of the HBT technology exhibited a bandwidth of up to 16 GHz, together with a high gain [76]. Such receivers can be used at bit rates of more than 20 Gb/s. Indeed, a high-sensitivity OEIC receiver module was used in 1995 at a bit rate of 20 Gb/s in a 1.55-μm lightwave system [77]. Even a decision circuit can be integrated within the OEIC receiver by using the HBT technology [78].

A third approach to InP-based OEIC receivers integrates a waveguide or an MSM photodetector with the FET or HEMT amplifiers [80]–[89]. A bandwidth of 15 GHz was realized for such an OEIC in 1995 by using modulation-doped FETs [81]. By 2000, such receivers exhibited bandwidths of more than 45 GHz with the use of waveguide photodiodes. Figure 7.21 shows the epitaxial-layer structure of such an OEIC receiver [84]. All epitaxial layers were grown on a semi-insulating (S.I.) InP substrate using the MOCVD technique. The ridge waveguide for the photodiode is 6 μm wide and was formed using the technique of reactive ion etching. It was coated with polyimide to reduce the parasitic capacitance. The entire receiver chip was 2.5 mm long and

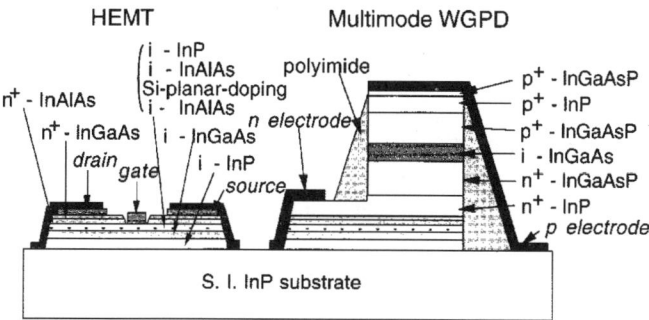

Figure 7.21: Epitaxial-layer structure of an OEIC receiver made by integrating a waveguide photodetector (WGPD) with an HEMT. (After Ref. [84]; ©2000 IEEE.)

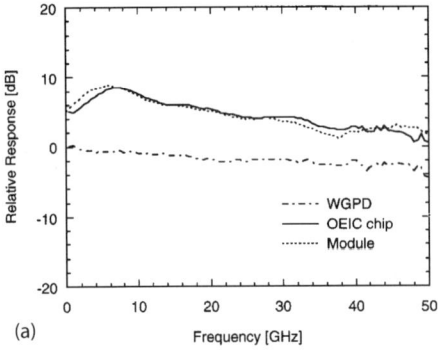
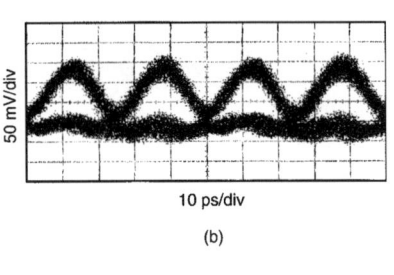

Figure 7.22: (a) Frequency response and (b) eye diagram at 40 Gb/s for an OEIC receiver made with a waveguide photodetector (WGPD). (After Ref. [84]; ©2000 IEEE.)

1.2 mm wide. It exhibited a responsivity of 0.62 A/W and a bandwidth of more than 40 GHz in the 1.55-μm wavelength region. Such a high bandwidth was realized using a distributed design for the electronic preamplifier integrated with the photodiode.

Figure 7.22(a) shows the measured frequency response [84] of the OEIC chip (solid curve) together with that of the waveguide photodiode alone (dashed curve). The integration of a photodiode with the HEMT improved the electric response but reduced the bandwidth. Nevertheless, this OEIC chip has a bandwidth of 46.5 GHz. The dotted line in this figure shows the frequency response of the packaged module. Since the solid and dashed lines nearly overlap, it is evident that packaging did not affect the bandwidth significantly. The OEIC chip also exhibited a clear eye opening at bit rates of up to 50 Gb/s. The eye diagram of an RZ signal at 40 GB/s is shown in Figure 7.22(b).

Packaging of optical receivers is another important issue [90]–[94]. It is common to package the receiver with a fiber pigtail so that it can be connected to the transmission line using a connector. The efficiency with which light from an optical fiber is coupled into the photodetector should be high since only a small amount of optical power is typically available at the end of an optical transmission line. Moreover, any optical feedback from the front facet of the photodetector should be minimized because unintentional reflections fed back into the transmission fiber can affect the system performance adversely. In practice, the fiber tip is cut at an angle to reduce the optical feedback. Several different techniques have been used to produce packaged optical receivers capable of operating at bit rates as high as 10 Gb/s. In one approach, an InGaAs APD was bonded to the Si-based IC by using the flip-chip technique [90]. Efficient fiber–APD coupling was realized by using a slant-ended fiber and a microlens monolithically fabricated on the photodiode. The fiber ferrule was directly laser-welded to the package wall with a double-ring structure for mechanical stability. The resulting receiver module withstood shock and vibration tests and had a bandwidth of 10 GHz.

A practical problem for high-speed OEIC receivers designed using a waveguide photodiode is that the size of the waveguide mode does not match with the fiber-mode size. In one approach, this problem is solved by integrating a spot-size converter with

7.5 Integrated Optical Receivers

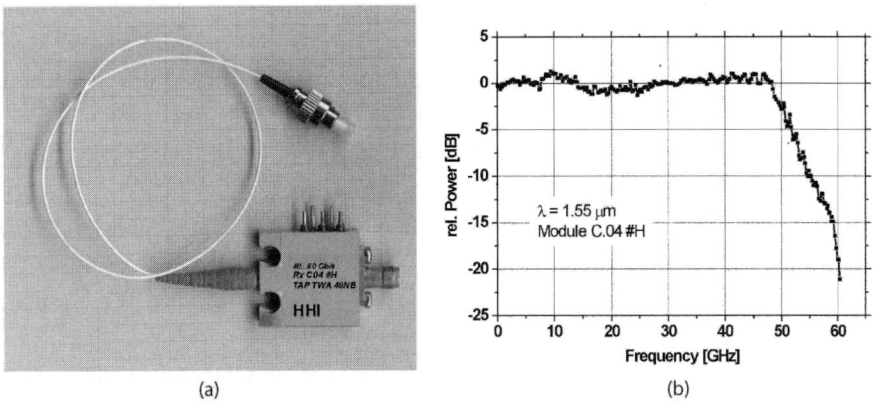

Figure 7.23: (a) A packaged OEIC receiver module made using a waveguide photodiode. The frequency response is shown on the right. (After Ref. [89]; ©2002 IEEE.)

the waveguide [89]. The OEIC chip is coupled to the fiber by placing the fiber tip directly against the tapered waveguide facet (the so-called butt coupling) and fixing it in place using an ultraviolet-curable resin. Figure 7.23(a) is a photograph of a packaged device with the fiber pigtail. The responsivity of the device was measured to be about 0.6 A/W and it remained unchanged after a 10-minute-long vibration test. The frequency response of such an OEIC receiver is shown in Figure 7.23(b). The response is flat to within 1 dB up to 47 GHz, and the 3-dB bandwidth exceeds 50 GHz. Such a device can operate at bit rates as high as 60 Gb/s.

Another approach to packaging makes use of a *planar lightwave circuit* platform containing silica waveguides on a silicon substrate (see Chapter 4). In one experiment, an InP-based OEIC receiver with two channels was flip-chip bonded to the platform [91]. The resulting module could detect two 10-Gb/s channels with negligible crosstalk. GaAs ICs have also been used to fabricate a compact receiver module capable of operating at a bit rate of 10 Gb/s [92]. By 2000, fully packaged 40-Gb/s receivers were available commercially [94].

For some applications such as Gigabit Ethernet, local-area networks, and optical interconnects, the bit rate is typically below 10 Gb/s but the receiver should be packaged to realize a low cost. Moreover, such receivers should be able to perform well over a wide temperature range extending from −40 to 85°C. The operating wavelength is typically near 850 nm because relatively inexpensive optical transmitters can be made in this spectral region using VCSELs. Silicon-based photodetectors are quite suitable for such lightwave systems. Indeed, several silicon-based technologies have been used to design low-cost OEIC receivers [95]–[102]. Among them, the use of complementary metal oxide–semiconductor (CMOS) technology has attracted considerable attention because its use allows one to take advantages of the enormous advances made by the computing industry in recent years. Such OEICs have been made using a feature size as small as 130 nm and can operate at bit rates as high as 8 Gb/s [102].

Problems

7.1 Calculate the responsivity of a photodetector at 1.3 and 1.55 μm if its quantum efficiency is 80%. Why is the photodetector more responsive at 1.55 μm?

7.2 Define the rise time and relate it to the rise time of an RC circuit containing a resistor in series with a capacitor.

7.3 Relate the bandwidth of a photodetector to the transit time and the RC time constant of the circuit.

7.4 What is the main difference between p–n and p–i–n photodiodes? Why do p–i–n photodiodes perform better?

7.5 What is meant by a resonant-cavity photodiode? What are the advantages offered by the resonant cavity?

7.6 What is meant by a waveguide photodiode? Sketch the design of such a device and discuss the advantages offered by the waveguide design.

7.7 Photons at a rate of 10^{10}/s are incident on an APD with responsivity of 6 A/W. Calculate the quantum efficiency and the photocurrent at the operating wavelength of 1.5 μm for an APD gain of 10.

7.8 Show by solving Eqs. (7.3.1) and (7.3.2) that the multiplication factor M is given by Eq. (7.3.5) for an APD in which electrons initiate the avalanche process. Treat α_e and α_h as constants.

7.9 Sketch the designs of SAM, SAGM, and SAGCM APD structures and discuss the advantages offered by them compared with the basic APD design.

7.10 How does an MSM photodetector work? Sketch a suitable design of such a photodetector and discuss the function of each part.

7.11 Draw a block diagram of a digital optical receiver showing its various components. Explain the function of each component. How is the signal used by the decision circuit related to the incident optical power?

7.12 The raised-cosine pulse shape of Eq. (7.5.6) can be generalized to generate a family of such pulses by defining

$$h_{\text{out}}(t) = \frac{\sin(\pi B t)}{\pi B t} \frac{\cos(\pi \beta B t)}{1 - (2\beta B t)^2},$$

where the parameter β varies between 0 and 1. Derive an expression for the transfer function $H_{\text{out}}(f)$ given by the Fourier transform of $h_{\text{out}}(t)$. Plot $h_{\text{out}}(t)$ and $H_{\text{out}}(f)$ for $\beta = 0, 0.5$, and 1.

References

[1] T. P. Lee and T. Li, in *Optical Fiber Telecommunications I*, S. E. Miller and A. G. Chynoweth, Eds., Academic Press, San Diego, CA, 1979, Chap. 18.

[2] R. G. Smith and S. D. Personick, in *Semiconductor Devices for Optical Communications*, H. Kressel, Ed., Springer, New York, 1980.

References

[3] S. R. Forrest, in *Optical Fiber Telecommunications II*, S. E. Miller and I. P. Kaminow, Eds., Academic Press, San Diego, CA, 1988, Chap. 14.

[4] B. L. Kasper, in *Optical Fiber Telecommunications II*, S. E. Miller and I. P. Kaminow, Eds., Academic Press, San Diego, CA, 1988, Chap. 18.

[5] S. B. Alexander, *Optical Communication Receiver Design*, Vol. TT22, SPIE Press, Bellingham, WA, 1995.

[6] R. J. Keyes, *Optical and Infrared Detectors*, Springer, New York, 1997.

[7] G. J. Brown, Ed., *Photodetectors Materials & Devices III*, SPIE Press, Bellingham, WA, 1998.

[8] M. J. Digonnet, Ed., *Optical Devices for Fiber Communication*, SPIE Press, Bellingham, WA, 1999.

[9] R. S. Tucker, A. J. Taylor, C. A. Burrus, G. Eisenstein, and J. M. Westfield, *Electron. Lett.* **22**, 917 (1986).

[10] K. Kishino, S. Ünlü, J. I. Chyi, J. Reed, L. Arsenault, and H. Morkoç, *IEEE J. Quantum Electron.* **27**, 2025 (1991).

[11] C. C. Barron, C. J. Mahon, B. J. Thibeault, G. Wang, W. Jiang, L. A. Coldren, and J. E. Bowers, *Electron. Lett.* **30**, 1796 (1994).

[12] I.-H. Tan, J. Dudley, D. I. Babić, D. A. Cohen, B. D. Young, E. L. Hu, J. E. Bowers, B. I. Miller, U. Koren, and M. G. Young, *IEEE Photon. Technol. Lett.* **6**, 811 (1994).

[13] I.-H. Tan, C.-K. Sun, K. S. Giboney, J. E. Bowers E. L. Hu, B. I. Miller, and R. J. Kapik, *IEEE Photon. Technol. Lett.* **7**, 1477 (1995).

[14] Y.-G. Wey, K. S. Giboney, J. E. Bowers, M. J. Rodwell, P. Silvestre, P. Thiagarajan, and G. Robinson, *J. Lightwave Technol.* **13**, 1490 (1995).

[15] K. Kato, S. Hata, K. Kwano, J. Yoshida, and A. Kozen, *IEEE J. Quantum Electron.* **28**, 2728 (1992).

[16] K. Kato, A. Kozen, Y. Muramoto, Y. Itaya, N. Nagatsuma, and M. Yaita, *IEEE Photon. Technol. Lett.* **6**, 719 (1994).

[17] K. Kato and Y. Akatsu, *Opt. Quantum Electron.* **28**, 557 (1996).

[18] T. Takeuchi, T. Nakata, K. Fukuchi, K. Makita, and K. Taguchi, *IEICE Trans. Electron.* **E82C**, 1502 (1999).

[19] K. Kato, *IEEE Trans. Microwave Theory Tech.* **47**, 1265 (1999).

[20] K. S. Giboney, R. L. Nagarajan, T. E. Reynolds, S. T. Allen, R. P. Mirin, M. J. W. Rodwell, and J. E. Bowers, *IEEE Photon. Technol. Lett.* **7**, 412 (1995).

[21] H. Ito, T. Furuta, S. Kodama, and T. Ishibashi, *Electron. Lett.* **36**, 1809 (2000).

[22] G. E. Stillman and C. M. Wolfe, in *Semiconductors and Semimetals*, Vol. 12, R. K. Willardson and A. C. Beer, Eds., Academic Press, San Diego, CA, 1977, pp. 291–393.

[23] H. Melchior, in *Laser Handbook*, Vol. 1, F. T. Arecchi and E. O. Schulz-Dubois, Eds., North-Holland, Amsterdam, 1972, pp. 725–835.

[24] R. J. McIntyre, *IEEE Trans. Electron. Dev.* **13**, 164 (1966); **19**, 703 (1972).

[25] J. C. Campbell, A. G. Dentai, W. S. Holden, and B. L. Kasper, *Electron. Lett.* **19**, 818 (1983).

[26] B. L. Kasper and J. C. Campbell, *J. Lightwave Technol.* **5**, 1351 (1987).

[27] L. E. Tarof, *Electron. Lett.* **27**, 34 (1991).

[28] L. E. Tarof, J. Yu, R. Bruce, D. G. Knight, T. Baird, and B. Oosterbrink, *IEEE Photon. Technol. Lett.* **5**, 672 (1993).

[29] J. Yu, L. E. Tarof, R. Bruce, D. G. Knight, K. Visvanatha, and T. Baird, *IEEE Photon. Technol. Lett.* **6**, 632 (1994).

[30] C. L. F. Ma, M. J. Deen, and L. E. Tarof, *IEEE J. Quantum Electron.* **31**, 2078 (1995).

[31] K. A. Anselm, H. Nie, C. Lenox, P. Yuan, G. Kinsey, J. C. Campbell, and B. G. Streetman, *IEEE J. Quantum Electron.* **34**, 482 (1998).

[32] T. Nakata, I. Watanabe, K. Makita, and T. Torikai, *Electron. Lett.* **36**, 1807 (2000).

[33] F. Capasso, in *Semiconductor and Semimetals*, Vol. 22D, W. T. Tsang, Ed., Academic Press, San Diego, CA, 1985, pp. 1–172.

[34] I. Watanabe, S. Sugou, H. Ishikawa, T. Anan, K. Makita, M. Tsuji, and K. Taguchi, *IEEE Photon. Technol. Lett.* **5**, 675 (1993).

[35] T. Kagawa, Y. Kawamura, and H. Iwamura, *IEEE J. Quantum Electron.* **28**, 1419 (1992); *IEEE J. Quantum Electron.* **29**, 1387 (1993).

[36] S. Hanatani, H. Nakamura, S. Tanaka, T. Ido, and C. Notsu, *Microwave Opt. Tech. Lett.* **7**, 103 (1994).

[37] I. Watanabe, M. Tsuji, K. Makita, and K. Taguchi, *IEEE Photon. Technol. Lett.* **8**, 269 (1996).

[38] I. Watanabe, T. Nakata, M. Tsuji, K. Makita, T. Torikai, and K. Taguchi, *J. Lightwave Technol.* **18**, 2200 (2000).

[39] A. R. Hawkins, W. Wu, P. Abraham, K. Streubel, and J. E. Bowers, *Appl. Phys. Lett.* **70**, 303 (1997).

[40] Y. Kang, P. Mages, A. R. Clawson, P. K. L. Yu, M. Bitter, Z. Pan, A. Pauchard, S. Hummel, and Y. H. Lo, *IEEE Photon. Technol. Lett.* **14**, 1593 (2002).

[41] Z. L. Liau and D. E. Mull. *Appl. Phys. Lett.* **56**, 737 (1990).

[42] C. Lenox, H. Nie, P. Yuan, G. Kinsey, A. L. Homles, B. G. Streetman, and J. C. Campbell, *IEEE Photon. Technol. Lett.* **11**, 1162 (1999).

[43] C. Cohen-Jonathan, L. Giraudet, A. Bonzo, and J. P. Praseuth, *Electron. Lett.* **33**, 1492 (1997).

[44] G. S. Kinsey, C. C. Hansing, A. L. Holmes, Jr., B. G. Streetman, J. C. Campbell, and A. G. Dentai, *IEEE Photon. Technol. Lett.* **12**, 416 (2000).

[45] T. Nakata, T. Takeuchi, I. Watanabe, K. Makita, and T. Torikai, *Electron. Lett.* **36**, 2033 (2000).

[46] G. S. Kinsey, J. C. Campbell, and A. G. Dentai, *IEEE Photon. Technol. Lett.* **13**, 842 (2001).

[47] P. Yuan, C. C. Hansing, K. A. Anselm, C. V. Lenox, H. Nie, A. L. Holmes, B. G. Streetman, and J. C. Campbell, *IEEE J. Quantum Electron.* **36**, 198 (2000).

[48] S. Wang, R. Sidhu, X. G. Zheng, X. Li, X. Sun, A. L. Holmes, and J. C. Campbell, *IEEE Photon. Technol. Lett.* **13**, 1346 (2001).

[49] S. Wang, J. B. Hurst, F. Ma, R. Sidhu, X. Sun, X. G. Zheng, A. L. Holmes, A. Huntington, L. A. Coldren, and J. C. Campbell, *IEEE Photon. Technol. Lett.* **14**, 1722 (2002).

[50] S. Wang, F. Ma, X. Li, R. Sidhu, X. Zheng, X. Sun, A. L. Holmes, and J. C. Campbell, *IEEE J. Quantum Electron.* **39**, 375 (2003).

[51] J. Burm, K. I. Litvin, D. W. Woodard, W. J. Schaff, P. Mandeville, M. A. Jaspan, M. M. Gitin, and L. F. Eastman, *IEEE J. Quantum Electron.* **31**, 1504 (1995).

[52] J. B. D. Soole and H. Schumacher, *IEEE J. Quantum Electron.* **27**, 737 (1991).

[53] J. H. Kim, H. T. Griem, R. A. Friedman, E. Y. Chan, and S. Roy, *IEEE Photon. Technol. Lett.* **4**, 1241 (1992).

[54] R.-H. Yuang, J.-I. Chyi, Y.-J. Chan, W. Lin, and Y.-K. Tu, *IEEE Photon. Technol. Lett.* **7**, 1333 (1995).

[55] O. Vendier, N. M. Jokerst, and R. P. Leavitt, *IEEE Photon. Technol. Lett.* **8**, 266 (1996).

References

[56] M. C. Hargis, S. E. Ralph, J. Woodall, D. McInturff, A. J. Negri, and P. O. Haugsjaa, *IEEE Photon. Technol. Lett.* **8**, 110 (1996).

[57] A. Bartels, E. Peiner, G.-P. Tang, R. Klockenbrink, H.-H. Wehmann, and A. Schlachetzki, *IEEE Photon. Technol. Lett.* **8**, 670 (1996).

[58] Y. G. Zhang, A. Z. Li, and J. X. Chen, *IEEE Photon. Technol. Lett.* **8**, 830 (1996).

[59] W. A. Wohlmuth, J. W. Seo, P. Fay, C. Caneau, and I. Adesida, *IEEE Photon. Technol. Lett.* **9**, 1388 (1997).

[60] E. Dröge, E. H. Bottcher, S. Kollakowski, A. Strittmatter, D. Bimberg, O. Reimann, and R. Steingruber, *Electron. Lett.* **34**, 2241 (1998).

[61] A. Umbach, T. Engel, H. G. Bach, S. van Waasen, E. Dröge, A. Strittmatter, W. Ebert, W. Passenberg, R. Steingruber, W. Schlaak, G. G. Mekonnen, G. Unterbörsch, and D. Bimberg, *IEEE J. Quantum Electron.* **35**, 1024 (1999).

[62] J. W. Shi, K. G. Gan, Y. J. Chiu, Y. H. Chen, C. K. Sun, Y. J. Yang, and J. E. Bowers, *IEEE Photon. Technol. Lett.* **16**, 623 (2001).

[63] R. G. Swartz, in *Optical Fiber Telecommunications II*, S. E. Miller and I. P. Kaminow, Eds., Academic Press, San Diego, CA, 1988, Chap. 20.

[64] K. Kobayashi, in *Optical Fiber Telecommunications II*, S. E. Miller and I. P. Kaminow, Eds., Academic Press, San Diego, CA, 1988, Chap. 11.

[65] T. Horimatsu and M. Sasaki, *J. Lightwave Technol.* **7**, 1612 (1989).

[66] O. Wada, H. Hamaguchi, M. Makiuchi, T. Kumai, M. Ito, K. Nakai, T. Horimatsu, and T. Sakurai, *J. Lightwave Technol.* **4**, 1694 (1986).

[67] M. Makiuchi, H. Hamaguchi, T. Kumai, O. Aoki, Y. Oikawa, and O. Wada, *Electron. Lett.* **24**, 995 (1988).

[68] K. Matsuda, M. Kubo, K. Ohnaka, and J. Shibata, *IEEE Trans. Electron. Dev.* **35**, 1284 (1988).

[69] H. Yano, K. Aga, H. Kamei, G. Sasaki, and H. Hayashi, *J. Lightwave Technol.* **8**, 1328 (1990).

[70] H. Hayashi, H. Yano, K. Aga, M. Murata, H. Kamei, and G. Sasaki, *IEE Proc.* **138**, Pt. J, 164 (1991).

[71] H. Yano, G. Sasaki, N. Nishiyama, M. Murata, and H. Hayashi, *IEEE Trans. Electron. Dev.* **39**, 2254 (1992).

[72] Y. Akatsu, M. Miyugawa, Y. Miyamoto, Y. Kobayashi, and Y. Akahori, *IEEE Photon. Technol. Lett.* **5**, 163 (1993).

[73] K. Takahata, Y. Muramoto, H. Fukano, K. Kato, A. Kozen, O. Nakajima, and Y. Matsuoka, *IEEE Photon. Technol. Lett.* **10**, 1150 (1998).

[74] S. Chandrasekhar, L. M. Lunardi, A. H. Gnauck, R. A. Hamm, and G. J. Qua, *IEEE Photon. Technol. Lett.* **5**, 1316 (1993).

[75] E. Sano, M. Yoneyama, H. Nakajima, and Y. Matsuoka, *J. Lightwave Technol.* **12**, 638 (1994).

[76] H. Kamitsuna, *J. Lightwave Technol.* **13**, 2301 (1995).

[77] L. M. Lunardi, S. Chandrasekhar, C. A. Burrus, and R. A. Hamm, *IEEE Photon. Technol. Lett.* **7**, 1201 (1995).

[78] M. Yoneyama, E. Sano, S. Yamahata, and Y. Matsuoka, *IEEE Photon. Technol. Lett.* **8**, 272 (1996).

[79] E. Sano, K. Kurishima, and S. Yamahata, *Electron. Lett.* **33**, 159 (1997).

[80] W. P. Hong, G. K. Chang, R. Bhat, C. K. Nguyen, and M. Koza, *IEEE Photon. Technol. Lett.* **3**, 156 (1991).

[81] P. Fay, W. Wohlmuth, C. Caneau, and I. Adesida, *Electron. Lett.* **31**, 755 (1995).

[82] P. Fay, M. Arafa, W. A. Wohlmuth, C. Caneau, S. Chandrasekhar, and I. Adesida, *J. Lightwave Technol.* **15**, 1871 (1997).

[83] G. G. Mekonnen, W. Schlaak, H. G. Bach, R. Steingruber, A. Seeger, T. Enger, W. Passenberg, A. Umbach, C. Schramm, G. Unterborsch, and S. van Waasen, *IEEE Photon. Technol. Lett.* **11**, 257 (1999).

[84] K. Takahata, Y. Muramoto, H. Fukano, K. Kato, A. Kozen, S. Kimura, Y. Imai, Y. Miyamoto, O. Nakajima, and Y. Matsuoka, *IEEE J. Sel. Topics Quantum Electron.* **6**, 31 (2000).

[85] N. Shimizu, K. Murata, A. Hirano, Y. Miyamoto, H. Kitabayashi, Y. Umeda, T. Akeyoshi, T. Furuta, and N. Watanabe, *Electron. Lett.* **36**, 1220 (2000).

[86] S. Pradhan, P. Bhattacharya, and W. K. Liu, *Electron. Lett.* **38**, 987 (2002).

[87] B. Mason, S. Chandrasekhar, A. Ougazzaden, C. Lentz, J. M. Geary, L. L. Buhl, L. Peticolas, K. Glogovsky, J. M. Freund, L. Reynolds, G. Przybylek, F. Walters, A. Sirenko, J. Boardman, T. Kercher, M. Rader, J. Grenko, D. Monroe, and L. Ketelsen, *Electron. Lett.* **38**, 1196 (2002).

[88] A. Leich, V. Hurm, J. Sohn, T. Feltgen, W. Bronner, K. Kohler, H. Walcher, J. Rosenzweig, and M. Schlechtweg, *Electron. Lett.* **38**, 1706 (2002).

[89] H. G. Bach, A. Beling, G. G. Mekonnen, and W. Schlaak, *IEEE J. Sel. Topics Quantum Electron.* **8**, 1445 (2001).

[90] Y. Oikawa, H. Kuwatsuka, T. Yamamoto, T. Ihara, H. Hamano, and T. Minami, *J. Lightwave Technol.* **12**, 343 (1994).

[91] T. Ohyama, S. Mino, Y. Akahori, M. Yanagisawa, T. Hashimoto, Y. Yamada, Y. Muramoto, and T. Tsunetsugu, *Electron. Lett.* **32**, 845 (1996).

[92] Y. Kobayashi, Y. Akatsu, K. Nakagawa, H. Kikuchi, and Y. Imai, *IEEE Trans. Microwave Theory Tech.* **43**, 1916 (1995).

[93] K. Emura, *Solid-State Electron.* **43**, 1613 (1999).

[94] M. Bitter, R. Bauknecht, W. Hunziker, and H. Melchior, *IEEE Photon. Technol. Lett.* **12**, 74 (2000).

[95] D. H. Hartman, M. K. Grace, and C. R. Ryan, *J. Lightwave Technol.* **3**, 729 (1985).

[96] C. L. Schow, J. D. Schaub, R. Li, J. Qi, and J. C. Campbell, *IEEE Photon. Technol. Lett.* **11**, 120 (1999).

[97] R. Li, J. D. Schaub, S. M. Csutak, and J. C. Campbell, *IEEE Photon. Technol. Lett.* **12**, 1046 (2000).

[98] T. K. Woodward, A. V. Krishnamoorthy, R. G. Rozier, and A. L. Lentine, *Electron. Lett.* **36**, 1489 (2000).

[99] O. Qasaimeh, Z. Ma, P. Bhattacharya, and E. T. Croke, *J. Lightwave Technol.* **18**, 1548 (2000).

[100] J. D. Schaub, R. Li, S. Csutak, and J. C. Campbell, *J. Lightwave Technol.* **19**, 272 (2001).

[101] H. Zimmermann, T. Heide, and A. Ghazi, *IEEE Photon. Technol. Lett.* **13**, 711 (2001).

[102] S. M. Csutak, J. D. Schaub, W. E. Wu, R. Shimer, and J. C. Campbell, *J. Lightwave Technol.* **20**, 2002 (2002).

Chapter 8

WDM Components

In principle, single-mode fibers used for optical communication systems have enormous capacity and can transport information at bit rates exceeding 10 Tb/s. In practice, however, the bit rate was limited to 10 Gb/s or less until 1995 by the speed of electronic components and by the limitations imposed by the dispersive and nonlinear effects occurring inside optical fibers. Since then, transmission of multiple optical channels over the same fiber through wavelength-division multiplexing (WDM) has extended the system capacity to beyond 1 Tb/s. As discussed in this chapter, many new passive and active components have been developed for WDM lightwave systems. Section 8.1 focuses on the most basic component—a tunable optical filter—that is needed for selecting a specific channel. Section 8.2 is devoted to multiplexers and demultiplexers that are required for all WDM systems for combining and splitting individual channels. Add–drop multiplexers capable of adding or dropping a specific channel are considered in Section 8.3. The focus of Section 8.4 is WDM devices that integrate multiple active and passive optical components such as lasers, demultiplexers, amplifiers, modulators, and detectors on the same chip to extend the functionality while reducing the overall cost at the same time.

8.1 Optical Filters

All optical filters modify the spectrum of incident light [1]. They can be classified into two broad categories depending on whether optical interference or diffraction is the underlying physical mechanism. Each category can be further subdivided according to the scheme adopted. Figure 8.1 shows four examples (a–d) of the schemes that can be used for making *tunable* optical filters, which not only select a specific channel within a WDM signal but also allow the wavelength of the selected channel to be changed. The desirable properties of a tunable optical filter include: wide tuning range to maximize the number of channels that can be selected, negligible crosstalk to avoid interference from adjacent channels, fast tuning speed to minimize the access time, small insertion loss, insensitivity to signal polarization, stability against environmental changes (humidity, temperature, vibrations, etc.), and last but not least, low cost.

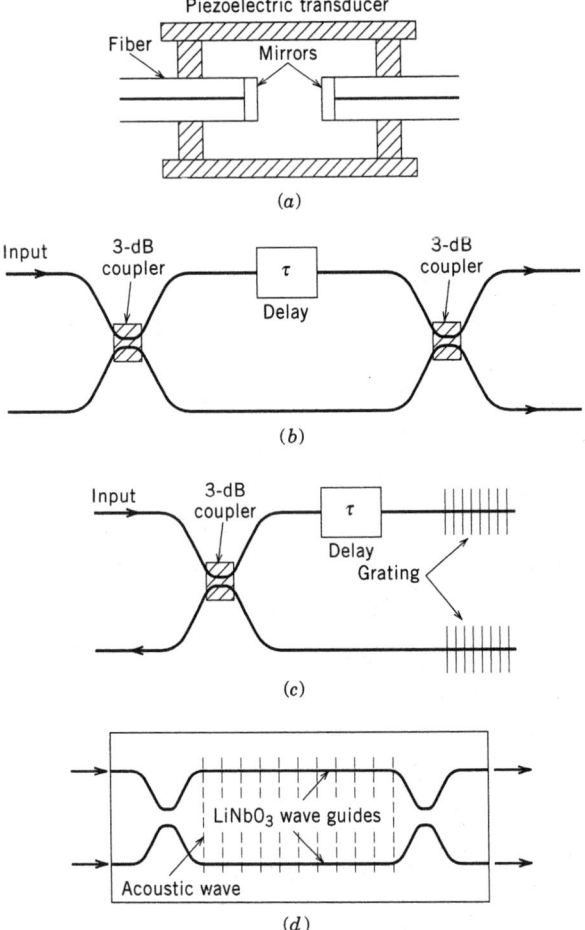

Figure 8.1: Four kinds of filters based on various interferometric and diffractive devices: (a) Fabry–Perot filter; (b) Mach–Zehnder filter; (c) Michelson filter; and (d) acousto-optic filter.

Mathematically, an optical filter is characterized by its transfer function $H_f(\omega)$. It modifies the incident optical field $A_{\text{in}}(t)$ such that the output field $A_{\text{out}}(t)$ is given by

$$A_{\text{out}}(t) = \frac{1}{2\pi} \int_{-\infty}^{\infty} H_f(\omega) \tilde{A}_{\text{in}}(\omega) \exp(-i\omega t)\, d\omega, \tag{8.1.1}$$

where $\tilde{A}_{\text{in}}(\omega)$ is the Fourier transform of the input field as defined in Eq. (1.5.17). In general, $H_f(\omega)$ is complex and can be written as

$$H_f(\omega) = |H_f(\omega)| \exp[i\phi(\omega)] \approx |H_f(\omega)| \exp[i(\phi_0 + \phi_1 \omega + \tfrac{1}{2}\phi_2 \omega^2 + \cdots)], \tag{8.1.2}$$

where $\phi_m = d^m \phi / d\omega^m$ ($m = 0, 1, \ldots$) is evaluated at the optical carrier frequency ω_0. The constant phase ϕ_0 can be ignored in practice. If we substitute Eq. (8.1.2) in Eq.

8.1 Optical Filters

(8.1.1) it is easy to see that the ϕ_1 term introduces a time delay that corresponds to the time light takes while passing through the filter. The ϕ_2 and other higher-order terms govern the dispersion characteristics of the optical filter. The modulus of transfer function, $|H_f(\omega)|$, governs the bandwidth of the filter. It is a single-peak function for some filters (such as a Bragg grating) but may contain multiple peaks for other filters, typically repeating periodically.

8.1.1 Fabry–Perot Filters

As discussed in Section 2.3.2, a Fabry–Perot (FP) resonator in its simplest form consists of a cavity formed with two high-reflectivity mirrors. The mirrors can also be made using fiber gratings, resulting in an all-fiber FP filter (see Figure 2.13). Figure 8.1(a) shows another fiber-based scheme for constructing a tunable FP filter. In this device, the wavelength of the selected channel is tuned by changing the width of an air gap between the two fibers. The filter bandwidth must be large enough to transmit the desired channel but, at the same time, small enough to block the neighboring channels.

The transfer function of an FP filter whose mirrors have the same reflectivity R_m can be obtained from Eq. (2.3.7) using $H_f(\omega) = A_t/A_i$ and is given by

$$H_f(\omega) = \frac{(1-R_m)e^{i\pi}}{1-R_m \exp(i\omega\tau_r)}, \tag{8.1.3}$$

where $\tau_r = 2L_f/v_g$ is the round-trip time inside the filter. The condition $\omega\tau_r = 2\pi m$, where m is an integer, determines the frequencies of the longitudinal modes associated with an FP resonator of length L_f. The FP filter constitutes an example of a multipeak filter because its transmission varies periodically with frequency and becomes maximum whenever the input frequency v coincides with one of the longitudinal modes of the resonator, that is, $v = v_m = m/\tau_r$. Figure 2.14 shows the multipeak transfer function of an FP filter for several values of mirror reflectivities. The sharpness of each peak is controlled by the finesse F_R, which increases rapidly as R_m approaches 1. The frequency spacing between two successive transmission peaks, or the *free spectral range*, is related to the round-trip time τ_r as

$$\Delta v_L = 1/\tau_r = v_g/(2L_f). \tag{8.1.4}$$

Such a device can act as a tunable optical filter if its length is controlled electronically using a piezoelectric transducer as shown in Figure 8.1(a).

The filtering action of an FP filter can be understood from Figure 8.2. If the filter is designed to pass a single channel, the combined bandwidth of the multichannel signal, $\Delta v_{sig} = N\Delta v_{ch} = NB/\eta_s$, must be less than Δv_L, where N is the number of channels, Δv_{ch} is the channel spacing, $\eta_s = B/\Delta v_{ch}$ is the spectral efficiency, and B is the bit rate. At the same time, the filter bandwidth Δv_{FP} (the width of the transmission peak in Figure 8.2) should be large enough to pass the entire frequency contents of the selected channel. Typically, $\Delta v_{FP} \sim B$. The number of channels is thus limited by

$$N < \eta_s(\Delta v_L/\Delta v_{FP}) = \eta_s F_R, \tag{8.1.5}$$

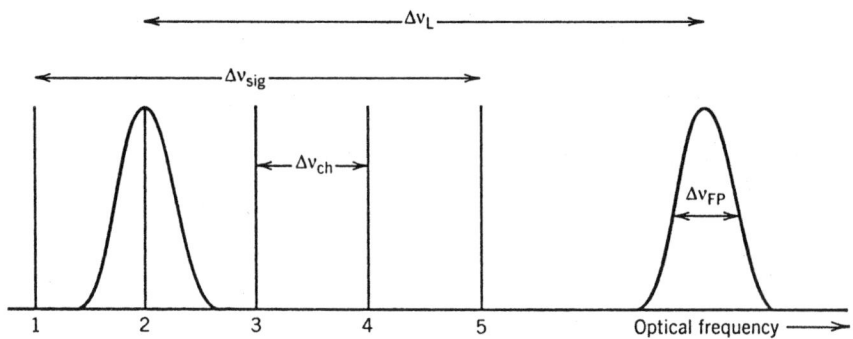

Figure 8.2: Channel selection through a tunable optical filter.

where $F_R = \Delta v_L/\Delta v_{FP}$ is the finesse of the FP filter [2]. If internal losses are negligible, the finesse is given by Eq. (2.3.9) and is determined solely by the reflectivity R_m of the two mirrors.

Equation (8.1.5) provides a remarkably simple condition for the number of channels that an FP filter can resolve. As an example, if $\eta_s = \frac{1}{3}$, an FP filter with 99% reflecting mirrors can select up to 104 channels. Channel selection is made by changing the filter length electronically. The length needs to be changed by only a fraction of the wavelength to tune the filter. The filter length L_f itself is determined from Eq. (8.1.4) together with the condition $\Delta v_L > \Delta v_{sig}$. As an example, for a 10-channel WDM signal with 0.8-nm channel spacing, $\Delta v_{sig} \approx 1$ THz. If $n_g = 1.5$ is used for the group index, L_f should be smaller than 100 μm. Such a short length together with the requirement of high mirror reflectivities underscores the complexity of the design of FP filters for WDM applications.

An all-fiber design of FP filters shown in Figure 8.1(a) uses an air gap between two optical fibers as a resonant cavity. The two fiber ends forming the gap are coated to act as high-reflectivity mirrors [3]. The entire structure is enclosed in a piezoelectric chamber so that the gap length can be changed electronically for tuning and selecting a specific channel. An advantage of fiber-based FP filters is that they can be integrated within a transmission system without incurring coupling losses. The number of channels is typically limited to below 100 ($F \approx 155$ for the 98% mirror reflectivity) but can be increased using two FP filters in tandem. Although tuning is relatively slow (~ 1 ms) because of the mechanical nature of the tuning mechanism, it suffices for some applications.

Tunable FP filters can also be made using several other materials such as liquid crystals, dielectric thin films, and semiconductor waveguides. Liquid-crystal-based filters make use of the anisotropic nature of liquid crystals that makes it possible to change the refractive index electronically [4]–[6]. An FP cavity is still formed by enclosing the liquid-crystal material within two high-reflectivity mirrors, but tuning is done by changing the refractive index rather than the cavity length. Such FP filters can provide high finesse ($F_R \sim 300$) with a bandwidth of about 0.2 nm [4]. They can be tuned electrically over 50 nm, but switching time is typically ~ 1 ms or more when

8.1 Optical Filters

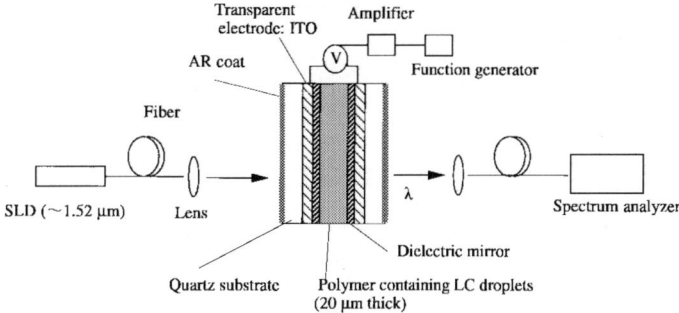

Figure 8.3: Schematic of a tunable optical filter made using nanosize droplets of a liquid crystal; SLD: semiconductor laser diode, ITO: indium tin oxide, AR: antireflection, LC: liquid crystal. (After Ref. [6]; ©1999 IEEE.)

nematic liquid crystals are used. It can be reduced to below 10 μs by using smectic liquid crystals [5].

All liquid-crystal filters suffer from polarization sensitivity and work only when the input field is polarized along a specific direction. This problem is solved in practice using a polarization-diversity technique in which the input field is first separated into two orthogonally polarized branches, filtered in each branch using an optical filter, and then recombined at the output end. In a novel approach, nanometer-size droplets of a liquid crystal (diameter <100 nm) are used to solve the polarization problem [6]. Figure 8.3 shows the device structure schematically. Mirrors are made by coating two quartz plates with a thin film of indium-tin oxide, providing a reflectivity of about 99% in the spectral region near 1.55 μm: The same film acts as a transparent electrode. The other side of each quartz plate is coated with an antireflection layer to avoid parasitic reflections. A polymer film containing liquid-crystal droplets is sandwiched between the two quartz plates. Droplets are formed by exposing the spin-coated liquid-crystal film containing a UV-curable polymer to intense UV light. The polarization-insensitive nature of such a filter results from the random orientation of the droplets.

Thin dielectric films are commonly used for making narrowband interference filters [1]. The basic idea is quite simple. A stack of suitably designed thin films acts as a high-reflectivity mirror. If two such mirrors are separated by a spacer dielectric layer, an FP cavity is formed that acts as an optical filter. The bandpass response can be tailored by using a multicavity filter formed with multiple thin-film mirrors separated by several spacer layers. Such FP filters are hard to tune because the thickness of the spacer layer is fixed and its refractive index is difficult to change except thermally. The use of semiconductors solves this problem to a large extent.

Several different designs have been used to realize FP filters with semiconductor materials. Silicon-based FP filters can be tuned using the thermo-optic effect [7]. For faster tuning rates, an electronic scheme should be used. In one approach, an In-GaAsP/InP ridge waveguide is etched to form two grooves using chemically assisted ion-beam etching [8]. Dielectric coatings are then deposited onto the etched facets to form two high-reflectivity mirrors. The refractive index of the waveguide material can

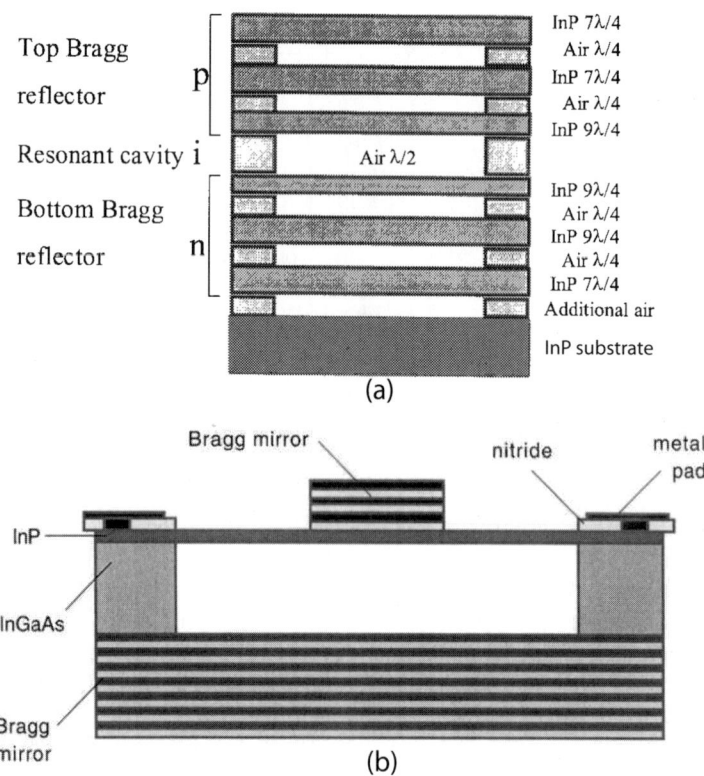

Figure 8.4: Schematic structure of two micromachined FP filters made using semiconductor layers with an adjustable air gap. (After Refs. [14] and [15]; ©1999 IEEE.)

be changed by injecting an external current, making it possible to tune the FP filter electronically. In essence, the structure is similar to that used for semiconductor lasers except that the waveguide is made transparent at the operating wavelength and remains passive even when electric current is injected.

Micromechanical tuning can also be used for semiconductor-based FP filters, and several designs have been proposed for making such devices [9]–[19]. Figure 8.4 shows two commonly used structures fabricated using micromachining techniques. In design (a), an air gap of thickness $\lambda/2$ is surrounded by two distributed Bragg reflectors (DBRs). Normally, one needs more than 40 alternating layers for forming a high-reflectivity DBR when semiconductors such as InGaAsP and InP are used. However, the number of layers required is reduced to just a few when InP layers are alternated with $\lambda/4$-thick air gaps [14]. Even when the thickness of InP layers is $7\lambda/4$, the 5-layer DBR shown in Figure 8.4(a) can provide 99.6% reflectivity over a 230-nm bandwidth. Such an FP filter exhibited in 1999 a tuning range of 62 nm with 0.6-nm bandwidth in the 1.55-μm spectral region when a voltage of up to 14 V was applied. In the structure shown in Figure 8.4(b), the top Bragg mirror (with a size of 30×30 μm^2) is suspended on four cantilever beams and can be moved by as much as 50 nm using electrostatic

8.1 Optical Filters

attraction [15]. The performance of micromachined FP filters has improved to the extent that by 2003 they were able to provide a tuning range of 140 nm with a maximum voltage of only 3.2 V [19].

8.1.2 Mach–Zehnder Filters

Mach–Zehnder (MZ) interferometers can also be used for making a tunable optical filter. As discussed in Section 2.3.4 and shown in Figure 8.1(b), an MZ interferometer can be constructed simply by connecting the two output ports of a 3-dB coupler to the two input ports of another 3-dB coupler. The first coupler splits the input signal equally into two parts, which acquire different phase shifts (if the arm lengths are made different) before they interfere at the second coupler. Since the relative phase shift is wavelength-dependent, the transmission through the bar (or cross) port also depends on the signal wavelength. This wavelength dependence converts an MZ interferometer into an optical filter.

The transfer function of the MZ filter can be obtained from Eq. (2.3.21) based on the transfer-matrix approach [20]. It can also be obtained following the method discussed in Section 4.5.1 and adding the contribution of all possible paths that an input signal can take while reaching the output end [1]. Following the second method, the transfer function through the bar and cross ports can be written as

$$H_b(\omega) = \sqrt{\rho_1 \rho_2} \exp(i\omega\tau) + i^2 \sqrt{(1-\rho_1)(1-\rho_2)}, \tag{8.1.6}$$

$$H_c(\omega) = i\sqrt{\rho_1(1-\rho_2)} \exp(i\omega\tau) + i\sqrt{\rho_2(1-\rho_1)}, \tag{8.1.7}$$

where ρ_1 and ρ_2 represent the bar-port transmissivity for the two couplers and $\tau = n_f \Delta L/c$ is the relative delay introduced when the two arms of the MZ interferometer differ by ΔL. Similar to the case of FP filters, the frequency response of an MZ filter is also periodic as both H_b and H_c are periodic functions of ω. As a simple example, consider the case when the MZ filter is made using two 3-dB couplers so that $\rho_1 = \rho_2 = \frac{1}{2}$. It is easy to see from Eqs. (8.1.6) and (8.1.7) that

$$|H_b(\omega)|^2 = \sin^2(\omega\tau/2), \qquad |H_c(\omega)|^2 = \cos^2(\omega\tau/2). \tag{8.1.8}$$

When Y-junction couplers are used in place of 3-dB directional couplers, the MZ filter has only single input and output ports. In this case, the transfer function is given by $|H_f(\omega)|^2 = \cos^2(\omega\tau/2)$. An optical filter with a sinusoidal transfer function is not sharp enough to be useful for selecting a specific channel in a WDM system. A much improved spectral response can be realized by cascading multiple MZ interferometers in series. Figure 8.5 shows an example of a filter with three cascaded MZ interferometers. In general, one can form a cascaded chain of M interferometers with different relative delays for each member of the chain. The transfer function of such a MZ chain can be obtained using the transfer matrices in Eq. (4.5.1). It can also be written as a sum over 2^M paths that the input signal may take in such a chain. In the specific case of 3-dB couplers, the transfer function can be written in the simple form

$$|H_f(\omega)|^2 = \prod_{m=1}^{M} \cos^2(\omega\tau_m/2), \tag{8.1.9}$$

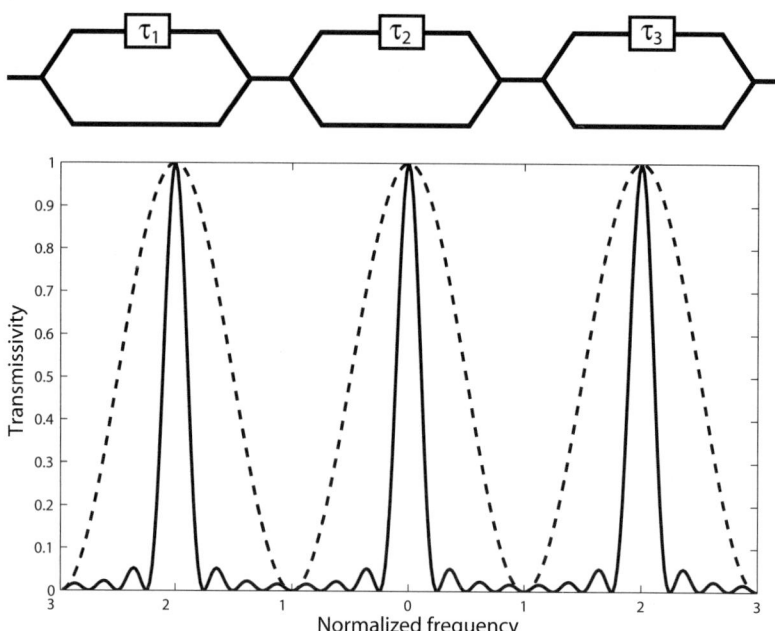

Figure 8.5: A three-stage MZ filter and its transfer function $|H_f(\omega)|^2$ plotted as a function of $\omega\tau/\pi$. The dotted curve shows for comparison the case of a single-stage MZ filter.

where the delays τ_m ($m = 1\text{–}M$) are M adjustable parameters. Figure 8.5 shows the transfer function of an MZ filter for $M = 3$ for the choice $\tau_1 = \tau$, $\tau_2 = 2\tau$, and $\tau_3 = 4\tau$. The $M = 1$ case is also shown for comparison. It is evident that the transmission peaks can be made quite narrow by increasing M.

In practice, the choice of delays depends on the specific application. A commonly used technique for WDM applications implements the relative delays such that each MZ stage blocks the alternate channels successively. This scheme requires $\tau_m = (2^m \Delta \nu_{\text{ch}})^{-1}$ for a channel spacing of $\Delta \nu_{\text{ch}}$. The resulting transmissivity of a 10-stage MZ chain has channel selectivity as good as that offered by an FP filter having a finesse of 1,600. Moreover, such a filter is capable of selecting closely spaced channels.

An all-fiber optical filter in the form of an MZ chain can be built by connecting multiple fiber couplers. However, a much more compact device can be fabricated in the form of a *planar lightwave circuit* [21]–[27] using the silica-on-silicon technology (see Section 4.3.3). Figure 8.6 shows the layout of such a device schematically. It consists of 24 cascaded MZ interferometers with asymmetric arm lengths. A chromium heater is deposited on one arm of some MZ interferometers to provide thermo-optic control of the optical phase. Such planar lightwave circuits are quite compact (with a size of ∼5 × 7 cm^2) but exhibit rather large insertion losses (>8 dB). The main advantage is that their dispersion characteristics can be controlled by changing the arm lengths and the number of MZ interferometers. Tuning is realized by changing the refractive index in one arm of each MZ interferometer (through temperature changes). Since the tuning

8.1 Optical Filters

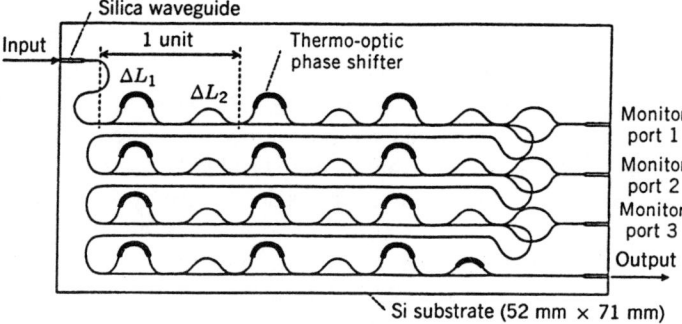

Figure 8.6: A planar lightwave circuit made using a chain of asymmetric MZ interferometers. (After Ref. [21]; ©1996 IEEE.)

mechanism is thermal, the switching time is ∼1 ms.

Tuning can be made much faster using LiNbO$_3$ waveguides because the optical path length of each MZ arm can be changed by changing the refractive index of the material electronically. Such a device is similar to a LiNbO$_3$ modulator except that the MZ interferometer is unbalanced because its two arms have different lengths [28]. Another alternative is to make use of semiconductor waveguides grown on GaAs or InP substrates. Such MZ devices can be made quite compact (with a size of ∼1 mm or less) and they can be coupled with other components such as optical amplifiers. Figure 8.7 shows an example of a GaAs-based notch filter fabricated by integrating a microring in one arm of a MZ interferometer [29]. The GaAs core of the waveguide is 0.5 μm thick and is sandwiched between two Al$_{0.5}$Ga$_{0.5}$As cladding layers. The MZ arms are separated by 10 μm and are less than 0.1 mm long. The transmission of such a notch filter drops to nearly zero periodically at well-defined frequencies. It can be used in any spectral region beyond 850 nm because photon energy is then less than the bandgap energy. The main drawback of such filters is large insertion losses occurring when the waveguide is coupled to an optical fiber.

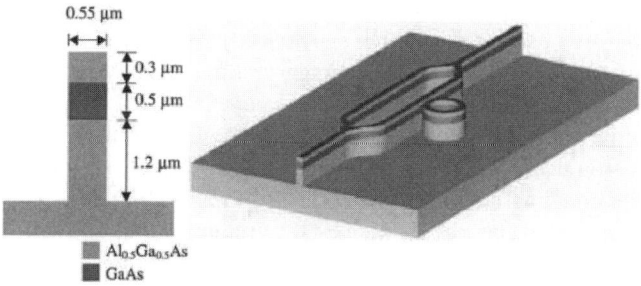

Figure 8.7: Schematic of a GaAs-based notch filter fabricated by integrating a microring in one arm of a MZ interferometer. The waveguide design is shown on the left. (After Ref. [29]; ©2000 IEEE.)

8.1.3 Bragg-Grating Filters

A separate class of tunable optical filters makes use of the wavelength selectivity provided by a Bragg grating. As discussed in Section 2.2, Bragg gratings can be made within the core of an optical fiber and act as a narrowband optical filter because they reflect light if its wavelength falls within the stop band [30]. Thus, a fiber grating acts as a reflection filter whose central wavelength is determined by the Bragg wavelength λ_B related to the grating period Λ as $\lambda_B = 2\bar{n}\Lambda$, where $\bar{n}$ is the effective index of the fiber mode. Figure 2.8 shows the reflectivity spectrum of a 7.5-cm-long fiber grating. It exhibits a bandwidth of about 25 GHz (0.2 nm in the 1.55-μm spectral region). Such a grating can reflect a single channel when a WDM signal with a channel spacing of 40 GHz or more is incident on it. The filter bandwidth can be tailored by changing the grating strength or by chirping the grating period slightly.

The reflective nature of fiber gratings requires in practice the use of an *optical circulator* if the selected channel needs to be transmitted rather than reflected. Figure 2.21(b) shows how an optical circulator can convert a fiber grating into a transmission filter. Several other schemes employ the wavelength selectivity of a fiber grating for providing transmission-based filters [30]. In one approach, fiber gratings are used as mirrors of an FP filter, resulting in transmission filters whose free spectral range can vary over a wide range of 0.1–10 nm [31]. Another scheme based on two Bragg gratings is shown schematically in Figure 8.1(c). If two identical Bragg gratings are fabricated near the two output ports of a 3-dB fiber coupler, the channel whose wavelength coincides with the Bragg wavelength of the gratings will appear at the second input port when a multichannel signal is launched from one input port [32]. Since this channel has been physically separated from the input signal, the whole device acts as an optical filter. The reason behind the physical separation is related to the $\pi/2$ phase shift introduced at the coupler each time light passes over to the cross port. One can think of such a device as a Michelson interferometer built with two wavelength-selective mirrors. Other kinds of interferometers discussed in Chapter 2, such as the Sagnac and MZ interferometers, can also be used to realize grating-based optical filters. Most of such schemes can also be implemented in the form of a planar lightwave circuit by forming silica waveguides on a silicon substrate.

The use of a circulator or an interferometer can be avoided by introducing a $\lambda/4$ phase shift in the middle of the grating. Such a phase shift opens a narrow transmission peak within the stop band of the grating and converts a fiber grating into a narrowband transmission filter [33]. Phase-shifted Bragg gratings have been fabricated using a variety of techniques [34]–[38]. It is possible to open multiple transmission peaks by creating multiple phase shift regions along the length of a grating. Figure 8.8 shows the three transmission peaks within the stop band of a grating of length L_g using $\kappa L_g = 5$, where κ is the coupling coefficient given in Eq. (2.2.10), in which three $\lambda/4$ phase shift regions are spaced apart by $L_g/4$ along the grating length. The peak positions can be altered by changing the amount of phase shift as well as the location of phase-shift regions [33].

Another method for creating a transmission peak within the stop band is to write two gratings with slightly different periods over the same fiber section, creating a Moiré fringe pattern [39]. Physically, the two gratings have different phase characteristics and

8.1 Optical Filters

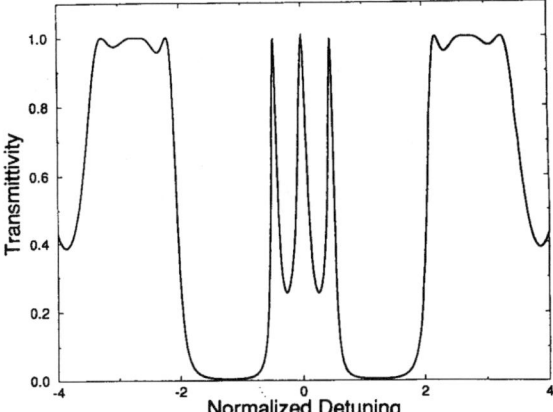

Figure 8.8: Narrow transmission peaks within the stop band of a fiber grating designed with three $\lambda/4$ phase shifts and $\kappa L_g = 3$. (After Ref. [33]; ©1994 IEEE.)

can thus become out of phase for some specific wavelengths, resulting in a transmission peak at those wavelengths [30]. Another technique makes use of a chirped grating because the stop band can be made quite wide (10 nm or more) for such gratings by varying the grating period along its length. Physically, such a wide stop band results from a superposition of many mini stop bands centered at the local Bragg wavelength associated with each small grating section. Thus, a simple way to open a transmission window at a specific location within the stop band of a chirped grating consists of blanking out a section of the grating that corresponds to that specific Bragg wavelength. This scheme is quite effective in creating multiple transmission windows at preset wavelengths and can provide an optical filter with a multipeak transfer function [40].

A Bragg grating does not exhibit periodic filtering characteristics because it has a single stop band centered at the Bragg wavelength. This property can be changed by making a superstructure or sampled grating discussed in Section 5.2.3 within the context of tunable semiconductor lasers. The basic idea can be understood from the concept of phase-shifted gratings. A single phase shift opens one transmission window within the stop band. Multiple windows can be formed by implementing multiple phase-shift regions. Since a phase-shift region is just a small section of the grating over which $\kappa = 0$ (no grating), one can make a grating in which such regions occur at periodic intervals over the entire length of the grating. Such devices are doubly periodic and are called superstructure gratings because they contain regions over which κ alternates from 0 to a finite value [40]–[42].

The periodic nature of κ results in a transfer function with multiple spectral peaks. This can be seen mathematically by writing the refractive index within the sampled grating as (see Section 2.2.4)

$$n(z) = \bar{n} + \text{Re}[\delta n_1(z) \exp(2\pi i z/\Lambda)], \tag{8.1.10}$$

where $\bar{n}$ is the mode index in the absence of the grating, Λ is the grating period, and

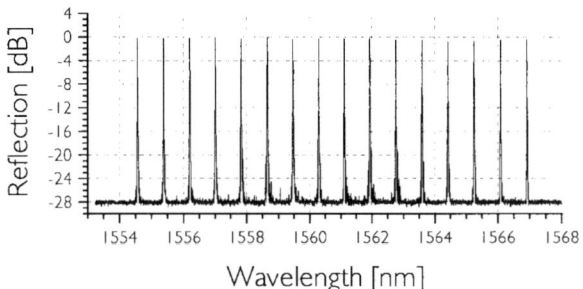

Figure 8.9: Reflection spectrum of a 10-cm-long sampled fiber grating designed to contain 16 peaks with 100 GHz spacing. (After Ref. [43]; ©1998 IEEE.)

δn_1 represents the amplitude of the grating. As discussed in Section 2.2.4, the coupling parameter κ is directly proportional to δn_1. In a superstructure grating $\delta n_1(z)$ varies along the grating length in a periodic manner. Expanding it in a Fourier series, we can write Eq. (8.1.10) as

$$n(z) = \bar{n} + \text{Re}\left(\sum_m F_m \exp[2i(\beta_B + m\beta_s)z]\right), \qquad (8.1.11)$$

where $\beta_B = \pi/\Lambda$ and $\beta_s = \pi/\Lambda_s$. It is common to chirp the grating period Λ to enlarge the stop band. The sampling period Λ_s associated with amplitude modulation determines the location of periodically located reflection peaks within the stop band. The frequency spacing $\Delta\nu$ between the two neighboring peaks is related to Λ_s as $\Delta\nu = c/(2\bar{n}\Lambda_s)$. In practice, such devices are made by placing an amplitude mask before the ultraviolet light creates an interference pattern within the fiber core [40]–[42].

Sampled gratings have evolved considerably from the simple design in which the grating amplitude δn_1 is varied in a digital fashion between two fixed values. The basic idea consists of choosing the amplitude and phase of the periodic modulation function $\delta n_1(z)$ such that the transmission characteristics of the sampled grating can be tailored to specific applications [43]–[49]. As early as 1998, it was shown that the use of a periodically repeating "sinc" function $(\sin x/x)$ for $\delta n_1(z)$ produces a much more desirable transfer function for sampled gratings. More specifically, the stop band of the grating acquires a rectangular shape and ensures that all reflection peaks have the same height. Figure 8.9 shows the reflection spectrum of a 10-cm-long sampled grating designed with this approach [43]. The sampling period $\Lambda_s \approx 1$ cm was chosen to provide 100 GHz spacing among peaks. Filter characteristics can be further improved for certain applications by using a pure phase-modulation sampling function in which only the phase of $\delta n_1(z)$ is modulated in a periodic function [48].

The spectral response of a fiber grating can be tuned by stretching or compressing the grating so that the grating period Λ, and the sampling period Λ_s in the case of a sampled grating, may be changed [50]–[54]. Tuning over a range of 40 nm was realized in 2002 by using the compression technique [53]. Figure 8.10(a) shows how the grating is mounted on a hybrid-material substrate that is bent to induce a com-

8.1 Optical Filters

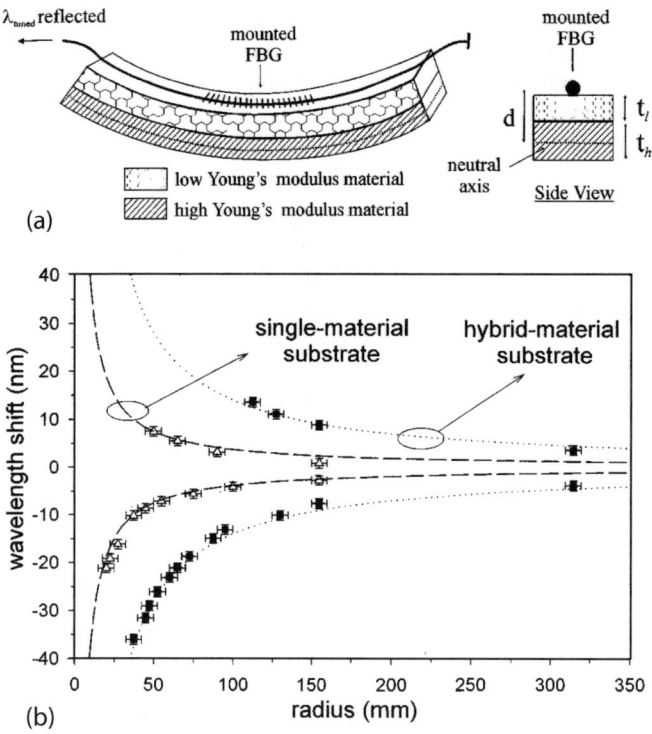

Figure 8.10: (a) Fiber Bragg grating mounted on a hybrid-material substrate and (b) observed shift in its Bragg wavelength as a function of bending radius. (After Ref. [53]; ©2002 IEEE.)

pressive strain on the grating. Figure 8.10(b) shows shift in the wavelength observed as a function of the bending radius. By 2003, a tuning range of 90 nm was demonstrated by combining an axial compressive strain with a tensile strain using the same beam-bending technique [54]. An alternative to stretching and compressing consists of changing the refractive index of the grating thermally [55]–[58]. Temperature changes shift the Bragg wavelength linearly through the refractive index variations induced by the thermo-optic effect. Typically, the Bragg wavelength shifts at a rate of 10 pm/°C, and temperature changes of ∼100°C are required to shift the Bragg wavelength by 1 nm or so. Clearly, the tuning range is limited when thermal effects are used to tune the grating. A nonuniform magnetic field can also be used for tuning over a few nanometers [59].

Several other techniques have been developed for making optical filters based on fiber gratings [60]–[63]. In one approach, a fiber grating is combined with an FP interferometer through a circulator, resulting in two cascaded optical filters [61]. Such a scheme benefits from the desirable properties of both filters while bypassing their limitations. Tuning range can be quite large for such a combination because FP filters can be tuned over a wide range. In another scheme, tilted gratings are used with a two-mode fiber to design a FP filter with a narrowband transmission peak [62]. Still another scheme makes use of two overlapping chirped fiber gratings to form an FP filter [63]. Such a device exhibits multiple transmission peaks, which correspond to FP resonances, within a wide stop band associated with the chirped gratings. The same type of transfer function can be realized by using an asymmetric Michelson interferometer [60], similar to that shown in Figure 8.1(c). Two chirped gratings in this configuration can provide a bandpass filter with periodic narrow transmission peaks [30].

Tunable optical filters can also be made using the well-developed DFB-laser technology. In this approach, InGaAsP waveguides are used with a built-in grating whose Bragg wavelength is tuned electrically through the electro-optic effect [64]. A phase-control section, similar to that used for multisegment distributed Bragg-reflector (DBR) lasers, can also be employed to tune such DBR filters. Multiple gratings fabricated on an InP substrate and integrated with an InGaAsP waveguide can provide tunability over a wide range [65]. Such filters can be tuned quickly (in a few nanoseconds) and can be designed to provide net gain since one or more amplifiers can be integrated with the filter. They can also be integrated with the receiver, as they use the same semiconductor material. These two properties of InGaAsP-based filters make them quite attractive for WDM applications.

8.1.4 Acousto-Optic Filters

In another class of tunable filters, the grating is formed dynamically using a high-frequency acoustic wave. Such filters, called *acousto-optic filters*, exhibit a wide tuning range (>100 nm) and have found applications in WDM systems [66]–[72]. An example of such a filter is shown in Figure 8.1(d). The physical mechanism behind the operation of acousto-optic filters is the *photoelastic effect* discussed in Section 6.2.3 in the context of optical modulators. Briefly, an acoustic wave creates an index grating by inducing periodic changes in the refractive index that can diffract an optical beam whenever the Bragg condition is satisfied. The wavelength selectivity stems from this acoustically induced grating.

Acousto-optic tunable filters can be made by using bulk components as well as planar waveguides, and both kinds are available commercially. In both cases, a piezoelectric transducer creates the acoustic wave, which diffracts a channel at a specific wavelength and thus separates it from the remaining incident channels. In the case of a bulk device, the filtered channel comes out in a direction different than that of the incident multichannel signal. In contrast, when the acousto-optic interaction occurs inside a single-mode waveguide, a collinear geometry must be used. In this case, the TE mode of a single-mode waveguide at a wavelength λ is converted into the TM mode by the acoustic wave when the Bragg condition $\beta_{TM} - \beta_{TE} = k_a$ is satisfied, where $k_a = 2\pi/\Lambda_a$ is the propagation constant for an acoustic wave of wavelength Λ_a. Using $\beta = 2\pi \bar{n}/\lambda$, we can write this condition as $\Lambda_a = \lambda/(\Delta n)$, where Δn is the difference in the effective mode indices associated with the TE and TM modes. Noting that beat length is defined as $l_b = \lambda/|\Delta n|$, we see that light is converted from TE to TM mode by the acoustic wave whenever the beat length equals the acoustic wavelength.

For WDM applications, the LiNbO$_3$-waveguide technology is often used since it can produce compact, polarization-independent, acousto-optic filters with a bandwidth of about 1 nm and a tuning range over 100 nm [68]. The basic design, shown schemat-

8.1 Optical Filters

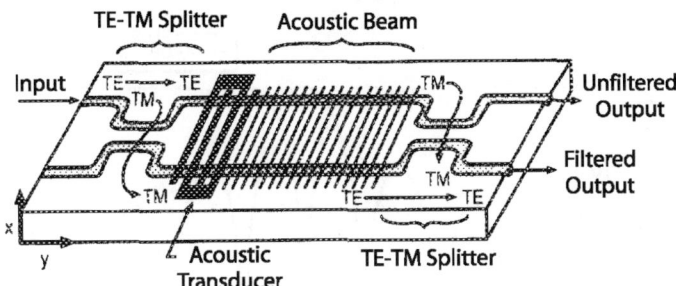

Figure 8.11: Schematic of an acousto-optic tunable filter based on LiNbO$_3$ technology. (After Ref. [66]; ©1990 IEEE.)

ically in Figure 8.11, uses two polarization splitters, two LiNbO$_3$ waveguides forming two arms of a MZ interferometer, and an interdigital transducer for exciting a surface acoustic wave. All components are integrated on the same LiNbO$_3$ substrate using the waveguide technology discussed in Section 4.3.2. Conceptually, one can think of such a device as a $\lambda/2$ wave plate sandwiched between two polarizers. The wave plate acts as a narrowband polarization converter at a wavelength set by the acoustic wave.

The operation of an acousto-optic filter can be understood from Figure 8.11 as follows. The input WDM signal is split by the polarization splitter into its orthogonally polarized components, which travel as TE and TM modes on the upper and lower LiNbO$_3$ waveguides, respectively. A surface acoustic wave of a fixed frequency is excited by the transducer and propagates along the waveguide length. The channel whose wavelength λ satisfies the Bragg condition, $\lambda = (\Delta n)\Lambda_a$, is converted in the upper waveguide from the TE to TM mode by the acoustically induced index grating. At the second splitter, this channel is transferred to the lower branch and it exits from the cross port. All other channels maintain their original polarization state in both waveguides and thus come out of the bar port after they are combined at the second polarization splitter. Figure 8.12 shows the transfer function of an acousto-optic filter by plotting the output from the bar and cross ports as a function of detuning from the center wavelength for which the Bragg condition is exactly satisfied. The filter bandwidth is a fraction of 1 nm, but it exhibits multiple side lobes that may lead to crosstalk. The side lobes can be suppressed, and the passband can be flattened, through suitable design modifications [68]. A packaged and pigtailed device in which two filters were integrated on the same substrate [69] exhibited in 1998 low insertion losses of 4.1 dB, minimal polarization dependence (about 0.1 dB), and side-lobe suppression of >30 dB. Its tuning range exceeded 40 nm in the 1,550-nm spectral region.

The TE–TM index difference Δn is about 0.07 for LiNbO$_3$ waveguides. In the spectral region near 1.55 μm, the acoustic wavelength Λ_a should be about 22 μm from the Bragg condition $\lambda = (\Delta n)\Lambda_a$. This value corresponds to a frequency of about 170 MHz if we use the acoustic velocity of 3.75 km/s for LiNbO$_3$. An acoustic wave of such frequency can easily be generated. The tuning is realized by changing electronically the acoustic-wave frequency (and hence the wavelength Λ_a) such that a different channel

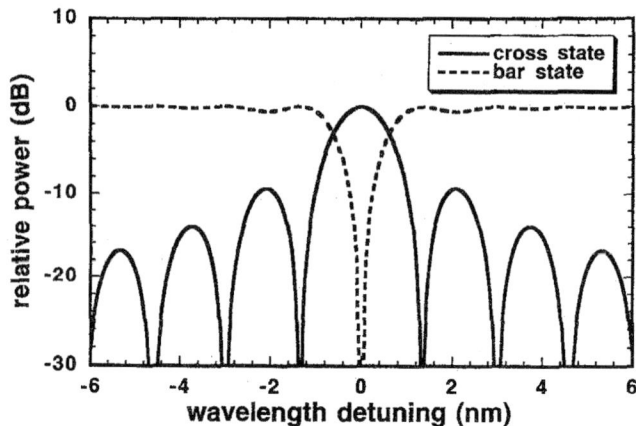

Figure 8.12: Transfer function of an acousto-optic tunable filter. Solid and dashed lines show the fraction of input power existing from the cross and bar ports, respectively. (After Ref. [68]; ©1996 IEEE.)

satisfies the Bragg condition. Tuning is relatively fast because of its electronic nature and can be accomplished in a switching time of less than 10 μs. Acousto-optic tunable filters are also suitable for wavelength routing and optical cross-connect applications in dense WDM systems [68].

The main drawback of any planar-waveguide approach is related to coupling losses incurred when the waveguide is coupled to optical fibers at the input and output ends. Such losses can be avoided using an all-fiber approach for making acousto-optic filters [73]–[79]. The basic idea is to couple two modes of the fiber through an acoustically induced index grating so that light at a specific wavelength is selectively transferred from the fundamental HE_{11} mode to a higher-order mode or to a cladding mode. The Bragg condition remains the same, $\Lambda_a = \lambda/(\Delta n)$, but Δn now corresponds to the difference in the mode indices for the two modes involved.

Figures 8.13(a) and (b) show two schemes for making an all-fiber acousto-optic tunable filter. In both cases, a piezoelectric transducer produces acoustic vibrations that are amplified and transmitted to the optical fiber using a horn. In the case of a tapered fiber, the light of a specific wavelength that can be tuned by varying the acoustic frequency is transferred to a higher-order mode that is not coupled into the single-mode buffered fiber and is thus physically separated from the reaming channels [71]. In the case of a two-mode fiber, a mode-selective coupler is used to separate the filtered channel so that it appears at a different port of the coupler [79]. In another scheme, acoustic wave transfers mode power to a cladding mode and is then transferred back to the core after a "core block" that removes all remaining channels from the core [78]. This technique cancels the frequency shift induced by an acoustically induced moving grating. In effect, the device works as a tunable bandpass filter that transmits a specific channel but blocks all others.

8.2 Multiplexers and Demultiplexers

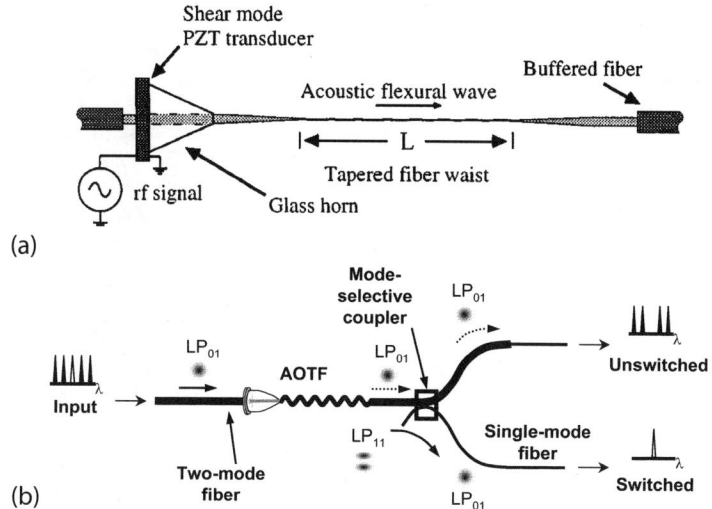

Figure 8.13: Two schemes for making an all-fiber acousto-optic tunable filter based on a (a) tapered fiber and (b) two-mode fiber. (After Refs. [71] and [79]; ©2002 IEEE.)

8.2 Multiplexers and Demultiplexers

Multiplexers and demultiplexers are used to combine or separate individual channels and are thus the essential components of all WDM systems. Similar to the case of optical filters, demultiplexers must incorporate a wavelength-selective mechanism in order to separate individual channels with different carrier wavelengths. They can be classified into two broad categories. Diffraction-based demultiplexers use an angularly dispersive element, such as a diffraction grating, which disperses incident light spatially into various wavelength components. Interference-based demultiplexers make use of devices such as optical filters and directional couplers. In both cases, the same device can be used as a multiplexer or a demultiplexer, depending on the direction of propagation, because of the inherent reciprocity of optical wave propagation in dielectric media.

8.2.1 Grating-Based Demultiplexers

Grating-based demultiplexers use the phenomenon of Bragg diffraction from an optical grating [80]–[85]. Figures 8.14(a) and (b) show the design of two such demultiplexers. The input WDM signal is focused onto a reflection grating, which separates various wavelength components spatially, and a lens focuses them onto individual fibers. Use of a graded-index lens simplifies alignment and provides a relatively compact device. The focusing lens can be eliminated altogether by using a concave grating. For a compact design, the concave grating can be integrated within a silicon slab waveguide.

A problem with such grating demultiplexers is that their bandpass characteristics depend on the dimensions of the input and output fibers. In particular, the core size of

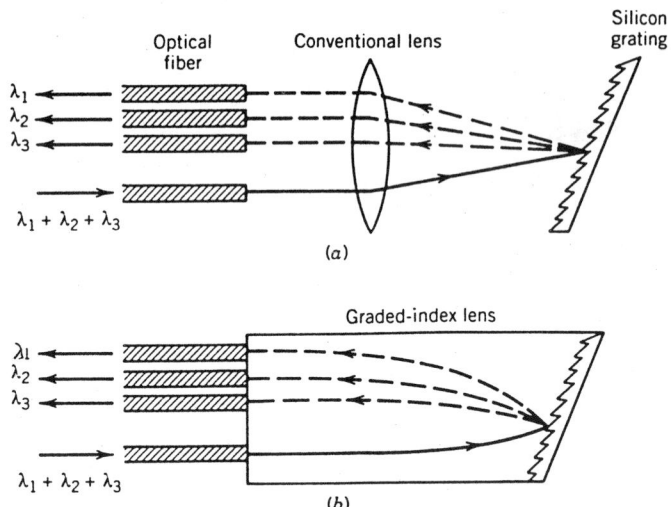

Figure 8.14: Grating-based demultiplexer making use of (a) a conventional lens and (b) a graded-index lens.

output fibers must be large to ensure a flat passband and low insertion losses. For this reason, most early designs of multiplexers used multimode fibers. In a 1991 design, a microlens array was used to solve this problem and to demonstrate a 32-channel multiplexer for single-mode fiber applications [81]. The fiber array was produced by fixing single-mode fibers in V-shaped grooves etched into a silicon wafer. The microlens transforms the relatively small mode diameter of fibers ($\sim$10 μm) into a much wider diameter (about 80 μm) just beyond the lens. This scheme provides a multiplexer that can work with channels spaced by only 1 nm in the wavelength region near 1.55 μm while accommodating a channel bandwidth of 0.7 nm.

In a different approach, multiple elliptical Bragg gratings are etched using the silica-on-silicon technology [80]. The idea behind this approach is simple. If the input and output fibers are placed at the two foci of the elliptical grating, and the grating period Λ is adjusted to a specific wavelength λ_0 by using the Bragg condition $2\Lambda n_{\text{eff}} = \lambda_0$, where n_{eff} is the effective index of the waveguide mode, the grating would selectively reflect that wavelength and focus it onto the output fiber. Multiple gratings need to be etched, as each grating reflects only one wavelength.

Because of the complexity of such a device, a single concave grating etched directly onto a silica waveguide is more practical. In a 1998 experiment such a grating was capable of demultiplexing up to 120 channels with a wavelength spacing of 0.3 nm [84]. Figure 8.15 shows the demultiplexer design schematically. In this device, the input and output waveguides were integrated with a concave diffraction grating on the same silicon substrate. The grating was designed such that its two neighboring teeth formed a retro-reflection prisms. The angle of each reflecting facet was varied across the grating such that the angle of incidence was always 45°. The waveguides had a 2-μm SiON core that was sandwiched between two SiO_2 cladding layers. A reactive-ion etching

8.2 Multiplexers and Demultiplexers

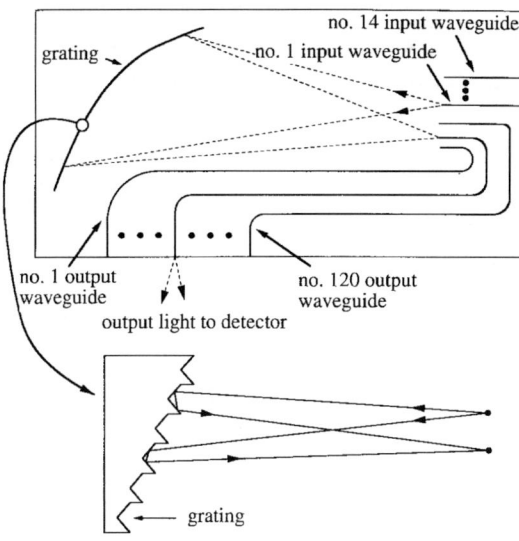

Figure 8.15: Grating-based demultiplexer built on a silicon chip using a concave diffraction grating. (After Ref. [84]; ©1998 IEEE.)

technique was used to form 14 input ridge waveguides, 120 output ridge waveguides, and a concave diffraction grating. The output waveguides were spaced 10 μm apart at the point where they collected light, but their spacing increased to 250 μm at the output edge of the chip. The entire chip was only 1 cm wide and 3.4 cm long, resulting in a compact device. Its main drawback was a relatively large fiber-to-fiber insertion loss (>20 dB) for each channel because of unavoidable coupling and diffraction losses.

8.2.2 Filter-Based Demultiplexers

Filter-based demultiplexers use the phenomenon of optical interference to select the wavelength. Demultiplexers based on the MZ filter have attracted the most attention [86]–[88]. Similar to the case of a tunable optical filter, one or more MZ interferometers are combined to form a demultiplexer. A single symmetric MZ interferometer with a ring resonator in one of its arms was fabricated with the silica-on-silicon technology as early as 1987 [87]. Such a device is similar to that shown in Figure 8.7. Its demultiplexing nature can be understood as follows. As shown in Section 2, ring resonators act as an all-pass filter but it modifies the phase of passing light in a periodic fashion in accordance with Eq. (2.3.4). The MZ interferometer converts such periodic phase changes into amplitude changes. The combination of two exhibits periodic transmission peaks at frequencies corresponding to the longitudinal modes of the ring resonator.

An alternate design combines several MZ interferometers to demultiplex a multichannel signal. Figure 8.16 illustrates the basic idea by showing the layout of a four-channel multiplexer [86]. It combines three MZ interferometers but does not cascade

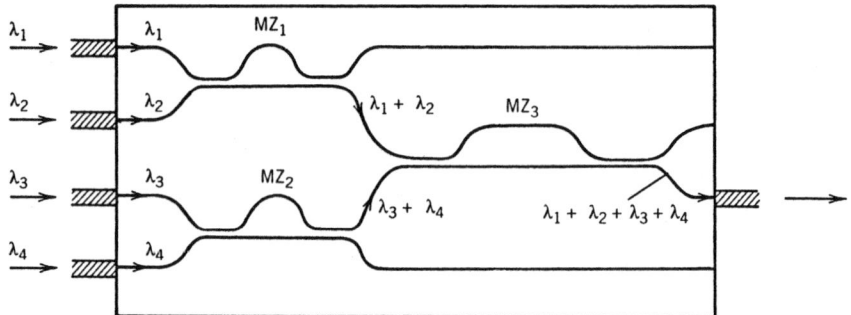

Figure 8.16: Layout of an integrated four-channel waveguide multiplexer based on Mach–Zehnder interferometers. (After Ref. [86]; ©1988 IEEE.)

them, as was done in Figure 8.7. Rather, two of them are placed in parallel. One arm of each MZ interferometer is made longer than the other to provide a wavelength-dependent phase shift between the two arms. The path-length difference is chosen such that the total power from two input ports at different wavelengths appears at only one output port. The same device works as a demultiplexer when a single 4-channel input is launched from the right end. The whole structure can be fabricated on a silicon substrate using SiO_2 waveguides in the form of a planar lightwave circuit. The basic idea can be extended to demultiplex a much larger number of channels in the form 2^M, where M is an integer.

The grating technology can be applied to form Bragg gratings directly on a planar silica waveguide. This approach has attracted attention since it permits the integration of Bragg gratings within planar lightwave circuits. In one demonstration of this idea, two standard Bragg gratings were incorporated in two arms of an asymmetric MZ interferometer [115]. Two chromium heaters were also integrated to change the path length so that the MZ output could be switched to a different port. Such a device can demultiplex an input signal with three channels so they appear at the three remaining ports.

Fiber Bragg gratings can also be used for making all-fiber demultiplexers. In one approach, a $1 \times N$ fiber coupler is converted into a demultiplexer by forming a phase-shifted grating at the end of each output port, and thus opening a narrowband transmission window (~ 0.1 nm) within the stop band [33]. The position of this window is varied by changing the amount of phase shift so that each arm of the $1 \times N$ fiber coupler transmits only one channel. Figure 8.17 shows the basic idea behind such an all-fiber demultiplexer schematically.

It is possible to construct multiplexers using multiple directional couplers. The basic scheme is similar to that shown in Figure 8.16 but simpler as MZ interferometers are not used. Furthermore, an all-fiber multiplexer made with fiber couplers avoids coupling losses that occur whenever light is coupled into or out of an optical fiber. A fused biconical taper can also be used for making fiber couplers [90]. Multiplexers based on fiber couplers can be used only when channel spacing is relatively large (>10 nm) and are thus suitable mostly for coarse WDM applications.

8.2 Multiplexers and Demultiplexers

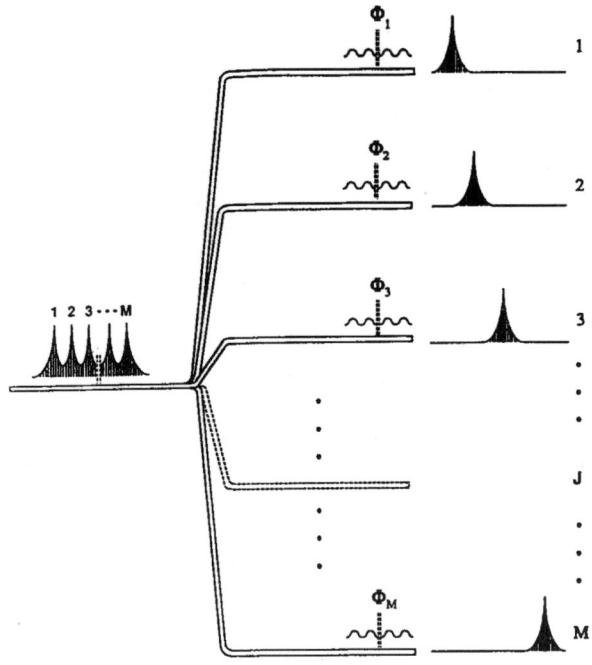

Figure 8.17: An all-fiber demultiplexer containing a phase-shifted grating in each arm of a $1 \times N$ fiber coupler. (After Ref. [33]; ©1994 IEEE.)

8.2.3 Waveguide-Grating Demultiplexers

As discussed in Section 4.5.4, a *phased array* of optical waveguides acts as a grating. Such an *arrayed waveguide grating* (AWG) can be fabricated using the silia, InP, or $LiNbO_3$ waveguides [91]–[100]. AWGs are normally designed with multiple input and output ports and have a variety of applications. They serve as a demultiplexer because a WDM signal launched into one of the input ports is separated into individual channels by the waveguide grating. The silica-on-silicon technology is used most commonly for making such demultiplexers in the form of a planar lightwave circuit [96]. However, the use of InP waveguides permits integration of the demultiplexer with a transmitter or receiver.

Figure 8.18 shows the schematic of a waveguide-grating demultiplexer, also known as a phased-array demultiplexer [91]. Physically, free-propagation regions act as a lens while the waveguide array acts as a grating. Thus, such a device is equivalent to focusing the input WDM signal onto a bulk grating, which diffracts each channel into a different direction while the second lens focuses it onto a different spot at the output end.

The operation of an AWG demultiplexer can be understood from Figure 8.18 as follows. The incoming WDM signal is coupled into an array of planar waveguides after passing through the first star coupler (see Section 4.5.6). In each waveguide, the WDM signal experiences a different phase shift because of different lengths of waveguides

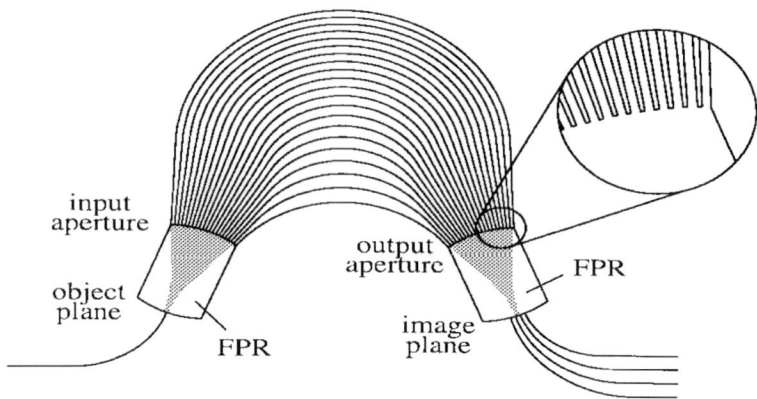

Figure 8.18: Schematic of a waveguide-grating demultiplexer consisting of an array of waveguides between two free-propagation regions (FPR). (After Ref. [91]; ©1996 IEEE.)

such that the phase difference between two neighboring waveguides is constant. This feature tilts the wavefront at the output end of the grating array. Moreover, the tilt is wavelength-dependent because the phase shift depends on the mode-propagation constant. As a result, different channels focus to different output waveguides when the light exiting from the array passes through the second star coupler. The net result is that the WDM signal is demultiplexed into individual channels.

AWG demultiplexers fabricated on a silicon substrate using silica waveguides were developed during the 1990s [94]–[100]. By 1996, such devices were able to resolve up to 128 channels with spacing as small as 25 GHz. The number of channels could be increased to 256 by 2000 by employing silica waveguides with a relatively large core-cladding index difference of 1.5% wile maintaining the 25-GHz channel spacing. Figure 8.19 shows the transmission spectra from 256 output waveguides for such a

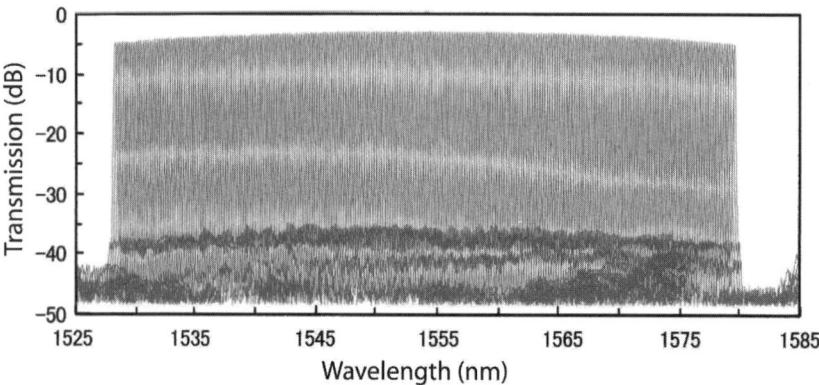

Figure 8.19: Transmission spectra of an AWG demultiplexer designed to demultiplex 256 channels with 25-GHz spacing. (After Ref. [100]; ©2002 IEEE.)

8.3 Optical Add–Drop Multiplexers

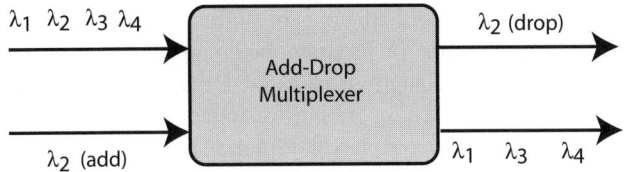

Figure 8.20: A generic add–drop multiplexer with four ports.

demultiplexer [100]. This device was designed with 712 waveguides within the central array such that the length of each grating increased by 27.7 μm from one end to another. The chip size was only 7.5×5.5 cm^2. Transmission losses on the chip ranged from 2.7 to 4.7 dB. It was possible to fabricate 400-channel demultiplexers by using an array with 1,175 waveguides, but the chip size increased to 12.4×6.4 cm^2 and the chip losses ranged from 3.8–6.4 dB. When coupling losses are included, the fiber-to-fiber insertion loss for AWG demultiplexers can exceed 10 dB. A combination of several suitably designed AWGs can increase the number of channels to more than 1,000 while maintaining the 10-GHz resolution [99].

The performance of multiplexers is judged mainly by the amount of insertion loss for each channel. The performance criterion for demultiplexers is more stringent. First, the performance of a demultiplexer should be insensitive to the polarization of the incident WDM signal. Second, a demultiplexer should separate each channel without any leakage from the neighboring channels. In practice, some power leakage is likely to occur, especially in the case of dense WDM systems with small interchannel spacing. Such power leakage is referred to as crosstalk and should be quite small (less than -20 dB) for a satisfactory system performance.

8.3 Optical Add–Drop Multiplexers

Optical add–drop multiplexers (OADMs) are needed for all-optical networks in which a specific channel at a certain wavelength must be dropped or added at a certain location while preserving the integrity of all other channels. Although sometimes referred to as an add–drop filter, such a component differs from optical filters in two respects. First, it should separate one channel from the rest and send it to a different port. Second, it should allow the possibility of adding the same-wavelength signal from yet another port. Thus, such a device should have at least four ports in contrast with optical filters that may have only two ports. Figure 8.20 shows the basic idea schematically with a generic black box. During the 1990s several kinds of add–drop multiplexers were developed using a variety of techniques. In this section we focus on a few of them.

8.3.1 Directional Couplers with Gratings

The wavelength selectivity of Bragg gratings is often used to make add–drop multiplexers [30]. However, since a fiber grating has only two ports, it should be combined with some other four-port device such as a directional coupler. In one approach, a Bragg grating is fabricated within one or both cores of a directional coupler [101]–[108]. The grating can be designed to assist or frustrate the coupling between the two cores for a specific channel whose wavelength satisfies the Bragg condition. Such a device is referred to as the grating-assisted or grating-frustrated directional coupler depending on the scheme employed. Fiber couplers are often used for making OADMs because their use results in an all-fiber device. A more compact device can be fabricated using semiconductor or silica waveguides although it invariably suffers from higher insertion losses.

In a grating-frustrated directional coupler [101], two cores are made identical and the coupler length is chosen such that it transfers light to the neighboring core in the absence of any grating (the so-called 100% coupler). A Bragg grating is fabricated in the second core. The presence of this grating frustrates the transfer of the channel whose wavelength falls within the stop band of a long-period grating. As a result, all channels of a WDM signal appearing from one input port (port 1) go to the cross output port (port 4) except the channel whose wavelength is close to the Bragg wavelength of the grating. This channel remains in the same core and appears at the bar output port of the coupler (port 3) in the form of a dropped channel. A channel at the same wavelength can be added to the WDM signal by injecting it from the unused port 2.

In a grating-assisted directional coupler [103], two cores are made dissimilar so that the device is unable to couple light propagating in one core to the neighboring core, except when its wavelength falls within the stop band of a Bragg grating. The grating occupies the second core and is designed such that it not only couples the light into the second core but also reflects it. As a result, all channels of a WDM signal entering from port 1 go to the bar port 3, except a specific channel whose wavelength is close to the Bragg wavelength of the grating. This channel transfers to the neighboring core and appears at the unused input port 2 and serves as the dropped channel. A channel at the same wavelength can be added to the WDM signal by injecting it from output port 4. This device can also be made using a long-period grating. The dropped channel in this case appears at port 4 because the grating transfers it to the second core but does not reflect it.

In another scheme [105], the coupler is made symmetric with identical cores and a Bragg grating is fabricated in both cores (or a single Bragg grating extends over both core regions). The grating does not cover the whole coupling region but is localized in its central part. In this case, all channels cross over to the second core, but a specific channel whose wavelength falls within the stop band of the grating is reflected by the grating. As a result, all channels of a WDM signal entering from port 1 go to the cross port 4, except the channel whose wavelength is close to the Bragg wavelength of the grating. This channel is reflected back and appears at the unused input port 2. The bar output port 3 of the coupler is now used for adding the channel at the Bragg wavelength.

Although fiber-based OADMs based on a directional coupler with an integrated grating exhibit ultralow insertion losses, they suffer from several problems such as the appearance of up to three peaks in the reflection spectrum of the grating [107], resulting from the coupling between the even–even, odd–odd, and even–odd eigenmodes of the coupler. Although these undesirable peaks can be suppressed with a suitable design, the fabrication of such devices is not a simple task [106].

8.3 Optical Add–Drop Multiplexers

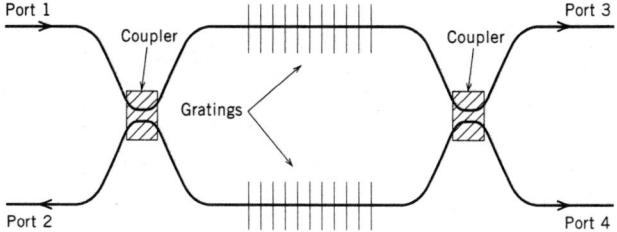

Figure 8.21: An add–drop multiplexer designed using a Mach–Zehnder interferometer and two identical Bragg gratings.

8.3.2 Mach–Zehnder Interferometer with Gratings

It is not necessary to fabricate a Bragg grating within the directional coupler. A practical scheme for making OADMs uses a MZ interferometer made with two 3-dB fiber couplers, the same device discussed earlier within the context of optical filters. The add–drop functionality is added by two Bragg gratings inserted in the two arms of the MZ interferometer [109]–[118]. The gratings should be identical in all respects and must be positioned symmetrically. Figure 8.21 shows such a device schematically. It can be fabricated using fiber couplers or one can make use of planar-waveguide technology.

The operation of such an OADM can be understood from Figure 8.21. Assume that the WDM signal is incident on port 1 of the device. The channel, whose wavelength λ_g falls within the stop band of the two identical Bragg gratings, is totally reflected. It appears at port 2 because of the π phase shift induced by the 3-dB coupler during a round trip. The remaining channels are not affected by the gratings and appear at port 4, again because of the π phase shift induced by the two 3-dB couplers. The same device can add a channel at the wavelength λ_g if the signal at that wavelength is injected from port 3. If the add and drop operations are performed simultaneously, it is important to make the gratings highly reflecting (close to 100%) to minimize the crosstalk.

Although semiconductor [109] and silica [110] waveguides were used in the initial demonstrations of MZ-based OADMs, by 1994 such a device was fabricated using fiber couplers [111]. The MZ interferometer should be perfectly balanced for such OADMs. A technique known as ultraviolet trimming is used in practice to ensure equal optical phase shifts in the two arms of the MZ interferometer. Even then, such a fiber-based device suffers from temperature-induced index fluctuations because of its interferometric nature. Two arms of the MZ interferometer are kept close to each other (5 mm apart or so) to keep them at nearly the same relative temperature and to minimize the environmental effects.

In practice, the performance of an OADM may deviate from the ideal behavior if (i) the MZ interferometer is not balanced perfectly, (ii) the directional couplers are not perfect 3-dB couplers, and (iii) the Bragg grating is not 100% reflecting. The performance is judged by the amount of insertion losses, drop-off efficiency, and crosstalk level [116]. Insertion losses are negligible for an all-fiber device, although they may become substantial if planar waveguides are used to form the MZ interferometer. How-

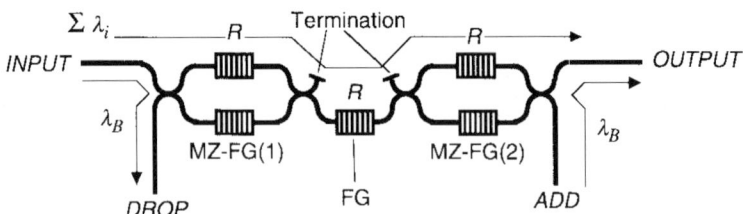

Figure 8.22: Add–drop multiplexer designed by cascading two MZ interferometers, each containing two identical fiber gratings (FG) reflecting the channel that need to be dropped or added. The fifth identical grating in the middle helps to reduce the crosstalk. (After Ref. [116]; ©1998 IEEE.)

ever, if the grating is not 100% reflecting and a differential phase shift exists in the two arms of the MZ interferometer, a part of the incident channel powers would leak to undesirable ports, producing not only some loss in the power of the channels that are transmitted but also generating crosstalk at the drop-off channel wavelength when a new signal at that wavelength is added. It is easy to estimate the magnitude of such crosstalk using the theory of Section 2.3.4 and optimize the OADM appropriately. As eaarly as 1995, all-fiber OADMs exhibited a drop-off efficiency of more than 99%, while keeping the crosstalk level below 1% [114].

For actual WDM applications, a crosstalk level of -20 dB is not sufficient and it should be reduced to below -40 dB. This can be realized by cascading two MZ devices in series [116] in a fashion shown in Figure 8.22. The first OADM drops off the channel at λ_B. Any power at this wavelength leaked through the first OADM will be reflected by the middle fiber grating before the WDM signal enters the second OADM stage. The two gratings in this section will further reduce the power at λ_B before the new signal is added by this section. It turns out that the coherent crosstalk at this wavelength can be reduced to below -40 dB even when gratings are only 95% reflecting.

A MZ chain similar to that used for making MZ filters can also be used for OADMs through serial cascading. However, in contrast with the case of MZ filters, the relative delay τ_m in Eq. (8.1.9) is kept the same for each MZ interferometer. The resulting device is sometimes referred to as a *resonant coupler* because it resonantly couples out a specific wavelength channel to one output port while the remainder of the channels appear at the other output port. Its performance can be optimized by controlling the coupling ratios of various directional couplers [112]. Although resonant couplers can be implemented in an all-fiber configuration using fiber couplers, the silica-on-silicon waveguide technology provides a compact alternative for designing such OADMs [113].

8.3.3 Optical Circulator with Gratings

Another class of OADMs makes use of optical circulators in combination with one or more fiber gratings [119]–[124]. Such a device is simple in design and can be made by connecting two ends of a fiber grating to two 3-port optical circulators. The channel

8.3 Optical Add–Drop Multiplexers

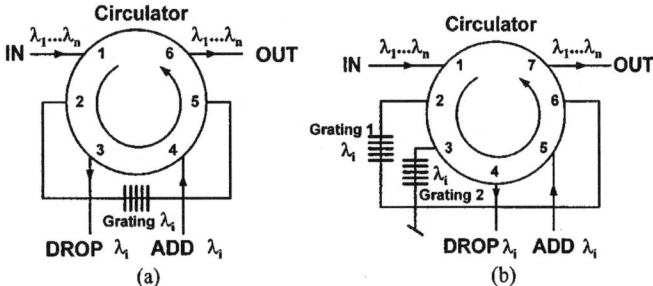

Figure 8.23: (a) Two designs of add–drop multiplexers using a single optical circulator in combination with fiber gratings. (After Ref. [121]; ©2001 IEEE.)

reflected by the grating appears at the unused port of the input-end circulator. The same-wavelength channel can be added by injecting it from the output-end circulator.

Since a circulator has considerable insertion loss, it would be better if the OADM is made using a single optical circulator. Such a configuration can be realized provided the circulator has more than three ports. Figure 8.23 shows two such schemes [121]. Scheme (a) uses a six-port circulator. The WDM signal entering from port 1 exits from port 2 and passes through a Bragg grating that reflects the channel whose wavelength falls within the stop band. The reflected channel appears at port 3 that acts as the "drop" port. The remaining channels re-enter the circulator at port 5 and leave the device from port 6. The channel to be added enters from port 4, exits from port 5, gets reflected by the grating, and is thus combined with other remaining channels. The in-band crosstalk occurs only if the grating reflectivity is less than 100%. It is possible to realize gratings with more than 99.9% reflectivity by increasing the coupling coefficient such that $\kappa L > 4$.

Out-of-band crosstalk occurs when the dropped channel is accompanied by other channels that fall outside the stop band of the grating but still experience some reflection. Typically, this crosstalk level is below 20 dB. It can be reduced further by employing the scheme of Figure 8.23(b). It uses two identical gratings to reduce the

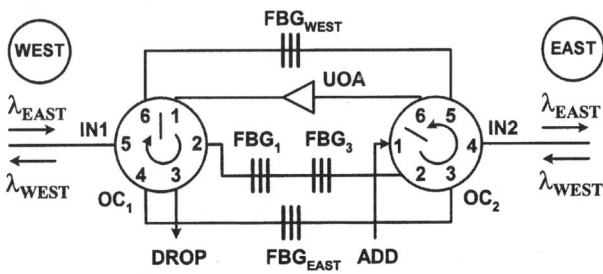

Figure 8.24: A bidirectional add–drop multiplexer made using two optical circulators in combination with four fiber gratings. A sigle optical amplifiers amplifies the WDM signal in both directions. (After Ref. [124]; ©2003 IEEE.)

crosstalk level. The WDM signal entering from port 1 is first reflected by the grating attached to port 2. The reflected signal is reflected a second time by the Bragg grating attached to port 3. Any leakage of channels other than the desired one is cut down drastically during two successive reflections. The dropped channel now appears at port 4. Many other variants are possible depending on the design criterion.

It is possible to construct a bidirectional OADM by combining two circulators with multiple Bragg gratings [119]. One can even insert an optical amplifier within such a device to amplify the channels while adding and dropping a channel at a selected wavelength. Figure 8.24 shows such a bidirectional OADM schematically [124]. It uses two 6-port optical circulators (with the path $6 \rightarrow 1$ blocked) in combination with four fiber gratings and a single optical amplifier. The four gratings are designed with different stop bands. The gratings marked FBG_{EAST} and FBG_{WEST} reflect the WDM signals traveling in the east and west directions, respectively. In contrast, the gratings marked FBG_1 and FBG_3 reflect the channels that need to be dropped (or added) in the east and west directions. One can follow the paths taken by the east- and west-going signals to see how a single amplifier amplifies both of them while dropping (and adding) the desired channel in both directions.

8.3.4 Microring Resonators

As discussed in Section 2.3.1, a ring resonator acts as an optical filter and can provide the wavelength selectivity required for building OADMs. This approach is particularly suitable for making compact OADMs because semiconductor-waveguide technology can be used to make a microring with a diameter of ~ 10 μm [125]–[131]. Moreover, a semiconductor optical amplifier can be integrated with such a device to compensate for insertion losses.

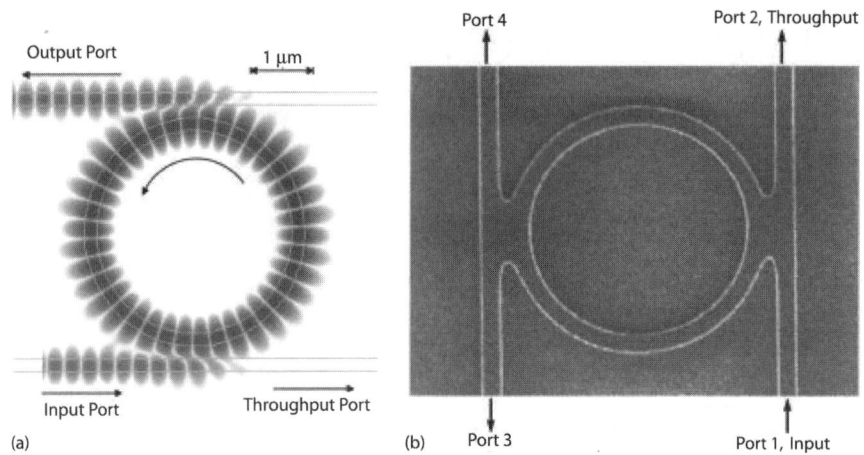

Figure 8.25: (a) Numerical simulations showing how a microring coupled to two waveguides laterally can act as an OADM. (b) Implementation of such a device using silica waveguides. (After Ref. [126]; ©1998 IEEE.)

8.3 Optical Add–Drop Multiplexers

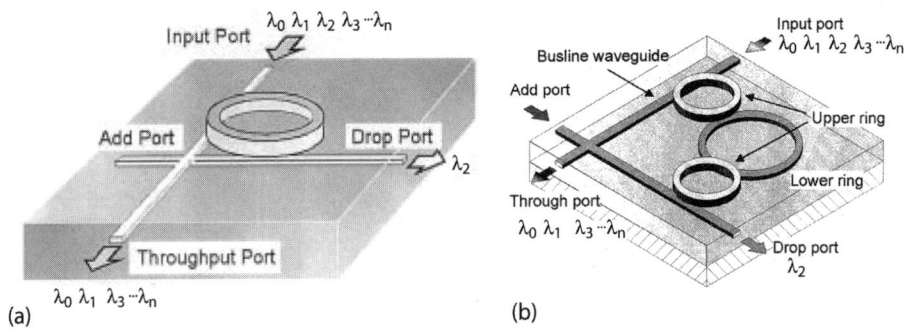

Figure 8.26: Two schemes (a and b) used for coupling a microring to linear waveguides vertically. The three-ring design provides a larger free spectral range and higher finesse. (After Refs. [129] and [130]; ©2002 IEEE.)

Figure 8.25(a) shows the basic idea behind a microring-based OADM. The microring is coupled to two planar waveguides whose ends provide the four ports required for the device to function. When a WDM signal is launched into the lower waveguide, a specific channel whose wavelength coincides with one of the longitudinal modes of the ring resonator is transferred to the upper waveguide by the microring and exits from the outport port of this waveguide, while the remaining channels stay in the first waveguide. The same wavelength channel can be added by launching it from the input port of the upper waveguide. Since light propagates inside the ring in a unidirectional fashion, the input and output ports are reversed for the two waveguides. Such an OADM can be fabricated using the silica or silicon waveguides. Figure 8.25(b) is a SEM photo of a device in which a 1-μm-thick SiO_2 layer acts as a cladding to the 0.2-μm-thick layer of polycrystalline silicon [126]. The central silicon ring has a radius of 3 μm and is coupled to the two planar waveguides on opposite ends. In spite of the fusing of the ring and the waveguide at the junctions, such a microring exhibited a Q value of about 250.

The microring can also be coupled to the two linear waveguides vertically [129]. Moreover, the two waveguides do not have to be parallel; they can even be perpendicular to each other. Figure 8.26 shows two such designs schematically. The vertical-stacking scheme (a) improves the device performance because it allows a more precise control of the coupling between the microring and linear waveguides. The reduced coupling losses allow fabrication of microrings with high finesse ($\sim$50). By reducing the radius of the microring to close to 10 μm, one can realize a resonance bandwidth of < 0.5 nm with a free spectral range of more than 10 nm. The three-ring design (b) permits one to tailor the shape of the resonance peak [130]. The scheme works in the same way as a chain of serially coupled MZ interferometers. The transfer function of the three-ring chain is obtained by multiplying the transfer functions of the three individual rings and is thus much closer to a "rect" function compared with the single-ring case. At the same time, the free spectral range is much larger because the wavelength of the selected channel has to correspond to the longitudinal mode of all three rings. As a result, the three-ring design allows one to achieve finesse of 1,000 or more.

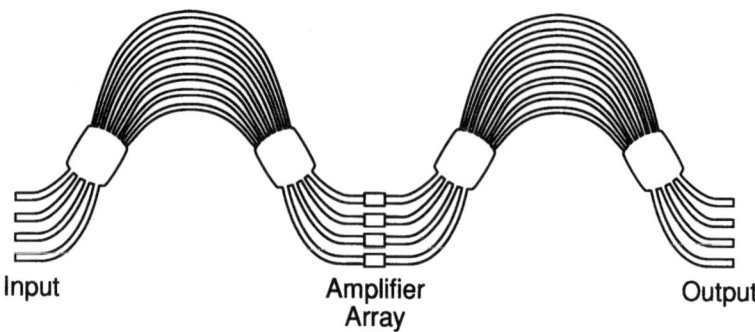

Figure 8.27: A tunable channel-dropping device made using two AWGs with an amplifier array between them. (After Ref. [133]; ©1994 IEEE.)

8.3.5 Tunable Add–Drop Multiplexers

Several other schemes have been used to make OADMs with a variety of optical components [132]–[142]. Among these, schemes that allow tuning of the wavelength of the dropped channel are most suitable for optical network applications. Even though such tunable devices make use of planar waveguides and are fabricated using semiconductor or silica waveguides, and thus suffer from insertion losses, their tunability makes them attractive for WDM systems.

In one approach shown schematically in Figure 8.27, two identical AWG demultiplexers fabricated using InGaAsP waveguides (on an InP substrate) are connected in series such that an array of semiconductor optical amplifiers connects each output port of one AWG with the corresponding input port of another [133]. The first AWG separates the incident WDM channels and directs them to different ports. The gain of amplifiers is adjusted such that only the channel to be dropped experiences amplification when passing through the device, while the other channels are attenuated considerably because of insertion and other types of losses. The second AWG multiplexes all channels so that the output appears from a single port, but this output consists mainly of the channel that was intended to be dropped. The wavelength of the dropped channel can be changed electronically by changing the currents injected into the amplifier array. The channels can be added using Y-junction couplers at the input end of each amplifier.

Another approach is based on tunable acousto-optic filters [28], discussed in Section 8.1.4. As seen in Figure 8.11, such a four-port device makes use of LiNbO$_3$ waveguides and works by changing the polarization of a specific channel through an acoustically induced index grating. The wavelength λ_d of the channel to be dropped (or added) can be altered by changing the wavelength Λ of the acoustic wave as the two are related as $\lambda_d = \Lambda |\Delta n|$, where $|\Delta n|$ is the birefringence or difference in the effective mode indices associated with the TE and TM modes of the LiNbO$_3$ waveguide.

Electro-optic tuning can also be used to realize LiNbO$_3$-based OADMs. Figure 8.28 shows the structure of such an OADM schematically. The device design is similar to that used in Figure 8.11, except that the grating is produced not acoustically but by applying strain along the waveguide length in a periodic fashion. This is accomplished

8.4 WDM Transmitters and Receivers

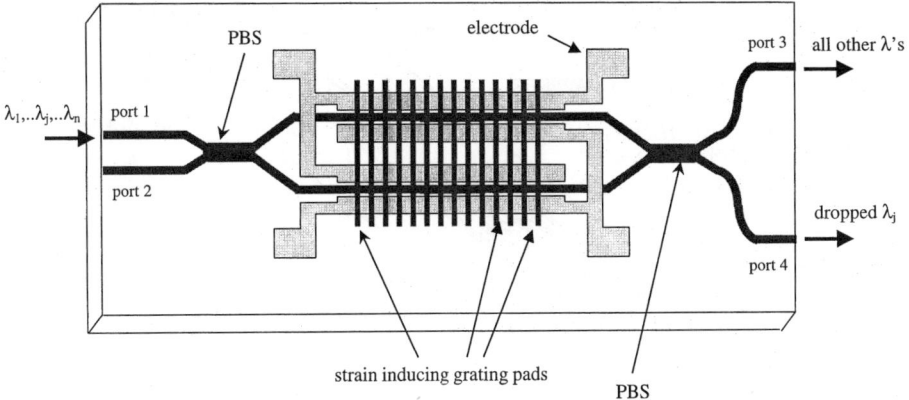

Figure 8.28: A rapidly tunable OADM made using two LiNbO$_3$ waveguides processing the TE and TM components separately. The period of the strain-induced grating remains fixed, but the birefringence is changed electrically for tuning the device. (After Ref. [142]; ©2003 IEEE.)

in practice by depositing a dielectric (typically SiO$_2$) film on the surface of the waveguides and placing metal contacts on top of it with a fixed spatial period. The strain results from a thermal-expansion mismatch between the substrate and film. Polarization conversion occurs for the channel whose wavelength satisfies the Bragg condition $\lambda_d = \Lambda|\Delta n|$, where Λ is the period of the grating and $|\Delta n|$ is the birefringence in the waveguide. Tuning is realized by changing the voltage, which changes the strain on the waveguide and thus modifies Δn. Thus, the tuning mechanism is quite different than an acousto-optic OADM in which the grating period Λ is changed but Δn is kept fixed. The main advantage of electro-optic tuning is that the wavelength of the dropped channel can be changed on a time scale shorter than 1 μs.

The use of the silica-on-silicon technology also permits tuning an OADM, but such a device makes use of the relatively slow thermo-optic effect [139]. The basic idea is discussed in Section 8.1.2 within the context of MZ filters and is similar to that shown in Figure 8.6. It consists of cascading several asymmetric MZ interferometers with a built-in chromium heater in one arm of each MZ interferometer. The relative phase shift between the two arms can be altered by changing the refractive index in one of the arms thermally. Such a change allows one to tune the wavelength of the channel that should be dropped or added by the device.

8.4 WDM Transmitters and Receivers

WDM systems require one narrow-linewidth source such as a distributed feedback (DFB) semiconductor laser (see Section 5.2) for each channel. The wavelengths or the carrier frequencies of two neighboring channels should be different by an amount known as the channel spacing, with a value typically falling in the range of 25–100 GHz. The laser wavelengths are chosen in practice to match precisely frequency grid standardized by the International Telecommunication Union (ITU).

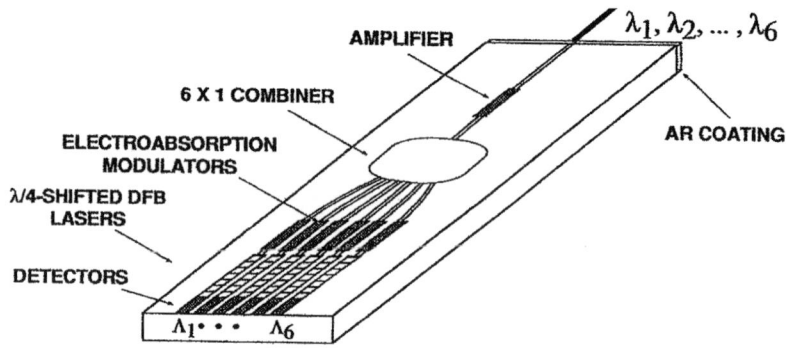

Figure 8.29: Schematic of a WDM transmitter made by integrating multiple DFB lasers, electroabsorption modulators, and back-facet detectors. The output of all lasers is combined and amplified before being coupled into a fiber. (After Ref. [152]; ©1997 Academic.)

The use of individual optical transmitters containing a fixed-wavelength DFB laser for each channel becomes impractical when the number of channels becomes large. Two solutions are possible. In one approach, tunable semiconductor lasers with a tuning range of 10 nm or more are employed. The use of such lasers reduces inventory and maintenance problems because a single device can be used for many channels by setting the operating wavelength appropriately. As discussed in Section 5.2.3, several different techniques can be used for making tunable semiconductor lasers. Multisection DBR lasers can be tuned up to 40 nm while exhibiting a narrow line width (<2 MHz), low intensity noise (below −140 dB/Hz) while coupling up to 20 mW of power into the fiber [143]–[147].

The number of lasers required can be still quite large for dense WDM systems. In an alternative approach, multiwavelength transmitters capable of generating light at several fixed wavelengths simultaneously are used. Although such WDM transmitters attracted some attention in the 1980s, it was only during the 1990s that monolithically integrated WDM transmitters, operating near 1.55 μm with a channel spacing of 1 nm or less, were developed using the InP-based optoelectronic integrated-circuit (OEIC) technology [148]–[161].

8.4.1 Optoelectronic Integrated Circuits

In integrated WDM transmitters, the output of several DFB or DBR semiconductor lasers, independently tunable through Bragg gratings, is combined using passive waveguides [148]–[156]. Figure 8.29 shows the basic idea schematically. Several phase-shifted DFB lasers tuned to operate at specific wavelengths are integrated with individual electroabsorption modulators so that the laser output can be modulated with channel data. Their output is combined using a $N \times 1$ multiplexer, such as a star coupler or a multimode-interference (MMI) coupler with one output port. A built-in amplifier boosts the power of the multiplexed signal to increase the transmitted power.

Considerable progress has been made in realizing integrated WDM transmitters. In a 1993 device, the WDM transmitter not only integrated 16 DBR lasers with 0.8-

8.4 WDM Transmitters and Receivers

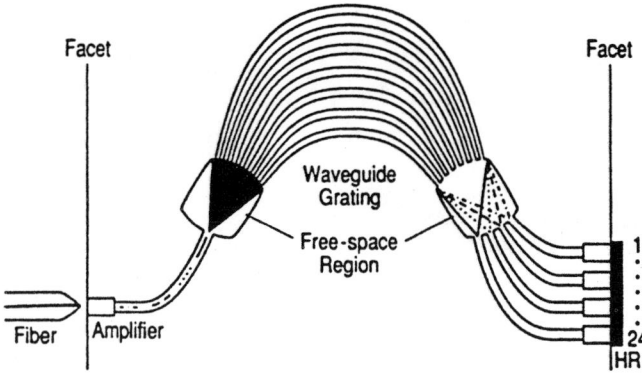

Figure 8.30: Schematic of a WDM laser made by integrating an AWG inside the laser cavity. (After Ref. [157]; ©1996 IEEE.)

nm wavelength spacing, but an electroabsorption modulator was also integrated with each laser [150]. In a 1996 device, 16 gain-coupled DFB lasers were integrated, and their wavelengths were controlled by changing the width of the ridge waveguides and by tuning over a 1-nm range using a thin-film heater [151]. In a different approach, sampled gratings with different periods are used to precisely tune the wavelengths of an array of DBR lasers [153]. The complexity of such devices makes it difficult to integrate more than 16 lasers on the same chip. The vertical-cavity surface-emitting laser (VCSEL) technology provides a unique approach to WDM transmitters since it can be used to produce a two-dimensional array of lasers covering a wide wavelength span at a relatively low cost [154]; it is well suited for local-area network and data-transfer applications.

An AWG, integrated within the laser cavity, can also provide sources operating at several wavelengths simultaneously [157]–[161]. The AWG is used for multiplexing the output of several optical amplifiers or DBR lasers. In a 1996 demonstration of the basic idea, simultaneous operation at 18 wavelengths (spaced apart by 0.8 nm) was realized using an intracavity AWG [157]. Figure 8.30 shows the laser design schematically. Spontaneous emission of the amplifier located on the left side is demultiplexed into 18 spectral bands by the AWG through the technique of spectral slicing. The amplifier array on the right side selectively amplifies the set of 18 wavelengths, resulting in a laser emitting at all wavelengths simultaneously. A 16-wavelength transmitter with 50-GHz channel spacing was built in 1998 by this technique [158]. By 1999, it was possible to design a WDM transmitter emitting light simultaneously at 40 wavelengths that were digitally tunable [159].

In a somewhat different approach, the AWG was not a part of the laser cavity but was used to multiplex the output of 10 DBR lasers, all produced on the same chip in an integrated fashion [160]. AWGs fabricated with the silica-on-silicon technology can also be used with this approach, although they cannot be integrated on the InP substrate. Since integrated devices are preferable from a system standpoint, it is better if the AWG is integrated within the laser cavity. A coupled-cavity approach has also

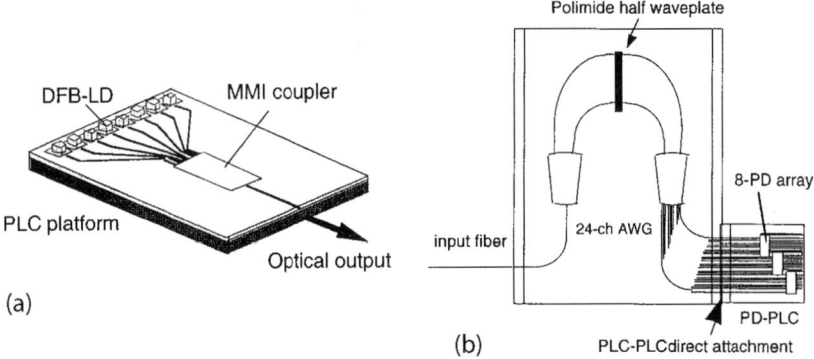

Figure 8.31: Multichannel transmitter (a) and receiver (b) modules fabricated using hybrid integration on a PLC platform; LD: laser diode; MMI: multimode interference; PD: photodiode. (After Ref. [98]; ©2000 IEEE.)

been used in which the cavities of two InGaAsP lasers, each containing a four-channel AWG were coupled using an MMI coupler [161]. The resulting compact device (with a size of 3.2×4.5 mm^2) was capable of emitting 16 wavelengths that were digitally tunable.

A hybrid approach can also be used in which multiple semiconductor lasers, fabricated using the InGaAsP/InP technology, are integrated on a photonic-lightwave-circuit (PLC) platform that is made with silica-on-silicon technology [98]. Figures 8.31(a) and (b) show the basic concept schematically. The output of eight DFB lasers, integrated individually on a PLC platform, is multiplexed using an MMI coupler fabricated with silica waveguides. If an electroabsorption modulator is integrated with each laser, the output consists of eight data-encoded channels.

On the receiver end, multichannel WDM receivers have been developed because their use simplifies the system design and reduces overall cost [162]. Monolithic receivers integrate a photodiode array with a demultiplexer on the same chip. Typically, a planar concave-grating demultiplexer or an AWG is integrated with the photodiode array. Even electronic amplifiers can be integrated within the same chip. The design of such monolithic receivers is similar to the transmitter shown in Figure 8.30, except that no cavity is formed and the amplifier array is replaced with a photodiode array. Such a WDM receiver was first fabricated in 1995 by integrating an eight-channel WGR (with 0.8-nm channel spacing), eight p–i–n photodiodes, and eight preamplifiers using heterojunction-bipolar transistor technology [163].

Similar to the case of transmitters, the hybrid PLC technology can be used to make WDM receivers [98]. An example is shown in Figure 8.31(b), where an AWG fabricated on a PLC platform using silica-on-silicon technology is attached to another PLC containing the photodiode array using a bonding technique. The second PLC contains three 45° micromirrors [164] that reflect the signal in the vertical direction before it reaches a photodiode array fabricated using InGaAsP/InP technology.

8.4 WDM Transmitters and Receivers

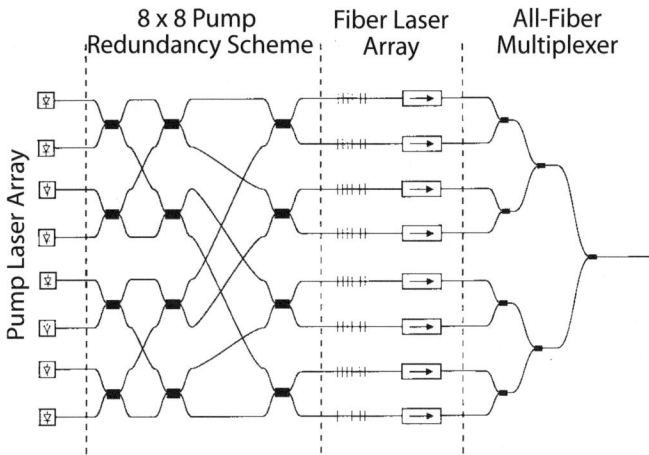

Figure 8.32: Schematic of an eight-channel, all-fiber, transmitter module with built-in pump redundancy. (After Ref. [166]; ©1999 IEEE.)

8.4.2 Fiber-Laser Transmitters

Although semiconductor DFB lasers are used almost exclusively as optical sources for WDM applications, it is possible to replace them with all-fiber DFB lasers (see Section 3.4.2) that make use of a fiber grating for fixing the output wavelength. Moreover, fiber lasers can be designed to operate at multiple wavelengths simultaneously, and are thus suitable as a WDM transmitter [165]–[170].

Figure 8.32 shows an eight-channel, all-fiber, transmitter module used in a 1999 experiment [166]. It employs twelve 3-dB fiber couplers such that each laser within the eight-laser array pumps all DFB fiber lasers. A single 8×8 star coupler can also be used in place of 12 fiber couplers. Such a pumping scheme provides a safeguard against the failure of one or two pump lasers. Eight DFB fiber lasers are separated in frequency by 100 GHz so that their wavelengths fall on the ITU grid. All lasers operate in the same polarization mode; the power in the orthogonal polarization was reduced by more than 40 dB. With 50 mW of power from each pump laser, each channel had an output power of about 5 dBm. The outputs of the eight lasers were combined using an all-fiber multiplexer consisting of seven 3-dB splitters, with a total insertion loss of 10 dB. A modulator can be placed after each laser before combining the outputs. This scheme can be extended to increase the number of channels. Moreover, a single pump laser can provide power to more than one fiber laser. For example, a 16-channel transmitter with only eight pump lasers can be built by placing additional fiber couplers just before the DFB lasers. The measured intensity noise was quite low (RIN < -160 dB/Hz) in such transmitters because pump-power fluctuations are averaged by the pump-redundancy scheme shown in Figure 8.32.

In an alternative scheme, a ring-cavity fiber laser is employed [169]. As shown schematically in Figure 8.33, the ring cavity contains one or more erbium-doped fiber amplifiers (EDFAs) together with a frequency shifter and an optical filter with periodic

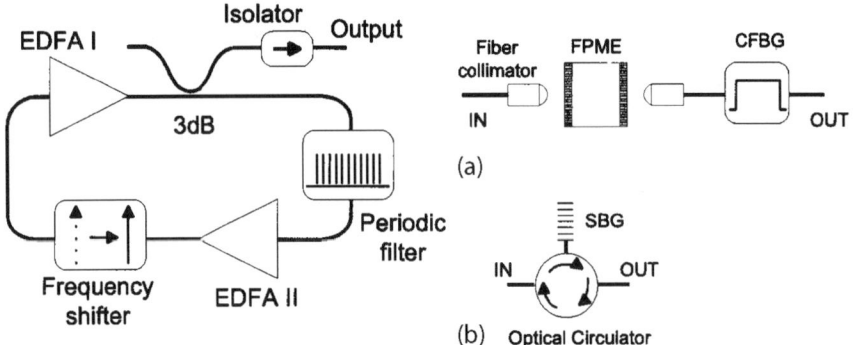

Figure 8.33: Schematic of a fiber-laser emitting light at multiple equally spaced wavelengths. The periodic filter can be (a) an FP filter or (b) a sampled grating. (After Ref. [169]; ©2000 IEEE.)

transmission peaks (such as a FP filter or a sampled grating). The EDFAs provide gain over a wide bandwidth approaching 5 THz or so. The frequency shifter such as an acousto-optic modulator shifts the frequency of intracavity light as it propagates through it (typically by 100 MHz or so). Such a periodically incurring frequency shift ensures that the fiber laser does not operate over a narrow bandwidth but spans the entire bandwidth of EDFAs. The periodic filter provides the maximum transmission at a comb of frequencies, which can be chosen to coincide with the ITU grid. The combination of a frequency shifter and a periodic filter forces the laser to emit light at the frequencies that coincide with the transmission peaks of the optical filter.

The periodic filter can be designed using several different techniques. Two designs are shown in Figure 8.33. In design (a), a FP filter is combined with a chirped fiber grating. The role of the grating is to limit the spectral band over which the FP filter provides transmission. In design (b), a sampled fiber grating is used in combination with an optical circulator. An AWG can also be used for this purpose. WDM transmitters emitting light at 16 distinct wavelengths have been made with this technique, although power is not generally uniform across all channels. Semiconductor optical amplifiers (SOAs) have also been used in place of EDFAs to make WDM transmitters. In one experiment, a fiber-ring laser containing two SOAs and a fiber FP filter provided output at 52 channels with 50-GHz spacing and nearly equal power levels [171]. It should be stressed that all such schemes provide multichannel CW output. A demultiplexer must be used to separate the channels before data can be imposed on individual channels using optical modulators.

8.4.3 Spectrally Sliced WDM Transmitters

A unique approach to WDM sources exploits the technique of spectral slicing for realizing WDM transmitters and is capable of providing more than 1,000 channels. In this approach, the output of a wide-bandwidth optical source is sliced spectrally using

8.4 WDM Transmitters and Receivers

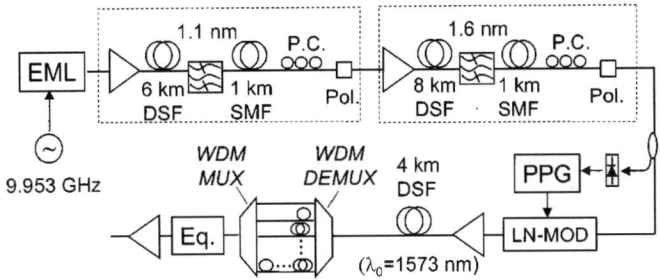

Figure 8.34: Schematic of a 40-channel WDM transmitter based on spectral slicing of the supercontinuum generated using the nonlinear effects in optical fibers. EML: electroabsorption-modulated laser; DSF: dispersion-shifted fiber; SMF: single-mode fiber; P.C.: polarization controller; LN-MOD: lithium-niobate modulator; PPG: pulse-pattern generator; and Eq: spectrum equalizer. (After Ref. [182]; ©2000 IEEE.)

a periodic optical filter such as an AWG. Several different types of optical sources have been used for this purpose.

As one might expect, the simplest solution is to use a broadband source (such as an LED, a superluminescent laser diode, or an optical amplifier with amplified spontaneous emission) and filter its output using a periodic filter so that the resulting device emits light in several equally spaced spectral bands [172]–[176]. One can also employ the amplified spontaneous emission of an optical amplifier for this purpose. However, such WDM transmitters cannot be used at high bit rates because of the incoherent nature of the original source. Nevertheless, they are quite suitable for local-area applications.

The most appropriate approach for high-speed WDM transmitters is to begin with a coherent optical source but broaden its spectrum while maintaining the coherence as much as possible [177]–[188]. In one implementation of this basic idea [177]–[179], picosecond pulses from a gain-switched semiconductor laser, or a mode-locked fiber laser (see Section 3.4.3), are first broadened spectrally to a bandwidth as large as 200 nm through supercontinuum generation by exploiting the nonlinear effects occurring inside optical fibers [190]. A mode-locked or directly modulated semiconductor laser can be used in place of the fiber laser.

Figure 8.34 shows the layout of a 40-channel transmitter in which an electroabsorption modulator, integrated within the semiconductor laser, is driven at about 10 GHz to produce 21.5-ps pulses at a center wavelength of 1,554 nm [182]. These pulses are amplified, propagated through a dispersion-shifted fiber ($D = -2$ ps/nm/km) that chirps them though self-phase modulation (SPM), filtered using a bandpass filter, and compressed inside a standard single-mode fiber. The bandpass filter is used to select the spectral region over which the chirp is mostly linear. The second stage is used to produce a train of 2.7-ps seed pulses, which are launched into 4-km-long dispersion-shifted fiber (with a zero-dispersion wavelength of 1573 nm) in which supercontinuum is generated. Most of the spectral broadening results from SPM in the normal-dispersion regime. A WDM demultiplexer with a channel spacing of 50 GHz is used to slice 40 channels from the broad supercontinuum with wavelengths in the range of

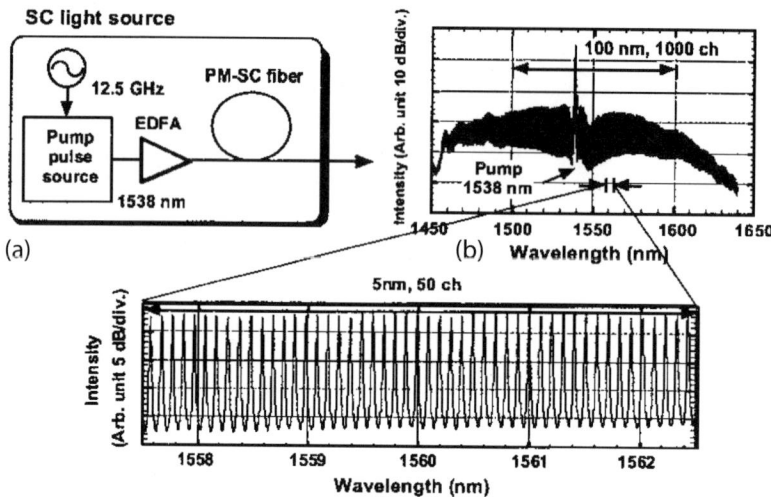

Figure 8.35: Schematic of a 1,000-channel WDM transmitter (a) and its output spectrum (b). The bottom trace show the enlarged spectrum over a 5-nm range. SC: supercontinuum; PM: polarization maintaining. (After Ref. [187]; ©2002 IEEE.)

1,546–1,562 nm. The LiNbO$_3$ modulator and delay lines were used to decorrelate the data patterns for different channels. In real WDM systems, an individual modulator will be used for each channel between the demultiplexer and multiplexer. Power levels for various channels can be equalized using an integrated dynamic equalizer.

Spectral slicing of the supercontinuum generated using a gain-switched semiconductor laser was first used in 1993 [177]. By 1994, this technique was used to build a 40-channel transmitter using a birefringent periodic optical filter [178]. In later experiments, AWGs built using silica-on-silicon technology were utilized to produce a much large number of WDM channels with a channel spacing of less than 1 nm. In a 2000 experiment, this technique produced 1,000 channels with 12.5-GHz channel spacing [184]. Figure 8.35(a) shows the the design of such a transmitter together with its multichannel spectrum (b). The output of a mode-locked semiconductor laser in the form of 4.3-ps pulses was amplified by an EDFA and then broadened spectrally in a polarization-maintaining fiber. In another experiment, 150 channels with 25-GHz channel spacing were realized within the C band covering the range of 1,530–1,560 nm [186]. The SNR of each channel exceeded 28 dB, indicating that the source was suitable for dense WDM applications. By 2003, the supercontinuum technique was used to produce a transmitter that delivered 50-GHz spaced optical carriers (on the ITU grid) over a spectral range from 1,425–1,675 nm covering the S, C, and L bands [189]. Even though channels powers were not uniform over the entire spectral range, this experiment shows clearly the potential of the supercontinuum technique.

The generation of supercontinuum is not necessary if a mode-locked laser producing femtosecond pulses is employed, and such ultrashort pulses have been used to produce WDM transmitters [191]–[195]. The spectral width of femtosecond pulses is quite large to begin with and can be further enlarged to 50 nm or more by chirping them using 10–15 km of standard single-mode fiber. Spectral slicing of the chirped output by a demultiplexer then provides many WDM channels, each of which can be modulated independently. This technique also permits simultaneous modulation of all channels using a single modulator before the demultiplexer if the modulator is driven by a suitable electrical bit stream composed by using the technique of time-division multiplexing in the electrical domain. A 32-channel WDM source was demonstrated in 1996 by using this method [191]. Since then, the technique has been used to make WDM transmitters with more than 1,000 channels [193].

Problems

8.1 Explain how a WDM signal is affected by an optical filter using the concept of the transfer function. What role does the phase of the transfer function play during the filtering process?

8.2 A Fabry–Perot filter of length L has equal reflectivities R for the two mirrors. Derive an expression for the transmission spectrum $T(v)$, considering multiple round trips inside the cavity containing air. Use it to show that the finesse is given by $F = \pi\sqrt{R}/(1-R)$.

8.3 A Fabry–Perot filter is used to select 100 channels spaced apart by 0.2 nm. What should the length and the mirror reflectivities of the filter be? Assume a 10-Gb/s bit rate, a refractive index of 1.5 and an operating wavelength of 1.55 μm.

8.4 The action of a directional coupler is governed by the matrix equation $\mathbf{E}_{out} = \mathbf{T}\mathbf{E}_{in}$, where $\mathbf{T}$ is the 2×2 transfer matrix and $\mathbf{E}$ is a column vector whose two components represent the input (or output) fields at the two ports. Assuming that the total power is preserved, show that the transfer matrix $\mathbf{T}$ must have the form

$$\mathbf{T} = \begin{pmatrix} \sqrt{\rho} & i\sqrt{1-\rho} \\ i\sqrt{1-\rho} & \sqrt{\rho} \end{pmatrix},$$

where ρ is the fraction of the power remaining in the bar port.

8.5 Explain how a Mach–Zehnder interferometer works. Using the transfer matrix of a 3-dB fiber coupler, find the transfer functions for the bar and cross ports when the signal experiences an additional delay of τ in the longer arm of the interferometer.

8.6 Prove that the transmission through a chain of M Mach–Zehnder interferometers built using 3-dB couplers is given by $T(v) = \prod_{m=1}^{M} \cos^2(\pi v \tau_m)$, where τ_m is the relative delay in the mth branch.

8.7 The reflection coefficient of a fiber grating of length L is given by

$$r_g(\delta) = \frac{i\kappa \sin(qL)}{q\cos(qL) - i\delta \sin(qL)},$$

where $q^2 = \delta^2 - \kappa^2$, $\delta = (\omega - \omega_B)(\bar{n})/c$ is the detuning from the Bragg frequency ω_B, and κ is the coupling coefficient. Plot the reflectivity spectrum using

$\kappa = 8$ cm^{-1} and $\bar{n} = 1.45$, and a Bragg wavelength of 1.55 μm for $L = 3, 5$, and 8 mm. Estimate the grating bandwidth in GHz in the three cases.

8.8 Write a computer program for calculating the transmissivity of phase-shifted gratings and reproduce Figure 8.8 when the grating has three $\lambda/4$ phase-shift regions, spaced apart by $L_g/4$. Assume that $\kappa L_g = 3$ for the grating of length L_g.

8.9 Explain how an acousto-optic filter fabricated using LiNbO$_3$ waveguides works. Derive the Bragg condition for such a filter and calculate the acoustic frequency for selecting a 1,552-nm channel when the waveguide birefringence is 0.07.

8.10 Sketch a design for an acousto-optic filter made using optical fibers. Explain how this filter would select a specific channel at a certain wavelength.

8.11 Sketch the design of a 8×8 demultiplexer with as few couplers as possible. Explain how such a device works.

8.12 Explain how an array of planar waveguides can be used for demultiplexing WDM channels. Use diagrams as necessary.

8.13 Use two fiber couplers and two fiber gratings to design an add–drop multiplexer. Explain how such a device functions.

8.14 Use a single Bragg grating and a single optical circulator to design an add–drop multiplexer. Explain clearly how such a device functions.

8.15 How can the nonlinear effects occurring inside an optical fiber be used for making WDM transmitters? Sketch the design of such a transmitter.

References

[1] C. M. Madsen and J. H. Zhao, *Optical Filter Design and Analysis*, Wiley, New York, 1999.

[2] M. Born and E. Wolf, *Principles of Optics*, 7th ed., Cambridge University Press, New York, 1999.

[3] J. Stone and L. W. Stulz, *Electron. Lett.* **23**, 781 (1987).

[4] K. Hirabayashi, H. Tsuda, and T. Kurokawa, *IEEE Photon. Technol. Lett.* **3**, 741 (1991); *J. Lightwave Technol.* **11**, 2033 (1993).

[5] A. Sneh and K. M. Johnson, *J. Lightwave Technol.* **14**, 1067 (1996).

[6] S. Matsumoto, K. Hirabayashi, S. Sakata, and T. Hayashi, *IEEE Photon. Technol. Lett.* **11**, 442 (1999).

[7] M. Iodice, G. Cocorullo, F. G. Della Corte, and I. Rendina, *Opt. Commun.* **183**, 415 (2000).

[8] H. K. Tsang, M. W. K. Mak, L. Y. Chan, J. B. D. Soole, C. Youtsey, and I. Adesida, *J. Lightwave Technol.* **17**, 1890 (1999).

[9] E. C. Vail, M. S. Wu, G. S. Li, L. Eng, and C. J. Chnag-Hasnain, *Electron. Lett.* **31**, 228 (1995).

[10] M. C. Larson, B. Pezeshki, and J. S. Harris, *IEEE Photon. Technol. Lett.* **7**, 382 (1995).

[11] J. Peerlings, A. Dehé, A. Vogt, M. Tilsch, C. Hebeler, F. Langenhan, P. Meissner, and H. L. Hartnagel, *IEEE Photon. Technol. Lett.* **9**, 1235 (1997).

[12] P. Tayebatti, P. D. Wang, D. Vakhshoori, and R. N. Sacks, *IEEE Photon. Technol. Lett.* **10**, 394 (1998).

[13] D. Sadot and E. Boimovich, *IEEE Commun. Mag.* **38** (12), 50 (1998).

[14] R. Le Dantec, T. Benyattou, G. Guillot, A. Spisser, C. Seassal, J. L. Leclercq, P. Viktorovitch, D. Rondi, and R. Blondeau, *IEEE J. Sel. Topics Quantum Electron.* **5**, 111 (1999).

[15] S. Greek, R. Gupta, and K. Hjort, *J. Microelectromech. Sys.* **8**, 328 (1999).

[16] J. Pfeiffer, J. Peerlings, R. Riemenschneider, R. Genovese, M. Aziz, E. Goutain, H. Kunzel, W. Gortz, G. Bohm, M. C. Amann, P. Meissner, and H. L. Hartnagel, *Mat. Sci. Semicond. Process.* **3**, 409 (2000).

[17] M. Aziz, J. Pfeiffer, M. Wohlfarth, C. Luber, S. Wu, and P. Meissner, *IEEE Photon. Technol. Lett.* **12**, 1522 (2000).

[18] M. Strassner, C. Luber, A. Tarraf, and N. Chitica, *IEEE Photon. Technol. Lett.* **14**, 1548 (2002).

[19] S. Irmer, J. Daleiden, V. Rangelov, C. Prott, F. Römer, M. Strassner, A. Tarraf, and H. Hillmer, *IEEE Photon. Technol. Lett.* **15**, 434 (2003).

[20] G. P. Agrawal, *Applications of Nonlinear Fiber Optics*, Academic Press, San Diego, CA, 2001.

[21] K. Takiguchi, K. Okamoto, S. Suzuki, and Y. Ohmori, *IEEE Photon. Technol. Lett.* **6**, 86 (1994).

[22] Y. Hibino, F. Hanawa, H. Nakagome, M. Ishii, and N. Takato, *J. Lightwave Technol.* **13**, 1728 (1995).

[23] S. Mino, K. Yoshino, Y. Yamada, T. Terui, M. Yasu, and K. Moriwaki, *J. Lightwave Technol.* **13**, 2320 (1995).

[24] M. Kawachi, *IEE Proc.* **143**, 257 (1996).

[25] K. Takiguchi, K. Okamoto, and K. Moriwaki, *J. Lightwave Technol.* **14**, 2003 (1996).

[26] Y. P. Li and C. H. Henry, *IEE Proc.* **143**, 263 (1996).

[27] K. Okamoto, *Opt. Quantum Electron.* **31**, 107 (1999).

[28] E. L. Wooten, L. Stone, E. W. Miles, and E. M. Bradley, *J. Lightwave Technol.* **14**, 2530 (1996).

[29] P. P. Absil, J. V. Hryniewicz, B. E. Little, R. A.Wilson, L. G. Joneckis, and P. T. Ho, *IEEE Photon. Technol. Lett.* **12**, 398 (2000).

[30] R. Kashyap, *Fiber Bragg Gratings*, Academic Press, San Diego, CA, 1999.

[31] G. E. Town, K. Sugden, J. A. R. Williams, I. Bennion, and S. B. Poole, *IEEE Photon. Technol. Lett.* **7**, 78 (1995).

[32] F. Bilodeau, K. O. Hill, B. Malo, D. C. Johnson, and J. Albert, *IEEE Photon. Technol. Lett.* **6**, 80 (1994).

[33] G. P. Agrawal and S. Radic, *IEEE Photon. Technol. Lett.* **6**, 995 (1994).

[34] J. Canning and M. G. Sceats, *Electron. Lett.* **30**, 1244 (1994).

[35] R. Kasyap, P. E. McKee, and D. Armes, *Electron. Lett.* **30**, 1977 (1994).

[36] R. Zengerle and O. Leminger, *J. Lightwave Technol.* **13**, 2354 (1995).

[37] L. Wei and W. Y. Lit, *J. Lightwave Technol.* **15**, 1405 (1997).

[38] F. Bhakti and P. Sansonetti, *J. Lightwave Technol.* **15**, 1433 (1997).

[39] S. Legoubin, E. Fertein, M. Douay, P. Bernage, P. Niay, F. Bayon, and T. Georges, *Electron. Lett.* **27**, 1945 (1991).

[40] M. C. Farries, K. Sugden, D. C. J. Reid, I. Bennion, A. Molony, and M. J. Goodwin, *Electron. Lett.* **30**, 891 (1994).

[41] B. J. Eggleton, P. A. Krug, L. Poladian, and F. Ouellette, *Electron. Lett.* **30**, 1620 (1994).
[42] M. Ibsen, J. Hübner, J. E. Pedersen, R. Kromann, L. U. A. Andersen, and M. Kristensen, *Electron. Lett.* **32**, 2233 (1996).
[43] M. Ibsen, M. K. Durkin, M. J. Cole, and R. I. Laming, *IEEE Photon. Technol. Lett.* **10**, 842 (1998).
[44] W. H. Loh, F. Q. Zhou, and J. J. Pan, *IEEE Photon. Technol. Lett.* **11**, 1280 (1999).
[45] X. F. Chen, C. C. Fan, Y. Luo, S. Z. Xie, and S. Hu, *IEEE Photon. Technol. Lett.* **12**, 1501 (2000).
[46] C. Y. Lin, G. W. Chern, and L. A. Wang, *J. Lightwave Technol.* **19**, 1212 (2001).
[47] C. R. Doerr, R. Pafchek, and L. W. Stulz, *IEEE Photon. Technol. Lett.* **14**, 334 (2002).
[48] J. E. Rothenberg, H. Li, Y. Li, J. Popelek, Y. Sheng, Y. Wang, R. B. Wilcox, and J. Zweiback, *IEEE Photon. Technol. Lett.* **14**, 1309 (2002).
[49] H. Lee and G. P. Agrawal, *IEEE Photon. Technol. Lett.* **15**, 1091 (2003).
[50] G. A. Ball and W. W. Morey, *Opt. Lett.* **19**, 1979 (1994).
[51] A. Iocco, H. G. Limberger, R. P. Salathe, L. A. Everall, K. E. Chisholm, J. A. R. Williams, and I. Bennion, *J. Lightwave Technol.* **17**, 1217 (1999).
[52] T. Inui, T. Komukai, and M. Nakazawa, *Opt. Commun.* **190**, 1 (2001).
[53] C. S. Goh, S. Y. Set, and K. Kikuchi, *IEEE Photon. Technol. Lett.* **14**, 1306 (2002).
[54] C. S. Goh, M. R. Mokhtar, S. A. Butler, S. Y. Set, K. Kikuchi, and M. Ibsen, *IEEE Photon. Technol. Lett.* **15**, 557 (2003).
[55] J. Lauzon, S. Thibault, J. Martin, and F. Ouellette, *Opt. Lett.* **19**, 2027 (1994).
[56] J. A. Rogers, B. J. Eggleton, J. R. Pedrazzani, and T. A. Strasser, *Appl. Phys. Lett.* **74**, 3131 (1999).
[57] K. M. Feng, J. X. Chai, V. Grubsky, D. S. Starodubov, M. I. Hayee, S. Lee, X. Jiang, A. E. Willner, and J. Feinberg, *IEEE Photon. Technol. Lett.* **11**, 1041 (1999).
[58] B. J. Eggleton, A. Ahuja, P. S. Westbrook, J. A. Rogers, P. Kuo, T. N. Nielsen, and B. Mikkelsen, *J. Lightwave Technol.* **18**, 1418 (2000).
[59] J. Mora, B. Ortega, M. V. Andres, J. Capmany, J. L. Cruz, D. Pastor, and S. Sales, *IEEE Photon. Technol. Lett.* **15**, 951 (2003).
[60] R. Kashyap, *Opt. Commun.* **153**, 14 (1998).
[61] B. Ortega, J. Capmany, and J. L. Cruz, *J. Lightwave Technol.* **17**, 1241 (1999).
[62] D. Johlen, P. Klose, H. Renner, and E. Brinkmeyer, *J. Lightwave Technol.* **18**, 1575 (2000).
[63] R. Slavik, S. Doucet, and S. LaRochelle, *J. Lightwave Technol.* **21**, 1059 (2003).
[64] T. Numai, S. Murata, and I. Mito, *Appl. Phys. Lett.* **53**, 83 (1988); **54**, 1859 (1989).
[65] J. P. Weber, B. Stoltz, and M. Dasler, *Electron. Lett.* **31**, 220 (1995).
[66] D. A. Smith, J. E. Baran, K. W. Cheung, and J. J. Johnson, *Appl. Phys. Lett.* **56**, 209 (1990).
[67] J. L. Jackel, J. E. Baran, A. d'Alessandro, and D. A. Smith, *IEEE Photon. Technol. Lett.* **7**, 318 (1995).
[68] D. A. Smith, A. d'Alessandro, J. E. Baran, D. J. Fritz, J. L. Jackel, and R. S. Chakravarthy, *J. Lightwave Technol.* **14**, 2044 (1996).
[69] H. Herrmann, K. Schafer, and C. Schmidt, *IEEE Photon. Technol. Lett.* **10**, 120 (1998).
[70] B. Beche, H. Porte, J. P. Goedgebuer, and C. Fontaine, *IEEE J. Quantum Electron.* **35**, 820 (1999).
[71] T. E. Dimmick, G. Kakarantzas, T. A. Birks, and P. S. J. Russell, *IEEE Photon. Technol. Lett.* **12**, 1210 (2000).

References

[72] J. Sapriel, D. Charissoux, V. Voloshinov, and V. Molchanov, *J. Lightwave Technol.* **20**, 892 (2002).
[73] T. A. Birks, P. St. J. Russell, and C. N. Pannell, *IEEE Photon. Technol. Lett.* **6**, 725 (1994).
[74] D. Ostling and H. E. Engan, *Opt. Lett.* **20**, 1247 (1995).
[75] T. A. Birks, P. St. J. Russell, and D. O. Culverhouse, *J. Lightwave Technol.* **14**, 2619 (1996).
[76] H. S. Kim, S. H. Yun, I. K. Kwang, and B. Y. Kim, *Opt. Lett.* **22**, 1476 (1997).
[77] T. Jin, Q. Li, J. Zhao, K. Cheng, and X. Liu, *IEEE Photon. Technol. Lett.* **14**, 1133 (2002).
[78] D. A. Satorius, T. E. Dimmick, and G. L. Burdge, *IEEE Photon. Technol. Lett.* **14**, 1324 (2002).
[79] H. S. Park, K. Y. Song, S. H. Yun, and B. Y. Kim, *J. Lightwave Technol.* **20**, 1864 (2002).
[80] C. H. Henry, R. F. Kazarinov, Y. Shani, R. C. Kistler, V. Pol, and K. J. Orlowsky, *J. Lightwave Technol.* **8**, 748 (1990).
[81] D. R. Wisely, *Electron. Lett.* **27**, 520 (1991).
[82] M. Fallahi, K. A. McGreer, A. Delage, I. M. Templeton, F. Chatenoud, and R. Barber, *IEEE Photon. Technol. Lett.* **5**, 794 (1993).
[83] K. A. McGreer, *IEEE Photon. Technol. Lett.* **8**, 553 (1996).
[84] S. J. Sun, K. A. McGreer, and J. N. Broughton, *IEEE Photon. Technol. Lett.* **10**, 90 (1998).
[85] F. N. Timofeev, E. G. Churin, P. Bayvel, V. Mikhailov, D. Rothnie, and J. E. Midwinter, *Opt. Quantum Electron.* **31**, 227 (1999).
[86] B. H. Verbeek, C. H. Henry, N. A. Olsson, K. J. Orlowsky, R. F. Kazarinov, and B. H. Johnson, *J. Lightwave Technol.* **6**, 1011 (1988).
[87] K. Oda, N. Tokato, H. Toba, and K. Nosu, *J. Lightwave Technol.* **6**, 1016 (1988).
[88] N. Takato, T. Kominato, A. Sugita, K. Jinguji, H. Toba, and M. Kawachi, *IEEE J. Sel. Areas Commun.* **8**, 1120 (1990).
[89] Y. Hibino, T. Kitagawa, K. O. Hill, F. Bilodeau, B. Malo, J. Albert, and D. C. Johnson, *IEEE Photon. Technol. Lett.* **8**, 84 (1996).
[90] B. S. Kawasaki, K. O. Hill, and R. G. Gaumont, *Opt. Lett.* **6**, 327 (1981).
[91] M. K. Smit and C. van Dam, *IEEE J. Sel. Topics Quantum Electron.* **2**, 251 (1996).
[92] R. Mestric, M. Renaud, M. Bachamann, B. Martin, and F. Goborit, *IEEE J. Sel. Topics Quantum Electron.* **2**, 257 (1996).
[93] H. Okayama, M. Kawahara, and T. Kamijoh, *J. Lightwave Technol.* **14**, 985 (1996).
[94] K. Okamoto, in *Photonic Networks*, G. Prati, Ed., Springer, New York, 1997.
[95] C. Dragone, *J. Lightwave Technol.* **16**, 1895 (1998).
[96] A. Kaneko, T. Goh, H. Yamada, T. Tanaka, and I. Ogawa, *IEEE J. Sel. Topics Quantum Electron.* **5**, 1227 (1999).
[97] P. Bernasconi, C. R. Doerr, C. Dragone, M. Cappuzzo, E. Laskowski, and A. Paunescu, *J. Lightwave Technol.* **18**, 985 (2000).
[98] K. Kato and Y. Tohmori, *IEEE J. Sel. Topics Quantum Electron.* **6**, 4 (2000).
[99] K. Takada, M. Abe, T. Shibata, and K. Okamoto, *IEEE Photon. Technol. Lett.* **13**, 577 (2001).
[100] Y. Hibino, *IEEE J. Sel. Topics Quantum Electron.* **8**, 1090 (2002).
[101] J. L. Archambault, P. S. J. Russell, S. Barcelos, P. Hua, and L. Reekie, *Opt. Lett.* **19**, 180 (1994).
[102] I. Baumann, J. Seifert, W. Novak, and M. Sauer, *IEEE Photon. Technol. Lett.* **8**, 1331 (1996).

[103] L. Dong, P. Hua, T. A. Birks, L. Reekie, and P. St. Russell, *IEEE Photon. Technol. Lett.* **8**, 1656 (1996).

[104] K. Bakhti, P. Sansonetti, C. Sinet, L. Gasca, L. Martineau, S. Lacroix, X. Daxhelet, and F. Gonthier, *Electron. Lett.* **33**, 803 (1997).

[105] A. S. Kewitsch, G. A. Rakuljic, P. A. Willems, and A. Yariv, *Opt. Lett.* **23**, 106 (1998).

[106] T. Erdogan, *Opt. Commun.* **157**, 249 (1998).

[107] C. Riziotis and M. N. Zervas, *J. Lightwave Technol.* **19**, 92 (2001).

[108] D. Mechin, P. Grosso, and D. Bosc, *J. Lightwave Technol.* **19**, 1282 (2001).

[109] C. M. Ragdale, T. J. Reid, D. C. J. Reid, A. C. Carter, and P. J. Williams, *Electron. Lett.* **28**, 712 (1992).

[110] R. Kashyap, G. D. Maxwell, and B. J. Ainslie, *IEEE Photon. Technol. Lett.* **5**, 191 (1993).

[111] T. J. Cullen, H. N. Rourke, C. P. Chew, S. R. Baker, T. Bricheno, K. C. Byron, and K. Fielding, *Electron. Lett.* **30**, 2160 (1994).

[112] M. Kuznetsov, *J. Lightwave Technol.* **12**, 226 (1994).

[113] H. H. Yaffe, C. H. Henry, M. R. Serbin, and L. G. Cohen, *J. Lightwave Technol.* **12**, 1010 (1994).

[114] F. Bilodeau, D. C. Johnson, S. Thériault, B. Malo, J. Albert, and K. O. Hill, *IEEE Photon. Technol. Lett.* **7**, 388 (1995).

[115] Y. Hibino, T. Kitagawa, K. O. Hill, F. Bilodeau, B. Malo, J. Albert, and D. C. Johnson, *IEEE Photon. Technol. Lett.* **8**, 84 (1996).

[116] T. Mizuochi, T. Kitayama, K. Shimizu, and K. Ito, *J. Lightwave Technol.* **16**, 265 (1998).

[117] S. Bethuys, L. Lablonde, L. Rivoallan, J. F. Bayon, L. Brilland, and E. Delevaque, *Electron. Lett.* **34**, 1250 (1998).

[118] D. Mechin, P. Yvernault, L. Brilland, and D. Pureur, *J. Lightwave Technol.* **15**, 1411 (2003).

[119] J. Kim and B. Lee, *IEEE Photon. Technol. Lett.* **12**, 561 (2000).

[120] Y. K. Chen, C. J. Hu, C. C. Lee, K. M. Feng, M. K. Lu, C. H. Chung, Y. K. Tu, and S. L. Tzeng, *IEEE Photon. Technol. Lett.* **12**, 1394 (2000).

[121] A. V. Tran, W. D. Zhong, R. C. Tucker, and R. Lauder, *IEEE Photon. Technol. Lett.* **13**, 582 (2001).

[122] A. V. Tran, W. D. Zhong, R. C. Tucker, and K. Song, *IEEE Photon. Technol. Lett.* **13**, 1100 (2001).

[123] I. Y. Kuo and Y. K. Chen, *IEEE Photon. Technol. Lett.* **14**, 867 (2002).

[124] A. V. Tran, C. J. Chae, and R. C. Tucker, *IEEE Photon. Technol. Lett.* **15**, 975 (2003).

[125] B. E. Little, S. T. Chu, H. A. Haus, J. Foresi, and J. P. Laine, *J. Lightwave Technol.* **15**, 998 (1997).

[126] B. E. Little, J. S. Foresi, H. A. Haus, E. P. Ippen, W. Greene, and S. T. Chu, *IEEE Photon. Technol. Lett.* **10**, 549 (1998).

[127] S. T. Chu, B. E. Little, W. Pan, T. Kaneko, S. Sato, and Y. Kokubun, *IEEE Photon. Technol. Lett.* **11**, 691 (1999); *IEEE Photon. Technol. Lett.* **11**, 1423 (1999).

[128] M. K. Chin, C. Youtsey, W. Zhao, T. Pierson, Z. Ren, S. L. Wu, L. Wang, Y. G. Zhao, and S. T. Ho, *IEEE Photon. Technol. Lett.* **11**, 1620 (1999).

[129] S. Suzuki, Y. Kokubun, M. Nakazawa, T. Yamamoto, and S. T. Chu, *J. Lightwave Technol.* **20**, 266 (2002).

[130] Y. Yanagase, S. Suzuki, Y. Kokubun, and S. T. Chu, *J. Lightwave Technol.* **20**, 1525 (2002).

[131] D. G. Rabus, M. Hamacher, U. Troppenz, and H. Heidrich, *IEEE Photon. Technol. Lett.* **14**, 1442 (2002).
[132] H. A. Haus and Y. Lai, *J. Lightwave Technol.* **10**, 57 (1992).
[133] M. Zirngibl, C. H. Joyner, and B. Glance, *IEEE Photon. Technol. Lett.* **6**, 513 (1994).
[134] L. Eldada, S. Yin, C. Poga, C. Glass, R. Blomquist, and R. A. Norwood, *IEEE Photon. Technol. Lett.* **10**, 1416 (1998).
[135] T. Augustsson, *J. Lightwave Technol.* **16**, 1517 (1998).
[136] C. R. Doerr, L. W. Stulz, M. Cappuzzo, E. Laskowski, A. Paunescu, L. Gomez, J. V. Gates, S. Shunk, and A. E. White, *IEEE Photon. Technol. Lett.* **11**, 1437 (1999).
[137] N. A. Riza and S. Yuan, *J. Lightwave Technol.* **17**, 1575 (1999).
[138] B. Liu, A. Shakouri, P. Abraham, and J. E. Bowers, *IEEE Photon. Technol. Lett.* **12**, 410 (2000).
[139] S. Rotolo, A. Tanzi, S. Brunazzi, D. DiMola, L. Cibinetto, M. Lenzi, G. L. Bona, B. J. Offrein, F. Horst, R. Germann, H. W. M. Salemink, and P. H. Baechtold, *J. Lightwave Technol.* **18**, 569 (2000).
[140] M. Raburn, B. Liu, Y. Okuno, and J. E. Bowers, *IEEE Photon. Technol. Lett.* **13**, 579 (2001).
[141] W. Jiang, Y. Sun, R. T. Chen, B. Guo, J. Horwitz, and W. Morey, *IEEE Photon. Technol. Lett.* **15**, 825 (2002).
[142] P. Tang, O. Eknoyn, and H. F. Taylor, *J. Lightwave Technol.* **21**, 236 (2003).
[143] I. Ishii, H. Tanobe, F. Kano, Y. Tohmori, Y. Kondo, and Y. Yoshikuni, *IEEE J. Quantum Electron.* **32**, 433 (1996).
[144] P.-J. Rigole, S. Nilsson, L. Bäckbom, B. Stalnacke, E. Berglind, J.-P. Weber, and B. Stoltz, *Electron. Lett.* **32**, 2352 (1996).
[145] H. Debrégeas-Sillard, A. Vuong, F. Delorme, J. David, V. Allard, A. Bodr, O. LeGouezigou, F. Gaborit, J. Rotte, M. Goix, V. Voiriot, and J. Jacquet, *IEEE Photon. Technol. Lett.* **13**, 4 (2001).
[146] D. A. Ackerman, J. E. Johnson, L. J. P. Ketelsen, L. E. Eng, P. A. Kiely, and T. G. B. Mason, in *Optical Fiber Telecommunications*, Vol. 4A, I. P. Kaminow and T. P. Lee, Eds., Academic Press, San Diego, CA, 2002, Chap. 12.
[147] Y. A. Akulova, G. A. Fish, P.-C. Koh, C. L. Schow, P. Kozodoy, A. P. Dahl, S. Nakagawa, M. C. Larson, M. P. Mack, T. A. Strand, C. W. Coldren, E. Hegblom, S. K. Penniman, T. Wipiejewski, and L. A. Coldren, *IEEE J. Sel. Topics Quantum Electron.* **8**, 1349 (2002).
[148] T. P. Lee, C. E. Zah, R. Bhat, W. C. Young, B. Pathak, F. Favire, P. S. D. Lin, N. C. Andreadakis, C. Caneau, A. W. Rahjel, M. Koza, J. K. Gamelin, L. Curtis, D. D. Mahoney, and A. Lepore, *J. Lightwave Technol.* **14**, 967 (1996).
[149] K. Sato, S. Seinke, Y. Kondo, and M. Yamamoto, *IEEE J. Quantum Electron.* **29**, 1805 (1993).
[150] M. G. Young, U. Koren, B. I. Miller, M. A. Newkirk, M. Chien, M. Zirngibl, C. Dragone, B. Tell, H. M. Presby, and G. Raybon, *IEEE Photon. Technol. Lett.* **5**, 908 (1993).
[151] G. P. Li, T. Makino, A. Sarangan, and W. Huang, *IEEE Photon. Technol. Lett.* **8**, 22 (1996).
[152] T. L. Koch, in *Optical Fiber Telecommunications*, Vol. 3B, I. P. Kaminow and T. L. Koch, Eds., Academic Press, San Diego, CA, 1997, Chap. 4.
[153] S. L. Lee, I. F. Jang, C. Y. Wang, C. T. Pien, and T. T. Shih, *IEEE J. Sel. Topics Quantum Electron.* **6**, 197 (2000).
[154] H. Li and K. Iga, *Vertical-Cavity Surface-Emitting Laser Devices*, Springer, New York, 2001.

[155] H. Hatakeyama, K. Kudo, Y. Yokoyama, K. Naniwae, and T. Sasaki, *IEEE J. Sel. Topics Quantum Electron.* **8**, 1341 (2002).

[156] S. W. Ryu, S. B. Kim, J. S. Sim, and J. Kim, *IEEE J. Sel. Topics Quantum Electron.* **8**, 1358 (2002).

[157] M. Zirngibl, C. H. Joyner, C. R. Doerr, L. W. Stulz, and H. M. Presby, *IEEE Photon. Technol. Lett.* **8**, 870 (1996).

[158] R. Monnard, A. K. Srivastava, C. R. Doerr, R. J. Essiambre, C. H. Joyner, L. W. Stulz, M. Zirngibl, Y. Sun, J. W. Sulhoff, J.L. Zyskind, and C. Wolf, *Electron. Lett.* **34**, 765 (1998).

[159] C. R. Doerr, C. H. Joyner, and L. W. Stulz, *IEEE Photon. Technol. Lett.* **11**, 1348 (1999).

[160] S. Menezo, A. Rigny, A. Talneau, F. Delorme, S. Grosmaire, H. Nakajima, E. Vergnol, F. Alexandre, and F. Gaborit, *IEEE J. Sel. Topics Quantum Electron.* **6**, 185 (2000).

[161] J. H. den Besten, R. G Broeke, M. van Geemert, J. J. M. Binsma, F. Heinrichsdorff, T. van Dongen, E. A. J. M. Bente, X. J. M. Leijtens, and M. K. Smit, *IEEE Photon. Technol. Lett.* **14**, 1653 (2002); *IEEE Photon. Technol. Lett.* **15**, 368 (2003).

[162] F. Tong, *IEEE Commun. Mag.* **36** (12), 42 (1998).

[163] S. Chandrasekhar, M. Zirngibl, A. G. Dentai, C. H. Joyner, F. Storz, C. A. Burrus, and L. M. Lunardi, *IEEE Photon. Technol. Lett.* **7**, 1342 (1995).

[164] H. Terui and K. Shuto, *J. Lightwave Technol.* **16**, 105 (1998).

[165] J. Hübner, P. Varming, and M. Kristensen, *Electron. Lett.* **33**, 139 (1997).

[166] M. Ibsen, M. N. Zervas, A. B. Grudinin, and R. I. Laming, *IEEE Photon. Technol. Lett.* **11**, 1114 (1999).

[167] M. Ibsen, E. Ronnekleiv, G. J. Cowle, M. N. Zervas, and R. I. Laming, *Electron. Lett.* **36**, 143 (2000).

[168] F. Liu, X. Zheng, R. J. S. Pedersen, P. Varming, A. Buxens, Y. Qian, and P. Jeppesen, *Electron. Lett.* **36**, 620 (2000).

[169] A. Bellemare, M. Karasek, M. Rochette, S. LaRochelle, and M. Tetu, *J. Lightwave Technol.* **18**, 825 (2000).

[170] L. B. Fu, R. Selvas, M. Ibsen, J. K. Sahu, J. N. Jang, S.-U. Alam, J. Nilsson, D. J. Richardson, D. N. Payne, C. Codemard, S. Goncharov, I. Zalevsky, and A. B. Grudinin, *IEEE Photon. Technol. Lett.* **15**, 655 (2003).

[171] N. Pleros, C. Bintjas, M. Kalyvas, G. Theophilopoulos, K. Yiannopoulos, S. Sygletos, and H. Avramopoulos, *IEEE Photon. Technol. Lett.* **14**, 693 (2002).

[172] M. H. Reeve, A. R. Hunwicks, W. Zhao, S. G. Methley, L Bickers, and S. Hornung, *Electron. Lett.* **24**, 389 (1998).

[173] N. S. K. Kwong, *IEEE Photon. Technol. Lett.* **4**, 996 (1992).

[174] J. S. Lee, Y. C. Chung, and D. J. DiGiovanni, *IEEE Photon. Technol. Lett.* **5**, 1458 (1993).

[175] W. T. Holloway, A. J. Keating, and D. D. Sampson, *IEEE Photon. Technol. Lett.* **9**, 1014 (1997).

[176] K. Y. Liou, K. K. Dreyer, E. C. Burrows, J. L. Zyskind, and J. W. Sulhoff, *IEEE Photon. Technol. Lett.* **10**, 270 (1998).

[177] T. Morioka, K. Mori, and M. Saruwatari, *Electron. Lett.* **29**, 862 (1993).

[178] T. Morioka, K. Mori, S. Kawanishi, and M. Saruwatari, *IEEE Photon. Technol. Lett.* **6**, 365 (1994).

[179] T. Morioka, K. Uchiyama, S. Kawanishi, S. Suzuki, and M. Saruwatari, *Electron. Lett.* **31**, 1064 (1995).

[180] T. Okuno, M. Onishi, and M. Nishimura, *IEEE Photon. Technol. Lett.* **10**, 72 (1998).

[181] J. J. Veselka and S. K. Korotky, *IEEE Photon. Technol. Lett.* **10**, 958 (1998).

References

[182] L. Boivin, S. Taccheo, C. R. Doerr, L. W. Stulz, R. Monnard, W. Lin, and W. C. Fang, *IEEE Photon. Technol. Lett.* **12**, 1695 (2000).
[183] Ö. Boyraz, J. Kim, M. N. Islam, F. Coppinger, and B. Jalali, *J. Lightwave Technol.* **18**, 2167 (2000).
[184] H. Takara, T. Ohara, K. Mori, K. Sato, E. Yamada, Y. Inoue, T. Shibata, M. Abe, T. Morioka, and K. I. Sato, *Electron. Lett.* **36**, 2089 (2000).
[185] F. Futami and K. Kikuchi, *IEEE Photon. Technol. Lett.* **13**, 73 (2001).
[186] E. Yamada, H. Takara, T. Ohara, K. Sato, T. Morioka, K. Jinguji, M. Itoh, and M. Ishii, *Electron. Lett.* **37**, 304 (2001).
[187] H. Takara, *Proc. Opt. Fiber Commun. Conf.*, Optical Society of America, Washington, DC, 2002, p. 314
[188] Ö. Boyraz and M. N, Islam, *J. Lightwave Technol.* **20**, 1493 (2002).
[189] K. Mori, K. Sato, H. Takara, and T. Ohara, *Electron. Lett.* **39**, 544 (2003).
[190] G. P. Agrawal, *Nonlinear Fiber Optics*, 3rd ed., Academic Press, San Diego, CA, 2001.
[191] M. C. Nuss, W. H. Knox, and U. Koren, *Electron. Lett.* **32**, 1311 (1996).
[192] L. Boivin, M. Wegmuller, M. C. Nuss, W. H. Knox, Y. Sun, A. K. Srivastava, J. W. Sulhoff, and C. Wolf, *IEEE Photon. Technol. Lett.* **11**, 466 (1999).
[193] B. C. Collings, M. L. Mitchell, L. Boivin, and W. H. Knox, *IEEE Photon. Technol. Lett.* **12**, 906 (2000).
[194] W. H. Knox, *IEEE J. Sel. Topics Quantum Electron.* **6**, 1273 (2000).
[195] L. Boivin and B. C. Collings, *Opt. Fiber Technol.* **7**, 1 (2001).

Chapter 9

Optical Switching

Multichannel lightwave systems can be used for constructing all-optical networks, provided it is possible to switch individual channels optically, rather than electronically. Several kinds of optical switches have been developed for realizing this goal [1]–[7]. Some of them function in the space domain such that all bits belonging to a specific-wavelength channel are routed to the same output port (circuit switching). Others operate in the time domain and can route individual bits or packets to different destinations (packet switching). This chapter focuses on space-domain switching schemes with emphasis on the device aspects; the next chapter is devoted to time-domain switching. The basic building blocks needed for space-domain switches are discussed in Section 9.1, with an emphasis on the technologies used to realize them. Section 9.2 is devoted to simple wavelength-based routers that switch the signal to a different port, depending on its wavelength. Wavelength converters that will be needed for such a scheme to work are discussed in Section 9.3. Section 9.4 focuses on optical cross-connects that are required for designing all-optical WDM networks.

9.1 Switching Technologies

Space-domain switching is accomplished using optical devices that can direct their input to two different output ports, depending on an electrical control signal. The simplest switch can be made by using a Y junction that has one input and two output ports (see Section 4.4.1). Normally, Y junctions serve as optical splitters and simply split the input signal into two parts that are spatially separated. However, if a mechanism is incorporated within the Y junction that will allow the input signal to exit selectively from one of the ports in a controlled fashion, a Y junction can be converted into an optical switch. A simple solution would be to insert a semiconductor optical amplifier (SOA) in the two output branches of a Y junction made using semiconductor waveguides. The SOA acts as an absorber in the absence of an electrical current, but can amplify the signal when a suitable amount of current is injected into it. Switches that work through selective absorption or amplification of the input signal are referred to as gate switches.

9.1 Switching Technologies

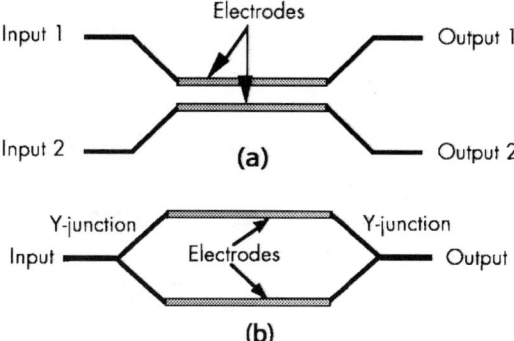

Figure 9.1: Schematic of an optical switch based on (a) a directional coupler and (b) a Mach–Zehnder interferometer. (After Ref. [8]; ©1996 IEEE.)

This section focuses on optical devices that can be switched between its two output ports using an electrical control signal. The physical mechanism behind switching can be electro-optic, acousto-optic, thermo-optic, or even electro-mechanical in nature, depending on the application requirements. The switching time can vary from a few milliseconds to <1 ns, depending on the scheme employed.

9.1.1 Electro-Optic Switches

The use of the electro-optic effect or electrorefraction, the same phenomenon discussed in Section 6.2.2 within the context of modulators, can provide relatively fast switching, that is, an input signal can be switched to different output ports on a time scale of ~1 ns. An electro-optic switching scheme can be implemented using a directional coupler, which acts as a 2×2 optical switch if it can direct an input signal toward different output ports in a controlled fashion. Figure 9.1(a) shows such a switch schematically. The electrodes deposited on top of the two waveguides in the central coupling region provide electrical control of the switch. By applying a static electric field across the waveguides, one can change the effective mode index in one or both waveguides: Changes in refractive indices affect the nature of coupling between the two waveguides and lead to optical switching. For example, if the directional coupler is designed to transfer all of its power to the cross port in the absence of any voltage, one can force the output to appear at the bar port by applying a certain voltage across one of the waveguides.

Directional-coupler switches can be made using any electro-optic material. The two most commonly used materials are $LiNbO_3$ and InP [8]. The $LiNbO_3$ technology was used as early as 1979 to realize a polarization-independent optical switch [9]. By 1988, it had been used to make working modules capable of switching among its eight input and eight output ports [10]. Figure 9.2 shows how such an 8×8 module can be made by interconnecting 12 directional couplers. By 1995, the operating characteristics of such 8×8 packaged switching modules were improved enough that they exhibited low crosstalk and excellent uniformity among all connecting ports [11]. An optical cross-

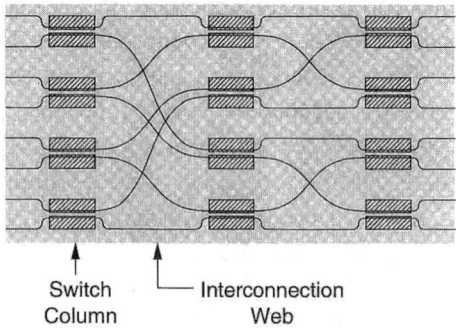

Figure 9.2: An 8 × 8 switching module designed with 12 interconnected directional couplers. (After Ref. [3]; ©1997 Academic.)

connect based on LiNbO$_3$ switches was used during the late 1990s for the MONET project [12].

The use of a four-port device is not a prerequisite for electro-optic switching. A Y junction with three ports can also act as a switch if the refractive index in the two output branches is controlled using the electro-optic effect [13]–[15]. Such a switch is sometimes referred to as a digital optical switch because it exhibits a more binary response than directional couplers and was developed during the 1990s for multiport switching modules [16]–[19]. Figure 9.3 shows the design (a) and digital switching characteristics (b) of a Y-junction electro-optic switch. As discussed in Section 4.4.1, the input waveguide splits in a Y junction into two parts that separate from each other at a small angle θ in a symmetric fashion. An external voltage is applied across the two output waveguides to change the effective mode index in each branch. When no voltage is applied, the symmetry ensures that the input power is split equally between the two output waveguides. When opposite voltages are applied to the two waveguides, index changes in the two arms can direct the output toward one of the selected ports.

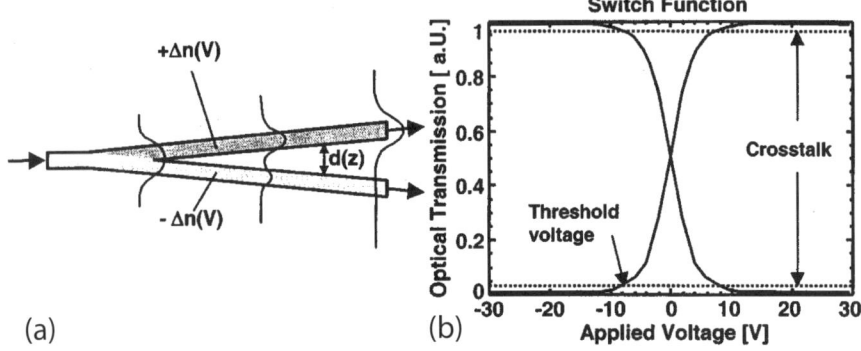

Figure 9.3: Design (a) and digital switching characteristics (b) of a Y-junction electro-optic switch. (After Ref. [19]; ©2002 IEEE.)

9.1 Switching Technologies

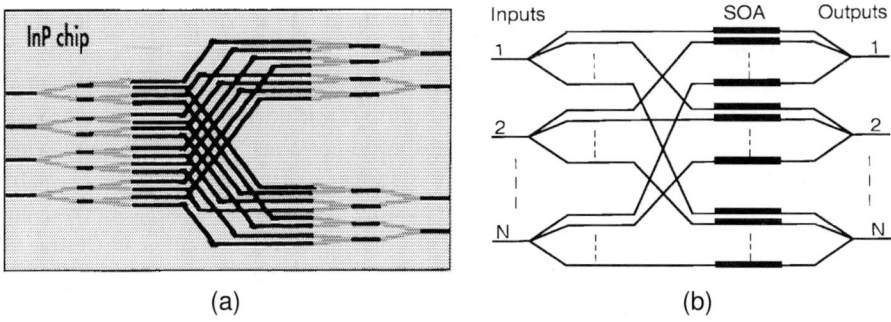

Figure 9.4: Examples of optical switches based on (a) Y-junction semiconductor waveguides and (b) SOAs with splitters. (After Ref. [8]; ©1996 IEEE.)

It is the voltage-induced asymmetry that converts a Y junction into a digital optical switch.

Mathematically, the coupling between two waveguides decreases exponentially as they separate from each other. Since the spacing d between them increases linearly with z, the coupling coefficient can be written as $\kappa = \kappa_0 \exp(-\gamma z)$, where γ is proportional to the angle θ and z is measured from the branching point. Coupled-mode theory in this case predicts that the power from one of the output ports changes with the applied voltage V as [20]

$$P_1(V) = \frac{P_{in} e^{-aV}}{1 + e^{-aV}}, \qquad (9.1.1)$$

where P_{in} is the input power and the constant a depends on the angle θ as well as on several other design parameters. This function is plotted in Figure 9.3(b). It shows how the power from an output port drops to nearly zero as V is increased above a threshold value V_{th} (typically $V_{th} < 10$ V). By reversing the sign of V, most of the input power can be forced to come out of this port.

The Mach–Zehnder (MZ) configuration, similar to that used for LiNbO$_3$ modulators, can also be used for making an electro-optic switch. The structure of such a device is shown in Figure 9.1(b) and is similar to that of Figure 4.19. In the absence of the applied voltage ($V = 0$), the MZ is balanced perfectly so that the launched power exits from the cross port. When $V = V_\pi$, where V_π is the voltage required to induce an additional phase shift of π, the input signal can be directed toward the bar port of the MZ interferometer. Although such MZ switches were fabricated as early as 1975 using semiconductor waveguides [21], silica waveguides were used for fabricating them during the 1990s. We discuss such silica-based switches in the subsection devoted to thermo-optic switches.

As mentioned earlier, GaAs or InP waveguides provide an alternative approach to making electro-optic switches. Similar to the case of LiNbO$_3$ waveguides, one can employ direction couplers, MZ interferometers, or Y junctions as the building blocks [8]. The InP-waveguide technology is most commonly used for such switches because it works well in the spectral region near 1.55 μm, where most lightwave systems operate. Figure 9.4(a) shows a 4×4 switch based on the Y junctions. Since InGaAsP waveguides can provide amplification, SOAs can be used for compensating insertion losses. SOAs themselves can be used for making optical switches. The basic idea is shown schematically in Figure 9.4(b), where SOAs act as a gate switch. Each input is divided into N branches using waveguide splitters, and each branch passes through an SOA, which either blocks light through absorption or transmits it while amplifying the signal simultaneously. Such devices have the advantage that all components can be integrated on the same InP substrate while providing low insertion losses, or even a net gain, because of the use of SOAs. They can also operate at high bit rates. Operation at a bit rate of 2.5 Gb/s was demonstrated in 1996 [22], but the bit rate can exceed 10 Gb/s for such devices.

9.1.2 Thermo-Optic Switches

Thermo-optic switches make use of the silica-on-silicon technology in which silica waveguides are deposited on top of a silicon substrate. As discussed in Section 4.3.3, this technology is quite advanced and has been used to fabricate a variety of complex devices known as planar lightwave circuits or PLCs [23]–[25]. It makes use of the MZ configuration for fabricating optical switches [26]–[29].

The PLC technology uses the thermo-optic effect to change the refractive index of silica. As seen in Figure 4.19, a thin film of chromium is deposited on top of the silica waveguide serving as an arm of the MZ interferometer. When a current is passed through this film, it heats the waveguide. The resulting temperature-induced changes in the refractive index provide an additional phase shift that can be used for optical switching. More precisely, if the MZ interferometer is balanced before heating, the input signal comes out of the cross port. However, it can be switched to the bar port by a thermally induced phase shift of π in one of the arms. As early as 1994, such optical switches were used to form a 8×8 switch module [26] and exhibited an average extinction ratio of 30 dB or so, with average insertion losses of about 7 dB.

For a dense WDM system, an extinction ratio of more than 35 dB is required at each optical switch to reduce crosstalk because switching may occur at multiple locations as the signals are routed through the network. Thus, it is important to increase the extinction ratio while maintaining low insertion losses. This problem was solved by

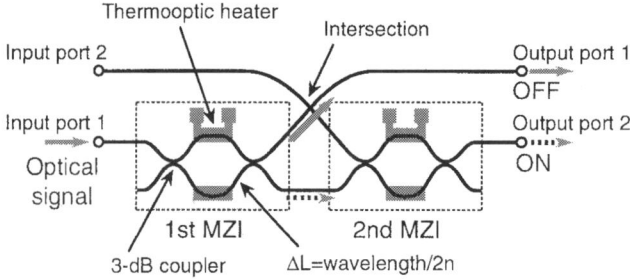

Figure 9.5: A thermo-optic switch with a high extinction ratio made by combining two MZ interferometers. (After Ref. [27]; ©1999 IEEE.)

9.1 Switching Technologies

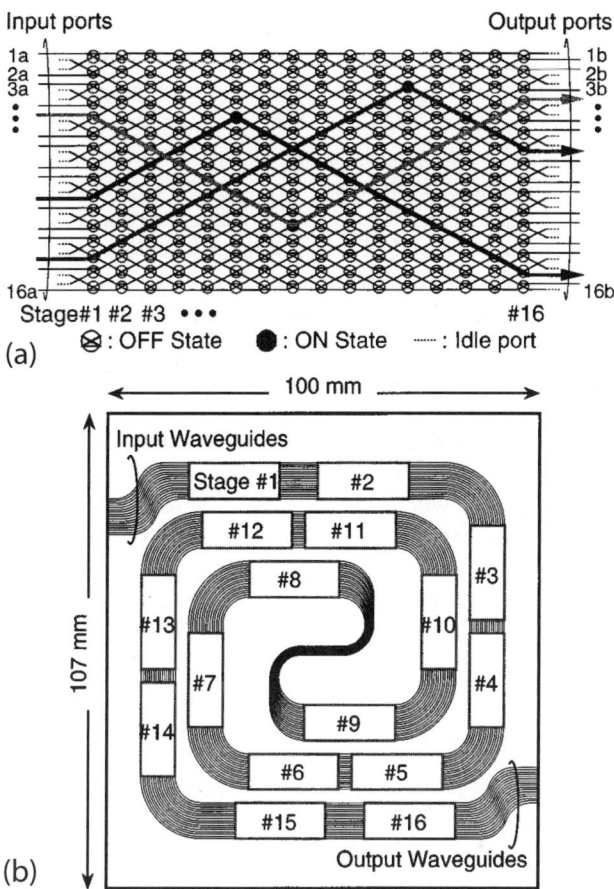

Figure 9.6: Configuration (a) and circuit layout (b) of a 16 × 16 switching fabric built using 256 thermo-optic switches, each containing two MZ interferometers. (After Ref. [28]; ©2001 IEEE.)

1999 by adopting a double-MZ design shown schematically in Figure 9.5. Such a switch combines two asymmetric MZ interferometers with thermo-optic heaters [27]. The relative phase shift of π between the two arms of each MZ interferometer ensures that the input signal entering from port 1 come out of the bar port (output port 1) when no external voltage is applied. This constitutes the "off" state of the switch. When an additional phase shift of π is added through thermo-optic heating, the signal switches to the cross port in each MZ interferometer and comes out of the output port 2 ("on" state). Since the signal as well as the leaked light responsible for crosstalk pass through two MZ interferometers in series, the extinction ratio is improved considerably; it exceeds 50 dB in practice.

Double-MZ switches were used by 2001 to build a 16 × 16 switching fabric that exhibited relatively high extinction ratios and low insertion losses across all switching paths [28]. Figure 9.6(a) shows schematically the configuration of such a strictly nonblocking switch. The switch was fabricated on a 6-in. silicon wafer and had the final dimensions of 10×10.7 cm^2. It was built using 16 stages, as shown in Figure 9.6(b). Each stage consisted of 16 double-MZ switching units, and the whole device employed 512 MZ interferometers. By changing the voltages allied to 256 switching nodes, one can switch the output from any input port to any output port. In practice, a phase-trimming technique is used to correct the phase errors that may occur during the fabrication of such a complex PLC. During phase trimming, the heating power is more than 10 times higher than that used for switching. As a result, the refractive index of the underlying waveguide increases permanently by a small amount, which is used to correct any phase error introduced during fabrication. The average insertion loss and average extinction ratio for such a device were measured to be 6.6 dB and 63 dB, respectively.

The consumption of electrical power is another relevant issue for thermo-optic switches. In one approach, the PLC structure is modified to reduce the power consumption. For example, the switching power could be reduced by a factor of about 4 using silicon trenches and heat-insulating grooves [29]. As an alternative, polymers are sometimes used in place of silica for making thermo-optic switches [30]. Most polymers have a thermo-optic coefficient that is more than 10 times larger compared with that of silica. The use of polymer waveguides reduces both the fabrication cost and power consumption. The switching time is $\sim$1 ms for all thermo-optic devices.

9.1.3 Micro-Mechanical Switches

When the switching time is not a limiting issue, even mechanical motion can be used for making a space-division switch. For example, a simple mirror acts as a switch if the output direction of the incident signal can be changed by tilting the mirror. The use of "bulk" mirrors is impractical as a large number of switches are needed for making even a 16×16 switching fabric. For this reason, the micro-electro-mechanical system (MEMS) technology is employed in practice for realizing compact switching devices [31]–[36]. It can be used to fabricate microscopic mirrors (or micromirrors) that can be rotated by applying an electrical signal.

Figure 9.7 shows the design (b) of a micromirror and how a two-dimensional array (a) of such free-rotating micromirrors can be used to fabricate a MEMS optical switch [31]. In this scheme, the entire matrix of micromirrors is integrated monolithically on a silicon chip. The input and output fibers are placed on two neighboring edges of the chip and are aligned with the location of the micromirrors precisely. An input signal from one of the fibers can be forced to couple into any output fiber by simply flipping the micromirror lying at the intersection of the input and output fibers. The switching time typically exceeds 5 ms for MEMS mirrors. Insertion losses depend on the size of the chip but are typically below 2 dB for an 8×8 switch. They increase to close to 3 dB for a 16×16 switch. With a 1-mm spacing between two neighboring micromirrors, the chip size is about 2×2 cm^2 for a 16×16 switch, but increases considerably as the the number of input and output ports increases.

The two-dimensional (2-D) geometry shown in Figure 9.7(a) employs N^2 micromirrors for a MEMS switch with N input and N output ports. This feature limits the

9.1 Switching Technologies

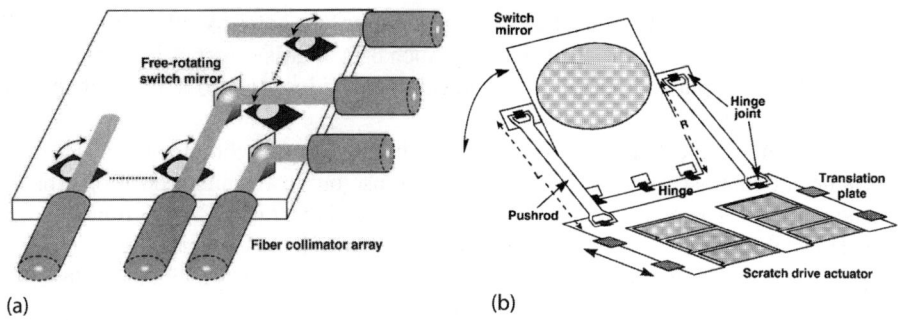

Figure 9.7: (a) A 2-D MEMS switch based on an array of micromirrors; (b) details of micromirror design. (After Ref. [31]; ©1999 IEEE.)

switch size in practice as the number of micromirrors exceeds 1,000 even for $N = 32$. Although multiple 2-D switches can be combined to increase effective size, a three-dimensional (3-D) configuration provides a better alternative. In this approach, shown schematically in Figure 9.8, the input and output fibers are located outside the switching-fabric plane, and two arrays of N micromirrors are used to switch an input signal to any one of the N output ports. This is possible because each micromirror is a two-axis gimbal mirror that can be used to steer the input beam in any direction in an analog fashion, in contrast with the digital operation of MEM mirrors in 2-D switches.

The operation of 3-D MEMS switches can be understood as follows. The optical beam from an input port is collimated and focused on the gimbal mirror that is associated with this port. The angle of this gimbal mirror is then adjusted electronically such that it redirects the optical beam to the mirror in the second MEMS mirror array that is associated with the desired output port. The second mirror directs the optical beam to the output fiber to complete the switching. A lens is sometimes placed between the two MEMS mirror arrays such that its focal length is just half the spacing between the arrays [36]. At the same time, the beam size is chosen such that the diffraction length is

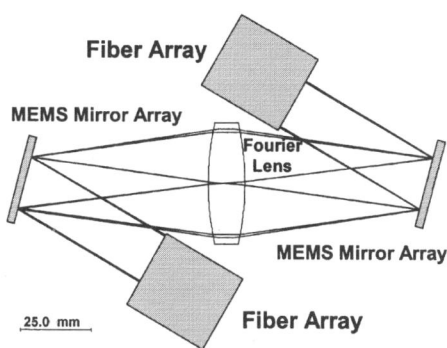

Figure 9.8: Schematic layout of a 3-D MEMS switch. The Fourier lens between the two MEMS mirror arrays is not necessary but its use improves performance. (After Ref. [36]; ©2003 IEEE.)

equal to the lens focal length. The use of the Fourier lens helps to reduce the crosstalk and has several other practical advantages. Such 3-D MEMs switches have been fabricated with 256 input and output ports and exhibited an insertion loss of less than 2 dB for all possible interconnections.

As one may expect, the electronic control circuitry is quite complex for 3-D MEMS switches in contrast with the 2-D switches that use only two mirror positions for the same purpose. The chief advantage of 3-D switches is that only $2N$ micromirrors are needed (rather than N^2) for fabricating an $N \times N$ switch. As a result, the number of input and outports can easily exceed 256. Indeed, 3-D MEMS switches with 1,100 input and 1,100 output ports were made by 2003 with only 4-dB maximum insertion loss between any two ports [36]. Such 3-D MEMS switches are likely to find applications in WDM networks [32]. They are relatively slow to reconfigure (switching time >5 ms) but that is not a major limitation for some applications.

9.1.4 Liquid-Crystal Switches

Another well-developed technology that can be used for fabricating optical switches is the liquid-crystal technology [37]–[41]. Liquid crystals are used extensively for making computer displays and spatial light modulators. When a thin layer of nematic liquid crystal is surrounded on each side by polarizers, incident light at any one spot can be transmitted or blocked, depending on the magnitude of an external voltage, and thus acts as an optical switch. The physical process behind such an optical switch is the electrically controlled birefringence that is used to change the state of polarization of the incident light in a birefringent medium such as a liquid crystal. The device is designed such that no light passes through in the absence of applied voltage (OFF state). With a suitable voltage, the birefringence is changed such that the incident light passes through the device (ON state).

A problem with the simple switch described above is its polarization sensitivity. This problem can be solved by splitting the input signal into orthogonally polarized components and switching each one separately, but only at the expense of increased complexity. In one approach, shown schematically in Figure 9.9(a), polarization beam splitters are used in combination with beam routers to realize a polarization-independent switch [37]. The operation of this device can be understood from the schematic shown in Figure 9.9(b). The input signal is first split into its two orthogonally components, each of which is processed by two different liquid-crystal switches.

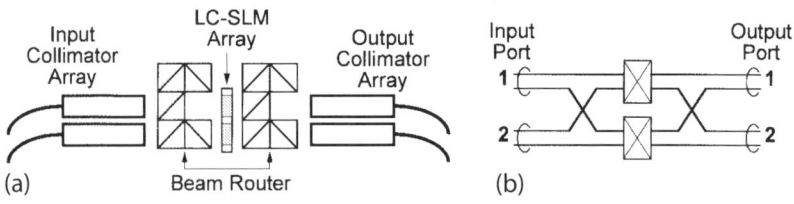

Figure 9.9: Design (a) and schematic (b) of a polarization-independent liquid-crystal switch. LC-SLM stands for liquid crystal-spatial light modulator. (After Ref. [37]; ©1998 IEEE.)

9.1 Switching Technologies

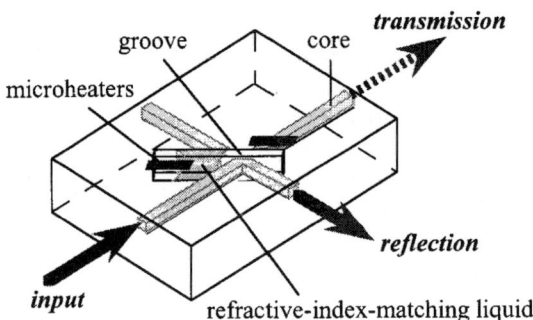

Figure 9.10: Schematic of a bubble switch based on total internal reflection from a thermally generated bubble. (After Ref. [45]; ©2001 IEEE.)

If both switches are in the OFF state, the signal entering from input port 1 are routed to output port 2. In contrast, when both switches are in the ON state, the same input signal appears on the output port 1. Thus, the combination of two polarization-dependent switches produces one polarization-independent switch. Using this combination as an elementary unit, one can construct 8×8 or higher-order crossbar-type switching fabric. A strictly nonblocking 32×32 switch was made as early as 1998 using this scheme. The insertion loss was in the range 9–10 dB for any path while the crosstalk remained below -25 dB.

9.1.5 Bubble Switches

The bubble technology makes use of the phenomenon of *total internal reflection* for optical switching [42]–[46]. Figure 9.10 shows the basic idea schematically. Two planar waveguides intersect in a groove filled with a liquid whose refractive index is chosen such that it nearly matches that of the waveguide mode. As a result, the input signal entering from a waveguide is transmitted straight through the waveguide under normal conditions. However, when a thermally generated bubble is moved to the intersection point, the optical signal undergoes total internal reflection such that the reflected beam is coupled into the other waveguide. Switching from one waveguide to another waveguide can be accomplished electrically using microheaters, but it is based on a relatively slow thermocapillary effect. The switching time of such an optical switch is close to 10 ms.

To fabricate a switching fabric with multiple input and output ports using such a switch, a two-dimensional array of optical waveguides is formed in such a way that they intersect inside liquid-filled channels [44]. The waveguide array is made in the form of a planar lightwave circuit using silica-on-silicon technology. The liquid-filled channels are etched chemically and are typically 15-μm-wide. The air bubble is generated using the inkjet technology used for printers. A 32×32 switch was made by 2000 using the bubble technology. It can be used to fabricate much larger switches using the well-known Clos architecture. This approach has the potential of realizing low-cost switches but it suffers from relatively large insertion losses. Measured losses for a 32×32 switch were as high as 9 dB, depending on how many times the signal was switched through total internal reflection. Recent technological improvements have reduced this value to about 2.5 dB [46]

A bubble switch is similar to a MEMS switch in its basic conception in the sense that both employ reflection for switching but it uses a bubble in place of a rotating mirror. Several other variations are possible. For example, electroholographic switches use a Bragg grating within a photorefractive medium (e.g., a $LiNbO_3$ crystal) for switching in place of a rotating mirror [47]. Incident light can be switched at any point by applying an electric field and creating a Bragg grating at that location. Because of the wavelength selectivity of the Bragg grating, only a single wavelength can be switched by such a device, in contrast with the MEMS or bubble switches that switch all wavelengths within the signal simultaneously. This feature increases the complexity of such a switching fabric.

9.2 Wavelength-Domain Routers

Wavelength-domain routers switch individual channels of an incoming WDM signal to different ports without using any of the switching mechanisms discussed in Section 9.1. The main attraction of such routers is that they act in a passive manner without any active element requiring electrical power. This occurs because switching is performed on the basis of channel wavelengths using a built-in Bragg grating. A WDM router is also called a *static router* since the routing topology is not dynamically reconfigurable.

Figure 9.11(a) shows the operation of a wavelength-domain router schematically in the case of five wavelengths. Each WDM signal consisting of five distinct wavelengths λ_1, λ_2, λ_3, λ_4, and λ_5 is demultiplexed into individual channels and directed toward five different output ports of the router. The channels are distributed in such a way that the WDM signal at each output port is composed of channels entering at different input ports. This operation results in a cyclic form of demultiplexing.

The most common design of a wavelength router uses an arrayed waveguide grating (AWG) and is similar to that of the demultiplexer shown in Figure 8.18. The only difference is that the AWG has the same number of input and output ports. Such a device, called the *waveguide-grating router* (WGR), is shown schematically in Figure 9.11(b). As discussed in Section 4.5.4, it consists of two star couplers such that the output ports of one star coupler are connected with the input ports of another star coupler through an array of planar waveguides [48]–[56]. Such a device represents a generalization of the MZ interferometer in the sense that a single input is divided coherently into M parts (rather than two), which acquire different phase shifts and interfere in the second star coupler such that they come out of N different ports depending on their wavelengths. The symmetric nature of an $N \times N$ WGR permits one to launch N WDM signals containing N different wavelengths simultaneously, and each WDM signal is demultiplexed to N output ports in a periodic fashion.

The physics behind the operation of a WGR requires careful consideration of the phase changes as different wavelength signals diffract through the free-propagation region inside star couplers and propagate through the waveguide array. The most important part of a WGR is the waveguide array designed such that the length difference

9.2 Wavelength-Domain Routers

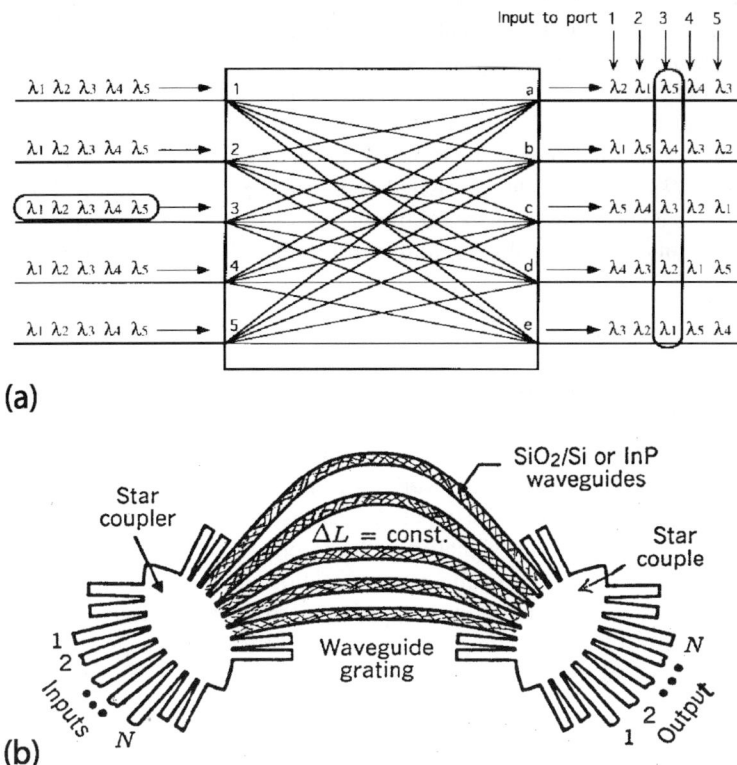

Figure 9.11: (a) Schematic illustration of a wavelength router and (b) its implementation using an AWG. (After Ref. [51]; ©1999 IEEE; reprinted with permission.)

ΔL between two neighboring waveguides remains constant from one waveguide to the next. The phase difference for a signal of wavelength λ, traveling from the pth input port to the qth output port through the mth waveguide (compared to the path connecting central ports), can be written as

$$\phi_{pqm} = (2\pi m/\lambda)(n_1 \delta_p + n_2 \Delta L + n_1 \delta'_q), \quad (9.2.1)$$

where n_1 and n_2 are the refractive indices in the regions occupied by the star couplers and waveguides, respectively. The lengths δ_p and δ'_q depend on the location of the input and output ports. When the condition

$$n_1(\delta_p + \delta'_q) + n_2 \Delta L = Q\lambda \quad (9.2.2)$$

is satisfied for some integer Q, the channel at the wavelength λ acquires phase shifts that are multiples of 2π while passing through different waveguides. As a result, all fields coming out of the M waveguides will interfere constructively at the qth port. Other wavelengths entering from the pth port will be directed to other output ports determined by the condition (9.2.2). Clearly, the device acts as a demultiplexer since a

WDM signal entering from the pth port is distributed to different output ports depending on the channel wavelengths.

The routing function of a WGR results from the periodicity of the transmission spectrum. This property is also easily understood from Eq. (9.2.2). The phase condition for constructive interference can be satisfied for many integer values of Q. Thus, if Q is changed to $Q+1$, a different wavelength will satisfy Eq. (9.2.2) and will be directed toward the same port. The frequency difference between these two wavelengths is the free spectral range (FSR), analogous to that of FP filters. For a WGR, it is given by

$$\text{FSR} = \frac{c}{n_1(\delta_p + \delta_q') + n_2 \Delta L}. \qquad (9.2.3)$$

Strictly speaking, FSR is not the same for all ports, an undesirable feature from a practical standpoint. However, when δ_p and δ_q' are designed to be relatively small compared with ΔL, FSR becomes nearly constant for all ports. In that case, a WGR can be viewed as N demultiplexers working in parallel with the following property. If the WDM signal from the first input port is distributed to N output ports in the order $\lambda_1, \lambda_2, \ldots, \lambda_N$, the WDM signal from the second input port will be distributed as $\lambda_N, \lambda_1, \ldots, \lambda_{N-1}$, and the same cyclic pattern is followed for other input ports.

The optimization of a WGR requires precise control of many design parameters for reducing the crosstalk and maximizing the coupling efficiency. Despite the complexity of the design, WGRs are routinely fabricated in the form of a compact commercial device (each dimension $\sim$1 cm) using either silica-on-silicon technology [53]. WGRs with 128 input and output ports were available by 1996 in the form of a planar lightwave circuit and were able to operate on WDM signals with a channel spacing as small as 0.2 nm while maintaining crosstalk below 16 dB. WGRs with 256 input and output ports have been fabricated using this same technology [25]. The primary performance criteria are low crosstalk and low insertion losses. By 2003, a 64×64 WGR module in the form of a silica-based planar lightwave circuit exhibited insertion losses in the range of 5.4–6.8 dB for all channels.

9.3 Wavelength Converters

The use of WDM routers results in optical networks that are not transparent in the wavelength domain. This problem can be solved if the wavelength of a blocked channel can be converted to a new wavelength without affecting the optical bit pattern of the channel. A wavelength converter is a simple device that changes the wavelength of the input channel to a new wavelength without modifying its data content. Many schemes were developed during the decade of the 1990s for realizing wavelength converters [57]–[65], and several others have been proposed since then [66]–[79]. It should be stressed that, strictly speaking, a wavelength converter works in the time domain and operates at a time scale of bit duration. However, since the entire channel is switched to a different output port, we include such devices in this chapter rather than Chapter 10.

A conceptually simple scheme uses an optoelectronic regenerator in the form of a receiver-transmitter pair, as shown schematically in Figure 9.12(a). The optical receiver converts the incident signal at the input wavelength λ_1 to an electrical bit pattern, which

9.3 Wavelength Converters

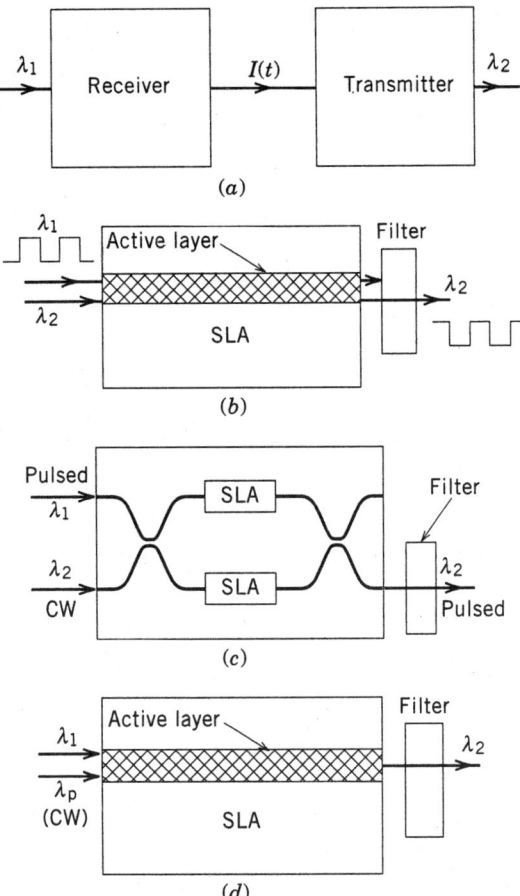

Figure 9.12: Four schemes for wavelength conversion: (a) optoelectronic regenerator; (b) gain saturation in a semiconductor laser amplifier (SLA); (c) phase modulation in a SLA placed in one arm of a Mach–Zehnder interferometer; (d) four-wave mixing inside a SLA.

is then used by a transmitter to generate the optical signal at the desired wavelength λ_2. Such a scheme is relatively easy to implement as it uses standard lightwave components. Its other advantages include insensitivity to input polarization, elimination of accumulated noise and other mechanisms of signal degradation, and the possibility of net amplification. Among its disadvantages are limited transparency to bit rate and data format, speed limited by electronics, and a relatively high cost, all of which stem from the optoelectronic nature of wavelength conversion.

9.3.1 Semiconductor-Based Devices

Several all-optical techniques for wavelength conversion make use of SOAs [57]–[61], devices covered in Section 5.5. The simplest scheme shown in Figure 9.12(b) is based on cross-gain saturation occurring when a weak field is amplified inside the SOA together with a strong field, and the amplification of the weak field is affected by the strong field. To use this phenomenon, the pulsed signal whose wavelength λ_1 needs to be converted is launched into the SOA together with a low-power CW beam at the wavelength λ_2 at which the converted signal is desired. Amplifier gain is mostly saturated by the λ_1 beam. As a result, the CW beam is amplified by a large amount during 0 bits (no saturation) but by a much smaller amount during 1 bits. Clearly, the bit pattern of the incident signal will be transferred to the new wavelength with reverse polarity such that 1 and 0 bits are interchanged. This technique has been used in many experiments and can work at bit rates as high as 40 Gb/s. It can provide net gain to the wavelength-converted signal and can be made nearly polarization-insensitive. Its main disadvantages are (i) relatively low on–off contrast, (ii) degradation due to spontaneous emission, and (iii) phase distortion because of frequency chirping that invariably occurs in SOAs (see Section 5.5). The use of an absorbing medium in place of the SOA solves the polarity reversal problem. An electroabsorption modulator (see Section 6.5) has been used for wavelength conversion with success [67]. It works on the principle of cross-absorption saturation. The device blocks the CW signal at λ_2 because of high absorption, except when the arrival of 1 bits at λ_1 saturates the absorption.

The contrast problem can be solved by using the MZ configuration of Figure 9.12(c) in which an SOA is inserted in each arm of a MZ interferometer [57]. The pulsed signal at the wavelength λ_1 is split at the first coupler such that most power passes through one arm. At the same time, the CW signal at the wavelength λ_2 is split equally by this coupler and propagates simultaneously in the two arms. In the absence of the λ_1 beam, the CW beam exits from the cross port (upper port in the figure). However, when both beams are present simultaneously, all 1 bits are directed toward the bar port because of the refractive-index change induced by the λ_1 beam. The physical mechanism behind this behavior is the cross-phase modulation (XPM). Gain saturation induced by the λ_1 beam reduces the carrier density inside one SOA, which in turn increases the refractive index only in the arm through which the λ_1 beam passes. As a result, an additional π phase shift can be introduced on the CW beam because of XPM, and the CW wave is directed toward the bar port during each 1 bit.

It should be evident from the preceding discussion that the output from the bar port of the MZ interferometer would consist of an exact replica of the incident signal with its wavelength converted to the new wavelength λ_2. An optical filter is placed in front of the bar port for blocking the original λ_1 signal. The MZ scheme is preferable over cross-gain saturation as it does not reverse the bit pattern and results in a higher on–off contrast simply because nothing exits from the bar port during 0 bits. In fact, the output from the cross port also has the same bit pattern but its polarity is reversed. Other types of interferometers (such as Michelson and Sagnac interferometers) can also be used with similar results. The MZ interferometer is often used in practice because it can be easily integrated on a single chip using InGaAsP/InP technology, resulting in a compact device.

Figure 9.13 shows an example of an MZ-based wavelength converter, monolithically integrated on a single InP chip [65] using multimode-interference (MMI) couplers. The input signal P_{in} is split and fed into the two MZ arms using an MMI coupler with a 60:40 splitting ratio and then combined back at the output end using another

9.3 Wavelength Converters

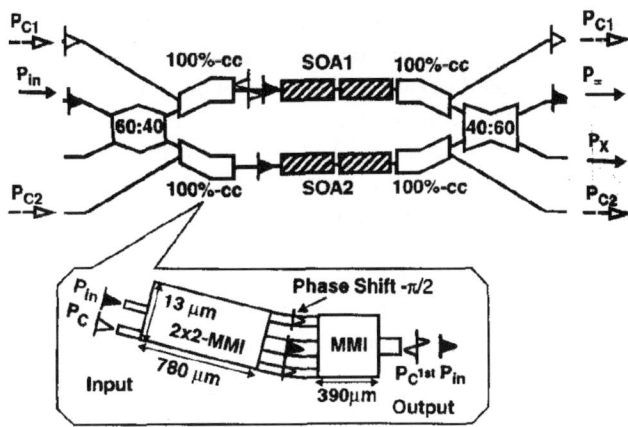

Figure 9.13: Schematic of a wavelength converter monolithically integrated on a single InP chip in the MZ configuration using two SOAs and several MMI couplers. Inset shows details of the MMI coupler used to convert control signals into first-order modes. (After Ref. [65]; ©1999 IEEE.)

MMI coupler with a 40:60 splitting ratio. Two control signals P_{c1} and P_{c2} are coupled into the two MZ arms using an especially designed MMI coupler (shown as an inset). This coupler converts control signals into first-order modes of the waveguide serving as an MZ arm while the signal propagates as the fundamental mode. At the output signal, the control signals are converted back into fundamental modes and they can be reused if needed. The SOA in each MZ is divided into two sections to enhance the switching speed. Such devices can operate at high bit rates (up to 80 Gb/s), offer a large contrast, and degrade the signal relatively little although spontaneous emission does affect the SNR. Its main disadvantage is a narrow dynamic range of the input power since the phase induced by the amplifier depends on it.

Another scheme employs the SOA as a nonlinear medium for four-wave mixing (FWM), a nonlinear phenomenon that can occur in many nonlinear media, including optical fibers [80]. As seen in Figure 9.12(d), its use requires an intense CW pump beam that is launched into the SOA together with the signal whose wavelength needs to be converted [57]. If v_1 and v_2 are the frequencies of the input signal and the converted signal, the pump frequency v_p is chosen such that $v_p = (v_1 + v_2)/2$. At the amplifier output, a replica of the input signal appears at the carrier frequency v_2 because FWM requires the presence of both the pump and signal. One can understand the process physically as the conversion of two pump photons of energy $2hv_p$ into two photons of energy hv_1 and hv_2. The nonlinearity responsible for the FWM in SOAs has its origin in fast intraband relaxation processes occurring at a time scale of 0.1 ps [81]. As a result, frequency shifts as large as 10 THz, corresponding to wavelength conversion over a range of 80 nm, are possible. For the same reason, this technique can work at bit rates as high as 100 Gb/s and is transparent to both the bit rate and the data format. Because of the gain provided by the amplifier, conversion efficiency can be quite high, resulting even in a net gain. An added advantage of this technique is the reversal of the frequency chirp since its use inverts the signal spectrum. The performance can be improved further using two SOAs in a tandem configuration.

The main disadvantage of any wavelength-conversion technique based on SOAs is that it requires a tunable laser source whose light should be coupled into the SOA, typically resulting in large coupling losses. An alternative is to integrate the functionality of a wavelength converter within a tunable semiconductor laser. Several such devices have been developed [57]. In the simplest scheme, the signal whose wavelength needs to be changed is injected into a tunable laser directly. The change in the laser threshold resulting from injection translates into modulation of the laser output, mimicking the bit pattern of the injected signal. Such a scheme requires relatively large input powers. Another scheme uses the low-power input signal to produce a frequency shift (typically, 10 GHz/mW) in the laser output for each 1 bit. The resulting frequency-modulated CW signal can be converted into amplitude modulation by using a MZ interferometer. Still another uses FWM inside the cavity of a tunable semiconductor laser, which also plays the role of the pump laser. A phase-shifted DFB laser provided wavelength conversion over a range of 30 nm with this technique [58]. A sampled grating within a distributed Bragg reflector has also been used for this purpose [62].

Wavelength converters can also be made using the $LiNbO_3$-waveguide technology. This semiconductor allows for the nonlinear effects such as difference-frequency generation that make use of the second-order susceptibility $\chi^{(2)}$. As early as 1993, this nonlinear effect in a $LiNbO_3$ waveguide was used for wavelength conversion [82]. The technique of periodic poling in which the sign of $\chi^{(2)}$ is inverted periodically along the waveguide length for realizing quasi-phase matching was used for efficient wavelength conversion. Such devices require the use of a single-mode pump laser operating in the spectral region near 780 nm with 50–100 mW of power. In practice, it is difficult to simultaneously couple a 780-nm pump and a 1.55-μm signal into the fundamental mode of the waveguide, although this problem can be solved using an integrated mode coupling structure. An alternative scheme makes use of two cascaded second-order nonlinear processes occurring in a periodically poled $LiNbO_3$ (PPLN) waveguide using a pump laser operating near 1.55 μm [64]. In such a wavelength converter, the pump at frequency ω_p is first up-converted to $2\omega_p$ through second-harmonic generation, which then generates wavelength-shifted output through difference-frequency generation. Both processes can be quasi-phase-matched using an appropriate grating period. Such a cascaded process mimics FWM, but is much more efficient than FWM based on third-order susceptibility. It is also less noisy than the FWM process occurring inside the active region of SOAs.

Figure 9.14(a) shows the optical spectrum recorded when four channels at wavelengths in the range of 1,552–1,558 nm were coupled into a PPLN waveguide together with a 1,562-nm pump with 110 mW power. The four peaks on the right side of the pump peak represent the four wavelength-converted channels. The conversion efficiency is about 5% for all channels but it can be increased by increasing pump power. The inset shows wavelength conversion of a single channel with 16% efficiency at a pump power of 175 mW. The conversion efficiency depends on the wavelength separation between the pump and signal, as shown in Figure 9.14(b). The bandwidth of the flat region over which conversion efficiency is nearly constant is more than 60 nm. Such a device can respond at femtosecond time scales and thus easily operate at bit

9.3 Wavelength Converters

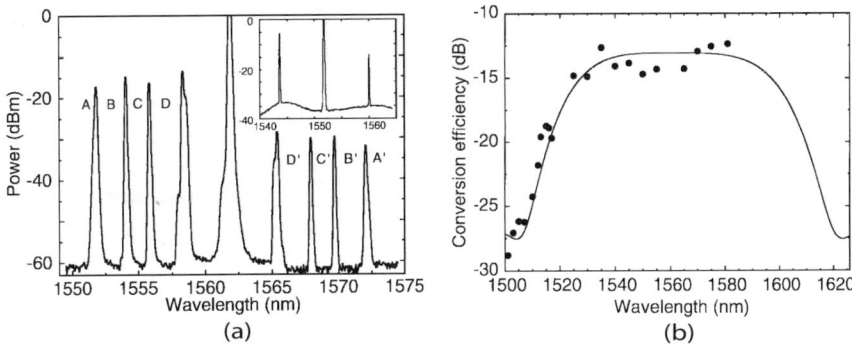

Figure 9.14: (a) Simultaneous wavelength conversion of four channels in a PPLN waveguide pumped at 1562-nm pump with 110-mW power. Inset shows 16% conversion efficiency realized at a pump power of 165 mW. (b) Conversion efficiencies measured as a function of signal wavelength are shown on the right together with the theoretical prediction. (After Ref. [64]; ©1999 IEEE.)

rates of 40 Gb/s or more. Wavelength conversion at 160 Gb/s was demonstrated in a 2000 experiment [69].

9.3.2 Fiber-Based Devices

Another class of wavelength converters makes use of optical fibers as a nonlinear medium. Both XPM and FWM can be employed for this purpose using the last two configurations shown in Figure 9.12. In the XPM case, the modulated signal is launched into a fiber of appropriate length together with a CW seed at the wavelength at which the signal needs to be converted. The signal is amplified considerably before launching so that it can act as a pump and thus impose a large enough XPM-induced phase shift in the time slots associated with the 1 bits. This phase shift is then converted into amplitude modulation using an interferometer. Although an MZ interferometer can be used in principle, its use is not practical because it is hard to maintain the linear phase shift constant in two arms of the MZ interferometer since the arm length typically exceeds 1 km because of the relatively weak nonlinearity of silica fibers.

The use of a Sagnac interferometer in the form of a nonlinear optical loop mirror solves this problem (see Section 2.3.3) and provided by 2000 wavelength converters capable of operating at bit rates of up to 40 Gb/s for both the return-to-zero (RZ) and nonreturn-to-zero (NRZ) formats [66]. The Sagnac loop was made using 3 km of dispersion-shifted fiber with the zero-dispersion wavelength at 1,555 nm and a dispersion slope of 0.08 ps/nm/km at this wavelength. The device reflected all 0 bits but 1 bits were transmitted through the fiber loop because of the XPM-induced phase shift. The on–off ratio between maximum and minimum transmission was measured to be 25 dB. The optical eye diagrams of the original as well as wavelength-converted signal were measured to judge the quality of wavelength conversion. Figure 9.15 shows the two eye patterns in the case of an RZ-format signal. It is evident that individual pulses remained almost unaffected during wavelength conversion.

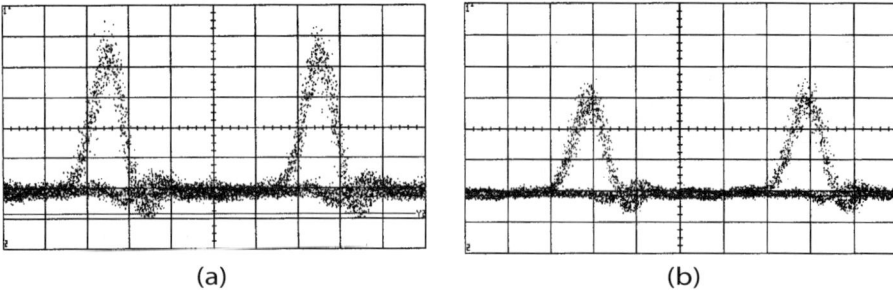

Figure 9.15: Optical eye diagrams (horizontal scale 20 ps/div) for the 1,545-nm signal (a) and wavelength-converted signal (b) at 1,557 nm. (After Ref. [66]; ©2000 IEEE.)

It is not essential to use an optical interferometer for XPM-based wavelength converters. As an alternative, one can pass the phase-modulated output of the fiber through a suitable optical filter. The basic idea behind this scheme is that the phase modulation of any optical carrier creates sidebands on each side of the carrier (similar to AM and FM sidebands). If an optical filter blocks the carrier but lets one of the sidebands pass, the output is a replica of the original bit stream at the carrier wavelength. Any optical filter with a bandwidth of ~ 1 nm (see Section 8.1) can be used for this purpose including fiber gratings or optical interferometers. In a 2000 experiment [68], the wavelength of a 40-Gb/s signal was shifted by several nanometers through XPM in a 10-km-long fiber. The XPM-induced phase modulation was converted into amplitude modulation by using a 4-m loop made of polarization-maintaining fiber. This loop formed a Sagnac interferometer that acted as a notch filter and suppressed the carrier while transmitting one or more sidebands.

The required fiber length can be reduced considerably by employing narrow-core fibers, sometimes referred to as highly nonlinear fibers even though it is the mode intensity that is enhanced in such fibers. In a 2001 experiment [70], wavelength conversion at a bit rate of 80 Gb/s was realized by using such a 1-km-long dispersion-shifted fiber designed to have a large value of the nonlinear parameter ($\gamma = 10.9$ W^{-1}/km). The zero-dispersion wavelength of the fiber was 1,552 nm and the dispersion slope was only 0.022 ps/nm^2/km at this wavelength. The 80-Gb/s data channel with the 1560-nm wavelength was first amplified to the 70-mW power level and then coupled into the highly nonlinear fiber together with a CW seed at the new wavelength that was varied in the range of 1,525–1,554 nm. A tunable optical filter with a 1.5-nm bandwidth was used at the fiber output to produce the wavelength-converted channel. Figure 9.16(a) shows the optical spectra just before and after the optical filter when the CW-seed wavelength was 1545.6 nm. Before the filter, the spectrum shows multiple sidebands generated through XPM but the seed spectrum is dominated by the carrier. After the filter, the carrier has been suppressed relative to the sidebands, resulting in a wavelength-converted signal with a bit stream identical to that of the original channel. The pulse width of the converted signal is shown in Figure 9.16(b) as a function of the wavelength. As seen there, the width remains nearly unchanged over a wide bandwidth. These results indicate that the information remains virtually intact during

9.3 Wavelength Converters

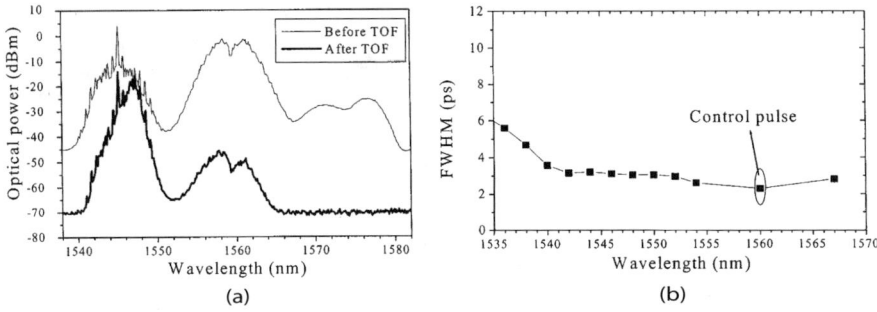

Figure 9.16: (a) Optical spectra measured before and after the tunable optical filter (TOF) for an XPM-based wavelength converter. (b) Pulse width as a function of converted wavelength. The width of the original data channel is marked as "control pulse." (After Ref. [70]; ©2001 IEEE.)

wavelength conversion. Indeed, bit-error rate-measurements showed negligible power penalty resulting from such a wavelength converter.

The use of FWM for wavelength conversion requires parametric amplification inside an optical fiber [80]. Such amplifiers have attracted considerable attention as wavelength converters, especially in the dual-pump configuration [72]–[79]. This configuration can provide nearly uniform gain over a wide bandwidth while allowing nearly polarization-independent operation of the device. The required pump power levels exceed 100 mW in practice. As a result, other nonlinear effects occurring inside the optical fiber used for FWM cannot be ignored. Stimulated Raman scattering (SRS) affects the device operation by transferring power from one pump to another, but this problem can be solved by adjusting pump powers at the input end. A second problem is related to stimulated Brillouin scattering (SBS) that hinders the propagation of high-power CW pumps.

The SBS problem is solved in practice by modulating the phase of each pump at high frequencies ($\sim$1 GHz) so that its spectrum is considerably broader than the SBS-gain bandwidth, resulting in a much higher SBS threshold. However, phase modulation of the two pumps needed for SBS suppression can lead to considerable spectral broadening of the wavelength-converted signal. This can be understood by noting that the complex amplitude A_i of the idler field during the FWM process has the form $A_i \propto A_{p1}A_{p2}A_s^*$, where A_{p1} and A_{p2} are the pump amplitudes while A_s is the signal amplitude [80]. Clearly, the phase of the idler can vary at the same frequency if the two pumps are modulated either in phase or in a random manner. The solution to this problem also follows from the preceding relation. If the two pumps are modulated such that their phases are always equal but opposite in sign, the product $A_{p1}A_{p2}$ will not exhibit any modulation [76]. As a result, even though the idler spectrum will be a mirror image of the signal spectrum, the bandwidth of the two spectra will be identical. A second approach makes use of binary phase modulation so that the phase of both pumps is modulated in the same direction but takes only two discrete values, namely 0 and π [79]. This approach works because the product $A_{p1}A_{p2}$ does not change under such a modulation scheme.

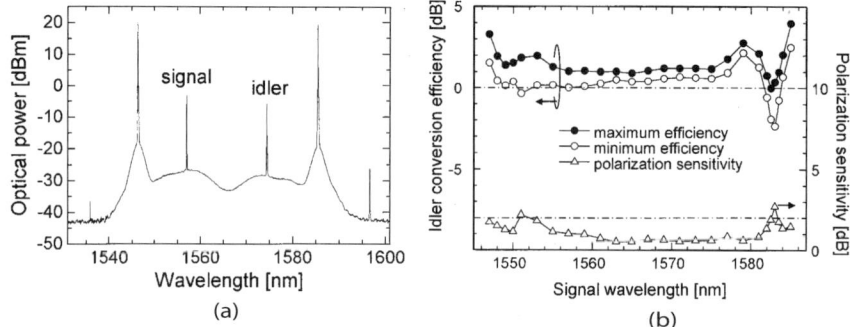

Figure 9.17: (a) Optical spectra measured for a FWM-based wavelength converter. Two dominant peaks correspond to two orthogonally polarized pumps. (b) Conversion efficiency and its polarization sensitivity as a function of signal wavelength. (After Ref. [79]; ©2003 IEEE.)

Figure 9.17(a) shows the optical spectrum recorded at the output when a 1,557-nm signal was launched inside a dual-pump wavelength converter [79]. The two pumps had power levels of 118 and 148 mW at wavelengths of 1,585.5 and 1,546.5 nm, respectively. The power was higher at the shorter wavelength to offset the Raman-induced power transfer to the longer-wavelength pump. FWM occurred inside a 1-km-long highly nonlinear fiber ($\gamma = 18$ W^{-1}/km). The zero-dispersion wavelength of the fiber was 1,566 nm with a dispersion slope of 0.027 ps/nm^2/km at this wavelength. The idler generated through FWM near 1,570 nm had the same bit pattern as the signal. Its average power was also comparable to that of the signal, indicating nearly 100% efficiency for such a wavelength converter. In fact, as shown in Figure 9.17(b), high efficiency could be maintained over a bandwidth of 40 nm or so. The efficiency varied somewhat with the signal state of polarization but variations were below 2 dB over a 30-nm range. Wavelength converters based on optical fibers can operate at bit rates as high as 1 Tb/s or more because of their much faster nonlinear response time compared with SOA-based converters.

9.4 Optical Cross-Connects

The development of wide-area WDM networks requires a dynamic wavelength routing scheme that can reconfigure the network while maintaining its nonblocking (transparent) nature. This functionality is provided by a device known as the *optical cross-connect* (OXC). It performs the same function as that provided by electronic digital switches in telephone networks. Such a device makes use of optical switches in place of electronic switches. However, the use of WDM technology makes the design of OXCs quite complex because individual channels must be demultiplexed before any switching can occur. The development and fabrication of such OXCs have remained a major topic of research since the advent of WDM systems [83]–[93]. In this section we discuss several schemes that have attracted considerable attention.

9.4 Optical Cross-Connects

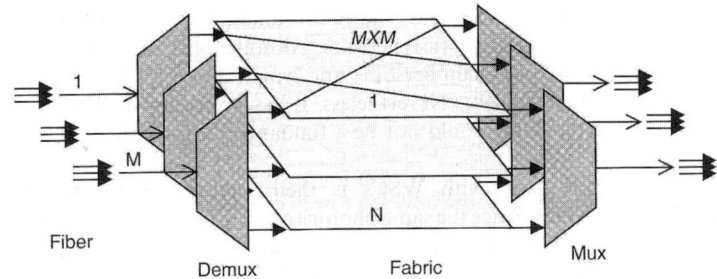

Figure 9.18: Schematic of a wavelength-selective cross-connect. Mux and Demux stand for multiplexers and demultiplexers, respectively. (After Ref. [91]; ©2002 Elsevier.)

9.4.1 Wavelength-Selective Cross-Connect

In a wavelength-selective OXC, shown schematically in Figure 9.18, demultiplexers split the WDM signal entering from all input ports into individual wavelengths and distribute each wavelength to a separate switching unit, each unit receiving all input signals at the same wavelength. Thus, for a wavelength-selective OXC interconnecting M input ports to M output ports such that each WDM signal consists of the same N distinct wavelengths, N switching units are needed. Each unit consists of a $M \times M$ switching fabric that can be configured to route the signals at a fixed wavelength in any desirable fashion. An extra input and output port can be added to the switching unit to allow dropping or adding of a local channel at that wavelength. Clearly, such a wavelength-selective OXC needs M multiplexers, M demultiplexers, and N switching units with M input and M output ports. Since the number of ports required on a single switching fabric is relatively small (it equals the number of incoming and outgoing fibers), such devices are relatively easy to fabricate and can be made using any technology discussed in Section 9.1. The only requirement is that the switching time be ∼5 ms or less.

The most common approach to designing wavelength-selective OXCs makes use of silica-on-silicon or InP technology because both multiplexers and demultiplexers can then be integrated with the switching fabric on a single chip. As early as 1996, a 16-channel OXC was demonstrated by integrating thermo-optic switches with arrayed waveguide gratings using silica-on-silicon technology [94]. By 2003, a 16×16 switching module had been developed in the form of a PLC chip that incorporated electrical driving circuits as well [95]. The module exhibited a relatively low insertion loss (5.6 dB on average), a high extinction ratio (>50 dB), and a low average PDL (about 0.11 dB). The switching time was less than 4.1 ms. This module is a promising component for realizing large photonic network components.

A strictly nonblocking 8×8 switching device has also been fabricated using MMI couplers made with silica-on-silicon technology [96]. Figure 9.19 shows schematically the architecture of such a switching fabric. It employs eight 1×2 MMI couplers, followed by sixteen 1×4 MMI couplers. On the output side, the signals are recombined by using sixteen 4×1 and eight 2×1 MMI couplers. Switching is performed by four-arm MZ interferometers, arranged in groups of four to minimize waveguide crossings

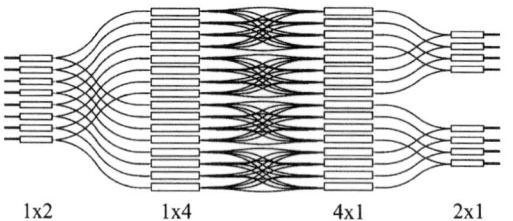

1x2   1x4   4x1   2x1

Figure 9.19: Schematic of a strictly nonblocking, 8×8 switching unit designed using several kinds of MMI couplers. (After Ref. [96]; ©2003 IEEE.)

and maintain relatively large crossing angles. Each MMI coupler introduces a small loss in the range of 1.2–1.4 dB. Since each connection employs four couplers, total insertion loss was close to 6 dB but the crosstalk level was below -34 dB on average.

The InP technology has also been used for making wavelength-selective OXCs [97]–[99]. Figures 9.20 shows the design of one such device schematically. A novel aspects of this device is that it incorporates an array of phase shifters between two AWGs for multiplexing and demultiplexing WDM channels [97]. By controlling the amount of phase shift, it is possible to connect any input channel to one of the output channels. Although the device shown in Figure 9.20 interconnects two input fibers to two output fibers, each carrying six WDM channels, the concept can easily be extended to larger-size OXCs.

A disadvantage of InP devices is that they are sensitive to the state of polarization of the input signal unless they are designed to be polarization-insensitive. A polarization-independent OXC was built on an InP substrate in 1998 using specially designed electro-optic MZ switches and AWGs [98]. Crosstalk was reduced in a later model of such a device using dilated or double-gate switches. Figure 9.21 shows the schematic of such a wavelength-selective OXC (with $M = 2$ and $N = 4$ in Figure 9.18) [99]. Each wavelength from the two input fibers is handled by a switch composed of four MZ interferometers and operating on the basis of phase shifts induced by the electro-optic effect. Switches were arranged to minimize the total number of crossings and to ensure that all interconnecting paths have the same number of crossings. Physical dimensions of the InP chip were only 11×6.5 mm^2. However, insertion losses for various inter-

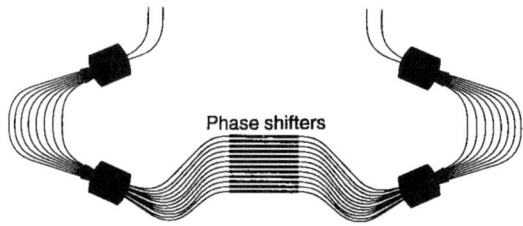

Figure 9.20: Schematic of an InP-based OXC in which switching is performed using phase shifters placed in between two AWGs. (After Ref. [97]; ©1998 IEEE.)

9.4 Optical Cross-Connects

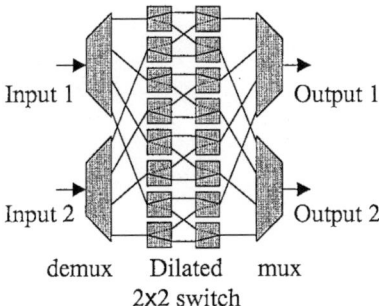

Figure 9.21: Schematic of an InP-based OXC designed using dilated MZ switches. (After Ref. [99]; ©1999 IEEE.)

connecting paths were in the range 15–17 dB. Such high losses have prevented InP devices from being competitive with the OXCs based on silica waveguides.

A new architecture shown in Figure 9.22 for wavelength-selective OXCs was used in a 2002 experiment [100]. It makes use of four circulators with two wavelength blockers and works as follows. Each WDM signal entering from two different input fibers is first split into two parts by a 3-dB coupler. Each part is sent to a different wavelength blocker surrounded by two optical circulators. The left blocker controls the bar state, while the right one controls the cross state. More precisely, if a WDM channel needs to be maintained in the bar state, it is passed by the left blocker but blocked by the right one. The opposite happens if that channel needs to be transferred to the cross state. The wavelength blockers work though the phase shifts introduced in one arm of the MZ switches in the same way as the device shown in Figure 9.20. Such an OXC was fabricated using silica-on-silicon technology with thermo-optic switches. It was capable of switching 128 channels in sets of eight. An advantage of such an architecture is that the OXC can be serviced without interrupting traffic.

The main advantage of a wavelength-selective OXC is its all-optical nature resulting from the absence of opto-electronic regenerators. As a result, wavelength-selective OXCs are transparent to both the format and the bit rate of the WDM signal. However, these advantages are overshadowed by the wavelength-blocking nature of a wavelength-selective OXC. To understand this problem, consider what happens if two

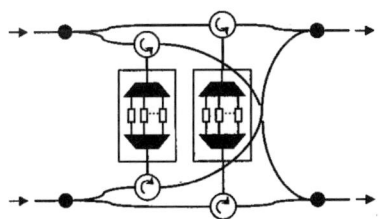

Figure 9.22: Architecture of a wavelength-selective OXC designed using four circulators and two wavelength blockers. (After Ref. [100]; ©2002 IEEE.)

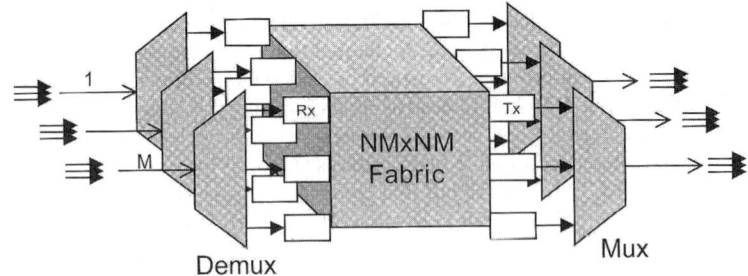

Figure 9.23: Schematic of a wavelength-interchanging cross-connect. Tx and Rx stand for transmitters and receivers, respectively. (After Ref. [91]; ©2002 Elsevier.)

channels entering from two input ports and having the same carrier wavelength are destined for the same output port. It is evident from the wavelength-selective OXC design shown in Figure 9.18 that only one channel can have the same wavelength at any output port. The two input channels can be unblocked only if the wavelength of one of the channels is changed at the transmitter end. Thus, wavelength-selective OXCs require the coordination of the wavelengths throughout the entire optical network, not an easy task in practice.

9.4.2 Wavelength-Interchanging Cross-Connect

The problems inherent in a wavelength-selective OXC can be solved by using a different architecture known as the wavelength-interchanging OXC. As the name suggests, the wavelength of any input channel can be interchanged within such a device using wavelength converters. Figure 9.23 shows the device configuration schematically. Similar to the case of a wavelength-selective OXC, all M input WDM signals are demultiplexed into N individual wavelengths. However, in contrast with a wavelength-selective OXC, channels at a given wavelength are not grouped together. Rather, a $NM \times NM$ optical or electrical switch is used to interconnect NM channels. In the case of an electrical switch, optical receivers convert all channels into electrical domain at the input end of the switch, while optical transmitters are used to regenerate the optical channel. Since transmitter wavelengths can be adjusted during this conversion, such a device does not block channels because of their identical wavelengths. By the same token, wavelength-interchanging OXCs making use of electronic switches are not transparent to the format and the bit rate of the optical bit stream. In spite of this bit-rate dependence, electrical switches are employed routinely in commercial WDM systems.

For all-optical networks, one needs wavelength-interchanging OXCs that use a single optical switch with NM input and NM output ports. Moreover, such wavelength-interchanging OXCs require wavelength converters that operate in the optical domain and can convert the wavelength of a channel while preserving the information carried by it. The main drawback of any OXC is the large number of components and inter-

9.4 Optical Cross-Connects

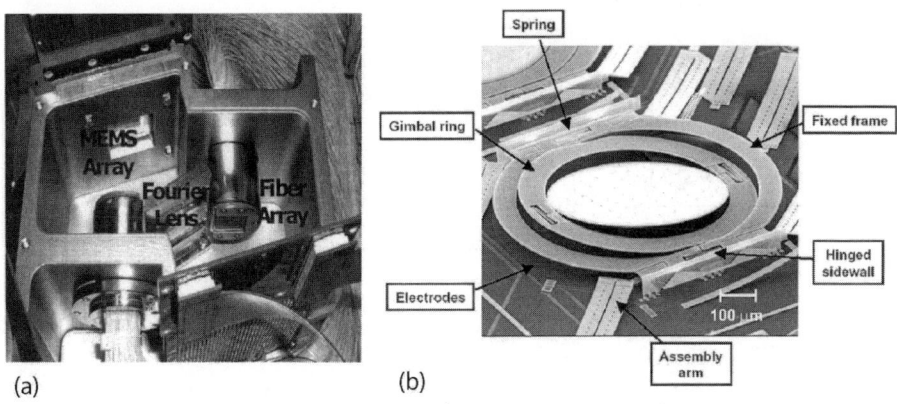

Figure 9.24: (a) A MEMs-based OXC and (b) a micromirror that rotates to make interconnection among input and output fibers. (After Refs. [101] and [102]; ©2003 IEEE.)

connections required that grows exponentially as the number of nodes and the number of wavelengths increase.

Although any switching technology of Section 9.1 can be employed for making OXCs in principle, the use of 3-D MEMS switches has attracted the most attention because it can be easily scaled to increase the the number of input and output ports. As mentioned earlier, this technology had provided by 2003 switching fabrics that could interconnect more than 1,000 input and output ports [36] while ensuring that the maximum loss does not exceed 4 dB between any two ports. Figure 9.24(a) is a photograph of such a device containing two MEMS chips with micromirrors that rotate

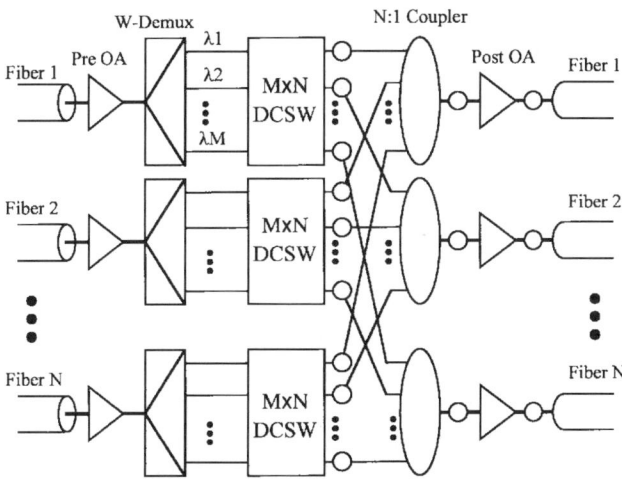

Figure 9.25: Optical-path cross-connect architecture based on delivery and coupling switches (DCSW) that connect M wavelengths to N fibers. (After Ref. [103]; ©1999 IEEE.)

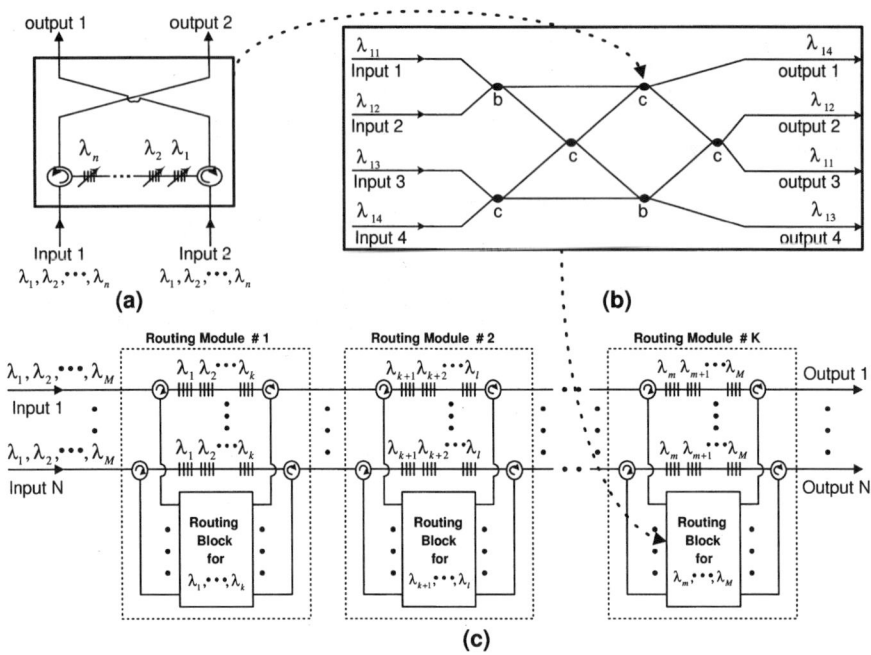

Figure 9.26: OXC Architecture based on tunable fiber Bragg gratings and optical circulators: (a) basic building block of a 2×2 switch; (b) 4×4 routing block made using six such units; (c) complete OXC interconnecting N fibers, each carrying M WDM channels. (After Ref. [106]; ©2003 IEEE.)

to make interconnections. A photograph of the micromirror appears in Figure 9.24(b).

It is possible to use multiple smaller-size switches utilizing an architecture referred to as optical-path cross-connect [88]. Figure 9.25 shows such an OXC schematically [103]. It makes use of N switches with M input and N output ports. Each switch accepts M WDM channels from one of the input fibers and switches them among N output fibers. The wavelength-contention problem is solved by incorporating wavelength converters within the switch. Such switches were made by 1998 with $N = 8$ and $M = 16$ using silica-on-silicon technology. Fiber-to-fiber insertion losses fall in the range of 12–14 dB for such OXCs.

Another architecture for OXCs makes use of fiber Bragg gratings in combination with optical circulators [104]–[106]. In one scheme, the basic building block of such OXCs is a 2×2 switch, as shown in Figure 9.26(a). Multiple WDM channels coming from two input ports are directed by optical circulators to a chain of tunable fiber gratings whose number equals the number of WDM channels. If a grating is tuned such that it reflects the channel falling within its stop band, that channel is reflected back and appears at the cross port. In contrast, if the grating is detuned so that the channel falls outside the stop band, that channel reaches the other circulator, which directs it to the bar port. By tuning each fiber grating inside the switch, each channel can be switched to either output port independently of the other channels.

Figure 9.26(b) shows how six of such 2×2 switches can be combined to produce a 4×4 routing block [106]. It is possible to make $N \times N$ routing blocks with a similar scheme. Finally, one can combine several such routing blocks to design an OXC, as shown in Figure 9.26(c), that can connect N input fibers to N output fibers while distributing M WDM channels from each fiber to output fibers in an arbitrary manner. An attractive feature of such an architecture is that it does not require demultiplexing and subsequent multiplexing of the WDM signals. Its modular nature also permits upgrading the number of wavelengths as needed.

Problems

9.1 Discuss how a directional coupler can be used for space-domain switching. Use diagrams liberally to make your explanation clear.

9.2 Discuss how a Mach–Zehnder interferometer can be used for space-domain switching. Use diagrams as necessary to make your explanation clear.

9.3 Consider a LiNbO$_3$ switch in the Mach–Zehnder configuration and estimate the voltage required for switching for 4-cm-long waveguides, assuming that the electrodes are separated by 10 μm. Use $r_{33} = 30.9$ pm/V and 2.2 for the mode index.

9.4 What is meant by a digital optical switch? Discuss how it can be made using a Y junction where a waveguide splits into two in a symmetric fashion.

9.5 Discuss the design of a thermo-optic switch made using two cascaded Mach–Zehnder interferometers. What is the main advantage of using two interferometers?

9.6 What is meant by 2-D and 3-D MEMS switches? Utilize diagrams to show how they are used in practice. How many MEMS mirrors will you need in the 2-D and 3-D cases for interconnecting 256 input fibers to 256 output fibers?

9.7 Explain how a liquid crystal can be used for optical switching. Do you expect such a switch to depend on the state of polarization of the input signal? If yes, how can you solve this problem?

9.8 What is meant by a bubble switch? Discuss how it can interconnect an array of input fibers to another array of output fibers.

9.9 Design an SOA-based wavelength converter using cross-gain saturation as the nonlinear mechanism. Explain how such a device works.

9.10 Design an SOA-based wavelength converter using cross-phase modulation as the nonlinear mechanism. Explain how such a device works.

9.11 Discuss how four-wave mixing inside an optical fiber can be used to make a wavelength converter. What are the main requirements on the fiber used to make such a device?

9.12 Describe the operation of wavelength-selective and wavelength-interchanging cross-connects and give an example of each kind.

References

[1] J. E. Midwinter, Ed., *Photonics in Switching*, Academic Press, San Diego, CA, 1993.
[2] A. Marrakchi, Ed., *Selected Papers on Photonic Switching*, SPIE Press, Bellingham, WA, 1996.
[3] E. J. Murphy, in *Optical Fiber Telecommunications*, Vol. 3B, I. P. Kaminow and T. L. Koch, Eds., Academic Press, San Diego, CA, 1997, Chap. 10.
[4] H. T. Mouftah, *Photonic Switching Technology*, IEEE Press, Piscataway, NJ, 1998.
[5] R. Ramaswami and K. N. Sivarajan, 2nd ed., *Optical Networks: A Practical Perspective*, Morgan Kaufmann, San Francisco, 2002.
[6] D. Y. Al-Salameh, S. K. Korotky, D. S. Levy, T. O. Murphy, S. H. Patel, G. W. Richards, and E. S. Tentarelli, in *Optical Fiber Telecommunications*, Vol. 4A, I. P. Kaminow and T. P. Lee, Eds., Academic Press, San Diego, CA, 2002, Chap. 7.
[7] G. I. Papadimitriou, C. Papazoglou, and A. S. Pomportsis, *J. Lightwave Technol.* **21**, 384 (2003).
[8] M. Renaud, M. Bachmann, and M. Ermann, *IEEE J. Sel. Topics Quantum Electron.* **2**, 277 (1996).
[9] R. C. Alferness, *Appl. Phys. Lett.* **35**, 748 (1979).
[10] R. C. Alferness, *IEEE J. Sel. Areas Commun.* **6**, 1117 (1988).
[11] E. J. Murphy, C. T. Kemmerer, D. T. Moser, M. R. Serbin, J. E. Watson, and P. L. Stoddard, *J. Lightwave Technol.* **13**, 967 (1995).
[12] W. T. Anderson, J. Jackel, G. K. Chang, H. Dai, X. Wei. M. Goodman, C. Allyn, M. Alvarez, and O. Clarke, et al., *J. Lightwave Technol.* **18**, 1988 (2000).
[13] H. Sasaki and R. M. de la Rue, *Electron. Lett.* **12**, 459 (1976).
[14] W. K. Burns, A. B. Lee, and A. F. Milton, *Appl. Phys. Lett.* **29**, 790 (1976).
[15] Y. Silberberg, P. Perlmutter, and J. E. Baran, *Appl. Phys. Lett.* **51**, 1230 (1987).
[16] H. Okkayama and M. Kawahara, *Electron. Lett.* **29**, 765 (1993).
[17] P. Granestrand, B. Lagerström, P. Svensson, H. Olofsson, J. E. Falk, and B. Stoltz, *IEEE Photon. Technol. Lett.* **6**, 71 (1994).
[18] M. N. Kahn, B. I. Miller, E. C. Burrows, and C. A. Burrus, *Electron. Lett.* **35**, 484 (1999).
[19] R. Krähenbhl, M. M. Howerton, J. Dubinger, and A. S. Greenblatt, *J. Lightwave Technol.* **20**, 92 (2002).
[20] W. K. Burns, M. M. Howerton, and R. P. Moeller, *J. Lightwave Technol.* **10**, 1408 (1992).
[21] W. E. Martin, *Appl. Phys. Lett.* **26**, 562 (1975); Y. Ohmachi and J. Noda, *Appl. Phys. Lett.* **27**, 544 (1975).
[22] E. Almström, C. P. Larsen, L. Gillner, W. H. van Berlo, M. Gustavsson, and E. Berglind, *J. Lightwave Technol.* **14**, 996 (1996).
[23] K. Kato and Y. Tohmori, *IEEE J. Sel. Topics Quantum Electron.* **6**, 4 (2000).
[24] T. Miya, *IEEE J. Sel. Topics Quantum Electron.* **6**, 38 (2000).
[25] Y. Hibino, *IEEE J. Sel. Topics Quantum Electron.* **8**, 1090 (2002).
[26] R. Nagase, A. Himeno, M. Okuno, K. Kato, K. Yukimatsu, and M. Kawachi, *J. Lightwave Technol.* **12**, 1631 (1994).
[27] T. Goh, A. Himeno, M. Okuno, H. Takahashi, and K. Hattori, *J. Lightwave Technol.* **17**, 1192 (1999).
[28] T. Goh, M. Yasu, K. Hattori, A. Himeno, M. Okuno, and Y. Ohmori, *J. Lightwave Technol.* **19**, 371 (2001).

References

[29] R. Kasahara, M. Yanagisawa, T. Goh, A. Sugita, A. Himeno, M. Yasu, and S. Matsui, *J. Lightwave Technol.* **20**, 993 (2002).

[30] L. Eldada and L. W. Shaklette, *IEEE J. Sel. Topics Quantum Electron.* **6**, 54 (2000).

[31] L. Y. Lin, E. L. Goldstein, and R. W. Tkach, *IEEE J. Sel. Topics Quantum Electron.* **5**, 4 (1999); *J. Lightwave Technol.* **18**, 482 (2000).

[32] D. J. Bishop, C. R. Giles, and G. P. Austin, *IEEE Commun. Mag.* **40** (3), 75 (2002).

[33] P. de Dobbelaere, K. Falta, L. Fan, and S. Gloeckner, *IEEE Commun. Mag.* **40** (3), 88 (2002).

[34] L. Y. Lin and E. L. Goldstein, *IEEE J. Sel. Topics Quantum Electron.* **8**, 163 (2002).

[35] P. de Dobbelaere, K. Falta, and S. Gloeckner, *IEEE Commun. Mag.* **41** (5), S16 (2003).

[36] J. Kim, C. J. Nuzman, B. Kumar, D. F. Lieuwen, J. S. Kraus, A. Weiss, C. P. Lichtenwalner, A. R. Papazian, R. E. Frahm, N. R. Basavanhally, D. A. Ramsey, V. A. Aksyuk, F. Pardo, M. E. Simon, V. Lifton, H. B. Chan, M. Haueis, A. Gasparyan, H. R. Shea, S. Arney, C. A. Bolle, P. R. Kolodner, R. Ryf, D. T. Neilson, and J. V. Gates, *IEEE Photon. Technol. Lett.* **15**, 1537 (2003).

[37] K. Noguchi, *J. Lightwave Technol.* **16**, 1473 (1998).

[38] Z. Z. Zhuang, Y. J. Kim, and J. S. Patel, *Appl. Phys. Lett.* **75**, 3008 (1999).

[39] S. Reinhorn, R. Oron, Y. Amitai, A. A. Friesem, K. Vinokur, and N. Pilossof, *Opt. Eng.* **38**, 1396 (1999).

[40] L. Sirleto, D. S. Hermann, G. Scalia, and L. Komitov, *Fiber Integ. Opt.* **21**, 277 (2002).

[41] A. d'Alessandro and R. Asquini, *Mol. Crys. Liq. Crys.* **398**, 207 (2003).

[42] J. E. Fouquet, S. Venkatesh, M. Troll, D. Chen, H. F. Wong, and P. W. Barth, Proc. LEOS Annual Meeting, IEEE Press, Piscataway, NJ, 1998, paper ThS1, pp. 169–170.

[43] M. Makihara, M. Sato, F. Shimokawa, and Y. Nishida, *J. Lightwave Technol.* **17**, 14 (1999).

[44] J. E. Fouquet, Tech. Dig. Opt. Fiber Commun. Conf., Optical Society of America, Washington, DC, 2000, Paper TuM1, pp. 204–206.

[45] T. Sakata, H. Togo, M. Makihara, F. Shimokawa, and K. Kaneko, *J. Lightwave Technol.* **19**, 1023 (2001).

[46] S. Venkatesh, J. E. Fouquet, R Haven, M. DePue, D. Seekola, H. Okano, and H. Uetsuka, Proc. LEOS Annual Meeting, IEEE Press, Piscataway, NJ, 2002, paper ME1, pp. 39–40.

[47] A. J. Agranat, Tech. Dig. Conf. Lasers and Electro-Optics, Optical Society of America, Washington, DC, 2002, paper CMH1, p. 37.

[48] M. K. Smit and C. van Dam, *IEEE J. Sel. Topics Quantum Electron.* **2**, 236 (1996).

[49] H. Okayama, M. Kawahara, and T. Kamijoh, *J. Lightwave Technol.* **14**, 985 (1996).

[50] C. Dragone, *J. Lightwave Technol.* **16**, 1895 (1998).

[51] A. Kaneko, T. Goh, H. Yamada, T. Tanaka, and I. Ogawa, *IEEE J. Sel. Topics Quantum Electron.* **5**, 1227 (1999).

[52] P. Bernasconi, C. R. Doerr, C. Dragone, M. Cappuzzo, E. Laskowski, and A. Paunescu, *J. Lightwave Technol.* **18**, 985 (2000).

[53] C. R. Doerr, in *Optical Fiber Telecommunications*, Vol. 4A, I. P. Kaminow and T. P. Lee, Eds., Academic Press, San Diego, CA, 2002, Chap. 9.

[54] K. Takada, M. Abe, T. Shibata, and K. Okamoto, *IEEE Photon. Technol. Lett.* **13**, 577 (2001).

[55] P. Bernasconi, L. Stulz, J. Bailey, M. Cappuzzo, E. Chen, L. Gomez, E. Laskowski, R. Long, and A. Wong-Foy, *IEEE J. Sel. Topics Quantum Electron.* **8**, 1115 (2002).

[56] S. Kamei, M. Ishii, M. Itoh, I. Shibata, Y. Inoue, and T. Kitagawa, *Electron. Lett.* **39**, 83 (2003).

[57] G.-H. Duan, in *Semiconductor Lasers: Past, Present, and Future*, G. P. Agrawal, Ed., AIP Press, Woodbury, NY, 1995, Chap. 10.

[58] H. Kuwatsuka, H. Shoji, M. Matsuda, and H. Ishikawa, *Electron. Lett.* **31**, 2108 (1995).

[59] T. Durhuus, B. Mikkelsen, C. Joergensen, S. Lykke Danielsen, and K. E. Stubkjaer, *J. Lightwave Technol.* **14**, 942 (1996).

[60] S. J. B. Yoo, *J. Lightwave Technol.* **14**, 955 (1996).

[61] C. Joergensen, S. L. Danielsen, K. E. Stubkjaer, M. Schilling, K. Daub, P. Doussiere, F. Pommerau, P. B. Hansen, H. N. Poulsen, A. Kloch, M. Vaa, B. Mikkelsen, E. Lach, G. Laube, W. Idler, and K. Wunstel, *IEEE J. Sel. Topics Quantum Electron.* **3**, 1168 (1997).

[62] H. Yasaka, H. Sanjoh, H. Ishii, Y. Yoshikuni, and K. Oe, *J. Lightwave Technol.* **15**, 334 (1997).

[63] A. Uchida, M. Takeoka, T. Nakata, and F. Kannari, *J. Lightwave Technol.* **16**, 92 (1998).

[64] M. H. Chou, I. Brener, M. M. Fejer, E. E. Chaban, and S. B. Christman, *IEEE Photon. Technol. Lett.* **11**, 653 (1999).

[65] J. Leuthold, P. A. Besse, E. Gamper, M. Dulk, S. Fischer, G. Guekos, and H. Melchior, *J. Lightwave Technol.* **17**, 1056 (1999).

[66] J. Yu, X. Zheng, C. Peucheret, A. T. Clausen, H. N. Poulsen, and P. Jeppesen, *J. Lightwave Technol.* **18**, 1001 (2000); *J. Lightwave Technol.* **18**, 1007 (2000).

[67] S. Hojfeldt, S. Bischoff, and J. Mork, *J. Lightwave Technol.* **18**, 1121 (2000).

[68] B. E. Olsson, P. Öhlen, L. Rau, and D. J. Blumenthal, *IEEE Photon. Technol. Lett.* **12**, 846 (2000).

[69] I. Brener, B. Mikkelsen, G. Raybon, R. Harel, K. Parameswaran, J. R. Kurz, and M. M. Fejer, *Electron. Lett.* **36**, 1788 (2000).

[70] J. Yu and P. Jeppesen, *IEEE Photon. Technol. Lett.* **13**, 833 (2001).

[71] S. Nakamura, Y. Ueno, and K. Tajima, *IEEE Photon. Technol. Lett.* **13**, 1081 (2001).

[72] J. Hansryd, P. A. Andrekson, M. Westlund, J. Li, and P. O. Hedekvist, *IEEE J. Sel. Topics Quantum Electron.* **8**, 506 (2002).

[73] B. N. Islam and Ö Boyraz, *IEEE J. Sel. Topics Quantum Electron.* **8**, 527 (2002).

[74] S. Radic and C. J. McKinstrie, *Opt. Fiber Technol.* **9**, 7 (2003).

[75] J. H. Lee, Z. Yusoff, W. Belardi, M. Ibsen,T. M Monro, and D. J. Richardson, *IEEE Photon. Technol. Lett.* **15**, 437 (2003).

[76] K. K. Y. Wong, K. Shimizu, M. E. Marhic, K. Uesaka, G. Kalogerakis, and L. G. Kazovsky, *Opt. Lett.* **28**, 692 (2003).

[77] S. Radic, C. J. McKinstrie, R. M. Jopson, J. C. Centanni, Q. Lin, and G. P. Agrawal, *Electron. Lett.* **39**, 838 (2003).

[78] T. Torounidis, H. Sunnerud, P. O. Hedekvist, and P. A. Andrekson, *IEEE Photon. Technol. Lett.* **15**, 1061 (2003).

[79] T. Tanemra and K. Kikuchi, *IEEE Photon. Technol. Lett.* **15**, 1573 (2003).

[80] G. P. Agrawal, *Nonlinear Fiber Optics*, 3rd ed., Academic Press, San Diego, CA, 2001.

[81] G. P. Agrawal, *J. Opt. Soc. Am. B* **5**, 147 (1988).

[82] C. Q. Xu, H. Okayama, and M. Kawahara, *Appl. Phys. Lett.* **63**, 3559 (1993).

[83] A. Jourdan, F. Masetti, M. Garnot, G. Soulage, and M. Sotom, *J. Lightwave Technol.* **14**, 1198 (1996).

[84] M. Koga, Y. Hamazumi, A. Watanabe, S. Okamoto, H. Obara, K. I. Sato, M. Okuno, and S. Suzuki, *J. Lightwave Technol.* **14**, 1106 (1996).

[85] J. M. Simmons, A. A. M. Saleh, E. L. Goldstein, and L. Y. Lin, *IEEE Photon. Technol. Lett.* **10**, 819 (1998).
[86] R. A. Soref and B. E. Little, *IEEE Photon. Technol. Lett.* **10**, 1121 (1998).
[87] M. W. Chbat, E. Grard, L. Berthelon, A. Jourdan, P. A. Perrier, A. Leclert, B. Landousies, A. Ramdane, N. Parnis, E. V. Jones, E. Limal, H. N. Poulsen, R. J. S. Pedersen, N. Flaaronning, D. Vercauteren, M Puleo, E. Ciaramella, G. Marone, R. Hess, H. Melchior, W. V. Parys, P. M. Demeester, P.J. Godsvang, T. Olsen, and D. R. Hjelme, *IEEE J. Sel. Areas Commun.* **16**, 1226 (1998).
[88] M. Koga, A. Watanabe, T. Kawai, K. Sato, and Y. Ohmori, *IEEE J. Sel. Areas Commun.* **16**, 1260 (1998).
[89] C. Dragone, *IEEE J. Sel. Topics Quantum Electron.* **6**, 1029 (2000).
[90] M. Lee, J. Yu, Y. Kim, C.-H. Kang, and J. Park, *IEEE J. Sel. Areas Commun.* **20**, 166 (2000).
[91] M. Zirnigbl, in *Optical Fiber Telecommunications*, Vol. 4A, I. P. Kaminow and T. P. Lee, Eds., Academic Press, San Diego, CA, 2002, Chap. 7.
[92] S. Sengupta, V. Kumar, and S. Saha, *IEEE Commun. Mag.* **41** (6), 60 (2003).
[93] S. Aisawa, A. Watanabe, T. Goh, Y. Takigawa, and M. Koga, *IEEE Commun. Mag.* **41** (9), 54 (2003).
[94] K. Okamoto, M. Okuno, A. Himeno, and Y. Ohmori, *Electron. Lett.* **32** 1471 (1996).
[95] T. Shibata, M. Okuno, T. Goh, T. Watanabe, M. Yasu, M. Itoh, M. Ishii, Y. Hibino, A. Sugita, and A. Himeno, *IEEE Photon. Technol. Lett.* **15**, 1300 (2003).
[96] M. P. Earnshaw, J. B. D. Soole, M. Cappuzzo, L. Gomez, E. Laskowski, and A. Paunescu, *IEEE Photon. Technol. Lett.* **15**, 810 (2003).
[97] C. R. Doerr, C. H. Joyner, L. W. Stulz, and R. Monnard, *IEEE Photon. Technol. Lett.* **10**, 117 (1998).
[98] C. G. P. Herben, C. G. M. Vreeburg, D. H. P. Maat, X. J. M. Leijtens, Y. S. Oei, F. H. Groen, J. W. Pedersen, P. Demeester, and M. K. Smit, *IEEE Photon. Technol. Lett.* **10**, 678 (1998).
[99] C. G. P. Herben, D. H. P. Maat, X. J. M. Leijtens, M. R. Leys, Y. S. Oei, and M. K. Smit, *IEEE Photon. Technol. Lett.* **11**, 1599 (1999).
[100] C. R. Doerr, L. W. Stulz, M. Cappuzzo, L. Gomez, A. Paunsecu, E. Laskowski, S. Chandrasekhar, and L. Buhl, *IEEE Photon. Technol. Lett.* **14**, 387 (2002).
[101] V. A. Aksyuk, S. Arney, N. R. Basavanhally, D. J. Bishop, C. A. Bolle, C. C. Chang, R. Frahm, A. Gasparyan, J. V. Gates, R. George, C. R. Giles, J. Kim, P. R. Kolodner, T. M. Lee, David T. Neilson, C. Nijander, C. J. Nuzman, M. Paczkowski, A. R. Papazian, F. Pardo, D. A. Ramsey, R. Ryf, R. E. Scotti, H. Shea, and M. E. Simon, *IEEE Photon. Technol. Lett.* **15**, 587 (2003).
[102] V. A. Aksyuk, F. Pardo, D. Carr, D. Greywall, H. B. Chan, M. E. Simon, A. Gasparyan, H. Shea, V. Lifton, C. Bolle, S. Arney, R. Frahm, M. Paczkowski, M. Haueis, Ronald Ryf, David T. Neilson, J. Kim, C. R. Giles, and D. Bishop, *J. Lightwave Technol.* **21**, 634 (2003).
[103] Y. Hamazumi, T. Kawai, M. Koga, and K. Sato, *IEEE Photon. Technol. Lett.* **11**, 370 (1999).
[104] Y.-K. Chen and C.-C. Lee, *J. Lightwave Technol.* **16**, 1746 (1998).
[105] E. Mutafungwa, *Opt. Fiber Technol.* **7**, 236 (2001).
[106] N. S. Moon and K. Kikuchi, *J. Lightwave Technol.* **21**, 703 (2003).

Chapter 10

Time-Domain Switching

The preceding chapter focused on space-domain switching in which a WDM channel is switched to another spatially separated port without affecting the information content of that channel. This chapter focuses on time-domain switching, a technique in which individual bits or packets of bits belonging to a specific channel are switched to different ports. The main difference between the two kinds of switching is that whereas whole-channel switching to a different port is acceptable even on a time scale of ~ 1 ms, time-domain switching must occur on the time scale of individual bits or packets, depending on the application. In the case of optical time-domain multiplexing, such switches must operate on a time scale shorter than 10 ps because individual bits arriving at a bit rate of 100 Gb/s or more are often directed toward different ports. In the case of packet-based networks, this time scale can be longer but rarely exceeds 100 ps. Most time-domain switches operate at a speed ranging from a few gigahertz to ~ 1 THz. Section 10.1 considers the basic building blocks needed for such fast optical switches, with an emphasis on the nonlinear phenomenon used to realize switching. Section 10.2 is devoted to optical flip-flops, while Section 10.3 focuses on several types of interferometric switches. Applications of time-domain switches are discussed in Section 10.4.

10.1 Nonlinear Switching Schemes

Time-domain switching requires an optical device that is capable of operating at high speeds and thus can respond at a time scale of a few picoseconds or shorter [1]–[10]. Such switches are sometimes referred to as an optical gate that can be opened for a short time using an external control. They are also referred to as an optical flip-flop, borrowing the terminology from electrical switches. In the optical domain, several nonlinear effects [11] such as optical bistability, cross-phase modulation (XPM), and four-wave mixing (FWM) are exploited for realizing time-domain switching. The switching speed depends in part on the nonlinear medium employed and on how fast this medium responds to an external control. Optical fibers can allow ultrafast switching on a time scale shorter than 1 ps, but require high optical powers and long fiber lengths because of

10.1 Nonlinear Switching Schemes

the relative weak nonlinearity of silica fibers. Semiconductor optical amplifiers (SOAs) exhibit stronger nonlinearities but typically respond on a nanosecond time scale. This section focuses on the basic nonlinear schemes to illustrate the physics behind time-domain switching.

10.1.1 Optical Bistability

Optical bistability is one of the main nonlinear phenomena exploited for making optical switches [12]. As the name implies, under certain conditions, the output of an optical device can have two discrete stable values for the same input. If the output can be switched between these two values through an external time-dependent control, the device acts as a time-domain switch. A simple device that exhibits optical bistability is a Fabry–Perot (FP) resonator containing a nonlinear medium [12]. A ring resonator can also be used for the same purpose. In fact, as early as 1983, a single-mode fiber was used as the nonlinear medium inside a ring cavity to realize an optically bistable device [13].

The origin of optical bistability in FP resonators can be understood from Eq. (2.3.8) for the transmissivity T_R. When the intracavity medium is nonlinear such that its refractive index changes with intensity I as $n = n_l + n_2 I$, the round-trip phase shift ϕ_R depends on input power. The nonlinear contribution to ϕ_R, which is also responsible for self-phase modulation (SPM), can be written as

$$\phi_{NL} = \gamma P_{av} L_R, \qquad (10.1.1)$$

where γ is the nonlinear parameter related linearly to n_2, P_{av} is the average intracavity power, and L_R is the round-trip length. For high-finesse resonators, transmitted power $P_t \approx (1 - R_m) P_{av}$, where R_m is the mirror reflectivity assumed to be the same for both mirrors. If we use this relation in Eq. (2.3.8) with $T_R = P_t / P_i$, the transmitted power is found to satisfy the transcendental equation

$$P_t \left\{ 1 + \frac{4R_m}{(1-R_m)^2} \sin^2\left[\frac{\phi_0}{2} + \frac{\gamma P_t L_R}{2(1-R_m)} \right] \right\} = P_i. \qquad (10.1.2)$$

It is clear from this equation that multiple values of P_t are possible at a fixed value of the incident power P_i because of the nonlinear phase shift. The number of solutions depends on the input power P_i. At low powers, only one solution exists. With increasing input power, the number of solutions increases from one to three, then to five and beyond. We focus on the case of three solutions since it requires the least input power.

Multiple solutions of Eq. (10.1.2) lead to dispersive optical bistability, a nonlinear phenomenon that has been observed using several different nonlinear media [12]. It occurs when the linear phase shift ϕ_0 does not correspond to a resonance of the device so that little light is transmitted at low power levels. The nonlinear phase shift brings the signal onto a resonance, resulting in higher transmission. However, the transmitted power P_t does not increase linearly with P_i, as is evident from the nonlinear nature of Eq. (10.1.2). Figure 10.1 shows the expected behavior for three values of detuning quantified by $\delta = (\omega - \omega_0)\tau_r$ in the vicinity of an FP resonance located at ω_0, where ω is the optical frequency of incident light and τ_r is the round-trip time in the cavity. Over

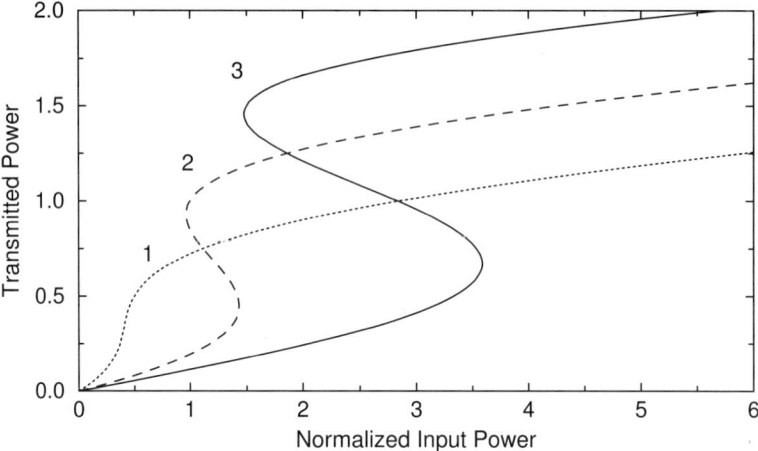

Figure 10.1: Bistable response of a fiber resonator with $R_m = 0.5$ for three values of detuning δ. Powers are normalized using $P_n = (\gamma L_R)^{-1}$.

a certain range of δ, three solutions of Eq. (10.1.2) produce the well-known S-shaped curve associated with optical bistability. The middle branch with a negative slope is always unstable [12]. As a result, the transmitted power jumps up and down at specific values of P_i in such a way that it exhibits hysteresis. The low-output state is referred to as the "off" state, while high-output state corresponds to the "on" state. Such a device can be switched on and off by changing input power, input wavelength, or other controls that change the initial detuning δ. Indeed, any mechanism that changes the linear refractive index of the intracavity material can be used for controlling such an optical switch.

Optical bistability has been observed using several different nonlinear media including atomic vapors, semiconductor waveguides, and optical fibers [12]. As early as 1978, a $LiNbO_3$ waveguide modulator was used for this purpose by coating its two cleaved ends with silver to form an FP cavity [14]. Waveguides formed using multiple quantum wells were used during the 1980s [15]. In the case of optical fibers, stimulated Brillouin scattering (SBS) hampers the observation of optical bistability when CW beams or relatively broad optical pulses are used. Bistability in a fiber-ring resonator was first seen in a 1983 experiment in which SBS was avoided using picosecond pulses [13]. In a later experiment, SBS was suppressed by placing an optical isolator inside the ring cavity that allowed the propagation of light in a single direction [16]. Bistable behavior was observed in this experiment at CW power levels below 10 mW. The nonlinear phase shift ϕ_{NL} at this power level was relatively small in magnitude (below 0.01 rad) but still large enough to induce bistability. An improved stabilization scheme was used in a 1998 experiment [17]. Figures 10.2(a) through (d) show the observed behavior at four values of the detuning δ. The experiment used mode-locked pulses (with a width of ~ 1 ps) emitted from a Ti:sapphire laser. The length of fiber-ring resonator (about 7.4 m) was adjusted precisely so that an entering laser

10.1 Nonlinear Switching Schemes

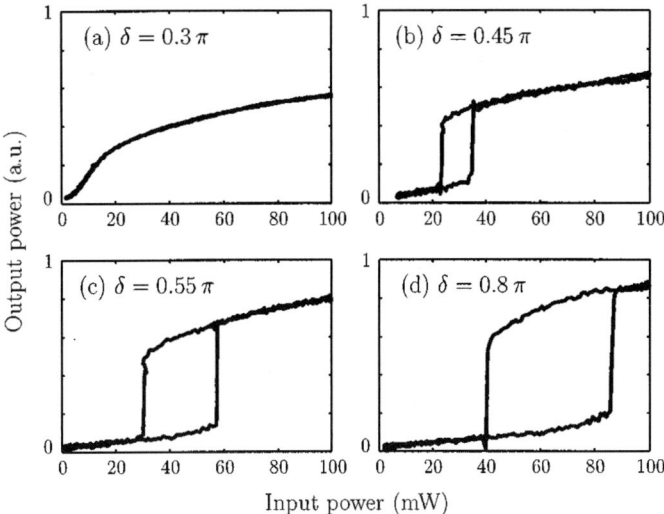

Figure 10.2: Hysteresis cycles observed in a fiber-ring resonator at four values (a–d) of detuning δ. (After Ref. [17]; ©1998 OSA.)

pulse overlapped in time with another pulse already circulating inside the cavity (synchronous pumping).

The use of semiconductor lasers as optically bistable devices attracted considerable attention during the 1990s [18]. The main advantage is that such lasers employ an FP cavity by design and that the active semiconductor waveguide of the laser can provide sufficient nonlinearity for bistability to occur. Moreover, an external holding beam is not needed since lasers generate this beam internally. Thus, it is sufficient to send a control signal, provided the laser exhibits bistable behavior in some range of applied current. Most semiconductor lasers are not intrinsically bistable but can be made so by integrating one or more saturable-absorber sections within the laser cavity [19]–[22]. Even an SOA can be used as a bistable device. Indeed, SOAs were used during the 1980s for observing bistability and for realizing all-optical flip-flops [23], [24]. Even though SOAs require an external holding beam, the power required is relatively low as SOAs also provide optical amplification.

FP cavities, although common, are not essential for bistability as long as there is a built-in mechanism that can provide optical feedback. Distributed feedback (DFB) from a Bragg grating formed inside a nonlinear medium can serve this purpose and lead to optical bistability [25]. One can employ a fiber grating or a planar waveguide with the built-in grating for making time-domain switches. DFB semiconductor lasers and SOAs are a natural candidate for such devices and they have been used for this purpose since the 1980s [26]–[29]. The physical mechanism responsible for optical bistability is the dependence of the refractive index on the carrier density; as the carrier density within the active region decreases in response to gain saturation, the refractive index increases, leading to a shift of the stop band associated with the Bragg grating. This nonlinear shift of the stop band is equivalent to changing detuning in Figure 10.2.

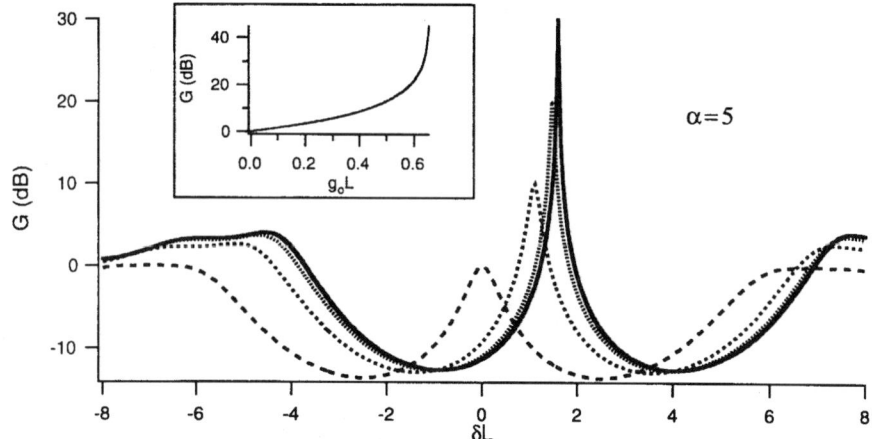

Figure 10.3: Gain spectra for a DFB amplifier of length L with a phase-shifted grating ($\kappa L = 3$); peak gain equals 0, 10, 20, and 30 dB for the four curves. Detuning δ is measured from the Bragg wavelength at transparency. Inset shows the peak gain as a function of $g_0 L$. (After Ref. [27]; ©1997 IEEE.)

The stop band can also be shifted by changing the SOA gain through current injection. Figure 10.3 shows how the gain spectrum of a phase-shifted DFB amplifier changes as the peak gain occurring at the Bragg wavelength is varied from 0 to 30 dB [27]. The shift in the gain peak is due to changes in the refractive index with the carrier density within the active layer.

10.1.2 Cross-Phase Modulation

Optical bistability is not an indispensable ingredient for making time-domain switches. The nonlinear phenomenon of XPM can be used for ultrafast switching even without a resonator, although it does require a Mach–Zehnder or Sagnac interferometer for converting the XPM-induced phase shift into amplitude changes. The basic idea behind this scheme has been discussed in Section 2.3. Figure 2.18 shows the MZ configuration schematically that can be realized in practice using optical fibers or semiconductor waveguides. A signal entering from one of the input ports is directed toward the cross port in the case of a symmetric MZ interferometer. However, if a control pulse propagates in only one of the MZ arms such that it introduces a nonlinear phase shift of π in that arm, a temporal slice of the input signal overlapping with the control pulse can be directed toward the bar port. Thus, such a device can select different time slices of the input signal in a controlled fashion and acts as a time-domain switch. The duration of the temporal slice depends on the response time of the nonlinear medium and can be shorter than 1 ps for optical fibers.

Although XPM-based switching was demonstrated by 1991 in an all-fiber MZ interferometer [30], its use is limited in practice because it is hard to maintain identical optical path lengths in the two MZ arms whose lengths typically exceed 1 km because

10.1 Nonlinear Switching Schemes

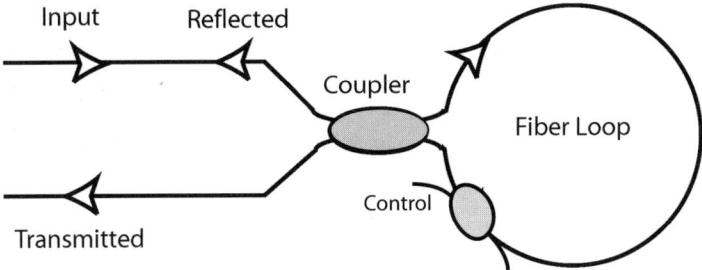

Figure 10.4: Configuration of a Sagnac loop in which a control beam is injected such that it produces much larger XPM-induced phase shift in one direction compared with the other.

of the relatively weak nonlinearity of silica fibers. For this reason, semiconductor waveguides are preferred for constructing an MZ interferometer. However, if an active waveguide is used, the switching speed can be limited to a few nanoseconds by the carrier lifetime. This problem was solved in 1992 when ultrafast all-optical switching was realized in the 1.55-μm spectral region using AlGaAs waveguides [31]. For this material, the photon energy for a 1.55-μm signal is slightly smaller than half the bandgap energy. As a result, two-photon absorption can be avoided. Because of the nonresonant nature of the nonlinearity, response time is in the femtosecond regime but high power levels are need for switching. More than 80% energy could be switched in the experiment even for 330-fs pulses. A directional coupler can also be used in place of the MZ interferometer [32].

For fiber-based devices, XPM can be used much more effectively by employing a Sagnac interferometer [33]–[39]. Such devices are used often for demultiplexing a high-bit-rate signal formed by multiplexing multiple channels in the time domain, a technique referred to as optical time-division multiplexing (OTDM). As discussed in Section 2.3.3, such interferometers employ a fiber loop that acts as a perfect mirror when the fiber coupler used to form the loop splits the input power equally (a 3-dB coupler) into counterpropagating directions. The reason for this behavior is related to the fact that the linear as well as nonlinear phase shifts are the same in both directions in a symmetric Sagnac loop. To make a time-domain switch, this symmetry is broken by injecting a control signal into the Sagnac loop such that it propagates in only one direction and induces a nonlinear phase shift on one of the counterpropagating waves through XPM. Figure 10.4 shows the scheme schematically. Physically, only the temporal slice of the input signal that overlaps with the control pulse and acquires an additional π phase shift is transmitted, while the rest of the input signal is reflected. As a result, such a scheme provides the ability to select different temporal slices of the signal, and each slice can be quite short because of the femtosecond response time of the fiber nonlinearity.

Even though the response time of fiber nonlinearity allows switching at femtosecond time scales, several other physical mechanisms limit the switching time to >1 ps. For example, wavelengths of the control and signals pulses are often different in practice. Because of group-velocity dispersion, the signal and control pulses travel at different speeds even when the two propagate in the same direction, and thus walk away from each other after some distance. We can estimate the relative phase shift between the counterpropagating waves after one round trip inside the loop using the analysis of Section 1.6.2. If we neglect the contribution of self-phase modulation, assuming that signal pulses are relatively weak, the phase difference $\Delta\phi$ between the counterpropagating fields is given by

$$\Delta\phi(t) = \phi_a^{NL} - \phi_b^{NL} = 2\gamma \int_0^L [P_c(t-\delta z) - P_c(t-\delta' z)]\,dz, \qquad (10.1.3)$$

where L is the loop length and $P_c(t)$ is the power profile of the control pulse. The parameters δ and δ' are related to the group velocities v_c and v_s of the control and signal as

$$\delta = \frac{1}{v_s} - \frac{1}{v_c}, \qquad \delta' = \frac{1}{v_s} + \frac{1}{v_c}, \qquad (10.1.4)$$

where the minus and plus signs occur when the control and signal propagate in the same and opposite directions, respectively.

Consider, for simplicity, the case of a Gaussian-shape control pulse so that $P_c(t) = P_0 \exp(-t^2/T_0^2)$ for a pulse of width T_0 and peak power P_0. The integrals in Eq. (10.1.3) can be carried out analytically in this case. In the counterpropagating direction, the relative speed of the signal and control is so large that only a small constant shift is introduced by XPM. Ignoring this shift, we obtain

$$\Delta\phi(t) = \frac{\gamma L E_c}{T_W}\left[\operatorname{erf}\left(\frac{T_W - t}{T_0}\right) - \operatorname{erf}\left(\frac{-t}{T_0}\right)\right], \qquad (10.1.5)$$

where $E_c = \sqrt{\pi} P_0 T_0$ is the control-pulse energy, $T_W = \delta L$ is the total walk off in a loop of length L, and $\operatorname{erf}(x)$ is the standard error function.

The loop transmissivity is obtained from Eq. (2.3.15) and is given by

$$T(t) = 1 - 2\rho(1-\rho)[1 + \cos(\Delta\phi)] = \sin^2[\Delta\phi(t)/2], \qquad (10.1.6)$$

if we use $\rho = \frac{1}{2}$ for a 3-dB fiber coupler. Figures 10.5(a) and (b) shows the transmission window for a Sagnac loop for two combinations of T_0 and T_W when the loop length L and pulse energy E_c are large enough to produce a maximum phase shift close to π. For short control-pulses, the switching window is nearly rectangular and its width is set by the walk-off time T_W.

The walk-off problem can be solved by using a fiber whose zero-dispersion wavelength lies between the pump and signal wavelengths such that the two waves have nearly the same group velocity. It can also be avoided by using orthogonally polarized control pulses at the same wavelength as that of the signal together with a polarization-maintaining fiber [36]. A relatively small group-velocity mismatch still occurs because of polarization-mode dispersion, but its effects are negligible for typical loop lengths. Moreover, it can be used to advantage by constructing a Sagnac loop in which the slow and fast axes of polarization-maintaining fibers are interchanged in a periodic fashion [37]. The parameter δ in such a fiber changes from positive to negative in alternating sections in such a way that the control and signal pulses nearly overlap throughout the fiber. The XPM-induced phase shift is enhanced considerably with this configuration.

10.1 Nonlinear Switching Schemes

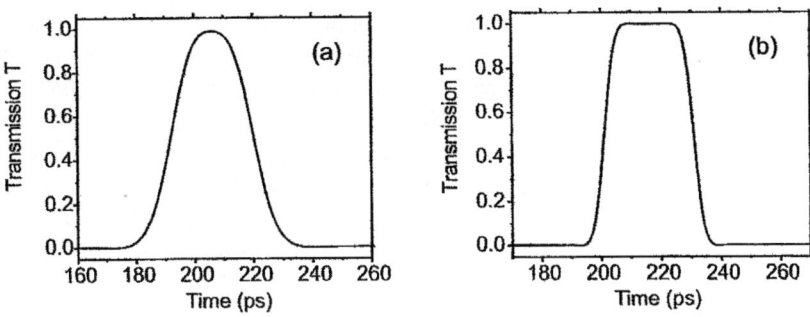

Figure 10.5: Switching window of a Sagnac-loop switch for (a) $T_0 = 30$ ps with $T_W = 10$ ps and (b) $T_0 = 10$ ps with $T_W = 30$ ps. (After Ref. [10]; ©2003 Elsevier.)

10.1.3 Four-Wave Mixing

Another nonlinear scheme for ultrafast switching, shown schematically in Figure 10.6, makes use of FWM inside a fast-responding nonlinear medium such as an optical fiber. It is similar to the schemes used for wavelength conversion and discussed in Section 9.3.2. In the switching context, FWM is used for demultiplexing individual channels from an OTDM bit stream. The OTDM signal is launched together with the control (at a different wavelength) into the nonlinear medium. The control beam consists of a periodic pulse train at the signal-channel bit rate and is referred to as the optical clock. This clock plays the role of the pump for the FWM process. In time slots in which the clock pulse overlaps with a signal bit, FWM produces a pulse at the idler wavelength. The pulse train at this new wavelength is an exact replica of the channel that needs to be demultiplexed. In effect, such a device copies selectively bits from one wavelength to another and thus can be used for wavelength conversion as well. An optical filter is used to separate the demultiplexed channel from the OTDM and clock signals.

FWM in optical fibers was first used in 1991 for time-domain switching although the 16-Gb/s signal employed relatively wide optical pulses ($\sim$20 ps) in that experiment [40]. In later experiments, the signal bit rate approached 1 THz, confirming that

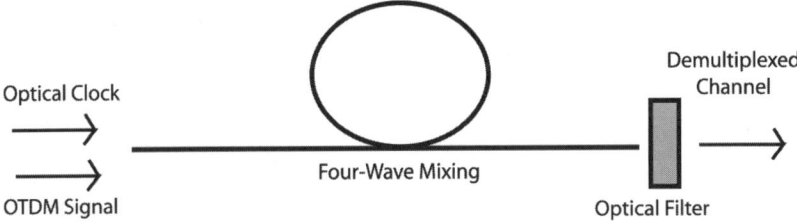

Figure 10.6: Schematic of FWM-based ultrafast switching. Clock pulses act as a pump and create copies of certain signal bits at the idler wavelength. An optical filter passes the idler but blocks other wavelengths.

optical fibers can be used to switch pulses shorter than 1 ps. A polarization-maintaining fiber is often used as the nonlinear medium for FWM because of its ability to preserve the state of polarization despite environmental fluctuations. As early as 1996, demultiplexing of 10-Gb/s channels from a 500-Gb/s OTDM signal was demonstrated using clock pulses of 1-ps duration [41]. This scheme can also amplify the demultiplexed channel through parametric amplification inside the same fiber.

The mathematical description of FWM is somewhat involved as it requires the solution of up to four coupled nonlinear equations [11]. However, when a single pump is used at power levels much higher than the signal, and depletion of the pump is ignored, the FWM process is governed by the following two coupled linear equations written in the Fourier domain:

$$\frac{dB_s}{dz} = 2i\gamma P_0 \exp(-i\kappa z) B_i^*, \tag{10.1.7}$$

$$\frac{dB_i^*}{dz} = -2i\gamma P_0 \exp(i\kappa z) B_s, \tag{10.1.8}$$

where B_s and B_i represent slowly varying spectral components of the signal and idler pulses, P_0 is the pump power, and κ represents the total phase mismatch given by

$$\kappa = \beta(\omega_s) + \beta(\omega_i) - 2\beta(\omega_p) + 2\gamma P_0, \tag{10.1.9}$$

where β is the propagation constant of the fiber mode. The signal, idler, and pump frequencies are assumed to satisfy the FWM relation $\omega_s + \omega_i = 2\omega_p$.

Equations (10.1.7) and (10.1.8) can be solved easily to study how the idler wave grows along the fiber length as a result of FWM. The idler power at the end of a fiber of length L is found to be [11]

$$P_i(L) = |B_i(L)|^2 = P_s(0)(1 + \kappa^2/4g^2)\sinh^2(gL), \tag{10.1.10}$$

where the parametric gain $g = [(\gamma P_0)^2 - (\kappa/2)^2]^{1/2}$. This solution shows that $P_i(L)$ can exceed input signal $P_s(0)$ when the phase-matching condition is nearly satisfied. In fact, $P_i(L)/P_s(0) = \sinh^2(\gamma P_0 L) \gg 1$ when $\kappa = 0$ and $\gamma P_0 L > 1$. Thus, FWM-based switches can provide significant gain while switching the signal to the idler wavelength.

The main limitation of fiber-based ultrafast switches stems from the weak fiber nonlinearity. Typically, fiber length should be 5 km or more for the device to function at practical power levels of the clock signal. This problem can be solved in two ways. In one approach, the required fiber length is reduced by up to a factor of 10 by using fibers especially designed such that the nonlinear parameter γ is enhanced because of the reduced spot size of the fiber mode [42]. Alternatively, a different nonlinear medium can be used in place of the optical fiber. The nonlinear medium of choice in practice is again the SOA. Since the carrier lifetime in SOAs typically exceeds 100 ps, it appears at first sight that FWM in SOAs will be limited to bit rates of below 10 Gb/s by this relatively slow response. However, it turns out that one can make use of intraband nonlinearities that can respond at time scales of 1 ps or less [43]. Several schemes have been developed in recent years that allow the use of SOAs as a time-domain switch at speeds as high as 250 GHz; they are discussed later in this chapter.

10.2 Optical Flip-Flops

Figure 10.7: Schematic of an all-optical flip-flop. Set and reset pulses turn the flip-flop on and off, respectively.

10.2 Optical Flip-Flops

As mentioned in Section 10.1.1, optical flip-flops constitute time-domain switches that can be turned on and off using an external control. Such devices attracted considerable attention during the 1980s because they mimic the functionality of electrical flip-flops and provide the most versatile solution for optical switching, optical memory, and optical logic elements [44]–[51]. We discuss in this section some of the recent advances related to optical flip-flops.

10.2.1 Semiconductor Lasers and Amplifiers

All flip-flops require an optical bistable device that can be switched between its two output states by changing an external control. Semiconductor lasers and SOAs are often used for making flip-flops because of their compact size, low power consumption, and potential for monolithic integration with other photonic devices. The external control can be electrical or optical for such devices. When an optical control is used, the device is referred to as an all-optical flip-flop. Figure 10.7 shows the basic idea behind such a device. The device output can be switched to the on state by sending an optical set signal in the form of a short pulse. At a later time, a reset pulse turns the flip-flop off. Unlike the switching schemes discussed in Section 10.3, the output remains on for the duration between the set and reset pulses. In this sense, a flip-flop retains memory of the set pulse and can be used as an optical memory element.

An InGaAsP semiconductor laser was used in a 1987 experiment [49] as an FP amplifier by biasing it at slightly below threshold (97% level). Two 1.53-μm lasers with a frequency difference of only 1 GHz were used for holding and control beams. The device could be switched on and off using these two beams but the switching time in this experiment was relatively long (>1 μs). In a 2000 experiment [29], a DFB laser was biased below threshold (97% level), and the resulting SOA was employed as an optically bistable device. The holding beam at 1,547 nm was tuned toward the longer-wavelength side of the Bragg resonance. Set and reset pulses were 15-ns-wide and were obtained from two InGaAsP lasers operating at 1,567 and 1,306 nm, respectively. The set pulse had a peak power of only 22 μW (0.33-pJ energy), while the peak power of reset pulses was close to 2.5 mW (36-pJ energy). Figure 10.8 shows a sequence of two set and two reset pulses (a) together with the output of the flip-flop (b). Such a device is capable of switching on a time scale comparable to the carrier lifetime ($\sim$1 ns).

The physical mechanism behind such a flip-flop is related to the shift in the stop band of the grating as the refractive index changes in response to variations in the

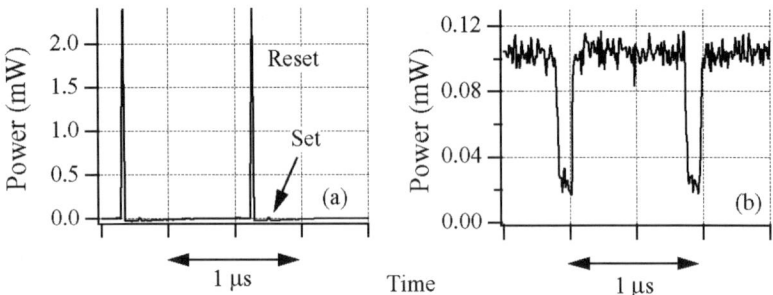

Figure 10.8: Operating characteristics of an optical flip-flop built using a DFB laser biased slightly below threshold. (a) Timing sequence of 1,567-nm set pulses (small peaks) and 1,306-nm reset pulses; (b) output power of 1,547-nm holding beam as a function of time. (After Ref. [29]; ©2000 OSA.)

carrier density (see Figure 10.3). The set pulse saturates the SOA gain, reduces the carrier density, and thus increases the effective refractive index $\bar{n}$ and shifts the Bragg wavelength to longer wavelengths as the two are related by $\lambda_B = 2\bar{n}\Lambda$, where Λ is the grating period. In contrast, the reset pulse is absorbed by the SOA, which increases the carrier density, and thus decreases the effective refractive index $\bar{n}$ and shifts the Bragg wavelength toward shorter wavelengths. The set-pulse wavelength must be within the gain bandwidth of the SOA so that it can saturate the amplifier. The exact wavelength of the reset pulse is not important as long as it is shorter than the holding-beam wavelength by an amount large enough that it falls outside the gain bandwidth and is thus absorbed by the amplifier. Thus, both control signals have a broad wavelength range of operation. The polarization of the reset pulse does not play any role. The dependence on the polarization of the set pulse can be reduced by designing the SOA appropriately. As control signals operate independently from the holding beam, they can propagate in a direction opposite to that of the holding beam; their role is only to change the carrier density. This transparency to the direction should be useful for system design.

Optical flip-flops can be constructed using several other designs [52]–[62]. In a 1995 experiment, flip-flop operation at 1.2 GHz was realized using a vertical-cavity surface-emitting laser (VCSEL) by injecting optical set and reset pulses with orthogonal polarizations [52]. The physical mechanism behind this flip-flop is related to polarization bistability. More specifically, the state of polarization of the output is switched from TE to TM by using set and reset pulses. In another experiment [54], an optical flip-flop was realized by switching between two modes of a semiconductor laser. FWM in photorefractive crystals can also be used for making an optical flip-flop when the feedback is provided by placing the crystal in a ring cavity [55]. However, the speed of such a device is limited by the response time of the photorefractive crystal.

Optical flip-flops based on two mutually synchronized semiconductor lasers were proposed in 1997 [56]. By 2001, two coupled semiconductor lasers were used for making flip-flops in which the output wavelength was switched between two values by selectively turning one of the lasers off [59]. Figure 10.9 shows the experimental scheme. Two lasers, each built using an SOA and two fiber Bragg gratings acting as

10.2 Optical Flip-Flops

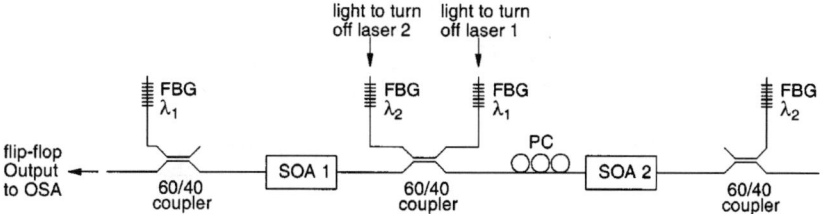

Figure 10.9: Optical flip-flop based on two coupled semiconductor lasers, each built using an SOA and two fiber gratings acting as mirrors. Each laser can be turned on and off by injecting light at a wavelength different than its own. (After Ref. [59]; ©2001 IEEE.)

mirrors, operate at different wavelengths, say, λ_1 and λ_2. One of the lasers is selectively turned off using the technique of *gain quenching* by injecting light at a wavelength different than that at which the laser operates in isolation. As a result, the output wavelength can be switched between λ_1 and λ_2 using optical controls.

An optical flop-flop in which two coupled lasers were integrated on the same chip has also been fabricated [58]. In this device shown schematically in Figure 10.10, a vertical-cavity surface-emitting laser (VCSEL) is integrated with an edge-emitting laser. The two lasers share the same active region and are mutually coupled through gain saturation since they compete for the gain in this shared region. The edge-emitting laser contains a short unbiased section that acts as a saturable absorber and makes it bistable. This laser is biased such that its output is relatively weak (off state). A set pulse injected into the saturable-absorber section switches the laser to the on state because it reduces cavity losses by saturating the absorber. The device can be turned off by injecting a reset pulse through the VCSEL if the pulse is intense enough to saturate the gain in the active region shared by the two lasers. As the intracavity intensity is reduced in response to lower gain, eventually it becomes too low to saturate the absorber; this results in increased cavity losses, and the device returns to the original

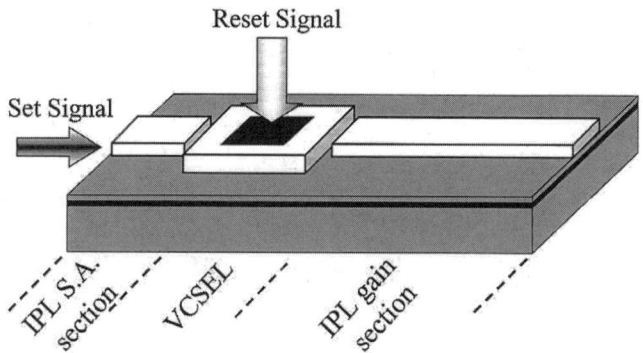

Figure 10.10: Schematic of an optical flip-flop in which a VCSEL is integrated with an in-plane laser (IPL) containing a short unbiased section acting as a saturable absorber (S.A.). Directions of set and reset pulses are also shown. (After Ref. [58]; ©2000 IEEE.)

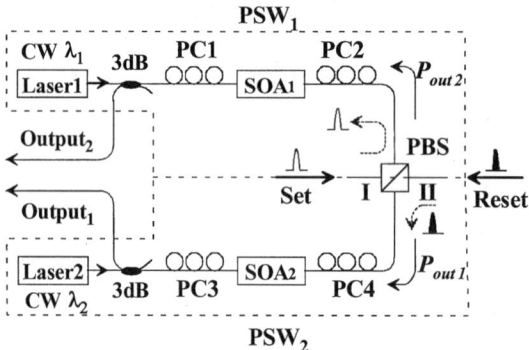

Figure 10.11: Optical flip-flop based on two polarization switches (PSW) built using SOAs whose gain-saturation characteristics are polarization-dependent; PC and PBS denote the polarization controller and polarization beam splitter, respectively. (After Ref. [62]; ©2003 IEEE.)

off state. This cycle can be repeated to switch the flip-flop on and off using set and reset pulses. Such a device is quite different from the DFB SOA discussed earlier as it does not make use of changes in the refractive index of the material.

A problem with the use of semiconductor lasers and SOAs is that their gain is polarization-dependent. This feature could be used to advantage for realizing a polarization switch. Figure 10.11 shows the configuration of an all-optical flip-flop designed using two such switches [62]. The SOAs were made with a strained active region so that their gain was significantly different for the TE and TM modes. Moreover, the TE–TM gain difference changed with the control power injected into each SOA because its gain-saturation characteristics were also polarization-dependent. Two CW lasers at wavelength λ_1 and λ_2 act as holding beams whose polarization can be adjusted using polarization controllers. The set and reset pulses act as controls and are used to saturate the SOA gain. The output at one of the wavelengths can be reduced to close to zero when the control power is large enough to saturate the SOA (about 0.5 mW). This drop in the output is a consequence of the polarization-dependent nature of the gain saturation. It produces refractive-index changes that are different for the TE and TM modes. As a result, the two polarization components of the probe acquire a relative phase difference. If this phase difference is an odd multiple of π, the output from the polarization beam splitter is suppressed.

10.2.2 Passive Semiconductor Waveguides

Passive semiconductor waveguides can also be used for constructing an all-optical flip-flop. Such devices cannot employ gain saturation as the nonlinear mechanism. Instead, they often operate below the bandgap of the semiconductor material and employ the optical Kerr effect to introduce intensity-dependent changes in the refractive index. A Bragg grating is also fabricated along the waveguide length to make the device bistable. Figure 10.12 shows the structure of such a bistable device fabricated on an InP substrate [57]. The bandgap of the 500-nm-thick InGaAsP waveguide corresponded to the

10.2 Optical Flip-Flops

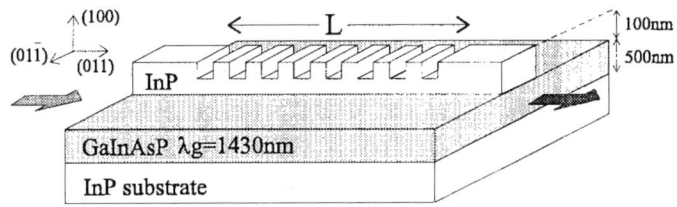

Figure 10.12: Optical flip-flop designed using a passive InGaAsP waveguide. The grating etched on the top InP cladding layer provides distributed feedback needed for bistable operation. (After Ref. [57]; ©1998 IEEE.)

1,430-nm wavelength to enure below-bandgap operation near 1.55 μm. A Bragg grating was etched onto the 100-nm-thick top InP layer for providing distributed feedback at the wavelength of the injected input signal. The grating periodicity and depth were 233 and 50 nm, respectively.

The physics behind such optical flip-flops is different from those employing SOAs with a Bragg grating. The device is biased at an input power level such that it is close to but below the switching threshold so that the output is relatively low (off state). It is switched on and off by increasing or decreasing the input power. Figure 10.13 shows the input (a) and output (b) of such a device. The required power levels are relatively large because of a weaker nonlinearity compared with the active devices based on gain saturation. For the device shown in Figure 10.12, the Kerr coefficient is estimated to be $n_2 = -4.5 \times 10^{-12}$ cm^2/W at wavelengths near 1,560 nm [57]. For a 3-mm-long grating, the set operation required an input powers of 27 mW; reset occurred when the power was lowered close to 10 mW. Set and reset pulses were 8-ns-wide in this experiment. However, because of the electronic nature of the Kerr nonlinearity, such optical flip-flops can respond at time scales of picoseconds or shorter. This is the main advantage of passive waveguides because the response time of active devices such as semiconductor lasers and SOAs is often limited to above 1 ns by the carrier lifetime.

For practical applications, an optical switch should be polarization-independent. Although a polarization-diversity scheme in which orthogonally polarized components

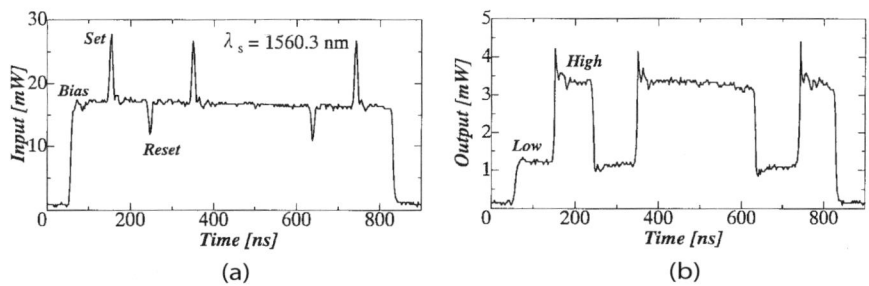

Figure 10.13: Time-dependent input (a) and output (b) traces showing the performance of an optical flip-flop built using a passive InGaAsP waveguide. (After Ref. [57]; ©1998 IEEE.)

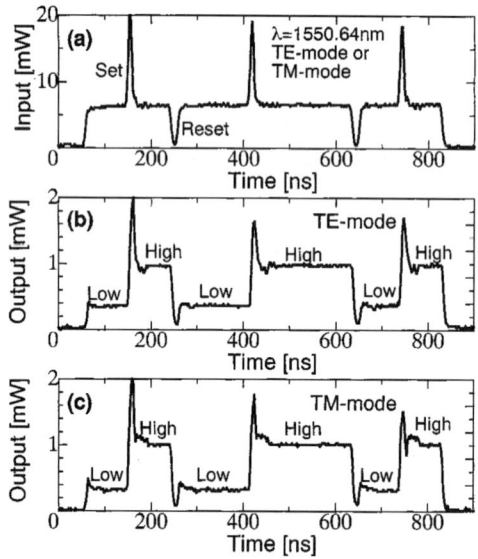

Figure 10.14: Experimental results for a polarization-independent optical flip-flop. (a) Sequence of set and reset pulses, (b) flip-flop output when the input signal excites the TE mode, and (c) flip-flop output when the input signal excites the TM mode. (After Ref. [60]; ©2002 IEEE.)

are processed separately can solve this problem, the resulting devices are too complex to be integrated monolithically. A compact, polarization-independent, all-optical flip-flop has been realized by using a passive semiconductor waveguide integrated with a vertically etched Bragg reflector [60]. The waveguide was designed such that the stop band of the Bragg grating changed little when the polarization of incident light was changed from the TE to TM mode. Figure 10.14 shows the sequence of set and reset pulses (a) and the flip-flop output when the input signal excites the TE mode (b) or the TM mode (c) of the waveguide. A virtually identical output for the TE and TM modes ensures that such a device operates in a polarization-independent manner.

10.3 Ultrafast Interferometric Switches

This section focuses on ultrafast switches that make use of an optical interferometer. As discussed in Section 10.1, such devices make use of the nonlinear phenomenon of XPM for optical switching and convert phase modulation into amplitude modulation using either a Sagnac interferometer in the form of a nonlinear optical loop mirror or a Mach–Zehnder interferometer made using directional couplers. Although an optical fiber was used as the nonlinear medium during the early 1990s, fiber-based devices often suffer from dispersive and environmental stability problems as they employ long fiber lengths to ensure a large enough XPM-induced phase shift. These problems can be solved using an SOA as a nonlinear medium but only at the expense of a slower

10.3 Ultrafast Interferometric Switches

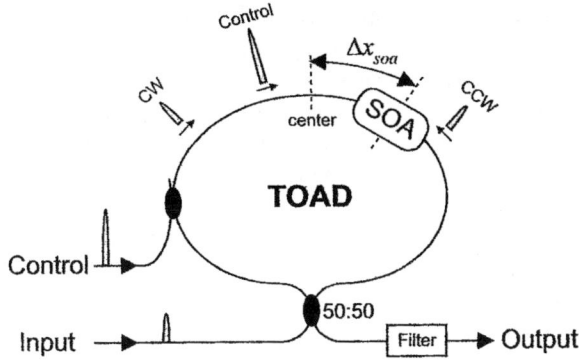

Figure 10.15: Schematic of a Sagnac-loop switch. The SOA is offset from the center of the loop by a small but controlled amount. (After Ref. [10]; ©2003 Elsevier.)

nonlinear response. In this section we consider several SOA-based interferometric devices and discuss their relative merits as an ultrafast optical switch.

10.3.1 Sagnac-Loop Switches

The physics behind time-domain switches built using a nonlinear optical loop mirror acting as a Sagnac interferometer has been discussed in Section 10.1.2. When optical fibers are used as the nonlinear medium for producing XPM-induced phase shift, dispersive effects limit the switching window, mainly because the signal and control pulses walk off from each other, owing to their different group velocities (see Figure 10.5). The walk-off problem can be solved if an SOA is placed within the Sagnac loop [63]–[65]. The loop length in this case can be 1 m or less because the loop is only used to propagate the input signal into counterpropagating directions after it has been split into two parts by a directional coupler.

As discussed earlier, SOAs produce a nonlinear phase shift when the refractive index of the active region changes in response to carrier density variations induced by gain saturation. The active nature of this process produces a relatively large nonlinear response but such nonlinear changes cannot occur at an ultrafast time scale because their speed is limited inherently by the carrier lifetime (typically in the range of 0.2–1 ns for most SOAs). It turns out that the speed limitation can be overcome to a large extent by a clever trick [65], [66]. The trick consists of placing the SOA such that it is shifted from the center of the loop by a small but precise distance. It is this shift that governs the temporal window over which switching occurs rather than the carrier lifetime. Figure 10.15 shows such a device schematically. It is referred to as a semiconductor laser amplifier in a loop mirror or SLALOM [64]. It is also called a terahertz optical asymmetric demultiplexer (TOAD) to emphasize the asymmetric placement of SOA and its operation at terahertz speeds [65].

In Figure 10.15, a 3-dB directional coupler splits the input pulse into two pulses propagating in the clockwise (CW) and counterclockwise (CCW) directions. The control pulse propagates only in the CW direction and is intense enough to saturate the

SOA, which is offset from the loop center in the CW direction by an amount Δx_{soa}. In the absence of the control pulse, the signal pulse acquires the same phase shift in both directions (as it is too weak to saturate the SOA), and it is totally reflected by the Sagnac loop. The control pulse is timed such that it arrives at the SOA after the CCW signal pulse but before the CW pulse. As a result, the XPM-induced phase shift is produced on the CW pulse, but not on the CCW pulse. If this differential phase shift equals π, the signal pulse is transmitted, rather than being reflected. Neighboring signal pulses do not experience a large differential phase shift and are thus reflected. After a time interval comparable to the carrier lifetime, the gain recovers, and the process can be repeated as long as control pulses are separated by the gain-recovery time. Such a device works well as a time-domain demultiplexer in which control pulses are injected at a relatively low rate corresponding to the single-channel bit rate even though the OTDM signal can have a repetition rate in excess of 100 GHz [10].

The switching time of the TOAD device is governed by the speed with which a short control pulse can saturate the SOA and change its refractive index. The temporal gain response of an SOA has been analyzed in Section 5.5.4 by solving the carrier-density rate equation. The phase shift imposed on the CW signal pulse arriving just after the control pulse is given by

$$\phi(t) = -\tfrac{1}{2}\beta_c h(t), \tag{10.3.1}$$

where $h(t)$ is the solution of Eq. (5.5.20) with $P_{\text{in}}(t)$ representing the total power passing through the amplifier at any time t. The relative phase shift produced by the amplifier is thus given by

$$\Delta\phi(t) = \tfrac{1}{2}\beta_c[h(t) - h(t+t_d)], \tag{10.3.2}$$

where $t_d = 2\Delta x_{\text{soa}}/v_g$ is the delay between the two counterpropagating pulses traveling at the group velocity v_g.

The power P_{out} transmitted through the loop can be calculated using the method of Section 2.3.3. If we assume a perfect 3-dB coupler ($\rho = \tfrac{1}{2}$), it is given by

$$P_{\text{out}}(t) = \tfrac{1}{4}P_{\text{in}}(t)[G_1(t) + G_2(t) - 2\sqrt{G_1(t)G_2(t)}\cos[\Delta\phi(t)], \tag{10.3.3}$$

where G_1 and G_2 represent the SOA gain for the CW and CCW directions, respectively. Figure 10.16(a) shows how phase shifts vary as a function of time for counterpropagating pulses when $t_d = 10$ ps and control pulses separated by 200 ps are injected into the loop in a periodic fashion [10]. Transmissivity of the Sagnac-loop switch, defined as $T_S = P_{\text{out}}/P_{\text{in}}$, is shown as a function of time in Figure 10.16(b). Since the differential phase shift $\Delta\phi$ between the counterpropagating pulses is large only for a duration t_d, loop transmission is large during this interval and drops to zero afterward. Clearly, the switching window of such a Sagnac loop can be controlled by changing the SOA location within the loop. Note that T_S exceeds 1 because of the amplification provided by the SOA as signal pulses pass through it.

The main advantage of using an SOA within the Sagnac loop is that it provides switching at relatively low energy levels of the control pulses. Typically, energy of control pulses should exceed 10% of the saturation energy E_{sat} given in Eq. (5.5.14) for imposing a relative phase shift of π on counterpropagating signal pulses [67]. For

10.3 Ultrafast Interferometric Switches

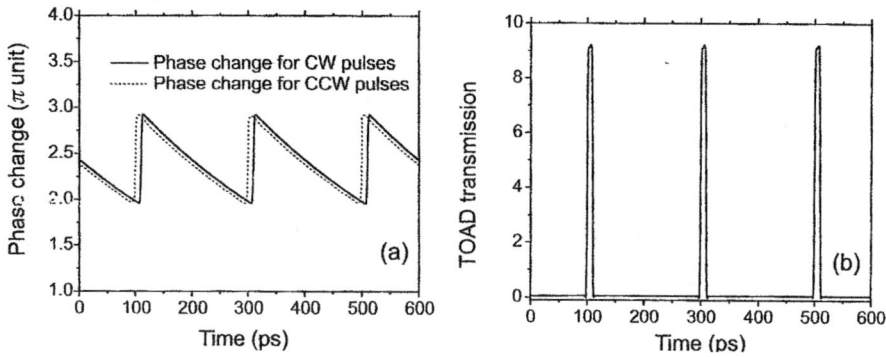

Figure 10.16: (a) Phase shifts as a function of time for counterpropagating pulses. (b) Resulting switching window of the Sagnac loop. (After Ref. [10]; ©2003 Elsevier.)

typical values of the SOA parameters, such phase shifts can be realized at energy levels below 1 pJ. A drawback of using SOAs is that they have a relatively high noise figure and thus reduce the signal-to-noise ratio at the output because of spontaneous emission added to the signal during its amplification.

10.3.2 Mach–Zehnder Switches

As discussed in Section 10.1.2, an MZ interferometer can be used in place of the Sagnac loop for ultrafast switching. In the simplest configuration, the control pulse is injected such that it propagates in only one of the arms of the MZ interferometer and thus adds a relative phase shift of π through XPM between the two parts of the signal pulse propagating through the two MZ arms. This approach was used in 1991 for optical switching [30]. However, it suffers from the stability problem because it is difficult to maintain identical linear phase shifts in the two long arms of a fiber-based MZ interferometer.

Similar to the case of a Sagnac loop, the use of an SOA as the nonlinear medium reduces the arm lengths drastically and thus can solve the stability problem. Moreover, such a device can be integrated on a single chip using semiconductor waveguides, resulting in a compact monolithic switch. For these reasons, the MZ geometry with SOAs has continued to attract considerable attention [68]–[76]. Figure 10.17 shows two basic schemes that are employed for making ultrafast MZ switches [10]. These are referred to as (a) colliding-pulse MZ and (b) symmetric MZ switching configurations.

In the case of a colliding-pulse MZ switch, both the signal and control pulses are split into two parts, and the two counterpropagate in each arm of the MZ interferometer. A relative phase shift is introduced by different locations of the SOA in the two arms, as seen in Figure 10.17(a). The SOAs are positioned such that the control pulse passes through the SOA before the signal pulse in one arm but after it in the other. This difference produces a relative phase shift because the carrier density and the refractive index are different in the second case, owing to gain saturation produced by the control pulse. In essence, this scheme mimics the Sagnac switch discussed earlier, and the

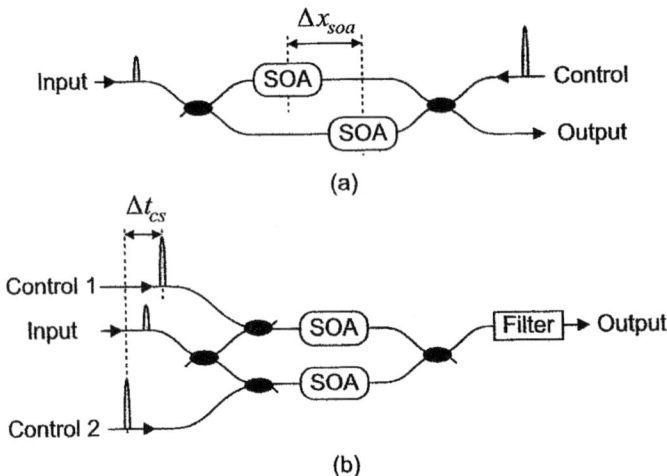

Figure 10.17: Two schemes used for making MZ switches with SOAs acting as nonlinear elements: (a) colliding-pulse MZ switch; (b) symmetric MZ switch. (After Ref. [10]; ©2003 Elsevier.)

relative spacing between the two SOAs sets the temporal window during which the device output is directed to a different port. Its main advantage is that control pulses need not be separated from signal pulses because of their counterpropagating nature.

Figure 10.18(a) shows the structure of a MZ switch fabricated in a monolithic form using InP-waveguide technology [74]. The MZ interferometer is built using two 3-dB multimode-interference (MMI) couplers that couple light into and out of two passive waveguides forming the MZ arms. Each of the MMI couplers is 570 μm long. Two 1.5-mm-long SOAs are integrated, one in each MZ arm, but they are offset by 1.5 mm; this distance sets the switching window to about 20 ps. Figure 10.18(b) shows the switching window together with the measured gain profiles of the two SOAs when 2-ps control pulses at a 2.5-GHz repetition rate are used to saturate the SOA gain. The dip in the gain corresponds to the arrival time of control pulses. It is followed by the long gain-recovery time but the gain is fully recovered before the next control pulse arrives. Thus, such a switch can respond at a time scale of 20 ps every 400 ps or so.

In the case of a symmetric MZ switch, the SOAs are located symmetrically in the two MZ arms. Control pulses are still injected in both arms, but they are launched independently and can be arranged to propagate in the forward or backward direction with respect to the signal pulses. An asymmetry is introduced by the relative delay between the two control pulses, as seen in Figure 10.17(b). The device works the same way as the colliding-pulse switch except that the switching window is now controllable externally by adjusting separation between the two control pulses. An optical filter is placed at the output port to block the control pulses when they copropagate with the signal pulses. This filter is not needed in the counterpropagating case.

Figure 10.19 shows the structure of the symmetric MZ switch fabricated in monolithic form using InP-waveguide technology [70]. The signal and control pulses are

10.3 Ultrafast Interferometric Switches

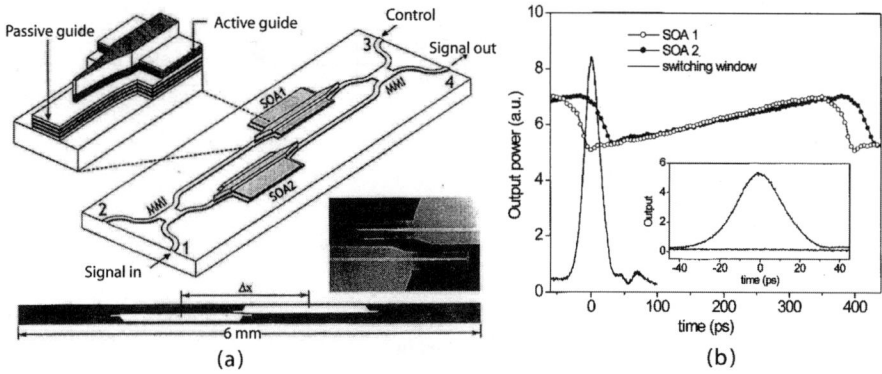

Figure 10.18: (a) Structure and microphotograph of a colliding-pulse MZ switch fabricated on an InP substrate. (b) Gain variations in the two SOAs and the resulting switching window; the inset shows the switching window on a magnified scale using a sampling oscilloscope. (After Ref. [74]; ©2001 IEEE.)

arranged to propagate in opposite directions in this device. The MZ interferometer is built using two 3-dB MMI couplers that couple light into two single-mode InGaAsP waveguides. The SOAs are designed to be polarization-insensitive and were 500 μm long. The entire device was grown on an InP substrate using an epitaxial growth technique. The switching window of the symmetric MZ switch is controlled by the relative delay Δt between the control pulses. Figure 10.19 shows how such a switch can be used for demultiplexing OTDM signals. Each control pulse saturates the SOA gain and changes the mode index slightly, but the differential phase shift in the two arms occurs only during a time interval Δt. The main advantages of the symmetric MZ switch is that the delay time can be adjusted externally and the same device can be used at different bit rates.

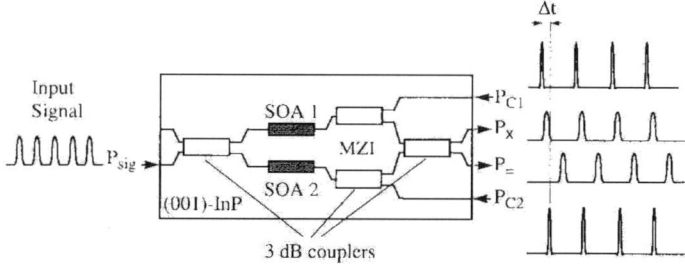

Figure 10.19: Layout of a monolithically integrated symmetric MZ switch fabricated on an InP substrate. The signal and control pulses show how such a device can be used as an OTDM demultiplexer. Two control pulses are delayed by a time interval Δt that sets the duration of the switching window. (After Ref. [70]; ©1998 IEEE.)

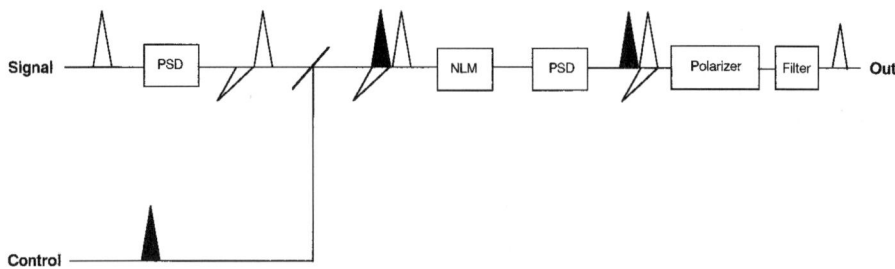

Figure 10.20: Schematic of a polarization-discriminating or UNI switch: PSD and NLM stand for polarization-sensitive delay and nonlinear medium, respectively. (After Ref. [81]; ©1998 OSA.)

10.3.3 Polarization-Discriminating Switches

Stability of a fiber-based MZ interferometer can be ensured by adopting a single-arm MZ design in which the same physical path is shared by two orthogonally polarized components of the signal pulse. The basic idea was first demonstrated in 1995 and has been used since then in a number of configurations [77]–[84]. Such a switch is sometimes referred to as polarization-discriminating as its operation is based on the state of polarization of the signal and control pulses. It is also called an ultrafast nonlinear interferometric (UNI) switch. It offers a number of advantages related to the single-arm nature of the underlying interferometer. The nonlinear element can be a passive waveguide [80] but SOAs are most often used in practice. The main difference from other SOA-based switches is that the relative delay between the orthogonally polarized signal pulses is produced by the birefringence of a polarization-maintaining fiber (PMF).

Figure 10.20 shows the design of a UNI switch. The signal pulse entering from the left of the device is split into its orthogonally polarized components using a PMF whose length L is chosen such that the two pulses are delayed by a fixed amount. The delay induced by the PMF depends on its birefringence and can be written as

$$\Delta t = |n_x - n_y| L/c. \tag{10.3.4}$$

For typical values of $\sim 10^{-4}$ of the refractive-index difference, a few meters of PMF can introduce relative delays of a few picoseconds. The two delayed signal pulses are then sent through a nonlinear medium (often an SOA). The position of the control pulse is adjusted such that it arrives in between the two components of the signal pulse. As a result, only the trailing component of the signal pulse acquires an additional XPM-induced phase shift. The two components of the signal are then synchronized by a second PMF whose fast and slow axes are reversed compared to the first PMF. A polarizer set at 45° forces interference between the two polarization components of the signal pulses and completes the XPM-induced switching. Control pulses are filtered at the output of the device. This filtering can be avoided by using a configuration in which signal pulses propagate in a direction opposite to that of the signal.

10.3 Ultrafast Interferometric Switches

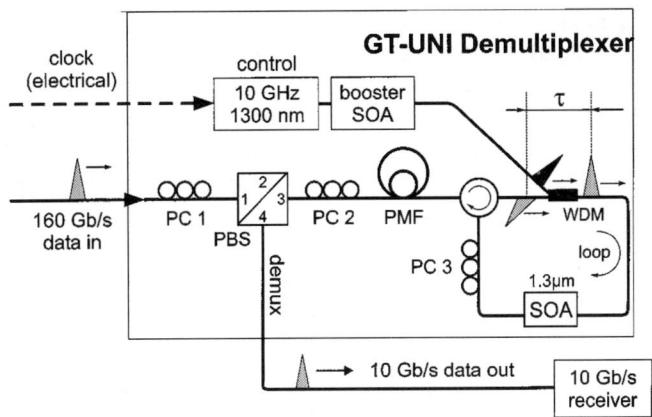

Figure 10.21: Schematic of a gain-transparent (GT) UNI switch with SOA inside a loop. The same PMF splits and recombines signal pulses. (After Ref. [82]; ©2001 IEEE.)

The main advantage of UNI geometry is its interferometric stability. Since all signals travel along the same path, they are exposed to identical environmentally induced fluctuations. As the output depends on the phase difference between the two polarization components of the signal pulse, the device is automatically immune to all environmental fluctuations. Indeed, UNI switches do not require any active stabilization in practice. For the same reason, any fluctuations in the linear refractive index of the nonlinear medium also cancel out and do not impact the device performance as long as such fluctuations occur over time scales longer than the signal-pulse width.

The use of folded geometry, as shown in Figure 10.21, can further improve the performance of UNI switches [82]. This configuration uses the same PMF for splitting and recombining signal pulses by placing the SOA inside a loop. Signal pulses are first polarized by the polarization controller PC-1 so that they leave the polarization beam splitter via port 3. The polarization controller PC-2 is used to align the polarization of the signal pulse at 45° with respect to the principal axes of the PMF, which splits it into two orthogonally polarized components and introduces a fixed relative delay between them. The XPM-induced phase shift on one of the components is induced by a polarization-insensitive SOA placed inside a loop; control pulses enter into this loop through a WDM coupler. A third polarization controller PC-3 placed after the SOA rotates the polarization such that the pulse pair is launched at −45° into the same PMF that was used to split the signal pulse. However, this PMF now reverses the delay and synchronizes the two polarization components.

As before, the two signal components recombine at the PBS. If both components encounter the same phase shifts in the loop containing the SOA, they leave the polarization beam splitter via port 1, and no signal is present at the output port 4. However, if a control pulse enters the SOA in between the two signal components, the trailing component experiences an additional XPM-induced phase shift. This relative phase shift switches that specific signal pulse to the output port 4. The use of the same PMF for splitting and recombining the signal components ensures that any random phase shifts occurring in the PMF will not affect the switching quality as long as such shifts occur on a time scale longer than the round-trip time.

The device shown in Figure 10.21 also makes use of an additional improvement realized though a gain-transparent SOA. The main idea behind gain transparency is to use control pulses whose wavelength is tuned close to the gain peak while the wavelength of signal pulses falls outside the SOA-gain spectrum in the transparent regime below the bandgap of the active material [85]. Since signal experiences virtually no gain (or loss), it does not suffer from the noise added by spontaneous emission. At the same time, control pulses saturate the gain and produce refractive-index changes at the signal wavelength. Other advantages offered by this scheme include low crosstalk, operation over a wide range of signal wavelengths, and negligible power variations among signal pulses at the device output.

10.3.4 Lithium-Niobate Switches

As we have seen in Chapter 6, LiNbO$_3$ waveguides can be used for making optical modulators capable of operating at 40 Gb/s and thus responding at picosecond time scales. It is thus not surprising that they can be employed to make time-domain switches. The main difference is that, rather than exploiting the electro-optic effect that was needed for optical modulators, such switches make use of the second-order nonlinear susceptibility $\chi^{(2)}$ associated with LiNbO$_3$ crystals.

The second-order nonlinear effects produce the frequency combinations $\omega_1 \pm \omega_2$ when two optical field at frequencies ω_1 and ω_2 are launched into a LiNbO$_3$ crystal, provided the phase-matching condition is satisfied. The sum-frequency generation corresponding to the combination $\omega_1 + \omega_2$ is used for switching because it creates the output in the spectral region near 775 nm when the signal and control pulses are in the 1.55-μm spectral region [86]–[88]. The phase-matching problem is solved using the periodic-poling technique in which the sign of $\chi^{(2)}$ is alternated in a periodic fashion along the waveguide length. This approach is known as quasi-phase matching (QPM).

All-optical switching in a periodically poled lithium-niobate (PPLN) waveguide was first observed in a 2000 experiment [86]. The 12-μm-wide waveguide was designed to support multiple transverse modes at the signal and control wavelengths for enhancing the conversion efficiency of the nonlinear process. An adiabatic taper reduced the width to 4 μm at the input end to facilitate coupling into the fundamental mode of the waveguide. The PPLN waveguide was 6 cm long and had a QPM period of 14.7 μm over a 5.55-cm-long region. As expected, the sum-frequency radiation was generated only when the signal and control were present simultaneously.

The performance of a PPLN switch can be improved by integrating it into one arm of a MZ interferometer [88]. The whole device can be integrated on a single chip. Figure 10.22 shows the device structure schematically. The QPM period was about 18 μm over the 5-mm-long section, over which the sum frequency $\omega_1 + \omega_2$ was generated through periodic inversion of ferroelectric domains. The whole device was only 11 mm long. This device also incorporated three thin-film heaters in the other arm for inducing a relative phase shift of π (through the thermo-optic effect) in the two MZ arms in a controlled fashion. This phase shift ensures the transmission of the sum-frequency radiation through the single output port of the device.

10.4 Applications

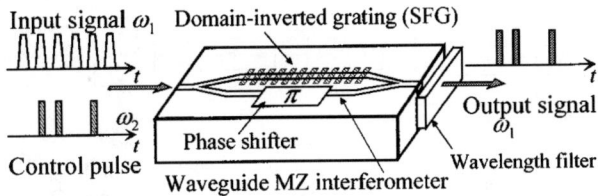

Figure 10.22: Schematic of a LiNbO$_3$ switch in which a PPNL waveguide is integrated in one arm of a MZ interferometer. (After Ref. [88]; ©2001 IEEE.)

The main drawback of such PPLN switches is that the output wavelength is in the spectral region near 775 nm even though the original signal occurs at a wavelength close to 1,550 nm. This limitation can be rectified by adopting a scheme in which two different $\chi^{(2)}$ processes are quasi-phase matched in the same device. Such a cascading nonlinear scheme has been used since the early 1990s [89] and can be adopted for the PPLN waveguide [90]–[92]. More specifically, the PPLN device first generates the second harmonic $2\omega_1$ at the control frequency, which in turn creates a new frequency $2\omega_1 - \omega_2$ through difference-frequency generation. Such a cascaded nonlinear process is equivalent to an effective FWM process and creates the output at a wavelength close to the signal and control wavelengths.

In principle, the response time of PPLN switches can be in the femtosecond regime because of the nonresonant nature of the underlying $\chi^{(2)}$ nonlinearity. In practice, dispersion-related effects limit the response to few picoseconds. In the case of the cascaded FWM-like process, the group-velocity mismatch between the fundamental and its second harmonic is about 350 ps/m [91]. The delayed second harmonic of the control pulse not only reduces the device efficiency but also creates crosstalk if signal pulses are too closely spaced. This crosstalk can be avoided by spacing signal pulses further apart but it limits the switch speed. One can enhance the efficiency and reduce the crosstalk to some extent by timing the control pulse such that it does not overlap with the signal pulse at the input end of the device but its second harmonic does overlap further along the device [92]. With such adjustments, PPLN devices have been used for demultiplexing OTDM signals at a bit rate of up to 160 Gb/s. We discuss demultiplexing in the next section devoted to applications of time-domain switches.

10.4 Applications

Time-domain switching can be used for a number of applications related to lightwave systems [6]–[10]. As already mentioned, their use for demultiplexing an OTDM signal is indispensable because each optical pulse is only a few picosecond wide at aggregate bit rates exceeding 100 Gb/s. They can also be used for extracting the optical clock in self-clocking OTDM networks. Another application is related to packet-switched optical networks that are gaining momentum with the growth of the Internet. In such systems, information is sent in the form of packets consisting of a few tens to hundreds of bits. This section considers these applications in more detail.

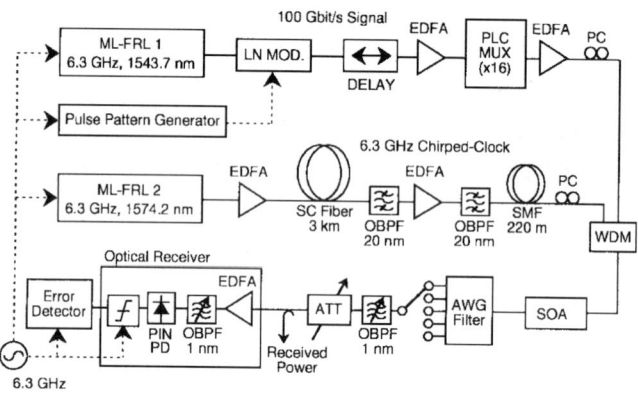

Figure 10.23: Experimental setup for time-domain demultiplexing using FWM in an SOA: ML-FRL, mode-locked fiber-ring laser; LN MOD, LiNbO$_3$ modulator; PC, polarization controller; SC, supercontinuum; OBPF, optical bandpass filter. (After Ref. [96]; ©1998 IEEE.)

10.4.1 Time-Domain Demultiplexing

An OTDM signal consists of a high-speed bit stream that is composed of several channels, each operating at a lower bit rate and interwoven with others in a periodic fashion. Thus, if ten channels, each operating at 10 Gb/s, are multiplexed in the time domain, every 11th bit of the composite 100-Gb/s bit stream belongs to the same channel. Demultiplexing a channel from such a high-speed OTDM signal requires optical switches that pick all the bits belonging to a specific channel and direct those bits to a different port. Such switches require control pulses at the single-channel bit rate (10 Gb/s in the preceding example) that is used to switch signal pulses selectively using a nonlinear phenomenon such as XPM, FWM, or bistability [93]. This regular pulse train is referred to as the optical clock.

As discussed in Section 10.1.3 and shown in Figure 10.6, one way to select individual pulses is to employ FWM in a nonlinear medium. In this scheme, the optical clock is chosen to be at a wavelength that is a few nanometers away from the signal wavelength. Only when the signal and control pulses are present simultaneously and overlap in the time domain, FWM creates an idler pulse at the new wavelength. An optical filter at this wavelength blocks the pump and signal pulses, resulting in an output bit stream that belongs to a specific channel. A different channel can be selected by adjusting the clock phase such that clock pulses overlap with the pulses belonging to this channel.

FWM inside an optical fiber was used as early as 1991 for demultiplexing OTDM channels [40]. This scheme has remained attractive in spite of long fiber lengths required mainly because of the ultrafast nature of the fiber nonlinearity that allows switching at femtosecond time scales, at least in principle. In a 1996 experiment, 10-Gb/s channels were demultiplexed from a 500-Gb/s OTDM signal using clock pulses of about 1-ps duration [41]. This scheme can also amplify the demultiplexed channel through parametric amplification inside the same fiber [94].

10.4 Applications

SOAs have also been used as a nonlinear medium for FWM [95]–[100]. This appears surprising at first because of a relatively slow carrier lifetime in SOAs (typically >100 ps). However, intraband nonlinearities can respond at time scales <1 ps [43], and they constitute the main nonlinear mechanism behind FWM in SOAs. Figure 10.23 illustrates the experimental setup used in 1998 for demultiplexing 6.3-Gb/s channels from a 100-Gb/s OTDM signal [96]. In this experiment, the 100-Gb/s OTDM signal was obtained by multiplexing sixteen 6.3-Gb/s channels in the time domain and consisted of 5-ps pulses at a wavelength of 1,543.7 nm. The optical clock at a wavelength of 1,559 nm used 33-ps pulses at the 6.3-GHz repetition rate that were chirped at a rate of 0.32 nm/ps. Figure 10.24 shows the OTDM signal (a) together with a single clock pulse (b). Both the signal and clock pulses were launched into an SOA module with 18-dB fiber-to-fiber gain. Their states of polarization were matched to that of the TE mode supported by the SOA to avoid any polarization effects. The optical spectrum recorded after the SOA is shown in Figure 10.24(c). It shows that FWM inside the SOA has generated an idler with multiple peaks that correspond to individual channels. This spectral shift of individual channels is a consequence of the 30-ps-wide chirped clock pulse. The clock pulse overlaps with multiple pulses belonging to different channels, but its wavelength is slightly different for each channel because of the linear chirp imposed on it. An optical filter in the form of an arrayed waveguide grating was used to filter individual channels.

In most FWM experiments, shorter clock pulses are used so that they overlap with only one signal pulse. The idler in this case contains only one demultiplexed channel. Such a device demultiplexes one channel at a time but different channels can be selected by adjusting the clock phase that amounts to shifting the position of clock pulses. The width of the clock pulse is an adjustable parameter that can be used to optimize the device performance [98]. The highest bit rate of the OTDM signal that has been demultiplexed using FWM in an SOA stands at 200 Gb/s [95]. This experiment also employed a polarization-independent FWM process using two PMFs. The first PMF separated the two polarization components of the signal temporally, while the second PMF synchronized them back by introducing the opposite delay. This technique is similar to that used in Figure 10.20.

XPM in an SOA provides an alternative to FWM for demultiplexing OTDM signals [10]. The Sagnac-loop switch has been used for this purpose since 1993 when it was realized that a short switching window of 10 ps or less can be obtained, even when the carrier lifetime of the SOA exceeds 100 ps, by shifting the SOA from the center of the loop [65]. Although the bit rate of the OTDM signal was only 50 Gb/s in the initial 1993 experiment, it was boosted to 250 Gb/s within a year [101]. The use of MZ interferometers provides a similar performance and also allows the possibility of monolithic integration. Demultiplexing of a 168-Gb/s OTDM signal (16 channels at 10.5 Gb/s) was realized in 2000 using such an integrated demultiplexer [73]. Figure 10.25 shows the device configuration schematically. In the hybrid-integration technique employed, the InP chip containing two SOAs was mounted on a silica-based planar lightwave circuit (PLC) that contained the silica waveguides forming the MZ interferometer. The entire device was only 2 cm long. The control pulses were 2.4 ps wide in this experiment and were delayed by 6 ps in the two MZ arms to open a 6-ps switching window.

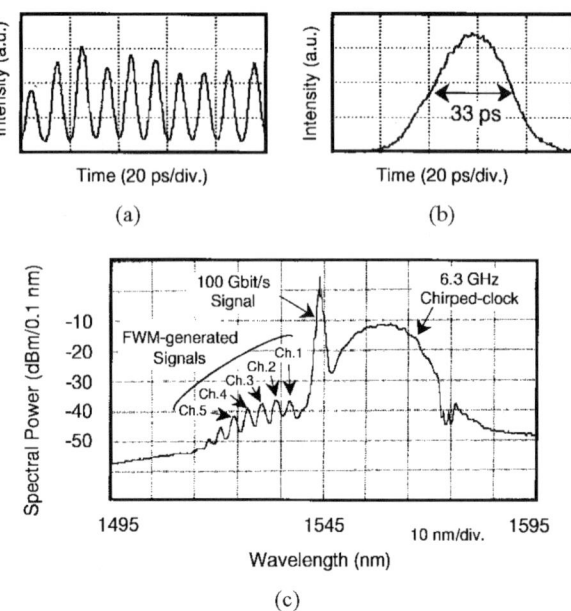

Figure 10.24: (a) OTDM signal, (b) clock pulse, and (c) optical spectrum showing FWM generated signals. (After Ref. [96]; ©1998 IEEE.)

The UNI switch with a single-arm MZ interferometer has also been used for demultiplexing applications [79]. Figure 10.21 shows the configuration of a 2001 experiment in which a 160-Gb/s OTDM signal (16 channels at 10 Gb/s) was demultiplexed using 1.3-ps control pulses [82]. In this device, a PMF splits each signal pulse into its orthogonally polarized components and introduces a relative delay between them. The switching window is thus set by the length of the PMF and can be adjusted during the experiment. Figure 10.26 shows how switching window changes as the PMF length is changed from 3–6 m. The window was 4.57 ps wide (FWHM) when the PMF fiber was 3 m long but changed to 8.71 ps for a 6-m-long fiber. This experiment also utilized a gain-transparent SOA by choosing the signal wavelength outside the gain spectrum. In fact, the SOA provided gain in the 1.3-μm region, while the OTDM signal was in the 1.55-μm region. Control pulses at 1.3 μm saturated the amplifier and changed the refractive index by a small amount for the signal pulses. In this mode of operation, the SOA does not act as an amplifier for the signal; its sole role is to provide XPM through cross-gain saturation.

All SOA-based demultiplexers suffer from a basic problem related to their active nature. Even though the amplifier gain can be saturated by control pulses on a time scale of ∼1 ps, it does not recover for a duration of 100 ps or more depending on the exact value of the carrier lifetime. Thus, the bit rate of a single channel, which determines the repetition rate of the clock or control pulses, should be 10 Gb/s or less for such switches to work even though the OTDM signal may exist at 100 Gb/s or more. The use of optical fibers or passive semiconductor waveguides solves this

10.4 Applications

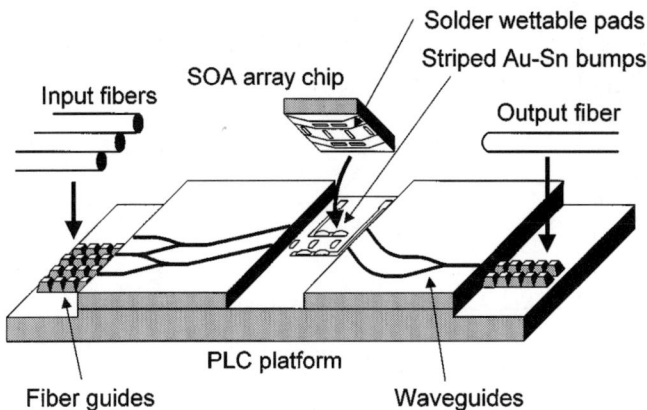

Figure 10.25: Configuration of a symmetric MZ demultiplexer built as a planar lightwave circuit. The InP chip containing two SOAs is mounted on top in this hybrid configuration. (After Ref. [73]; ©2000 IEEE.)

problem because such nonlinear media do not experience gain recovery.

XPM inside an optical fiber was used for demultiplexing in a 2001 experiment [102]. Control pulses are made so intense in this approach that they shift the spectrum of selected signal pulses through XPM by 1 or 2 nm when the two overlap inside a long optical fiber [11]. An optical filter is then used to select the demultiplexed channel at the shifted wavelength. The experiment used a 5-km-long fiber with its zero-dispersion wavelength at 1,543 nm. The 14-ps control pulses at a repetition rate of 10 GHz had a wavelength of 1,534 nm and were propagated with the 80-Gb/s OTDM signal at 1,538.5 nm. Figure 10.27 shows the optical spectra before and after the fiber together with the filtered spectrum. The group-velocity mismatch between the signal and control pulses plays a major role in such optical switches. In fact, the switching window

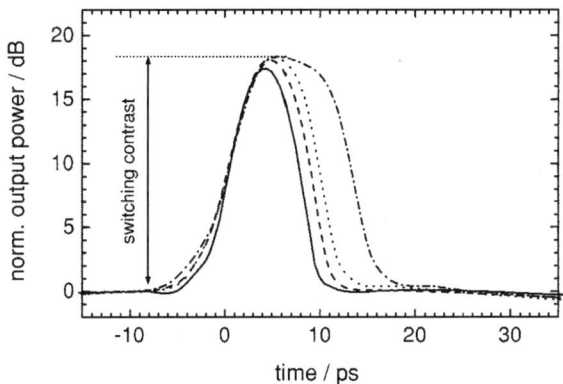

Figure 10.26: Switching window of a UNI switch for several PMF lengths. The length was 3, 3.5, 4, and 6 m from the innermost to the outermost curve. (After Ref. [82]; ©2001 IEEE.)

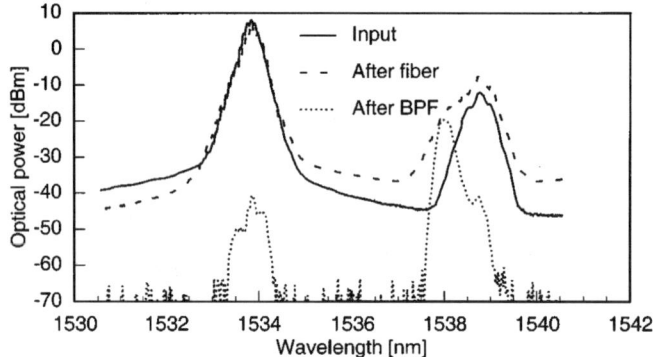

Figure 10.27: Optical spectra measured before (solid curve) and after (dashed curve) the fiber. The filtered spectrum is shown as a dotted curve. (After Ref. [102]; ©2001 IEEE.)

is determined by the initial offset between the two. Thus, only the group-velocity mismatch prevents such a demultiplexer to fully make use of the ultrafast nature of the fiber nonlinearity. This mismatch can also be reduced by locating the control and and signal pulses on the opposite sides of the zero-dispersion wavelength of the fiber. In addition, the use of highly nonlinear fibers not only can reduce the required average power of control pulses but also help with the problem of group-velocity mismatch as much shorter lengths are needed. An added benefit of this technique is that it can be used to demultiplex multiple channels simultaneously by simply employing multiple control pulses at different wavelengths [103].

The highest demultiplexing speed was realized in an experiment in which a passive InGaAsP waveguide was used as a nonlinear element [80]. This experiment employed a polarization-discriminating, single-arm, MZ interferometer, as shown in Figure 10.28 (similar to the UNI switch shown in Figure 10.20). The nonlinear waveguide was an InGaAsP bulk waveguide with the absorption edge near 1.7 μm. Control pulses were in the 1.55-μm region and thus were absorbed by the waveguide while copropagating signal pulses were below the bandgap and experienced little absorption. Carriers gen-

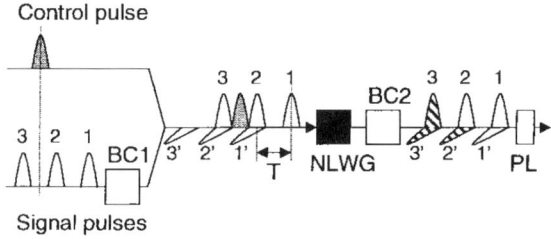

Figure 10.28: Schematic of an ultrafast MZ switch capable of responding on a time scale of 200 fs. The InGaAsP nonlinear waveguide (NLWG) provides such fast nonlinearity through the absorption of control pulses. (After Ref. [80]; ©1998 IEEE.)

10.4 Applications

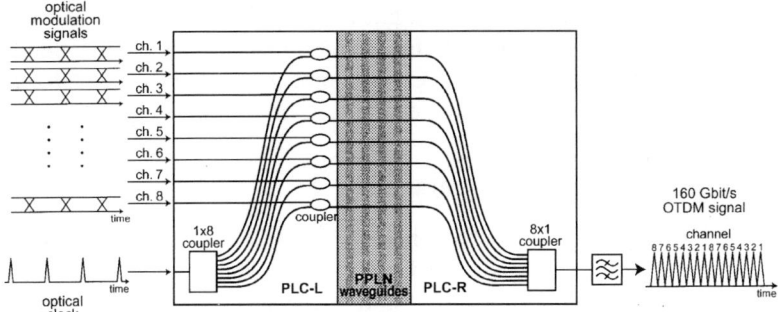

Figure 10.29: Simultaneous time-domain multiplexing of eight channels using a LiNbO$_3$ chip containing eight PPLN waveguides and sandwiched between two PLCs. (After Ref. [105]; ©2003 IEEE.)

erated through absorption of a control pulse change the refractive index and modify the signal phase. Even though the process appears similar to that used by an SOA, the device can respond at time scales shorter than 1 ps. In fact, 200-fs switching time was observed in this experiment, and the device was capable of demultiplexing a 1.5-THz OTDM signal. The reason that a passive waveguide can respond on such a fast time scale is related to the band-filling effects within the conduction band. Even though the nonlinearity itself is of the slowly relaxing type, it can respond relatively fast when the switch is turned on or off.

Another example of a passive waveguide is provided by LiNbO$_3$ switches that make use of the $\chi^{(2)}$ nonlinearity. As discussed earlier, two $\chi^{(2)}$ processes can be simultaneously quasi-phase-matched in a suitably designed PPLN waveguide such that they act as an effective FWM process and produce a new wave at the frequency $2\omega_1 - \omega_2$, where ω_1 and ω_2 are the carrier frequencies of the clock and signal pulses. Such a device has been used to demultiplex a 160-Gb/s OTDM bit stream [104]. It can also be integrated with one or more PLCs using a hybrid approach similar to that shown in Figure 10.25 and allows multiplexing or demultiplexing of multiple channels simultaneously [105].

Figure 10.29 shows the configuration of such a device used to multiplex eight 20-Gb/s channels simultaneously to form a 160-Gb/s OTDM signal. The middle LiNbO$_3$ chip contains eight PPLN waveguides and is sandwiched between two PLCs made using silica-on-silicon technology. Each PLC has a 1×8 coupler and eight waveguides of different lengths. The PLC on the input end also has eight integrated 2×1 couplers that allow the coupling of eight 20-Gb/s channels at the same frequency ω_2. The 4-cm-long PPLN waveguides employ an integrated mode filter with a taper for efficient coupling, similar to that used for the device shown in Figure 10.22. A single optical clock at the frequency ω_1 (with a repetition rate of 20 GHz) is split into eight optical clocks by the 1×8 coupler, each of which is combined with a 20-Gb/s NRZ-format signal into a PPLN waveguide. The cascaded nonlinear process creates an RZ-format signal at the frequency $2\omega_1 - \omega_2$ at the waveguide output. The second PLC then combines these RZ signals to form an OTDM signal at 160 Gb/s. The path-length difference between neighboring waveguides in this PLC corresponds to a delay of one bit-slot at 160 Gb/s so that multiplexing occurs automatically at the output. Such a device is useful because it performs NRZ-to-RZ conversion while multiplexing multiple channels in the time domain. Conversion between the RZ and NRZ formats can also be carried out using other time-domain switches [10].

10.4.2 Optical-Clock Recovery

Demultiplexing of an OTDM signal requires an optical clock in the form of a pulse train at the bit rate of channels that need to be demultiplexed. This clock should be generated from the OTDM signal itself, just as optical receivers recover an electrical clock from the incoming data. A self-clocking scheme is sometimes employed in which each clock pulse is explicitly transmitted with the signal using a bit slot within the OTDM bit stream. The alternative is to employ a mode-locked laser operating at the single-channel bit rate. However, pulses emitted from this laser still need to be synchronized with the OTDM data to ensure that signal and clock pulses overlap to an accuracy of 1 ps or so. Several techniques have been used for the purpose of synchronization [106]–[111]. Some among them employ a phase-locked loop and require time-domain switches not only in the clock recovery section but also within the demultiplexer.

The synchronization problem is much easier to solve in self-clocked OTDM networks because clock pulses are incorporated within the signal itself. However, one still needs to separate clock pulses from the data and direct them to a different port [10]. There are several possibilities. One can send clock pulses at a wavelength different than that of the OTDM signal and use an optical filter to recover the clock. However, this method suffers from the group-velocity mismatch; because of fiber dispersion, clock pulses travel at a slightly different speed and will shift from their original position, resulting in clock skew. As an alternative, one may use orthogonal polarization for clock pulses. Such a scheme can work only if PMFs are employed throughout the network and is not practical. The third possibility consists of keeping the same wavelength and polarization for clock pulses but making them much more intense than signal pulses. An intensity-based thresholding device can then separate the clock pulses, provided it responds on a time scale shorter than the width of pulses.

Figure 10.30 shows the basic idea (a) and the modified Sagnac loop (b) that was used for clock separation in a 1997 experiment [112]. The main difference between a Sagnac loop used for clock recovery and the one used for demultiplexing is that the directional coupler does not split the input signal into counterpropagating directions with equal powers. Rather, the coupler is designed to split the pulse with $\alpha \approx 0.2$ so that the clockwise-propagating pulse closer to the SOA is much weaker than that in the other direction. Consider what happens to an intense clock pulse as it passes through the loop. The weaker component passes first through the SOA, and its phase is not affected by the SOA because it cannot saturate the gain. The stronger component enters the SOA later and its phase changes through self-phase modulation (SPM) as it depletes the carrier density and saturates the SOA. If the relative phase shift between the two components of the clock pulse is arranged to be close to π, this pulse will be transmitted by the Sagnac loop. In contrast, much weaker data pulses do not experience a large relative phase shift and are thus reflected by the Sagnac loop. By properly choosing the splitting ratio α of the coupler, different gains for two counterpropagating components

10.4 Applications

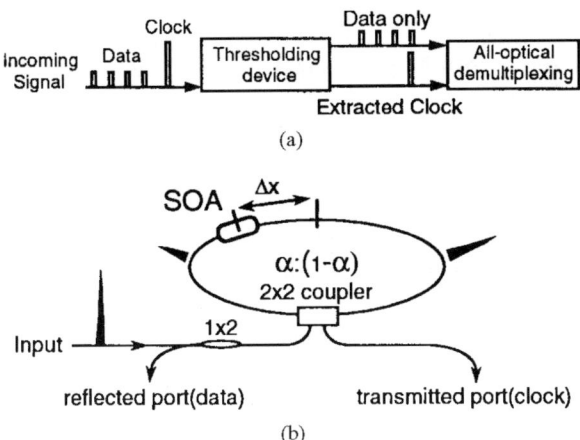

Figure 10.30: (a) Schematic illustration of the self-clocking scheme and (b) the Sagnac-loop switch used to separate the clock from the signal using an intensity-based thresholding scheme. (After Ref. [112]; ©1997 IEEE.)

of each clock pulse can balance the intensity difference caused by the uneven splitting at the coupler.

To avoid switching data pulses that immediately follow the clock pulse, they must arrive at the SOA after the switching window of duration T_w opened by the clock pulse has closed. Thus, for the proper operation of the device, it is important to position the SOA in the loop such that the offset Δx from the center of the loop satisfies the condition $T_s > T_w = 2\Delta x/v_g$, where v_g is the group velocity of the clock pulse and T_s is the separation between the clock pulse and the signal pulse that follows it. This condition can be satisfied by keeping one or more bit slots after the control pulse unoccupied depending on the bit rate of the ODTM signal.

10.4.3 Packet-Switched Networks

Packet-switched networks, such as the Internet, route information in the form of packets consisting of a few tens to hundreds of bits. Each packet begins with a header that contains all the destination information. When a packet arrives at a node, a router reads the header and sends it toward its destination. At bit rates of up to 10 Gb/s or so, electronic routers are fast enough that the switching can be performed in the electrical domain. However, future packet-switched networks are likely to operate at a bit rate of 100 Gb/s or more. Optical packet-switching techniques are almost a prerequisite for such systems and have been studied since the mid-1990s [113]–[122].

Figure 10.31 shows the layout of an all-optical cross-connect for packet-switched networks [116]. Similar to the case of a wavelength-selective cross-connect discussed in Section 9.4.1, each incoming WDM signal arriving at the cross-connect is first demultiplexed, and all channels at a fixed wavelength are grouped together. The packet-switching unit at each wavelength consists of three devices, each performing a separate

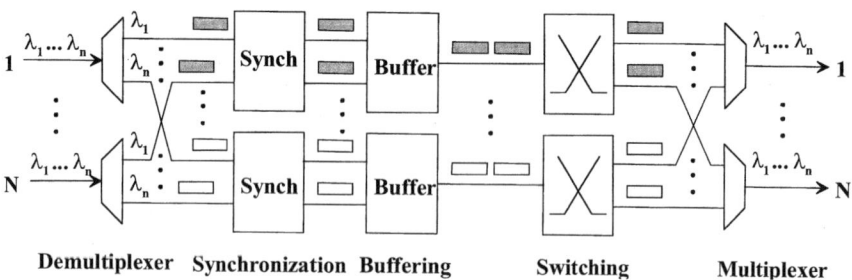

Figure 10.31: Generic node structure for all-optical packet-switched cross-connects. (After Ref. [116]; ©2003 IEEE.)

step. The first device synchronizes packets, the second buffers them to avoid packet collisions, and the third performs the switching operation. All three steps can be performed optically using time-domain switches. The switched packets are then reassembled in individual channels that are multiplexed to form the outgoing WDM signal.

In one implementation of such a cross-connect [116], the basic element is an all-optical packet switch with the layout shown in Figure 10.32. Optical power of the input packet is split into two branches using a directional coupler. One branch processes the header, while the other delivers the payload and simply contains a fiber-delay line to compensate for the latency in the header branch. Between the header and the payload, a few 0 bits are inserted that serve as the guard time. The entire switch is composed of three units. The header-processing unit is a time-domain switch in the form of a Sagnac loop. The flip-flop memory unit is implemented using two coupled lasers that switch the output between two wavelengths, say, λ_1 and λ_2. The third unit is simply a wavelength converter; it copies the initial data packet to the output wavelength from the flip-flop. Thus, the switch produces output at different wavelengths depending on the header information.

Figure 10.33 shows another configuration of an 1×2 all-optical packet switch that uses two TOAD devices in the form of Sagnac-loop switches [10]. In this case, clock bits are sent with the data using orthogonal polarization and can be separated from the

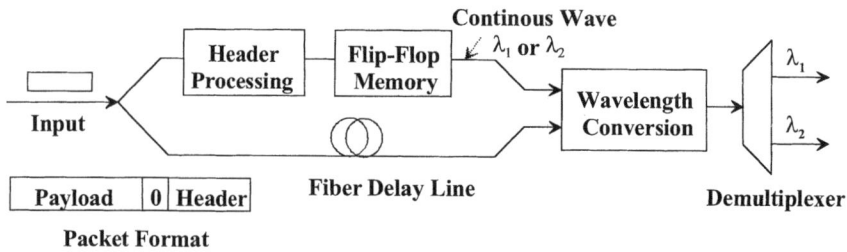

Figure 10.32: Design of an 1×2 all-optical packet switch producing its output at two different wavelengths depending on the header address. (After Ref. [116]; ©2003 IEEE.)

10.4 Applications

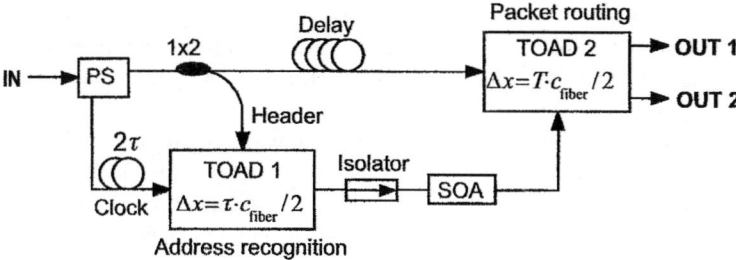

Figure 10.33: Design of an 1 × 2 packet switch based on two TOAD devices. It switches packets to different output ports depending on a single header bit. (After Ref. [10]; ©2003 Elsevier.)

data with the help of a polarization splitter (PS). This clock is used in the first TOAD with a narrow switching window to demultiplex a single header bit. This bit acts as a control for the second TOAD whose switching window remains open for the duration of the entire packet. When the header bit corresponds to 1, the packet is transmitted (output port 1) but if this bit is 0, the packet is reflected and appears at the output port 2 after passing through a circulator. The time delay τ in the first TOAD corresponds to the duration of a single bit within the packet but equals $T = N\tau$ for the second TOAD, where N is the number of bits within the packet.

A different approach to packet-switched networks makes use of the label-swapping technique. Such networks are referred to as all-optical label-swapping networks [115]. The basic idea behind them is to attach an optical label to each packet before it enters the network. The optical label contains the destination information. An optical multiplexing layer multiplexes "labeled packets" onto an available WDM channel. Routing of packets within the WDM network is performed at routers, which swap incoming labels with different ones using an internal routing table that assigns a different wavelength to each packet depending on its destination. With this approach, the header and the payload of each packet are never examined optically. Rather, wavelength swapping of optical labels at the routers is used to send each packet to its destination.

Several other schemes for packet-switched networks have been proposed. In one

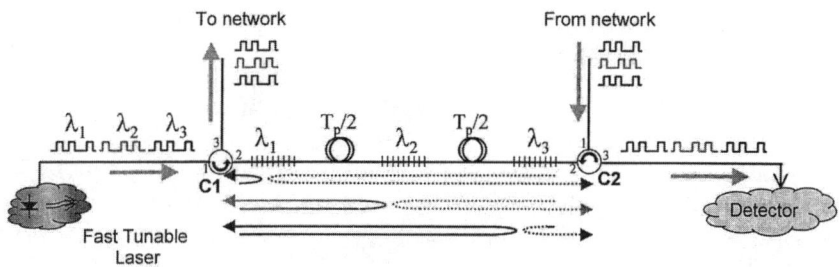

Figure 10.34: Layout of a device used to form composite packets consisting of multiple fixed-duration packets at different wavelengths. The same device can be used to demultiplex composite packets. (After Ref. [121]; ©2003 IEEE.)

scheme, optical switches based on arrayed waveguide gratings (AWGs) are employed [120]. The header is processed electronically, and this information is used to control the switching matrix. Synchronization and bufferings of incoming packet are handled in the optical domain using fiber-delay lines and 2×2 optical switches. Still in another approach, multiple optical packets at different wavelengths are combined serially to form a composite packet using fiber gratings and delay lines [121]. Figure 10.34 shows the basic idea schematically. In this scheme, individual packets of fixed length are generated serially at different wavelengths using a tunable laser. These packets are multiplexed and transmitted as a composite packet using a simple device consisting of an optical circulator followed by multiple fiber gratings spaced apart by fixed optical delay lines. AWGs can also be used in place of fiber gratings when the number of packets (and hence the wavelengths) is large. The same device can be used to demultiplex incoming packets, as seen in Figure 10.34. Packets can be added or dropped at each node using 2×2 switches. Packet-switched networks are still evolving and constitute a subject of intense research in 2003.

Problems

10.1 Consider a high-finesse Fabry–Perot resonator with a nonlinear medium whose refractive index varies with intensity as $n = n_0 + n_2 I$. Prove that the transmitted power is related to the input power as in Eq. (10.1.2).

10.2 Reproduce the bistability curves shown in Figure 10.1, using Eq. (10.1.2). Why do you need some minimum detuning from the Fabry–Perot resonance before bistability appears?

10.3 Explain how a Sagnac loop can be used for time-domain switching. Use diagrams as necessary.

10.4 Prove that the transmissivity of a Sagnac loop made using a 3-dB coupler is given by $T = \sin^2(\Delta\phi/2)$, where $\Delta\phi$ is the relative phase shift between the counterpropagating waves.

10.5 Explain how four-wave mixing inside a fiber can be used for time-domain switching. Use diagrams as necessary.

10.6 Solve Eqs. (10.1.7) and (10.1.8) and prove that the idler power generated through four-wave mixing is given by Eq. (10.1.10).

10.7 What is a flip-flop? How can you use a DFB amplifier as a flip-flop? Explain both the set and reset mechanisms for such a device.

10.8 Explain the physical mechanism that makes the switching window of a TOAD device to be much smaller than the carrier lifetime of the SOA inside the Sagnac loop.

10.9 Explain the operating principle behind a colliding-pulse Mach–Zehnder switch containing an SOA in each arm of the interferometer.

10.10 Explain the operating principle behind a symmetric Mach–Zehnder switch containing an SOA in each arm of the interferometer.

10.11 Discuss how an ultrafast nonlinear interferometer (UNI) works as a time-domain switch. How would you operate a UNI in the gain-transparency mode?

10.12 How can one employ periodically poled lithium niobate (PPLN) for time-domain switching? Explain the concept of cascaded nonlinearity in this context.

References

[1] J. E. Midwinter, Ed., *Photonics in Switching*, Academic Press, San Diego, CA, 1993.

[2] A. Marrakchi, Ed., *Selected Papers on Photonic Switching*, SPIE Press, Bellingham, WA, 1996.

[3] E. J. Murphy, in *Optical Fiber Telecommunications*, Vol. 3B, I. P. Kaminow and T. L. Koch, Eds., Academic Press, San Diego, CA, 1997, Chap. 10.

[4] M. Saruwatari, *IEEE J. Sel. Topics Quantum Electron.* **6**, 1363 (2000).

[5] L. H. Spiekman, in *Optical Fiber Telecommunications*, Vol. 4A, I. P. Kaminow and T. P. Lee, Eds., Academic Press, San Diego, CA, 2002, Chap. 14.

[6] P. Leclerc, B. Lavingne, and D. Chiaroni, in *Optical Fiber Telecommunications*, Vol. 4A, I. P. Kaminow and T. P. Lee, Eds., Academic Press, San Diego, CA, 2002, Chap. 15.

[7] G. P. Agrawal and D. N. Maywar, in *Nonlinear Photonic Crystals*, Springer Series in Photonics, Vol. 10, R. E. Slusher and B. H. Eggleton, Eds., Springer, New York, 2003, Chap. 13.

[8] G. I. Papadimitriou, C. Papazoglou, and A. S. Pomportsis, *J. Lightwave Technol.* **21**, 384 (2003).

[9] K. Vlachos, N. Pleros, C. Bintjas, G. Theophilopoulos, and H. Avramopoulos, *J. Lightwave Technol.* **21**, 1857 (2003).

[10] I. Glesk, B. C. Wang, L. Xu, V. Baby, and P. R. Prucnal, in *Progress in Optics*, Vol. 45, E. Wolf, Ed., Noth-Holland, Amsterdam, 2003, Chap. 2.

[11] G. P. Agrawal, *Nonlinear Fiber Optics*, 3rd ed., Academic Press, San Diego, CA, 2001.

[12] H. M. Gibbs, *Optical Bistability: Controlling Light with Light*, Academic Press, San Diego, CA, 1984.

[13] H. Nakatsuka, S. Asaka, H. Itoh, K. Ikeda, and M. Matsuoka, *Phys. Rev. Lett.* **50**, 109 (1983).

[14] P. W. Smith, I. P. Kaminow, P. J. Maloney, and L. W. Stulz, *Appl. Phys. Lett.* **33**, 24 (1978); *Appl. Phys. Lett.* **34**, 62 (1979).

[15] P. L. K. Wa, J. E. Sitch, N. J. Mason, and P. N. Robson, *Electron. Lett.* **21**, 26 (1985).

[16] R. M. Shelby, M. D. Levenson, and S. H. Perlmutter, *J. Opt. Soc. Am. B* **5**, 347 (1988).

[17] S. Coen, M. Haelterman, P. Emplit, L. Delage, L. M. Simohamed, and F. Reynaud, *J. Opt. Soc. Am. B* **15**, 2283 (1998).

[18] H. Kawaguchi, *IEEE J. Sel. Topics Quantum Electron.* **3**, 1254 (1997).

[19] C. Harder, K. Y. Lau, and A. Yariv, *IEEE J. Quantum Electron.* **18**, 1351 (1982).

[20] M. Ueno and R. Lang, *J. Appl. Phys.* **58**, 1689 (1985).

[21] H. F. Liu, Y. Hashimoto, and T. Kamiya, *IEEE J. Quantum Electron.* **24**, 43 (1988).

[22] G. H. Duan, P. Landais, and J. Jacquet, *IEEE J. Quantum Electron.* **30**, 2507 (1994).

[23] W. F. Sharfin and M. Dagenais, *Appl. Phys. Lett.* **48**, 321 (1986); *Appl. Phys. Lett.* **48**, 1510 (1986).

[24] M. J. Adams, H. J. Westlake, and M. J. O'Mahony, in *Optical Nonlinearities and Instabilities in Semiconductors*, H. Haug, Ed., Academic Press, San Diego, CA, 1988, Chap. 15.

[25] H. G. Winful, J. H. Marburger, and E. Garmire, *Appl. Phys. Lett.* **35**, 379 (1979).
[26] M. J. Adams and R. Wyatt, *Proc. Inst. Elect. Eng.*, **134**, 35, 1987.
[27] D. N. Maywar and G. P. Agrawal, *IEEE J. Quantum Electron.* **33**, 2029 (1997); *IEEE J. Quantum Electron.* **34**, 2364 (1998).
[28] D. N. Maywar and G. P. Agrawal, *Opt. Exp.* **3**, 440 (1998).
[29] D. N. Maywar and G. P. Agrawal, and Y. Nakano, *Opt. Exp.* **6**, 75 (2000).
[30] B. K. Nayar, N. Finlayson, N. J. Doran, S. T. Davey, W. L. Williams, and J. W. Arkwright, *Opt. Lett.* **16**, 408 (1991).
[31] K. Al-Hemyari, J. S. Aitchison, C. N. Ironside, G. T. Kennedy, R. S. Grant, and W. Sibbett, *Electron. Lett.* **28**, 1090 (1992).
[32] J. S. Aitchison, A. H. Kean, C. N. Ironside, A. Villeneuve, and G. I. Stegeman, *Electron. Lett.* **27**, 1709 (1991).
[33] M. C. Farries and D. N. Payne, *Appl. Phys. Lett.* **55**, 25 (1989).
[34] K. J. Blow, N. J. Doran, B. K. Nayar, and B. P. Nelson, *Opt. Lett.* **15**, 248 (1990).
[35] M. Jinno and T. Matsumoto, *IEEE Photon. Technol. Lett.* **2**, 349 (1990); *Electron. Lett.* **27**, 75 (1991).
[36] H. Avramopoulos, P. M. W. French, M. C. Gabriel, H. H. Houh, N. A. Whitaker, and T. Morse, *IEEE Photon. Technol. Lett.* **3**, 235 (1991).
[37] J. D. Moores, K. Bergman, H. A. Haus, and E. P. Ippen, *Opt. Lett.* **16**, 138 (1991); *J. Opt. Soc. Am. B* **8**, 594 (1991).
[38] M. Jinno, *J. Lightwave Technol.* **10**, 1167 (1992); *Opt. Lett.* **18**, 726 (1993); *Opt. Lett.* **18**, 1409 (1993).
[39] H. Bülow and G. Veith, *Electron. Lett.* **29**, 588 (1993).
[40] P. A. Andrekson, N. A. Olsson, J. R. Simpson, T. Tanbun-Ek, R. A. Logan, and M. Haner, *Electron. Lett.* **27**, 695 (1991).
[41] T. Morioka, H. Takara, S. Kawanishi, T. Kitoh, and M. Saruwatari, *Electron. Lett.* **32**, 832 (1996).
[42] J. Hansryd and P. A. Andrekson, *IEEE Photon. Technol. Lett.* **13**, 732 (2001).
[43] G. P. Agrawal, *Appl. Phys. Lett.* **51**, 302 (1987); *J. Opt. Soc. Am. B* **5**, 149 (1988).
[44] M. Kondo, Y. Ohta, M. Fujiwara, and M. Sakaguchi, *IEEE J. Quantum Electron.* **18**, 1759 (1982).
[45] J. M. Liu and Y. C. Chen, *Electron. Lett.* **21**, 236 (1985).
[46] K. Okumura, Y. Ogawa, H. Ito, and H. Inaba, *IEEE J. Quantum Electron.* **21**, 377 (1985).
[47] S. Suzuki, T. Terakado, K. Komatsu, K. Nagashima, A. Suzuki, and M. Kondo, *J. Lightwave Technol.* **4**, 894 (1986).
[48] K. Otsuka and K. Ikeda, *Opt. Lett.* **12**, 599 (1987).
[49] K. Inoue, *Opt. Lett.* **12**, 918 (1987).
[50] K. Otsuka, *Electron. Lett.* **24**, 800 (1988).
[51] L. L. Chern and J. K. McIver, *Opt. Lett.* **15**, 186 (1990).
[52] H. Kawaguchi, *Electron. Lett.* **31**, 1150 (1995).
[53] J. Zhou, M. Cada, G. P. Li, and T. Makino, *IEEE Photon. Technol. Lett.* **7**, 1125 (1995).
[54] B. B. Jian, *Electron. Lett.* **32**, 349 (1996).
[55] M. S. Petrovic, M. R. Belic, M. V. Jaric, and F. Kaiser, *Opt. Commun.* **138**, 349 (1997).
[56] T. Chattopadhyay and M. Nakajima, *Opt. Commun.* **138**, 320 (1997).
[57] K. Nakatsuhara, T. Mizumoto, R. Munakata, Y. Kigure, and Y. Naito, *IEEE Photon. Technol. Lett.* **10**, 78 (1998).
[58] F. Robert, D. Fortusini, and C. L. Tang, *IEEE Photon. Technol. Lett.* **12**, 465 (2000).

[59] M. T. Hill, H. de Waardt, G. D. Khoe, and H. J. S. Dorren, *IEEE J. Quantum Electron.* **37**, 405 (2001).

[60] S.-H. Jeong, H.-C. Kim, T. Mizumoto, J. Wiedmann, S. Arai, M. Takenaka, and Y. Nakano, *IEEE J. Quantum Electron.* **38**, 706 (2002).

[61] V. Van, T. A. Ibrahim, P. P. Absil, F. G. Johnson, R. Grover, and P.-T. Ho, *IEEE J. Sel. Topics Quantum Electron.* **8**, 705 (2002).

[62] H. J. S. Dorren, D. Lenstra, Y. Liu, M. T. Hill, and G. D. Khoe, *IEEE J. Quantum Electron.* **39**, 141 (2003).

[63] A. W. O'Neil and R. P. Webb, *Electron. Lett.* **26**, 2008 (1990).

[64] M. Eiselt, *Electron. Lett.* **28**, 1505 (1992).

[65] J. P. Sokoloff, P. R. Prucnal, I. Glesk, and M. Kane, *IEEE Photon. Technol. Lett.* **5**, 787 (1993).

[66] M. Eiselt, W. Pieper, and H. G. Weber, *J. Lightwave Technol.* **13**, 2099 (1995).

[67] K. I. Kang, T. G. Chang, I. Glesk, and P. R. Prucnal, *Appl. Opt.* **35**, 417 (1996); *Appl. Opt.* **35**, 1485 (1996).

[68] K. I. Kang, T. G. Chang, I. Glesk, P. R. Prucnal, and R. K. Boncek, *Appl. Phys. Lett.* **67** 605 (1995).

[69] E. Jahn, N. Agrawal, M. Arbert, H.-J. Ehrke, D. Franke, R. Ludwig, W. Pieper, H. G. Weber, and C. M. Weinert, *Electron. Lett.* **31** 1857 (1995).

[70] R. Hess, M. Caraccia-Gross, W. Vogt, E. Gamper, P. A. Besse, M. Duelk, E. Gini, H. Melchior, B. Mikkelsen, M. Vaa, K. S. Jepsen, K. E. Stubkjaer, and S. Bouchoule, *IEEE Photon. Technol. Lett.* **10**, 165 (1998).

[71] J. Leuthold, P.-A. Besse, E. Gamper, M. Dulk, S. Fischer, G. Guekos, and H. Melchior, *J. Lightwave Technol.* **17**, 1056 (1999).

[72] D. Wolfson, A. Kloch, T. Fjelde, C. Janz, B. Dagens, and M. Renaud, *IEEE Photon. Technol. Lett.* **12**, 332 (2000).

[73] S. Nakamura, Y. Ueno, K. Tajima, J. Sasaki, T. Sugimoto, T. Kato, T. Shimoda, M. Itoh, H. Hatakeyama, T. Tamanuki, and T. Sasaki, *IEEE Photon. Technol. Lett.* **12**, 425 (2000).

[74] P. V. Studenkov, M. R. Gokhale, J. Wei, W. Lin, I. Glesk, P. R. Prucnal, and S. R. Forrest, *IEEE Photon. Technol. Lett.* **13**, 600 (2001).

[75] C. Schubert, J. Berger, S. Diez, H. J. Ehrke, R. Ludwig, U. Feiste, C. Schmidt, H. G. Weber, G. Toptchiyski, S. Randel, and K. Petermann, *J. Lightwave Technol.* **20**, 618 (2002).

[76] R. P. Schreieck, M. H. Kwakernaak, H. Jäckel, and H. Melchior, *IEEE J. Quantum Electron.* **38**, 1053 (2002).

[77] K. Tajima, S. Nakamura, and Y. Sugimoto, *Appl. Phys. Lett.* **67**, 3709 (1995).

[78] N. S. Patel, K. L. Hall, and K. A. Rauschenbach, *Opt. Lett.* **21**, 1466 (1996).

[79] K. L. Hall and K. A. Rauschenbach, *Opt. Lett.* **23**, 1271 (1998).

[80] S. Nakamura, Y. Ueno, and K. Tajima, *IEEE Photon. Technol. Lett.* **10**, 1575 (1998).

[81] N. S. Patel, K. L. Hall, and K. A. Rauschenbach, *Appl. Opt.* **37**, 2831 (1998).

[82] C. Schubert, S. Diez, J. Berger, R. Ludwig, U. Feiste, H. G. Weber, G. Toptchiyski, K. Petermann, and V. Krajinovic, *IEEE Photon. Technol. Lett.* **13**, 475 (2001).

[83] B. S. Robinson, S. A. Hamilton, and E. P. Ippen, *IEEE Photon. Technol. Lett.* **14**, 206 (2002).

[84] C. Bintjas, K. Vlachos, N. Pleros, and H. Avramopoulos, *J. Lightwave Technol.* **21**, 2629 (2003).

[85] S. Diez, R. Ludwig, and H. G. Weber, *IEEE Photon. Technol. Lett.* **11**, 60 (1999).

[86] K. R. Parameswaran, M. Fujimura, M. H. Chou, and M. M. Fejer, *IEEE Photon. Technol. Lett.* **12**, 654 (2000).

[87] G. S. Kanter, P. Kumar, K. R. Parameswaran, and M. M. Fejer, *IEEE Photon. Technol. Lett.* **13**, 341 (2001).

[88] T. Suhara and H. Ishizuki, *IEEE Photon. Technol. Lett.* **13**, 1203 (2001).

[89] C. N. Ironside, J. S. Aitchison, and J. M. Arnold, *IEEE J. Quantum Electron.* **29**, 2650 (1993).

[90] H. Kanbara, H. Itoh, M. Asobe, K. Noguchi, H. Miyazawa, T. Yanagawa, and I. Yokohama, *IEEE Photon. Technol. Lett.* **11**, 328 (1999).

[91] Y. Fukuchi, T. Sakamoto, K. Taira, K. Kikuchi, D. Kunimatsu, A. Suzuki, and T. Ito, *IEEE Photon. Technol. Lett.* **14**, 1267 (2002).

[92] Y. Fukuchi and K. Kikuchi, *IEEE Photon. Technol. Lett.* **14**, 1409 (2002).

[93] K. Nonaka, F. Kobayashi, K. Kishi, T. Todokoro, Y. Itoh, C. Amano, and T. Kurokawa, *IEEE Photon. Technol. Lett.* **10**, 1484 (1998).

[94] P. O. Hedekvist, M. Karlsson, and P. A. Andrekson, *J. Lightwave Technol.* **15**, 2051 (1997).

[95] T. Morioka, H. Takara, S. Kawanishi, K. Uchiyama, and M. Saruwatari, *Electron. Lett.* **32**, 840 (1996).

[96] K. Uchiyama, S. Kawanishi, and M. Saruwatari, *IEEE Photon. Technol. Lett.* **10**, 890 (1998).

[97] I. Tomkos, I. Zacharopoulos, D. Syvridis, R. Calvani, F. Cisternino, and E. Riccardi, *IEEE Photon. Technol. Lett.* **11**, 1464 (1999).

[98] N. K. Das, Y. Yamayoshi, T. Kawazoe, and H. Kawaguchi, *J. Lightwave Technol.* **19**, 237 (2001).

[99] I. Shake, H. Takara, K. Uchiyama, I. Ogawa, T. Kitoh, T. Kitagawa, M. Okamoto, K. Magari, Y. Suzuki, and T. Morioka, *Electron. Lett.* **38**, 27 (2002).

[100] S. L. Jansen, H. Heid, S. Spalter, E. Meissner, C. J. Weiske, A. Schopflin, D. Khoe, and H. de Waardt, *Electron. Lett.* **38**, 978 (2002).

[101] I. Glesk, J. P. Sokoloff, and P. R. Prucnal, *Electron. Lett.* **30**, 339 (1994).

[102] B.-E Olsson and D. J. Blumenthal, *IEEE Photon. Technol. Lett.* **13**, 875 (2001).

[103] L. Rau, W. Wang, B.-E. Olsson, Y. Chiu, H.-F. Chou, D. J. Blumenthal, and J. E. Bowers, *IEEE Photon. Technol. Lett.* **14**, 1725 (2002).

[104] Y. Fukuchi, T. Sakamoto, K. Taira, and K. Kikuchi, *Electron. Lett.* **39**, 789 (2003).

[105] T. Ohara, H. Takara, I. Shake, K. Mori, S. Kawanishi, S. Mino, T. Yamada, M. Ishii, T. Kitoh, T. Kitagawa, K. R. Parameswaran, and M. M. Fejer, *IEEE Photon. Technol. Lett.* **15**, 302 (2003).

[106] O. Kamatani and S. Kawanishi, *IEEE Photon. Technol. Lett.* **8**, 1094 (1996).

[107] D. T. K. Tong, K. L. Deng, B. Mikkelsen, G. Raybon, K. F. Dreyer, and J. E. Johnson, *Electron. Lett.* **36** 1951 (2000).

[108] T. Yamamoto, L. K. Oxenlowe, C. Schmidt, C. Schubert, U. Feiste, J. Berger, R. Ludwig, and H. G. Weber, *Electron. Lett.* **37**, 509 (2001).

[109] T. Miyazaki, H. Sotobayashi, and W. Chujo, *IEEE Photon. Technol. Lett.* **14**, 1734 (2002).

[110] E. S. Awad, P. S. Cho, N. Moulton, and J. Goldhar, *IEEE Photon. Technol. Lett.* **15**, 126 (2003).

[111] T. Miyazaki and F. Kubota, *IEEE Photon. Technol. Lett.* **15**, 1008 (2003).

[112] K.-L. Deng, I. Glesk, K. I. Kang, and P. R. Prucnal, *IEEE Photon. Technol. Lett.* **9**, 830 (1997).

References

[113] A. Carena, M. D. Vaughn, R. Gaudino, M. Shell, and D. J. Blumenthal, *J. Lightwave Technol.* **16**, 2135 (1998).

[114] P. Toliver, I. Glesk, R. J. Runser, K.-Li Deng, Ben Y. Yu, and P. R. Prucnal, *J. Lightwave Technol.* **16**, 2169 (1998).

[115] D. J. Blumenthal, B.-E. Olsson, G. Rossi, T. E. Dimmick, L. Rau, M. Masanovic, D. Lavrova, R. Doshi, O. Jerphagnon, J. E. Bowers, V Kaman, L. A. Coldren, and J. Barton, *J. Lightwave Technol.* **18**, 2058 (2000).

[116] H. J. S. Dorren, M. T. Hill, Y. Liu, N. Calabretta, A. Srivatsa, F. M. Huijskens, H. de Waardt, and G. D. Khoe, *J. Lightwave Technol.* **21**, 2 (2003).

[117] N. Calabretta, Y. Liu, M. T. Hill, H. de Waardt, G. D. Khoe, and H. J. S. Dorren, *IEE Proc.* **150**, 219 (2003).

[118] S. J. B. Yoo, F. Xue, Y. Bansal, J. Taylor, Z. Pan, J. Cao, M. Jeon, T. Nady, G. Goncher, K. Boyer, K. Okamoto, S. Kamei, and V. Akella, *IEEE J. Sel. Areas Commun.* **21**, 1041 (2003).

[119] H. J. Chao, K.-L. Deng, and Z. Jing, *IEEE J. Sel. Areas Commun.* **21**, 1096 (2003).

[120] S. Bregni, A. Pattavina, and G. Vegetti, *IEEE J. Sel. Areas Commun.* **21**, 1113 (2003).

[121] M. Boroditsky, N. J. Frigo, C. F. Lam, K. F. Dreyer, D. A. Ackerman, J. E. Johnson, L. J. P. Ketelsen, A. Chen, and A. Smiljanic, *J. Lightwave Technol.* **21**, 1717 (2003).

[122] M. Y. Jeon, Z. Pan, J. Cao, Y. Bansal, J. Taylor, Z. Wang. V. Akella, K. Okamoto, S. Kamei, J. Pan, and S. J. B. Yoo *J. Lightwave Technol.* **21**, 2723 (2003).

Appendix A

System of Units

The international system of units (known as the SI, short for *Système International*) is used in this book. The three fundamental units in the SI are meter (m), second (s), and kilogram (kg). A prefix can be added to each of them to change its magnitude by a multiple of 10. Mass units are rarely required in this book. On the other hand, measures of distance required in this text range from nanometers (10^{-9} m) to kilometers (10^3 m), depending on whether one is dealing with planar waveguides or silica fibers. Similarly, time measures range from picoseconds (10^{-12} s) to a few minutes. Other common units in this book are Watt (W) for optical power and W/m^2 for optical intensity. They can be related to the fundamental units through energy because optical power represents the rate of energy flow (1 W = 1 J/s). The energy can be expressed in several other ways using $E = h\nu = k_B T = mc^2$, where h is the Planck constant, k_B is the Boltzmann constant, and c is the speed of light. The frequency ν is expressed in hertz (1 Hz = 1 s^{-1}). Of course, because of the large frequencies associated with optical waves, most frequencies in this book are expressed in GHz or THz.

In designing lightwave systems, optical powers can vary over several orders of magnitude as the signal travels from the transmitter to the receiver. Such large variations are handled most conveniently using decibel units, abbreviated dB, commonly used by engineers in many different fields. Any ratio R can be converted into decibels by using the general definition

$$R \text{ (in dB)} = 10 \log_{10} R. \tag{A.1}$$

The logarithmic nature of the decibel allows a large ratio to be expressed as a much smaller number. For example, 10^9 and 10^{-9} correspond to 90 dB and −90 dB, respectively. Since $R = 1$ corresponds to 0 dB, ratios smaller than 1 are negative in the decibel system. Furthermore, negative ratios cannot be written using decibel units.

The most common use of the decibel scale occurs for power ratios. For instance, the signal-to-noise ratio (SNR) of an optical or electrical signal is given by

$$\text{SNR} = 10 \log_{10}(P_S/P_N), \tag{A.2}$$

where P_S and P_N are the signal and noise powers, respectively. For example, loss of an optical fiber is expressed in decibel units in Section 1.3 because it corresponds to

Appendix A System of Units

a decrease in the optical power during transmission through the fiber and thus can be expressed as a power ratio. For example, if a 1-mW signal reduces to 1 μW after transmission over 100 km of fiber, the 30-dB loss over the entire fiber span translates into a loss of 0.3 dB/km. The same technique can be used to define the insertion loss of any passive optical component. For instance, a 1-dB loss of a fiber connector implies that the optical power is reduced by 1 dB (by about 20%) when the signal passes through the connector. The bandwidth of an optical filter is defined at the 3-dB point, corresponding to a 50% reduction in the signal power. The modulation bandwidth of semiconductor lasers in Section 5.3 is also defined at the 3-dB point.

Since losses of all passive components in a lightwave system are expressed in dB units, it is useful to express the transmitted and received powers also by using a decibel scale. This is achieved by using a derived unit, denoted as dBm and defined as

$$\text{power (in dBm)} = 10 \log_{10}\left(\frac{\text{power}}{1 \text{ mW}}\right), \tag{A.3}$$

where the reference level of 1 mW is chosen simply because typical values of the transmitted power are in that range (the letter m in dBm is a reminder of the 1-mW reference level). In this decibel scale for the absolute power, 1 mW corresponds to 0 dBm, whereas powers below 1 mW are expressed as negative numbers. For example, a 10-μW power corresponds to -20 dBm. Because of the logarithmic nature of the decibel scale, powers ranging over a wide range from 1 pW to 1 GW are expressed in dBm units over a much smaller range from -90 to 120 dBm.

Appendix B

Software Package

The back cover of the book contains a software package on a compact disk (CD) provided by RSoft, Inc. (Website: www.rsoftdesign.com). This state-of-the art software package should prove useful to readers of this book for solving problems provided at the end of each chapter. It also contains additional problems for each chapter that may help in understanding the difficult material. The CD contains an especially prepared version of several integrated computer-added design (CAD) modules that are marketed commercially by RSoft, under the trade names such as BeamPROP, Full-WAVE, BandSOLVE, GratingMOD, and LaserMOD, for modeling of various active and passive photonic components. The latest version of these modules are integrated using a photonics, computer-aided-design (CAD), layout tool. The CD also contains a few examples from two system-design software packages called OptSIM and LinSIM.

The CD should work on any personal computer (PC) and is intended for PCs running under Windows 2000 and Windows XP operating systems. The first step is to install the software package on the computer. The installation procedure should be straightforward for most users: Insert the CD in the CD-ROM drive and follow the instructions. If the installer does not start automatically for some reason, one may have to click on the "setup" program in the root directory of the CD. After the installation, the user simply has to click on the icon named "RSoft CAD-Layout" to start the program. Figure B.1 shows the CAD-layout window appearing when the program begins. The user should read the documentation files that are provided with the software to understand the purpose of various tools available in the two toolbars on the top and to the left of this window.

The examples that can be solved by the RSoft software are arranged in different folders using the same name and number as the book chapter so that the material can be easily identified. The folder marked "Optical Fibers" contains problems related to the material presented in Chapter 1 of the book. Several examples in this folder illustrate how the modes of different types of fibers can be calculated together with their dispersive properties. The user can display the mode profiles for step-index, graded-index, and plastic optical fibers. Even the modal properties of microstructure and photonic-crystal fibers can be calculated using the software package. Dispersive and nonlinear effects can be examined using several other examples contained in this folder. The next

Appendix B   Software Package   409

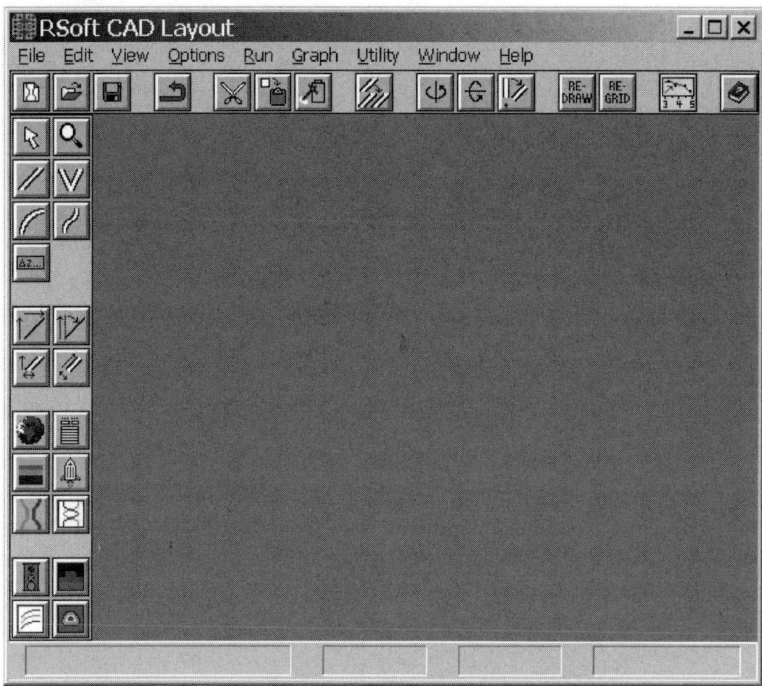

Figure B.1: RSoft CAD-layout window appearing when the software package on CD is started after installing it on a PC. It shows two toolbars on the top and to the left of the window in addition to the standard menu bar at the top and a status line at the bottom.

folder contains examples related to passive fiber components discussed in Chapter 2, including directional couplers and fiber gratings. The same pattern continues for other folders.

The CD includes a number of additional problems that go beyond the material covered in the book. These problems are grouped in a separate folder entitled Appendix. Most of them are related to two- and three-dimensional photonic crystals capable of exhibiting a photonic bandgap. Any radiation whose wavelength falls within this bandgap is totally reflected because of the existence of such a photonic bandgap. Such devices constitute a natural generalization of fiber gratings, which can be thought of as one-dimensional photonic crystals. The BandSolve and FullWave packages included with this CD can calculate the photonic band structure of these devices and should prove quite useful for understanding such new lightwave components.

Appendix C

Acronyms

Each scientific field has its own jargon, and the field of lightwave technology is not an exception. Although an attempt was made to avoid extensive use of acronyms, many still appear throughout the text. Each acronym is defined the first time it appears in a chapter so that the reader does not have to search the entire text to find its meaning. As a further help, we list all acronyms here, in alphabetical order.

AM	amplitude modulation
APD	avalanche photodiode
ASE	amplified spontaneous emission
ASK	amplitude-shift keying
AWG	arrayed-waveguide grating
BH	buried heterostructure
CPW	coplanar waveguide
CVD	chemical vapor deposition
CW	continuous wave
DBR	distributed Bragg reflector
DCF	dispersion-compensating fiber
DDF	dispersion-decreasing fiber
DFB	distributed feedback
EAM	electroabsorption modulator
EDFA	erbium-doped fiber amplifier
EDFL	erbium-doped fiber laser
FET	field-effect transistor
FM	frequency modulation
FP	Fabry–Perot
FSK	frequency-shift keying
FSR	free spectral range
FWHM	full-width at half-maximum
FWM	four-wave mixing
GRIN	gradient index
GVD	group-velocity dispersion

Appendix C   Acronyms

HBT	heterojunction-bipolar transistor
HEMT	high-electron-mobility transistor
IC	integrated circuit
ISI	intersymbol interference
ITU	International Telecommunication Union
LAN	local-area network
LEAF	large effective-area fiber
LED	light-emitting diode
LPE	liquid-phase epitaxy
MBE	molecular-beam epitaxy
MCVD	modified chemical vapor deposition
MEMS	micro-electro-mechanical system
MMI	multimode interference
MOCVD	metal-organic chemical vapor deposition
MONET	multiwavelength optical network
MPN	mode-partition noise
MQW	multiquantum well
MSM	metal–semiconductor–metal
MSR	mode-suppression ratio
MZ	Mach–Zehnder
NA	numerical aperture
NALM	nonlinear amplified-loop mirror
NLS	nonlinear Schrödinger
NOLM	nonlinear optical-loop mirror
NRZ	nonreturn to zero
OADM	optical add-drop multiplexer
OEIC	optoelectronic integrated circuit
OOK	on–off keying
OTDM	optical time-division multiplexing
OVD	outside-vapor deposition
OXC	optical cross-connect
PDG	polarization-dependent gain
PDM	polarization-division multiplexing
PLC	photonic lightwave circuit
PM	phase modulation
PMD	polarization-mode dispersion
PMF	polarization-maintaining fiber
PPLN	periodically poled lithium niobate
PSK	phase-shift keying
RF	radio frequency
RIN	relative intensity noise
RMS	root-mean-square
RZ	return to zero
SAGCM	separate absorption, grading, charge, and multiplication
SAGM	separate absorption, grading, and multiplication
SAM	separate absorption and multiplication

SBS	stimulated Brillouin scattering
SEM	scanning electron microscope
SI	Système International
SLM	single longitudinal mode
SNR	signal-to-noise ratio
SOA	semiconductor optical amplifier
SOI	semiconductor on insulator
SPM	self-phase modulation
SRS	stimulated Raman scattering
TDM	time-division multiplexing
TE	transverse electric
TM	transverse magnetic
TOAD	terahertz optical asymmetric demultiplexer
UNI	ultrafast nonlinear interferometer
UV	ultraviolet
VAD	vapor-axial deposition
VCSEL	vertical-cavity surface-emitting laser
VPE	vapor-phase epitaxy
WDM	wavelength-division multiplexing
WGR	waveguide-grating router
XPM	cross-phase modulation
YAG	yttrium aluminium garnet
YIG	yttrium iron garnet
ZDWL	zero-dispersion wavelength

Index

absorption
    excited-state, 97
    extrinsic, 24, 25
    free-carrier, 183, 213
    infrared, 26
    intrinsic, 24
    material, 24, 26
    multiphoton, 161
    two-photon, 58, 196, 371
    ultraviolet, 59
absorption coefficient, 226, 242, 254
absorption edge, 394
absorption spectrum, 58, 96, 225
acoustic velocity, 230, 301
acoustic wave, 40, 41, 230, 300
    surface, 301
adiabatic condition, 165
adiabatic taper, 388
Airy formula, 75
Airy function, 225
amplification factor, 95, 107, 121, 212, 216
amplifier
    bandwidth of, 96
    DFB, 370
    distributed, 110
    Fabry–Perot, 375
    hybrid, 109
    lumped, 113
    noise in, 100
    parametric, 118–128
    Raman, 44, 105–118
    semiconductor optical, 209–218, 374–378, 380–385
amplifier noise, *see* noise
amplitude mask, 298
amplitude-phase coupling, 196, 203, 234
angular momentum, 127
anisotropy, 156
annealing, 157, 161, 163
anticorrelation, 202

antireflection coating, 188, 209, 210, 245, 254, 272, 291
APD, 262–270
    bandwidth of, 265
    design of, 262, 269
    gain of, 263
    multiplication factor of, 264
    physical mechanism behind, 262
    reach-through, 266
    resonant-cavity, 269
    responsivity of, 264
    SAGCM, 267
    SAGM, 267
    SAM, 266
    staircase, 268
    superlattice, 268
    waveguide, 270
apodization technique, 68
arrayed waveguide grating, *see* grating
Auger recombination, 195
autocorrelation function, 200, 201
autocorrelation trace, 56
avalanche breakdown, 264
avalanche photodiode, *see* APD

band-filling effect, 395
bandgap, 152, 155, 179
    direct, 255
    indirect, 255, 259
bandwidth
    amplifier, 96, 102, 121
    APD, 266
    electrical, 113
    filter, 275, 289, 296, 300, 407
    front-end, 274
    gain, 96
    LED, 207, 209
    modulation, 198, 209
    modulator, 237, 242, 244, 245
    parametric amplifier, 121

photodetector, 256, 272
photodiode, 260, 261
Raman-amplifier, 105
Raman-gain, 44
RC, 236, 244, 270
semiconductor laser, 197
small-signal modulation, 197
bar port, 82, 293
barrier layer, 155, 245
barrier-enhancement layer, 271
beam splitter, 53
  polarization, 87
beam-pointing instability, 61
beat length, 20, 55, 300
Beer's law, 23
bending loss, 26
Bessel function, 16, 51, 54
biconical taper, 306
birefringence, 20, 32, 111, 175, 228, 316, 342, 386
  circular, 85
  degree of, 20, 33
  effects of, 36
  field-induced, 228
  modal, 20
  nonlinear, 134
  random, 21, 33
  strain-induced, 130
bistability, 367–370, 375–380
  dispersive, 367
  physical mechanism for, 369
  polarization, 376
bit rate–distance product, 4, 6
bit slot, 276, 277
Bloch formalism, 62
Bloch function, 172
Boltzmann constant, 207, 406
boundary condition, 16, 17, 52, 119, 143
Bragg condition, 57, 63, 174, 187, 230, 300, 301, 304, 310, 317
Bragg diffraction, 57, 187, 188
Bragg wavelength, 61, 69, 190, 296, 299, 300, 370, 375
Brillouin gain, 42
Brillouin scattering, 57
  spontaneous, 40
  stimulated, 40, 122, 353, 368
Brillouin shift, 41, 42
Brillouin zone, 172
bubble technology, 343

buffer layer, 151, 157, 235

carrier diffusion, 155
carrier heating, 196
carrier lifetime, 193, 195, 212, 371, 374, 381, 391
cavity
  all-fiber, 129
  design of, 128
  Fabry–Perot, 68, 128, 130, 133, 182, 245, 260, 269, 291
  figure-8, 129
  ring, 129, 130, 367
chain-transfer agent, 10
channel spacing, 215
characteristic temperature, 194
charge-bleed layer, 236
chemical etching, 188, 191, 240
chemical vapor deposition, 7, 161
  low-pressure, 160
  plasma-activated, 8
  plasma-enhanced, 160
chemical-vapor deposition
  metal-organic, 154
chirp, 234
  amplifier-induced, 217, 348
  modulator-induced, 248
chirp parameter, 234, 244, 245, 248
chromatic dispersion, *see* dispersion
chromium heater, 294, 306, 317, 338
chromophore, 165, 238, 240
circulator, 87, 296, 299, 322, 357, 400
  multiport, 87
  optical, 312
  polarization-independent, 87
  six-port, 313
  three-port, 87
clock pulse, 391, 396
clock recovery, 277
Clos architecture, 343
CMOS technology, 281
coherence
  spatial, 61
  temporal, 61
coherence length, 120
color center, 58
confinement factor, 22, 183, 193
connector loss, *see* loss
control pulse, 381–383, 388, 389, 391
core–cladding interface, 2, 4, 26

Index 415

correlation length, 33
coupled-cavity mechanism, 188
coupled-mode equations, 50, 62
    frequency-domain, 51, 63
    time-domain, 51, 63
coupled-mode theory, 50, 337
coupler
    3-dB, 54, 166, 293
    asymmetric, 50, 191
    directional, 49–56, 84, 166, 232, 306, 309, 335
    fiber, 49–56, 128, 294, 306, 311, 321, 371
    four-port, 166
    fused, 49
    grating-assisted, 167, 190, 310
    grating-frustrated, 310
    intermodal dispersion in, 55
    MMI, 163, 170, 318, 320, 348, 355, 384, 385
    mode-selective, 302
    power transfer in, 52, 53
    resonant, 84, 312
    star, 171, 307, 318, 321, 344
    supermodes of, 54, 55
    symmetric, 50, 52, 53, 310
    transfer matrix for, 54
    WDM, 129
    X, 167
coupling coefficient, 51, 54, 63, 69, 166, 296, 337
    effective, 52
    frequency dependence of, 56
    nonuniform, 71
coupling efficiency, 186, 205, 208, 261
coupling length, 53–56
critical angle, 2, 26, 143, 205
cross port, 54, 82, 293
cross-connect, 354–361
    optical-path, 360
    packet-switched, 397
    polarization-independent, 356
    wavelength-interchanging, 358
    wavelength-selective, 355
cross-gain saturation, 214, 215, 392
cross-phase modulation, 38, 51, 71, 80, 120, 127, 348, 351, 370, 380–389, 391, 393
crosstalk
    demultiplexer, 309
    electrical, 248
    FWM-induced, 40
    in-band, 114, 313
    interchannel, 40, 213, 214
    OADM, 312
    optical, 248
    out-of-band, 313
    Raman-induced, 44
Curie temperature, 157
cutoff condition, 18, 147
cutoff wavelength, 255

dark current, 256
decision circuit, 277
demultiplexer, 303–309
    all-fiber, 306
    AWG, 307, 316
    concave-grating, 320
    diffraction-based, 303
    fiber-based, 390
    filter-based, 305
    grating-based, 303
    interference-based, 303
    phased-array, 307
    SOA-based, 392
    terahertz optical asymmetric, 381
    waveguide-grating, 307
density of states, 154
depletion layer, 266
depletion width, 257, 259
depolarization, 111
detector, *see* photodetector
dielectric coating, 128
dielectric constant, 15
dielectric mirror, 128
dielectric tensor, 156
difference-frequency generation, 350, 389
diffusion coefficient, 155, 200
digital optical switch, *see* switch
dipole moment, 238
dipole relaxation time, 95
directional coupler, *see* coupler
dispersion, 27–34
    anomalous, 12, 102, 123, 133, 135
    chromatic, 15
    fourth-order, 122, 124
    grating-induced, 66
    group-velocity, 27–31, 55, 76, 184, 371
    intermodal, 56

material, 15, 28–29, 64, 183
modal, 3, 27, 33
multipath, 3, 4
normal, 12, 123, 133, 135
polarization-mode, 21, 32, 56, 80, 115, 128, 372
resonator-induced, 73
second-order, 35
third-order, 32, 35, 66, 124
waveguide, 28, 30, 65
dispersion compensation, 71, 74
dispersion length, 55
dispersion management, 80, 122
dispersion parameter, 28, 35
dispersion relation, 40, 64
dispersion slope, 32, 351, 354
distributed Bragg reflector, 131, 187, 261, 269, 292, 350
distributed feedback, *see* feedback
distributed feedback lasers, 186
fabrication of, 188
gain-coupled, 188, 319
linewidth saturation in, 203
multisection, 190
phase-shifted, 187
drift velocity, 257

EDFA, 321
amplification characteristics of, 97
C-band, 103
cascaded chain of, 102
gain spectrum of, 96
gain-clamped, 103
L-band, 103
multichannel amplification in, 102
noise figure of, 101
parallel configuration for, 104
pumping of, 94
S-band, 104
semiconductor lasers for, 94
spectral nonuniformity of, 102
two-stage, 103
effective core area, 22, 36, 42, 52, 106
effective-index method, 150
eigenvalue equation, 19, 145, 150
elasto-optic coefficient, 42
electro-optic coefficient, 157
electro-optic effect, 156, 169, 227, 233, 238, 335, 356, 388
electroabosrption, 225–226

electron–hole recombination, 180, 200
electron-beam lithography, 61, 188
electrorefraction, 227–229, 335
electrostriction, 40
energy-band diagram, 155
epitaxial growth, 154, 181, 188, 248
equalizer, 275
erbium ions
energy levels of, 94
fluorescence time of, 99
gain spectrum of, 96
erbium-doped fiber amplifier, *see* EDFA
error function, 372
evanescent wave, 50
excess noise factor, 269, 270
exciton, 226
extinction ratio, 233, 241, 244, 245, 338, 355
extrusion velocity, 10
eye diagram, 277, 280, 351

Fabry–Perot cavity, *see* cavity
Fabry–Perot interferometer, *see* interferometer
Fabry–Perot resonator, 74–76
finesse of, 75
free spectral range of, 76
Faraday effect, 85
Faraday rotator, 86, 136
fast axis, 21, 80, 372
feedback
cavity, 182
distributed, 122, 186, 317, 369, 379
negative, 275
optical, 188, 280, 369
Fermi level, 155, 180
ferroelectric domain, 388
fiber
birefringent, 20, 55
chalcogenide, 26
depressed-cladding, 7
design of, 6–13
dispersion properties of, 27
dispersion-compensating, 133
dispersion-decreasing, 30, 80
dispersion-flattened, 30
dispersion-shifted, 7, 30, 40, 80, 323, 351, 352
dry, 24
dual-core, 50, 56, 82

Index

fabrication of, 6–13
fluoride, 26, 104
fluorinated, 25
fluorophosphate, 94
geometrical-optics description of, 1
graded-index, 4–6, 9
highly nonlinear, 12, 122, 124, 127, 352, 354, 394
holey, 11
large effective-area, 37
loss of, 22–27
low-PMD, 34
microstructure, 11, 31
modes of, *see* fiber modes
multimode, 2–6, 18
nonlinear effects in, 36–44
parabolic-index, 4
PFBVE, 24
photonic-crystal, 11
photosensitive, 58
plastic, 6, 9–11, 24
PMMA, 24
polarization-maintaining, 21, 33, 135, 213, 324, 352, 372, 374, 386, 391
polycrystalline, 26
pulse propagation in, 34–36
reduced-slope, 32
reverse-dispersion, 32
silica, 6–9
single-mode, 18–22
standard, 80
step-index, 2–6, 16
tapered, 302
telecommunication, 7, 19
tellurite, 104
two-mode, 302
wave propagation in, 13–18
fiber amplifiers, 93–128
erbium-doped, *see* EDFA
parametric, *see* amplifier
Raman, *see* amplifier
fiber cable, 12
fiber coupler, *see* coupler
fiber gratings, *see* grating
fiber laser
cavity of, 128
coupled-cavity, 131
CW, 130–132
DFB, 321

distributed-feedback, 131
dual-frequency, 131
figure-8, 133
mode-locked, 132–136, 323
multiwavelength, 132, 321
ring-cavity, 321
tuning of, 130
fiber modes, 15–22
classification of, 18
effective index of, 18
eigenvalue equation for, 17
field distribution of, 20
fundamental, 20
hybrid, 18
linearly polarized, 18
propagation constant of, 18
spot size of, 21
transverse-electric, 18
transverse-magnetic, 18
field-effect transistor, 278
modulation-doped, 279
filter
acousto-optic, 300–302, 316
add–drop, 309–317
all-pass, 72, 305
AWG, 323
bandpass, 277, 300, 302, 323
birefringent, 324
Bragg, 296–300
comb, 132
Fabry–Perot, 289, 322
grating, 67–69, 296–300
high-pass, 277
interference, 291
liquid-crystal, 290
low-pass, 275
Mach–Zehnder, 168, 293–295, 305
mode, 395
multicavity, 291
notch, 295, 352
optical, 287–302
periodic, 322
raised-cosine, 276
surface-acoustic-wave, 277
tunable, 287, 300–302
finesse, 75, 289, 290, 294, 315
flame hydrolysis, 8, 158
flip-chip technique, 278, 280
flip-flop, 369
all-optical, 375

418   *Index*

    electrical, 375
    laser-based, 375
    optical, 375–380
    passive-waveguide, 378
    physical mechanism behind, 375
    SOA-based, 375
fluorescence time, 95, 99
four-wave mixing, 39, 57, 118–128, 214, 215, 349, 373, 389, 391
    efficiency of, 349
    polarization-independent, 391
Franz–Keldysh effect, 225
free spectral range, 76, 183, 210, 289, 315, 346
frequency chirping, 199, 225, 234, 248, 348
front end, 274
    bandwidth of, 274
    high-impedance, 274
    low-impedance, 274
    transimpedance, 275

gain
    APD, 264
    differential, 182
    distributed, 113
    optical, 94
    parametric, 120, 123
    polarization-dependent, 34, 115, 126
    Raman, 105, 120
    recovery of, 393
gain bandwidth, *see* bandwidth
gain coefficient, 95, 115, 181
gain margin, 187
gain quenching, 377
gain saturation, 121, 212, 215, 369, 376, 377, 381, 383, 396
gain spectrum, 96, 121
    parametric, 124
    Raman, 105
gain switching, 200
gain–bandwidth product, 267–269
gain-flattening technique, 103
gain-recovery time, 382
Gaussian distribution, 21, 116
Gaussian random process, 200, 202
gel effect, 10
Gigabit Ethernet, 10, 281
grating
    acoustically induced, 300, 301, 316
    apodized, 68, 69

Bragg, 58, 84, 131, 133, 296–300, 304, 306, 309, 311, 344, 360, 369, 376, 379
built-in, 186, 300, 369
bulk, 307
chirped, 61, 69, 190, 297, 322
concave, 303, 304
DFB-laser, 187
diffraction, 303
dispersion relation for, 64
elliptical, 304
external, 130, 188
fabrication of, 59–62
fiber, 56–71, 88, 99, 129, 131, 296–300, 306, 312, 360, 369, 376, 400
filtering by, 67–69
first-order, 63
index, 58, 61, 167
long-period, 62, 103, 168, 310
moving, 302
nonuniform, 69
phase-shifted, 131, 296, 306
reduced speed in, 65
sampled, 71, 190, 297, 319, 322, 350
stop band of, 64
superstructure, 71, 190, 297
tilted, 299
tunable, 360
waveguide, 173–175, 307, 319, 344, 400
grating period, 57, 59, 61, 187, 190, 304
GRIN lens, 86
group index, 183
group velocity, 27, 33, 51, 63, 65, 76, 372
group-velocity dispersion, *see* dispersion
group-velocity mismatch, 39, 80, 372, 389, 393, 396
GVD, *see* dispersion
GVD parameter, 27, 35, 51, 63, 74

header-processing unit, 398
Helmholtz equation, 15, 50, 62
heterostructure design, 180, 208, 260
holding beam, 369, 376
holographic technique, 188
    dual-beam, 59, 70
    single-beam, 59
    single-pulse, 60
homogeneous broadening, 96, 132

Index

hybrid integration, 320
hydrogen soaking, 59
hysteresis, 368

idler, 119, 120, 124, 354, 373, 374
impact ionization, 262, 270
impedance matching, 237, 262
impermeability tensor, 157, 227
index
    effective, 70, 150, 151, 296, 300, 304, 335
    fluctuations of, 311
    gradient, 86, 303
    mode, 18, 20, 70, 150, 187, 336
index ellipsoid, 228
inhomogeneous broadening, 96, 132
injection locking, 199
inkjet technology, 343
integrated circuits
    optoelectronic, 246, 278, 318
interdigited electrode, 271
interface scattering, 183
interfacial gel polymerization, 10
interferometer
    Fabry–Perot, 74–76, 132, 210, 289
    Mach–Zehnder, 81–84, 130, 168, 232, 293–295, 305, 306, 311, 317, 338, 348, 370, 380, 383
    Michelson, 84, 296, 300, 348
    phase-mask, 61
    polarization-discriminating, 386, 394
    Sagnac, 76–81, 133, 296, 348, 351, 371, 381
    ultrafast nonlinear, 386
International Telecommunication Union, 317
Internet, 10, 389, 397
intersymbol interference, 202, 275
intraband relaxation, 349
ion-beam etching, 291
ionization coefficient ratio, 264, 266
isolator, 86, 129, 368
    pigtailed, 86
    polarization-independent, 87
    polarization-sensitive, 134
    polarizing, 135
ITU grid, 317, 321

Jones matrix, 33
Jones vector, 36
junction heating, 195

KDP crystal, 228
Kerr effect, 378
Kramers–Kronig relation, 58, 244

label swapping, 399
Lambertian source, 205, 209
Langevin force, 112, 200
laser
    ArF, 161
    argon-ion, 58, 59, 94
    color-center, 109
    coupled, 376
    DBR, 131, 246, 318
    DFB, 122, 131, 225, 246, 247, 317, 350, 375
    dye, 94
    edge-emitting, 377
    excimer, 58, 60
    fiber, *see* fiber laser
    figure-8, 133
    He–Ne, 80
    InGaAsP, 320, 375
    KrF, 161
    mode-locked, 396
    modulator-integrated, 246, 247
    multiquantum-well, 155
    Nd:YAG, 79, 94
    Nd:YLF, 69
    quantum-well, 155
    semiconductor, *see* semiconductor lasers
    surface-emitting, 377
    Ti:sapphire, 161
    tunable, 350
laser threshold, 182
lattice constant, 152, 153, 181
lattice defects, 152
lattice mismatch, 245
LED, 204–209
    bandwidth of, 207
    broad-spectrum, 209
    Burrus-type, 208
    edge-emitting, 209
    modulation response of, 207
    P–I characteristics of, 204
    resonant-cavity, 209
    responsivity of, 206
    structures for, 208
    surface-emitting, 208
    temperature dependence of, 206
    transfer function of, 207

light-emitting diodes, *see* LED
LiNbO$_3$ technology, 335, 350, 388
linewidth enhancement factor, 196, 199, 203, 216
liquid crystal, 290, 342
    droplets of, 291
    nematic, 291
    smectic, 291
liquid-phase epitaxy, 154
Lorentzian spectrum, 41, 96, 203
loss
    bending, 26
    cavity, 183, 193
    connector, 13
    coupling, 213, 241, 290, 302, 306, 309, 315, 350
    diffraction, 128, 305
    fiber, 22–27, 406
    insertion, 87, 165, 173, 174, 225, 233, 240, 241, 244, 245, 294, 301, 305, 309, 313, 338, 356, 407
    internal, 183, 213, 290
    macrobending, 26
    microbending, 12, 26
    microwave, 237, 240
    polarization-dependent, 34, 87
    propagation, 160, 175, 241
    scattering, 150, 213
    splice, 13
    wavelength dependence of, 23, 24
    wavelength-selective, 130

Mach–Zehnder interferometer, *see* interferometer
Mach–Zehnder modulator, *see* modulator
magic T, 167
magneto-optic material, 85
Markoffian approximation, 200
material dispersion, *see* dispersion
Maxwell equations, 13, 17, 107, 118, 143, 149
Maxwellian distribution, 116
MEMS technology, 192, 340
Michelson interferometer, *see* interferometer
micro-electro-mechanical system, *see* switch
microlens, 304
micromachining, 293
micromirror, 340, 360
microring, 295, 314

Mie scattering, 26
mirror
    amplifying-loop, 79, 129, 134
    array of, 341
    Bragg, 261, 292
    DBR, 191
    dielectric, 128, 133, 261
    Faraday, 136
    fiber-loop, 77, 78, 129, 133, 371
    loop, 380, 381
    MEMS, 340
    nonlinear optical loop, 77, 351
    thin-film, 291
    wavelength-selective, 296
mode
    asymmetric, 165, 166
    bounded, 148
    cladding, 302
    even, 146
    fiber, *see* fiber modes
    fundamental, 19, 148
    guided, 15
    HE$_{11}$, 302
    higher-order, 163, 165, 302
    leaky, 15
    longitudinal, 75, 183, 186, 260, 289
    odd, 146
    radiation, 15
    super, 172
    symmetric, 166
    TE, 144, 145, 147, 213, 228, 300, 378
    TM, 147, 213, 228, 300, 378
    transverse, 261
    vibrational, 24
    waveguide, *see* waveguide modes, 304
mode locking, 199
    active, 136
    additive-pulse, 85, 133
    harmonic, 136
    hybrid, 136
    interferometric, 85
    passive, 84, 132–136
    saturable-absorber, 132
mode-partition noise, 202
mode-suppression ratio, 186, 187
modulation
    amplitude, 196, 224, 298
    chirp-free, 234
    cross-phase, 38
    direct, 224

Index 421

efficiency of, 233
external, 225
frequency, 224
large-signal, 198
phase, 196, 224, 232, 298, 352
schemes for, 224
self-phase, 36
sinusoidal, 197, 207
small-signal, 196
modulation bandwidth, 207
modulation format
    ASK, 224
    FSK, 224
    NRZ, 223
    on–off keying, 224
    PSK, 224
    RZ, 223
modulation instability, 102, 121
modulator
    acousto-optic, 230, 322
    amplitude, 136, 232
    digital, 233, 237
    electro-optic, 136, 228
    electroabsorption, 199, 242, 319, 348
    integrated with laser, 246
    $LiNbO_3$, 136, 199, 231–237, 324, 368
    Mach–Zehnder, 231–237
    MQW, 247
    multiquantum-well, 242
    phase, 136, 231
    polymer, 237–242
    push–pull, 234
    temporal response of, 233
    traveling-wave, 236, 245
Moiré pattern, 296
molecular-beam epitaxy, 154
MONET project, 336
MOPA configuration, 131
multimode interference, 163, 170
multiplexer
    add–drop, 309–317
    all-fiber, 306
    AWG, 319
    WDM, *see* demultiplexer
multiplication layer, 263, 270
network
    all-optical, 309, 334, 358
    label-swapping, 399
    local-area, 207, 281, 319

packet-switched, 99, 397, 399
telephone, 354
WDM, 209, 334, 342, 354, 399
wide-area, 354
noise
    amplifier, 100
    intensity, 117, 200, 318, 321
    intesnity, 200–202
    Langevin, 112
    laser, 200–204
    mode-partition, 202
    $1/f$, 204
    phase, 202–204
    pump, 117, 125
    quantum limit of, 126
    Rayleigh-induced, 114
    shot, 200
    spontaneous-emission, 100
    transfer of, 117
noise figure, 100, 101, 213, 383
    effective, 113
    parametric amplifier, 126
nonlinear effects, 36–44
nonlinear fiber-loop mirror, *see* mirror
nonlinear gain, 202
nonlinear length, 121
nonlinear parameter, 37, 52, 119, 352, 367
nonlinear polarization rotation, 134
nonlinear Schrödinger equation, 37, 39
nonlinearity
    carrier-induced, 215
    cascading, 389
    electro-optic, 233
    fiber, 134, 371, 374, 390
    intraband, 215, 374, 391
    Kerr, 379
    nonresonant, 371
    second-order, 395
    silica, 12
    third-order, 215
NRZ format, 223, 277, 351, 396
numerical aperture, 3, 9, 205, 208

on–off ratio, 233, 235, 244, 348
optical amplifiers, *see* amplifiers
optical bistability, *see* bistability
optical circulator, *see* circulator
optical clock, 373, 390
    recovery of, 396
optical cross-connect, *see* cross-connect

optical detector, *see* photodetector
optical fibers, *see* fiber
optical filter, *see* filter
optical isolator, *see* isolator
optical label, 399
optical receiver, *see* receiver
optical switching, *see* switching
optoelectronic integration, 278
overlap factor, 232
oxygen-deficient bonds, 58

p–i–n photodiode, *see* photodiode
p–n junction, 155, 180, 257
    reverse-biased, 257
p–n photodiode, 257
packet
    buffering of, 400
    header of, 397
    labeled, 399
    payload of, 398
    switching of, 397
    synchronization of, 400
parametric amplification, 118–128, 353, 390
    dual-pump, 123
    phase-sensitive, 126
    single-pump, 120
parametric fluorescence, 126
parametric gain, 374
parasitic capacitance, 271, 279
parasitic reflection, 291
paraxial approximation, 5
periodic poling, 350, 388
periodically poled lithium niobate, 388
phase conjugation, 40
phase errors, 61
phase mask, 61, 168
phase matching, 230, 232
    condition for, 121
phase mismatch, 119, 121, 127
phase shift
    differential, 382, 385
    nonlinear, 38, 78, 80, 108, 135, 367,
        368, 371, 381
    relative, 383
    SPM-induced, 78, 80, 135, 367
    voltage-induced, 232, 234
    wavelength-dependent, 306
    XPM-induced, 39, 77, 81, 135, 351,
        352, 370, 372, 381, 387
phase-locked loop, 396

phase-matching condition, 39, 57, 76, 120,
    121, 374, 388
phase-trimming technique, 340
phonon
    acoustic, 41
    optical, 43
    population of, 112
photodetector
    avalanche, *see* APD
    bandwidth of, 256
    dark current of, 256
    inverted MSM, 272
    MSM, 271–273
    quantum efficiency of, 254
    responsivity of, 254
    rise time of, 256
    traveling-wave, 262, 273
photodiode
    array of, 320
    p–i–n, 259–262
    p–n, 257
    reverse-biased, 257
    waveguide, 261, 280
photoelastic effect, 229, 300
photoelectric effect, 253
photolithography, 157, 160, 240
photon lifetime, 193
photonic bandgap, 11, 64, 71
photonic crystal, 11
photonic lightwave circuit, 320, 395
photonic switching, *see* switching
photorefractive crystal, 344, 376
photoresist, 188, 240
photosensitivity, 58–59, 161
piezoelectric effect, 236
piezoelectric transducer, 289, 300, 302
pigtail, 236, 280, 301
planar lightwave circuit, 158, 281, 294, 296,
    306, 307, 338, 343, 346, 391
Planck constant, 406
plastic
    PMMA, 10
    polystyrene, 10
plastic optical fiber, *see* fiber
PMD, 32
    compensaton of, 34
    first-order, 33
    pulse broadening induced by, 33
    second-order, 34
PMD parameter, 34, 116

Index

Pockels effect, 156, 227, 238
point symmetry, 227
polarization
    circular, 85, 127
    degree of, 117
    electric, 14
    linear, 127
    magnetic, 14
    nonlinear, 36
    orthogonal, 126
    principal states of, 33
    TE, 213
    TM, 213
polarization controller, 129, 135, 378, 387
polarization converter, 301
polarization diversity, 291
polarization-dependent gain, *see* gain
polarization-mode dispersion, *see* PMD
poling process, 238, 240
polycrystalline silicon, 315
polymer, 164
    fluorinated, 10
    organic, 10
    PMMA, 10
population inversion, 94, 102, 179, 181
population-inversion factor, 100
preamplifier, 274, 276
preform, 7, 9
principal axis, 33, 157, 228, 387
propagation constant, 17, 33, 34, 85, 143, 150, 300
proton exchange, annealed, 157
pump depletion, 121, 123
pumping
    backward, 95, 105, 107, 108, 114, 116, 117
    bidirectional, 95, 108
    EDFA, 94
    efficiency of, 94, 101
    forward, 95, 105, 108, 116, 117
pyroelectric effect, 236

quantum dot, 155
quantum efficiency, 269
    differential, 195
    external, 195, 205, 269
    internal, 195, 204
    photodetector, 254
    total, 195, 206
quantum limit, 130

quantum well, 133, 154, 209, 226, 243, 268
    modulation-doped, 199
    shallow, 249
    strained, 155, 196, 244
quantum wire, 155
quarter-wave plate, 229
quasi-phase matching, 350, 388, 395
quaternary compound, 152

Raman amplification
    broadband, 109
    distributed, 113
    multiple-pump, 109
    PMD effects on, 115
    single-pump, 107
    vector theory of, 115
Raman amplifier, *see* amplifier
Raman gain, 44, 105
Raman scattering
    spontaneous, 42, 111, 112
    stimulated, 42, 105, 353
Raman shift, 43, 111
Raman–Nath scattering, 61, 231
rare-earth element, 93
rate equation, 97, 193, 200
Rayleigh backscattering, 111, 113, 125
Rayleigh scattering, 25, 114
RC circuit, 256
RC time constant, 258
reactive ion etching, 61, 158, 160, 164, 240, 245, 279, 304
receiver
    components of, 274
    digital, 274
    front end of, 274
    integrated, 274–281
    linear channel in, 275
    OEIC, 278
    packaging of, 280
    WDM, 317–325
refractive index
    average, 57
    carrier-induced change in, 196
    definition of, 15
    effective, 376
    extraordinary, 157
    nonlinear part of, 37
    ordinary, 157
    periodic, 57, 62
    temperature-induced change in, 338

regenerator, 346
relative intensity noise, 117, 131, 201
relaxation oscillations, 197, 198, 201, 203
reset pulse, 375–377, 379
resonance
  electronic, 24
  vibrational, 24
resonator
  Fabry–Perot, 183, 289, 367
  fiber-ring, 72–74, 368
  ring, 72–74, 305, 314, 367
responsivity, 272
  APD, 264
  LED, 206
  photodetector, 254
rise time, 198, 256
router
  electronic, 397
  passive, 344
  static, 344
  waveguide-grating, 344
  WDM, 344
Rowland ghost gap, 71
RZ format, 223, 277, 351, 396

Sagnac interferometer, 76–81, 296
  nonlinear switching in, 79
  switching characteristics of, 79
  transmissivity of, 77
  unbalanced, 79
Sagnac loop, 77, 79, 80, 129, 351, 372, 396
  dispersion-imbalanced, 80
  SOA in, 381
  symmetric, 371
  XPM in, 80
saturable absorber, 84, 132, 134, 135, 369, 377
  superlattice, 133
saturable Bragg reflector, 133
saturation
  cross-absorption, 348
  cross-gain, 348
  gain, 348
saturation current, 155
saturation energy, 99, 216, 382
saturation power, 95, 99, 212
saturation velocity, 258
SBS threshold, 353
scanning electron microscope, 154, 160, 272
scattering matrix, 53

Schottky barrier, 271, 272
Schrödinger equation, 225, 250
second-harmonic generation, 58, 350, 389
selective-area-growth, 248
self-clocking scheme, 396
self-phase modulation, 36, 37, 51, 69, 71, 109, 120, 217, 249, 323, 372, 396
Sellmeier equation, 28
semiconductor lasers
  bistability in, 369
  broad-area, 184
  buried heterostructure, 185
  characteristics of, 192–196
  coupled-cavity, 188
  DFB, see distributed feedback lasers, 369
  EDFA pumping by, 94
  gain in, 180
  gain-switched, 323, 324
  index-guided, 184
  linewidth of, 202
  longitudinal modes of, 183
  mode-locked, 217, 324
  modulation response of, 196–200
  noise in, 200–204
  SNR of, 202
  structures for, 184
  surface-emitting, 191
  temperature sensitivity of, 194
  threshold of, 182
  transfer function of, 197
  tunable, 190, 318, 350
semiconductor materials, 152
semiconductor optical amplifiers, 209–218
  angled-facet, 211
  array of, 316
  bandwidth of, 210
  buried-facet, 211
  design of, 210
  Fabry–Perot, 210
  polarization sensitivity of, 213
  properties of, 212–218
  switching with, 80, 366–400
  tilted-stripe, 211
  traveling-wave, 210
  WDM applications of, 322
  window-facet, 211
set pulse, 375–377, 379
shot noise, see noise

Index

signal
  TE-polarized, 213
  TM-polarized, 213
  WDM, 102, 346
signal-to-noise ratio, 100, 202, 383
silica-on-silicon technology, 158, 172, 174, 294, 304, 307, 312, 317, 319, 320, 324, 338, 343, 346, 355, 395
silicon substrate, 296, 304, 308
silicon-on-insulator technology, 162
single-mode condition, 19, 147, 150, 185
slope efficiency, 130, 195
slow axis, 21, 80, 372
small-signal gain, 212
spatial hole burning, 196
spatial light modulator, 342
spectral efficiency, 289
spectral hole burning, 97, 132, 196
spectral slicing, 319, 322
spin coating, 164, 239, 291
splice loss, *see* loss
spontaneous emission, 97, 100, 155, 193, 200, 207, 319, 383
  amplified, 95, 101, 104, 323
spontaneous-emission factor, 100, 193, 213
spot size, 21
spot-size converter, 186, 280
squeezing, 126
Stark effect, 226, 242
Stark shift, 226
Stark splitting, 96
stimulated emission, 94, 97, 155, 179, 193
stitching errors, 61
stochastic process, 112
stop band, 67, 296, 298, 306, 360, 369, 375
  edge of, 66, 69
strain
  compressive, 245, 299
  tensile, 245, 299
strain tensor, 230
strip line, 240
sum-frequency generation, 388
super-Gaussian pulse, 37
supercontinuum, 323, 324
superlattice, 133, 268, 272
superluminescent laser diode, 323
supermodes, 55
surface acoustic wave, 231
susceptibility

linear, 14
nonlinear, 36
  second-order, 350, 388, 395
  third-order, 36, 350
switch
  bubble, 343
  digital, 336
  dilated, 356
  directional-coupler, 335
  double MZ, 339
  double-gate, 356
  electro-optic, 356
  electroholographic, 344
  electronic, 358
  gate, 334, 338
  interferometeric, 380–389
  $LiNbO_3$, 388
  liquid-crystal, 342
  Mach–Zehnder, 337, 383
  MEMS, 340, 359
    three-dimensional, 341
    two-dimensional, 340
  nonblocking, 343
  packet, 398
  polarization, 378
  polarization-discriminating, 386
  polarization-independent, 335, 342
  polymer-based, 340
  PPLN, 388
  Sagnac-loop, 381
  SLALOM, 381
  SOA-based, 338
  thermo-optic, 168, 357
  TOAD, 381, 398
  ultrafast, 380–389
  UNI, 386, 392, 394
  Y junction, 334
  Y-junction, 336
switching
  applications of, 389–400
  circuit, 334
  electro-optic, 335
  FWM-based, 373
  micro-electro-mechanical, 340
  nonlinear, 79, 82
  nonlinear schemes for, 366–374
  packet, 334, 397
  power required for, 79
  quasi-CW, 80
  space-domain, 334–344

speed of, 371
SPM-induced, 79
technologies for, 334–344
thermo-optic, 338, 355
threshold of, 79
time-domain, 334, 366–400
ultrafast, 366–374
XPM-induced, 80, 83, 370–372, 380–389
switching fabric, 355
switching time, 340, 375, 382, 395
switching window, 372, 381, 384, 391, 393, 397
symmetry group, 227
synchronous pumping, 369
Systeme International, 406

Talbot effect, 170
tap, optical, 54
Taylor series, 51, 56, 65
temperature gradient, 71
ternary compound, 152
thermal equilibrium, 155
thermal expansion, 317
thermo-optic effect, 168, 291, 294, 299, 338, 340, 388
thermo-optic heater, 339
thermocapillary effect, 343
thermoelectric cooler, 194
thin-film heater, 168, 319, 388
third-order dispersion, *see* dispersion
three-level system, 97
threshold condition, 183
threshold current, 194
time-division multiplexing, 247, 325, 371, 390–396
time-domain demultiplexing, 390–396
time-reversal symmetry, 85
total internal reflection, 3, 11, 26, 143, 205, 343
transfer function, 169, 197, 207, 276, 288, 289, 293
transfer matrix, 53, 72, 293
transform-limited pulse, 247
transimpedance, 275
transistor
    field-effect, 278
    heterojunction-bipolar, 278, 320
    high-electron-mobility, 278
transit time, 256, 257, 259, 266, 269, 270

transition cross section, 95, 97
transmission line, 236, 237
tuning
    electro-optic, 316
    electronic, 291, 295, 302
    micromechanical, 292
    thermal, 291, 294, 299
tuning range, 190
tunneling breakdown, 266
twin-amplifier configuration, 213
two-level system, 95

ultrashort pulse, 324
ultrasound, 230
ultraviolet trimming, 311

V parameter, 18, 54
vacuum permeability, 14
vacuum permittivity, 14
vapor deposition
    axial, 7
    chemical, 7
    outside, 7
vapor-phase epitaxy, 154
velocity matching, 237
Verdet constant, 86
Vernier effect, 190
vertical-cavity surface-emitting laser, 191, 319, 376
vibrational frequency, 24

wafer-bonding technique, 191
walk-off effect, 80, 372
wall-plug efficiency, 195, 206
wave equation, 14
wave-vector mismatch, 120
waveguide, 142–165
    AlGaAs, 371
    array of, 307, 343
    asymmetric, 142, 147
    bulge, 151
    buried, 151
    channel, 151
    coplanar, 236, 273
    core of, 142
    design of, 150
    electro-optic, 156
    InGaAsP, 300, 378
    InP, 337
    inverted-rib, 240

Index 427

    laser-written, 161
    $LiNbO_3$, 157, 229, 301, 350, 368, 388
    materials for, 152
    metal, 261
    modes of, *see* waveguide modes
    mushroom-mesa, 261
    planar, 142–165
    polymer, 164, 237
    PPLN, 388, 395
    rectangular, 149–151, 241
    rib, 150, 163, 240
    ridge, 150, 185, 279, 291, 319
    semiconductor, 152–156, 226, 314, 369, 371, 378
    silica, 158–162, 296, 304, 308, 338, 391
    silicon, 315
    silicon oxynitride, 159
    strip-loaded, 150
    symmetric, 142, 145
    tapered, 281
waveguide dispersion, *see* dispersion
waveguide grating, *see* grating
waveguide modes
    asymmetric, 147
    planar, 143–148
    rectangular, 149–151
    symmetric, 145
waveguide photodiode, 261
wavelength converter, 40, 346–354, 358
    dual-pump, 354
    fiber-based, 351–354
    FWM-based, 353
    semiconductor-based, 347–351
    XPM-based, 347, 352
WDM components, 287–325
WDM receiver, 317–325
    integrated, 318
    monolithic, 320
WDM transmitter, 317–325
    fiber-laser, 321
    integrated, 318
    spectrally sliced, 322

Y junction, 165, 293, 316, 334

zero-dispersion wavelength, 29, 39, 80, 121, 122, 124, 128, 323, 351, 372, 393

CUSTOMER NOTE: IF THIS BOOK IS ACCOMPANIED BY SOFTWARE, PLEASE READ THE FOLLOWING BEFORE OPENING THE PACKAGE.

This software contains files to help you utilize the models described in the accompanying book. By opening the package, you are agreeing to be bound by the following agreement:

This software product is protected by copyright and all rights are reserved by the author and John Wiley & Sons, Inc. You are licensed to use this software on a single computer. Copying the software to another medium or format for use on a single computer does not violate the U.S. Copyright Law. Copying the software for any other purpose is a violation of the U.S. Copyright Law.

This software product is sold as is without warranty of any kind, either express or implied, including but not limited to the implied warranty of merchantability and fitness for a particular purpose. Neither Wiley nor its dealers or distributors assumes any liability of any alleged or actual damages arising from the use of or the inability to use this software. (Some states do not allow the exclusion of implied warranties, so the exclusion may not apply to you.)

WILEY